UNDERSTANDING ALGEBRA

UNDERSTANDING ALGEBRA

REVISED EDITION

John Baley
Martin Holstege
Cerritos College

WCB
McGraw-Hill

Boston, Massachusetts   Burr Ridge, Illinois   Dubuque, Iowa
Madison, Wisconsin   New York, New York   San Francisco, California   St. Louis, Missouri

Photo Credits

The Adler Planetarium, Chicago, IL, 195
Mark Antman, The Image Works, 79, 165
Camerique, 97, 229, 343, 406
Alan Carey, The Image Works, 68
Chicago Park District, Chicago, IL, 164
Dan Chidester, The Image Works, 410
Data General, Rob Brown, 312
Field Museum of Natural History; GEO 84382; 94157; Chicago, IL, 541, 563
Gerald Fritz, Monkmeyer Press Photo Service, 356
George Gardner, The Image Works, 143
Jean-Claude LeJeune, 555
Mobile Oil Corporation, 46
Motor Vehicle Manufacturers Assoc., Detroit, MI, 268
NASA—National Aeronautics and Space Administration, 184, 307
Nebraska Dept. of Agriculture, Lincoln, NE, 415
New York Convention and Visitors Bureau, Inc., 556
New York Stock Exchange, Inc., 56
H. Armstrong Roberts, 40, 41, 59, 121, 306, 385, 493, 553, 569, 575, 594
Sears, Roebuck and Company, 1
James L. Shaffer, 397
Jack Spratt, The Image Works, 407
Thermometer Corporation of America, 67
United Press International, 564, 601
U.S. Navy; photo by PH1 Fel Barbante, 454

WCB/McGraw-Hill

*A Division of The **McGraw·Hill** Companies*

UNDERSTANDING ALGEBRA - *Revised Edition*

First Edition

4567890 QF   9987

Library of Congress Cataloging in Publication Data

Baley, John D.
    Understanding Algebra / John Baley, Martin Holstege.
      p.   cm.
    ISBN 0-07-003566-0
    1. Algebra.   I. Holstege, Martin.   II. Title.
QA154.2.B32 1987
512.9—dc19   87-20729
                                                                              CIP
This book is printed on acid-free paper.

Preface

This text is an in-depth second course in algebra. The book was designed to be understood and used by students. Although the development of algebra starts on page one, we, the authors, realize that many students may have completed the prerequisite course in elementary algebra a few years ago. Therefore, a review of elementary algebra has been blended into the early chapters of the text. This book is designed to develop students' skills with algebra so that they will be thoroughly prepared to continue their study of mathematics and science.

Features: Organization and Pedagogy

This book is divided into sixteen chapters to allow instructors greater flexibility in arranging course outlines to conform to academic calendars and student needs.

Each chapter includes:

Preview: This gives the student an overview of how the chapter relates to previous chapters and how the coming chapter will be developed.

Sections: Each chapter is divided into sections of material that approximate one hour of classroom lecture time.

Numerous Examples: Over 400 examples illustrated with graphics demonstrate the concepts presented in the text. Students are given one or more examples of each task they are expected to perform.

Ample Exercises: Over 4,800 problems, usually in matched odd-even sets, give the student the opportunity to apply the concepts and practice the skills taught in the text. Problem sets are a critical part of the learning process because they not only give the student needed practice, but they also show the student the results of subtle variations. By doing problems a student gets an opportunity to internalize mathematics and see the effects of changes in a parameter.

Applications: Whenever possible, this text uses applications in examples and problems to show students that algebra is a powerful branch of mathematics that is used in engineering, electronics, geology, optics, aviation, surveying, construction, forestry, navigation, and physics.

Key Ideas: Each chapter concludes with a summary of key ideas to aid students in organizing their knowledge and help them prepare for exams.

Review Tests: To further insure students that they have mastered the concepts of each chapter, every chapter has a review test that closely approximates the questions that are likely to be asked on an exam.

Special features of this text

Cumulative Reviews: There are cumulative reviews after Chapter 8 and at the end of the book. These reviews are an excellent opportunity for students to get an overview of the course as they study for midterm and final exams.

Conversational Bubbles: These bubbles are spread throughout the book but generally appear in the context of a worked example. Teachers will quickly recognize that these bubbles anticipate and verbalize the questions that students are likely to ask as they are learning the material. Answering bubbles provide the answers that experienced teachers are likely to give.

Highlighted Definitions, Properties, Theorems, and Rules: These features, which are essential to remember and understand, are highlighted in boxes throughout the text, both to draw the students' attention to important concepts and to make it easy for students to reference key ideas.

Pointers for Better Understanding: Fifteen special boxes are spread throughout the text to clarify ideas or give students helpful hints about how to write or visualize certain algebra problems. These pointers deal with many ideas that are needed to work algebra but are frequently never explicitly taught to students. How to write equations, how to deal with denominate numbers, and why you can't divide by zero are examples of this kind of pointer.

Help with Problem Solving: This book makes a special effort to help students develop the skills and understanding needed to apply algebra to practical situations. Throughout the text, it helps the students to:

    a. develop skills to express in precise mathematical terms ideas that are imprecisely expressed in English;

    b. critically analyze a problem and identify both relevant and irrelevant pieces of information in the problem;

    c. build a knowledge-base of formulas that can be applied in this book and throughout life;

    d. visualize or draw mental pictures based on information given in a problem.

Sections 1.5, 2.5, 4.2, 7.5, 10.4, and 14.4 are six complete sections devoted to building problem-solving skills. There are also six Pointers for Problem Solving throughout the book that help students build skills of problem organization (p. 172), chosing variables (p. 158), defining exactly what a variable represents (p. 163), model making (p. 166), using variables (p. 406), and visualization (p. 407).

Using Your Calculator: This is a series of special features that show students how to use a scientific calculator efficiently to solve problems. Each calculator feature tells students what they need to know, when they need to know it.

Logarithms: Logarithms are approached with a recognition that scientific calculators exist and are readily available. This approach both reduces the arithmetic involved and allows the student to concentrate on logarithms as functions. This leaves time to introduce many interesting applications of logarithms.

Class Testing

This book has been class tested through five revisions by over 1,500 students in both lecture and semi-independent mathematics classes. The authors are grateful to Professor Ray Battee and Sister Pat Peach, who helped with the class testing of the book. They are also thankful to the many students and instructional aides who provided useful feedback, which helped to improve this text.

Ancillaries

Student Solution Manual

A student solution manual with all the even-numbered problems worked step by step is available. This manual also provides additional hints and explanations about how the problems were solved.

Instructor's Materials

An instructor's printed test manual provides four forms of each chapter test, a midterm exam, a final exam, and suggested tests for Chapters 1–3, 4–6, 7–9, 10–12, and 13–15 to assist teachers who wish to cover multiple chapters on each test. In addition, a separate manual is available that contains answers to all the exercises.

Acknowledgments

The authors wish to thank Debra Pentecost for the level of care and attention she gave to preparing the preliminary editions for class testing. Her efforts have made the development of this book a much more pleasant task for all concerned.

The authors are also specially grateful to the reviewers from colleges throughout the country who carefully read, critiqued, and provided many excellent suggestions that have been incorporated into the book you are holding. The quality of this text has been improved by the efforts of:

Sharon Bird—*Richland College*

Katherine Blackburn—*East Tennessee State University*

Betsy Darken—*University of Tennessee, Chattanooga*

Doug Dawson—*Arizona State University*

Joan Dykes—*Edison Community College*

Tom Farmer—*Miami University*

Josephine Gervase—*Manchester Community College*

Virginia Hamilton—*Ball State University*

Louis Hoelzle—*Bucks County Community College*

William Keils—*San Antonio College*

Steven Marsden—*Glendale Community College*

Scott Newey—*American River College*

Don Poulson—*Mesa Community College*

Jane Ritter—*Stephen F. Austin University*

Ned Schillow—*Lehigh County Community College*

Erik Schreiner—*Western Michigan University*

Faye Thames—*Lamar University*

Jan Vanderer—*South Dakota State University*

Many thanks are due to the entire editorial and production team, particularly our editor Wayne Yuhasz, whose experienced guidance has helped shape every aspect of this book. His able and trusty assistants Anne Wightman and Louise Bush provided much appreciated coordination and support for the project. The production staff in Cambridge—Margaret Pinette, Project Manager; her assistant, Pamela Niebauer; Michael Weinstein, Production Manager; and his assistant, Susan Brown—have done an admirable and highly-professional job. We also thank Luck Jenkins and all the people at Quarasan for their services and support.

Contents

UNDERSTANDING ALGEBRA

1 Real Numbers and Their Properties

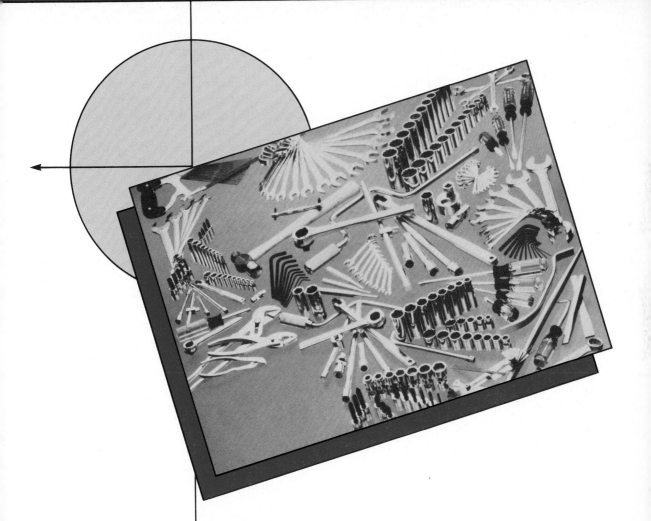

Contents

As you might expect, algebra is a series of ideas or concepts built into a system of knowledge. Like any system of knowledge it has a language to convey the ideas. It also has its own special vocabulary.

This chapter will start by building your algebra vocabulary with a series of definitions. Then it will list the properties of a mathematical system. All of algebra and most of mathematics are based in these definitions and properties.

Naturally the authors want you to treat them as important. However, it is our experience that students who have difficulty with this unit are trying to read too much into it. Try to accept the first two sections of this chapter just as formal statements of common sense.

The third and fourth sections of this chapter will develop rules for the operations of addition, subtraction, multiplication, and division as used in algebra. A study of problem solving follows. We will focus on *reasoning* rather than on *procedures* as we solve problems. Section 1.5 will show how to find the amount per unit of measure. This basic concept is a valuable tool in solving a wide range of problems.

■ 1.1   Introduction

Any system of knowledge is based upon certain undefined terms. These undefined terms are used to define other ideas in the system. A good system of knowledge has as few undefined terms as possible. One way to study algebra starts with the idea of a set. A **set** is simply a collection of objects. The objects in a set are called **elements.**

There must be some way to tell if a given element is part of the set. One way to determine if an element is in a set is to list all the members of the set. In mathematics, we frequently refer to sets of numbers. Sometimes it is convenient to list the members of the set. In such a listing the elements of the set are usually enclosed in braces. An example is the set containing the letters A, B, and C.

$$\{A, B, C\}$$

1.1A   Set Membership

The symbol $\in$ means "is an element of" and $\notin$ means "is not an element of."

1.1B   Equal Sets

Set A is **equal to** B if both sets contain the same elements.

If $A = \{1, 2, 3\}$ and $B = \{3, 2, 1\}$ then $A = B$ since $\{1, 2, 3\} = \{3, 2, 1\}$. The order of the elements in a set is not important.

Example 1 □ If $A = \{5, 10, 15, 20\}$, use $\in$ or $\notin$ to indicate membership in set A.

$$10 \in A \qquad 3 \text{\_\_\_\_} A$$

$\notin$

$\notin, \in$

$$7 \text{\_\_\_\_} A \qquad 15 \text{\_\_\_\_} A \qquad\qquad □$$

What am I supposed to do with the blanks?

Fill them in. Then check the left-hand column to be sure your answer was correct.

Why not just copy the answers, or better yet have the authors fill in the blanks?

Because learning works best if you're involved. That's why you should think of an answer before you look to the left to see if we agree with you. It doesn't work as well if you merely look in the left column and agree with us.

1.1C Definition: Variable

A variable is a symbol that can represent any element of a set of numbers.

The basic set of numbers we deal with is the set of natural numbers. These are the numbers that are used as a child learns to count.

1.1D Definition: Natural Numbers

The set of natural numbers is the set of counting numbers. They are

$$N = \{1, 2, 3, 4, 5, \ldots\}$$

(. . . means *continue with this pattern.*)

This set of numbers is also called the **counting numbers.**

1.1E Definition: Closure

A set is said to be closed under an operation if performing the operation with any two members of the set always yields a member of the set.

The set of counting numbers is closed under addition because the sum of any two counting numbers is another counting number.

Example 2 □   □

Both are counting numbers.

The result is a counting number.

Am I missing something? I don't see any big deal.

Like many mathematical ideas, closure is really a very simple concept. Don't try to make it complicated. Do notice that a set can be closed for one operation and not closed for another. Read on.

To show that the set of natural numbers is not closed for subtraction we need only give one counterexample. The answer to $3 - 5$ is not contained in

Chapter 1 ■ Real Numbers and Their Properties

the set of natural numbers. Therefore, the natural numbers are not closed for subtraction.

That means it is not always possible to do subtraction using only natural numbers.

Subtraction is always possible if you subtract the smaller number from the larger.

True, but you're imposing an extra condition. For a set to be self-contained or closed under subtraction you must be able to subtract any number from any other number in the set and get an answer that is in the set.

To define a set of numbers that is closed for subtraction, it is necessary to make a few more definitions.

1.1F   Definition: Identity Element for Addition

Zero is the identity element for addition. That is, if you add zero to any number, the result is identical to the original number.

$$4 + 0 = 4 \qquad 0 + 9 = 9$$

Zero isn't a natural number.

True. That's why we have to define whole numbers.

1.1G   Definition: Whole Numbers

The set of whole numbers is the set of natural numbers combined with the number zero. They are

$$W = \{0, 1, 2, 3, 4, \ldots\}$$

1.1H Definition: Negative Number

Every natural number n has a negative, $-n$, such that

$$n + (-n) = 0$$

For example,

$$3 + (-3) = 0 \qquad\qquad -2 + 2 = 0$$

The negative of a number can be viewed as the opposite of the number. Taking two steps forward can be represented as 2. Taking two steps backward can be represented as -2.

Example 3 □ Give the number you would use to represent each of the following.

-3
Going down three floors in an elevator _____

+15
A gain of 15 yards in football _____

-20
A debt of twenty dollars _____

0
The distance covered by a car that doesn't move _____

-6
The position of a car that backed up 6 feet _____

0
The location of a car that backed up 50 feet then drove forward 50 feet _____ □

The opposite of any number is called its **additive inverse.**

1.1I Definition: Additive Inverse

If two numbers have a sum of zero, they are called **additive inverses of each other.**

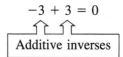

$$3 + (-3) = 0 \qquad\qquad -3 + 3 = 0$$

Additive inverses   Additive inverses

Example 4 □ Write the additive inverse of each number below.

-8
The additive inverse of 8 is _____

8
The additive inverse of -8 is _____

-5
The additive inverse of 5 is _____

-a
The additive inverse of a is _____

y
The additive inverse of $-y$ is _____

0
The additive inverse of 0 is _____ □

1.1J Definition: Integers

Integers are the set of numbers made up of the set of natural numbers, combined with zero and the negatives of the set of natural numbers. They are

$$I = \{\ldots -3, -2, -1, 0, 1, 2, 3 \ldots\}$$

Now we can define both addition and subtraction for integers.

1.1K Rule: Addition of Numbers with Like Signs

To add two positive numbers add as in arithmetic. To add two negative numbers add as in arithmetic, keeping the negative sign with the answer.

Example 5 □ Add the following.

9, −9,

−6, −10,

7, −7

$$3 + 6 = \underline{\hspace{1cm}}$$

$$-2 + (-4) = \underline{\hspace{1cm}}$$

$$4 + 3 = \underline{\hspace{1cm}}$$

$$(-3) + (-6) = \underline{\hspace{1cm}}$$

$$-5 + (-5) = \underline{\hspace{1cm}}$$

$$(-4) + (-3) = \underline{\hspace{1cm}} \quad □$$

To add a positive and a negative number we rely on the definition of additive inverses.

$$n + (-n) = 0$$

$$6 + (-6) = 0$$

Because these are additive inverses of each other, their sum is zero.

Example 6 □ To add $8 + (-6)$ think of 8 as $2 + 6$.

$$8 \quad + (-6) =$$

$$2 + 6 + (-6) =$$

$$2 + \quad 0 \quad = 2$$

Therefore,

$$8 + (-6) = +2 \qquad □$$

Example 7 □ Add the following.

$$-8 \quad + (+6) =$$

$$\overbrace{-2 + (-6)} + (+6) =$$

$$-2 + \quad \underbrace{\qquad} \atop 0 \quad = -2$$

> Think of -8
> as
> $(-2) + (-6)$.

□

Adding signed numbers by using additive inverses is a slow process. A faster method of adding numbers with unlike signs uses the idea of absolute value.

Magnitude

Magnitude is the value of a number if you ignore its sign, or how far the number is from zero on the number line. The magnitude of a number is called its **absolute value.**

1.1L Definition: Absolute Value

The absolute value of any number a is written $|a|$ and is given by the following rule:

if	$a \geq 0$	then	$|a| = a$
if	$a < 0$	then	$|a| = -a$

$$|3| = 3 \qquad |-3| = 3 \qquad |-212| = 212$$

Example 8 □ Find the absolute value of 12.
Since 12 is greater than zero, 12 is its own absolute value.

$$|12| = 12$$

□

Example 9 □ Find the absolute value of -12.
Since -12 is less than zero, its absolute value is the negative of -12, which is 12.

$$|-12| = -(-12)$$
$$|-12| = 12$$

□

We also need the formal definition of absolute value to determine the absolute value of a variable. Until we know if x represents a positive or negative number, $|x|$ and $|-x|$ can't be written in any simpler form.

Find the absolute value of each of the following.

1. $|13|$ **2.** $|-135|$ **3.** $|0|$ **4.** $|36|$ **5.** $|-13|$ **6.** $|-36|$

7. $|56|$ **8.** $|72|$ **9.** $|-46|$ **10.** $|-7|$ **11.** $|17|$ **12.** $|180|$

13. $|25|$ **14.** $|38|$ **15.** $|-12|$ **16.** $|-18|$ **17.** $|-42|$ **18.** $|-76|$

19. $|1001|$ **20.** $|869|$ **21.** $|750|$ **22.** $|850|$ **23.** $|-125|$ **24.** $|-98|$

1.1M   Rule: Addition of Numbers with Unlike Signs

To find the sum of two numbers with unlike signs, find the difference between their magnitudes and use the sign of the number with the larger magnitude.

Example 10 □ Add $2 + (-5)$.

-5

Negative

3

-3

The number with the larger absolute value is _____

The sign of the sum will be _____ .

5 minus 2 is _____ .

Therefore, the sum of $2 + (-5)$ is _____ .   □

Subtraction of Integers

The easy way to subtract is to avoid subtraction. To subtract a number we add its negative.

1.1N   Rule: Subtraction

To subtract a number, add its additive inverse.

$$a - b = a + (-b)$$

To see how this definition of subtraction makes sense, look at a problem for which you already know the answer.

Example 11 □ Subtract 12 − (+5).

Rewrite the subtraction as an addition.

$$12 - (+5) =$$
$$12 + (-5) = 7$$

The operation changed from subtraction to addition and the sign of the 5 changed from + to −.

□

Now let's subtract a negative number.

Example 12 □ Subtract 6 − (−4).

$$6 - (-4) =$$
$$6 + (+4) = +10$$

□

The sign of the 4 changed from − to +. +4 is the additive inverse of −4.

Yes, and the subtraction sign changed to addition.

Example 13 □ Subtract: 18 − (−5).

+5

The additive inverse of −5 is _____ .

Rewrite as an addition problem and add:

18, (+5), 23

_____ + _____ = _____ .

□

This method also works when you subtract a number from a negative number.

Example 14 □ Subtract: −4 − (+6).

$$-4 - (+6) =$$
$$-4 + (-6) = -10$$

□

The operation changed from subtraction to addition and the sign of the 6 changed from + to −.

Chapter 1 ■ Real Numbers and Their Properties

1.10 Summary of Rules for Addition and Subtraction of Signed Numbers

To add two positive numbers add as in arithmetic.

$$6 + 8 = 14$$

To add two negative numbers, add as in arithmetic, keeping the negative sign with the answer.

$$-6 + (-8) = -14$$

To add two numbers with unlike signs, find the difference between their absolute values and use the sign of the number with the larger absolute value.

$$6 + (-8) = -2$$

To subtract a number, add its additive inverse.

$$6 - (-8) = 6 + (+8)$$
$$= 14$$

Problem Set 1.1B

Simplify the following.

1. $9 + 5$
2. $9 + (-5)$
3. $9 - 5$
4. $9 - (-5)$
5. $-8 + 6$
6. $-8 + (-6)$
7. $-8 - 6$
8. $-8 - (-6)$
9. $0 + 9$
10. $0 + (-9)$
11. $0 - 9$
12. $0 - (-9)$
13. $|-9 - 10|$
14. $|-15| + |-5|$
15. $|-15| + (-5)$
16. $(-15) + |-5|$
17. $8 - (-8)$
18. $8 - |-8|$
19. $(-20) - (-9)$
20. $(-20) - |-9|$
21. $|-7| + 6$
22. $0 - |-8|$
23. $0 - (-8)$
24. $|-16| - |-12|$
25. $-9 - (-10)$
26. $7 + (-3)$
27. $-25 - 30$

28. $-40 - (-56)$

29. $(-12) + |-12|$

30. $(-12) + (-12)$

31. $|-28| + |-8|$

32. $(-50) + |-20|$

33. $|-18| + 7$

34. $|-27| - |-26|$

35. $17 - (-17)$

36. $17 + (-17)$

37. $19 - |-19|$

38. $19 + |-19|$

39. $(-12) + (-32)$

40. $(-12) - (-32)$

41. $|-12| - (-32)$

42. $|-12| + (-32)$

43. $|-12| - |-32|$

44. $|-12| + |-32|$

45. $|-6| + 7$

■ 1.2 Some Basic Properties of a Mathematical System

For some operations the order in which the numbers are written is important. At other times it is not. When the order in which the numbers are written is unimportant, the operation is said to be **commutative.**

Yes, No

Does $7 + 6 = 6 + 7$? _____   Does $7 - 6 = 6 - 7$? _____

Yes, No

Does $4 \cdot 2 = 2 \cdot 4$? _____   Does $4 \div 2 = 2 \div 4$? _____

Because the order of numbers does not matter when we add or multiply, we say that addition and multiplication are **commutative.** But the order of numbers does matter when we subtract or divide; therefore subtraction and division are not commutative.

1.2A Commutative Property for Addition and Multiplication

$$a + b = b + a \qquad \text{and} \qquad a \cdot b = b \cdot a$$

where a and b are variables that can stand for any number or algebraic expression.

Binary Operations

Most mathematical operations are binary operations, that is, they deal with only two elements at a time and yield a single result. If you wish to perform a binary operation with three elements, you have a problem.

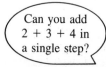

Can you add $2 + 3 + 4$ in a single step?

Nope.

How can you find the sum then?

You can add 2 and 3 to get 5; then add 4 to get 9.

Chapter 1 ■ Real Numbers and Their Properties

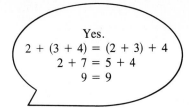

If it doesn't matter in which way the elements are grouped, the operation is said to be **associative.**

Example 15 □

Yes   Does $(8 + 4) + 2 = 8 + (4 + 2)$? _____

No   Does $(8 - 4) - 2 = 8 - (4 - 2)$? _____

Yes   Does $\quad (8 \cdot 4) \cdot 2 = 8 \cdot (4 \cdot 2)$? _____

No   Does $(8 \div 4) \div 2 = 8 \div (4 \div 2)$? _____   □

For addition or multiplication it does not matter how the elements are grouped. Therefore, we say that addition and multiplication are **associative.** How the elements are grouped does matter for subtraction and division; therefore, subtraction and division are not associative.

1.2B   Associative Property for Addition and Multiplication

$$(a + b) + c = a + (b + c) \quad \text{and} \quad (a \cdot b) \cdot c = a \cdot (b \cdot c)$$

where a, b, and c are variables that can stand for any number or algebraic expression.

Numbers that are added are called **terms.** Numbers that are multiplied are called **factors.**

Another important property involves multiplying by zero.

1.2C   Multiplication by Zero

$$a \cdot 0 = 0 \cdot a = 0$$

where a is a variable that can stand for any number or algebraic expression.

This says that if any factor in a product is zero, the product is zero.

$$0 \cdot 4 = 0 \qquad \text{and} \qquad -8 \cdot 0 = 0$$

There is a property that allows us to write some multiplication problems as addition and vice versa.

Notice that

Multiplication		Addition
$2(3 + 4)$	$=$	$2 \cdot 3 + 2 \cdot 4$
$2(7)$	$=$	$6 + 8$
14	$=$	14

This is called the **distributive property** of multiplication over addition.

1.2D   The Distributive Property

$$a(b + c) = a \cdot b + a \cdot c$$

It also works when the multiplier is on the right.

$$(b + c)a = b \cdot a + c \cdot a$$

where a, b, and c are variables that can stand for any number or algebraic expression.

Example 16 □ Use the distributive property to rewrite $4(10 + 2)$.

$$4(10 + 2) = 4 \cdot 10 + 4 \cdot 2 \qquad\qquad □$$

Example 17 □ Use the distributive property to rewrite the following.

4, 5 $3(4 + 5) = 3 \cdot \underline{\hspace{1cm}} + 3 \cdot \underline{\hspace{1cm}}$

9, 6 $9(6 + 3) = \underline{\hspace{1cm}} \cdot \underline{\hspace{1cm}} + 9 \cdot 3$

6, 6 $6(x + y) = \underline{\hspace{1cm}} \cdot x + \underline{\hspace{1cm}} \cdot y$

a, 8 $8(a + b) = 8 \cdot \underline{\hspace{1cm}} + \underline{\hspace{1cm}} \cdot b$

4, 3 $4(x + 3) = 4 \cdot x + \underline{\hspace{1cm}} \cdot \underline{\hspace{1cm}} \qquad\qquad □$

Use the distributive property to rewrite the following.

1. $5(3 + 6)$ **2.** $7(4 + 6)$ **3.** $8(3 + 1)$ **4.** $9(a + 2)$ **5.** $8(a + b)$

6. $9(x + 2)$ **7.** $4(x + y)$ **8.** $a(3 + 4)$ **9.** $a(b + 2)$ **10.** $x(a + 3)$

11. $x(a + b)$ **12.** $b(x + a)$

Problem Set 1.2B

Rewrite each of the following using only the property indicated.

1. $5a + x =$ _____ (Commutative property of addition)

2. $5a + x =$ _____ (Commutative property of multiplication)

3. $6 \cdot (a + 4) =$ _____ (Commutative property of multiplication)

4. $6 \cdot (a + 4) =$ _____ (Commutative property of addition)

5. $6 \cdot (a + 4) =$ _____ (Distributive property)

6. $2 \cdot a + 2 \cdot 4 =$ _____ (Commutative property of addition)

7. $2 \cdot a + 2 \cdot 4 =$ _____ (Commutative property of multiplication)

8. $2 \cdot (a + 4) =$ _____ (Distributive property)

9. $6 + (8 + c) =$ _____ (Associative property of addition)

10. $5 \cdot (9 + 4) =$ _____ (Commutative property of addition)

11. $6 \cdot (0 + 4) =$ _____ (Commutative property of multiplication)

12. $0 \cdot (a + 6) =$ _____ (Multiplication by zero)

13. $0 \cdot (a + 6) =$ _____ (Commutative property of multiplication)

14. $5 \cdot 6 + 5 \cdot 8 =$ _____ (Commutative property of addition)

15. $5 \cdot (a \cdot 4) =$ _____ (Associative property of multiplication)

16. $9 \cdot (x + 3) =$ _____ (Distributive property)

17. $9 \cdot (x + 3) =$ _____ (Commutative property of addition)

18. $(a + 2) + y =$ _____ (Associative property of addition)

19. $3 \cdot (a + b) =$ _____ (Distributive property)

Name the property for each statement.

20. $6 + x = x + 6$

21. $a \cdot (b \cdot 7) = (a \cdot b) \cdot 7$

22. $a \cdot (b \cdot c) = a \cdot (c \cdot b)$

23. $a + (5 + y) = (a + 5) + y$

24. $a + (5 + y) = a + (y + 5)$

25. $a + (5 + y) = (5 + y) + a$

26. $0 \cdot c = c \cdot 0$

27. $0 \cdot c = 0$

28. $c \cdot 0 = 0$

29. $0 \cdot (6 + w) = 0$

30. $0 \cdot (6 + w) = 0 \cdot (w + 6)$

31. $0 \cdot (6 + w) = 0 \cdot 6 + 0 \cdot w$

32. $12 \cdot (x + 5) = (x + 5) \cdot 12$

33. $12 \cdot (x + 5) = 12 \cdot (5 + x)$

34. $12 \cdot (x + 5) = 12 \cdot x + 12 \cdot 5$

35. $(x + 5) \cdot 12 = x \cdot 12 + 5 \cdot 12$

36. $5 \cdot 0 = 0$

37. $9 \cdot 0 = 0 \cdot 9$

38. $5 + x = x + 5$

39. $4 + 3 \cdot a = 4 + a \cdot 3$

40. $3 + (x + 6) = (3 + x) + 6$

41. $7 \cdot (2 + 4) = 7 \cdot 2 + 7 \cdot 4$

42. $0 \cdot (3 + 4) = 0 \cdot 3 + 0 \cdot 4$

43. $15 \cdot (x + 3) = 15 \cdot x + 15 \cdot 3$

44. $(x + 3) \cdot 15 = x \cdot 15 + 3 \cdot 15$

45. $3 \cdot 0 = 0 \cdot 3 = 0$

46. $(5 \cdot 4) \cdot 6 = 5 \cdot (4 \cdot 6)$

47. $(5 \cdot 4) \cdot 6 = (4 \cdot 5) \cdot 6$

48. $0 \cdot (6 + x) = (6 + x) \cdot 0$

49. $0 \cdot (6 + x) = 0 \cdot (x + 6)$

50. $(x + y) + 6 = (y + x) + 6$

51. $(x + y) + 6 = 6 + (x + y)$

■ 1.3   **Multiplication and Division**

With multiplication of signed numbers there are four cases to consider.

    I. positive number times a positive number
   II. positive number times a negative number
  III. negative number times a positive number
  IV. negative number times a negative number

Cases I and II   Multiplication by a Positive Number

When you multiply by a positive number, the proper sign for the product is clear because you can view multiplication as repeated addition.

$$3 \cdot (+2) = (+2) + (+2) + (+2) = +6$$

and

$$3 \cdot (-2) = (-2) + (-2) + (-2) = -6$$

Case III   Multiplication of a Positive Number by a Negative Number

$(-2)(+3)$ is a case where the first factor is negative. Because multiplication is commutative, we can change the order of the factors.

$$(-2)(+3) = (+3)(-2)$$

We know that $\quad (+3)(-2) = -6$

Therefore, $\quad (-2)(+3) = -6$

Case IV   Multiplication of a Negative Number by a Negative Number

To cover this case we need to maneuver a bit. Consider the example $(-2)(-3)$.

(-2)	$\cdot$	$0 = 0$	Multiplication by zero
(-2)	$\cdot[3 + -3] = 0$		Definition of additive inverse to replace 0
$(-2)(3) + (-2)(-3) = 0$			Distributive property
$-6 + (-2)(-3) = 0$			Replace $(-2)(3)$ with -6

Now $\qquad -6 + \quad ? \quad = 0$

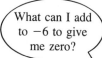 What can I add to -6 to give me zero?

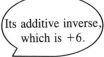 Its additive inverse, which is $+6$.

Therefore, the question mark must represent $+6$. Hence, $(-2)(-3) = +6$.

Summary of Rules for Multiplication of Signed Numbers

When two signed numbers are multiplied, the product is positive if the signs of the factors are alike, and the product is negative if the signs of the factors are different.

$$\left. \begin{array}{c} (\,+\,) \cdot (\,+\,) \\ (\,-\,) \cdot (\,-\,) \end{array} \right\} \Rightarrow + \qquad \left. \begin{array}{c} (\,+\,) \cdot (\,-\,) \\ (\,-\,) \cdot (\,+\,) \end{array} \right\} \Rightarrow -$$

Example 18 □

15, −12, −24

$5 \cdot 3 =$ _____ $4 \cdot (-3) =$ _____ $-6 \cdot 4 =$ _____

30, 0, 0

$(-5)(-6) =$ _____ $5 \cdot 0 =$ _____ $0 \cdot (-6) =$ _____

−ab, −ab, ab

$(a)(-b) =$ _____ $-a \cdot b =$ _____ $(-a)(-b) =$ _____

□

Exponents

An easy way to show repeated multiplication is to use an exponent.

1.3A   Definition of Powers

Base/Exponent/
Factor

The base is the factor that is to be multiplied repeatedly. The exponent is the number written to the right and a little above the base. The exponent indicates how many times the base is to be used as a factor.

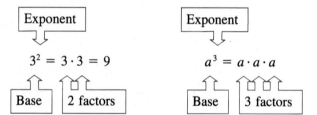

The expression 3^2 is read "three squared" or "the second power of 3" or "three to the second power." The expression a^3 is read "a cubed" or "the third power of a" or "a to the third power."

Example 19 □ $x^4 = x \cdot x \cdot x \cdot x$.

We must be careful of the sign when raising negative numbers to a power. For example $(-4)^2 = 16$ because $(-4)^2$ means $(-4)(-4) = 16$ but, $(-4)^3 = -64$ because $(-4)^3$ means $(-4)(-4)(-4) = -64$.   □

Example 20 □ Evaluate.

25, 25

$(5)^2 =$ _____ $(-5)^2 =$ _____

−25, −25

$-5^2 =$ _____ $-(-5)^2 =$ _____

$x^2, -x^3$

$(-x)^2 =$ _____ $(-x)^3 =$ _____

−1, 0

$(-1)^{243} =$ _____ $0^{64} =$ _____   □

Chapter 1 ■ Real Numbers and Their Properties

Multiplication of Expressions with Like Bases

We can multiply two powers of the same variables such as a^2 times a^3 if we use the definition of exponents.

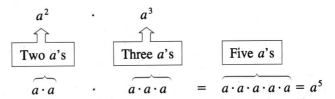

$$a^2 \cdot a^3$$

Two a's   Three a's   Five a's

$$\underbrace{a \cdot a} \cdot \underbrace{a \cdot a \cdot a} = \overbrace{a \cdot a \cdot a \cdot a \cdot a} = a^5$$

Rather than write out all the factors and then count the number of times the base is used, just add the exponents.

$$a^2 \cdot a^3 = a^{2+3} = a^5$$

1.3B   Multiplication of Powers with Like Bases

When multiplying powers with the same base, keep the base, and add the exponents.

$$a^x \cdot a^y = a^{x+y}$$

Example 21 □

$$x^3 \cdot x^4 = x^7 \qquad b^5 \cdot b^4 = b^9 \qquad y^3 \cdot y^2 = y^5 \qquad p^2 \cdot p^{100} = p^{102} \quad □$$

Example 22 □

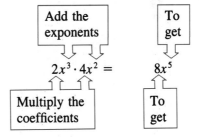

Add the exponents   To get

$$2x^3 \cdot 4x^2 = \qquad 8x^5$$

Multiply the coefficients   To get

□

Example 23 □

$$-3a^2b^2 \cdot 2a^2b^3 = -3a^2 \cdot b^2 \cdot 2a^2 \cdot b^3 \qquad \text{Associative property of multiplication}$$

$$= -3 \cdot 2 \cdot a^2 \cdot a^2 \cdot b^2 \cdot b^3 \qquad \text{Commutative property of multiplication}$$

$$= -6 \cdot a^4 \cdot b^5 \qquad \text{Multiplication law of powers} \qquad □$$

1.3C Definition: Quotient or Division

The quotient of $\dfrac{a}{b}$ is defined to be q if there exists a number q such that $b \cdot q = a$.

$$\frac{a}{b} = q \qquad \text{if and only if} \qquad b \cdot q = a$$

$$\frac{-6}{3} = -2 \qquad \text{because} \qquad -2 \cdot 3 = -6$$

▶ **Pointer for Better Understanding**

The definition of division explains why division by zero is impossible. Let's look for a quotient q that will satisfy the definition.

If $\qquad \dfrac{7}{0} = q \qquad$ then $\qquad q \cdot 0 = 7$.

BUT $\qquad\qquad\qquad q \cdot 0 = 0$.

Therefore it's impossible to define an answer to $\dfrac{7}{0}$.

In general, any attempt to define division by zero leads to a contradiction.

$$\frac{a}{0} = q \qquad \text{implies} \qquad q \cdot 0 = a$$

But $q \cdot 0 = 0$ not a. So assuming $\dfrac{a}{0}$ is a number leads to a contradiction.

$$\frac{0}{0} = q \qquad \text{implies} \qquad q \cdot 0 = 0$$

but 0 times any number is zero, so q can be any number, that is, $\dfrac{0}{0}$ is indeterminant.

In summary, you cannot divide by zero.

According to the definition of division above, each quotient is defined only if a number can be found that will serve as the quotient.

If we are limited to the set of integers, $\frac{5}{3}$ is undefined because there is no integer q such that $3 \cdot q = 5$.

To make division possible for more than a few special cases we must define another set of numbers.

1.3D Definition: Rational Numbers

The set of rational numbers, Q, is the set of all numbers that can be written as the ratio of two integers.

$q \in Q$ if and only if q can be written $q = \frac{a}{b}$ where a and b are integers and $b \neq 0$.

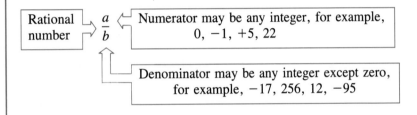

Some samples of rational numbers are

$$\frac{5}{3}, \; -\frac{5}{3}, \; \frac{0}{3}, \; \frac{4}{1}, \; 5, \; -\frac{5}{1}, \; \frac{7}{-1}, \; \frac{2159}{3456}, \; 0.25$$

This definition means that all integers are rational numbers.

With the definition of the set of rational numbers we now have a set of numbers that is closed for the basic operations of arithmetic—addition, subtraction, multiplication, and division.

Operations with Rational Numbers

Most of the procedures for operations with rational numbers will be familiar to you from arithmetic. However, as you read this section, pay attention to why the procedures of arithmetic work rather than simply to what you should do.

Since we use multiplication to define division, we start with multiplication.

1.3E Rule: Multiplication of Rational Numbers

The product of two rational numbers $\dfrac{a}{b}$ and $\dfrac{c}{d}$ is:

$$\frac{a}{b} \cdot \frac{c}{d} = \frac{a \cdot c}{b \cdot d}$$

Notice that the expression on the left is the product of two fractions. The expression on the right is a single fraction with a product in the numerator and another product in the denominator.

1.3F Definition: Identity Elements for Multiplication

$$\frac{a}{b} \cdot 1 = \frac{a}{b} \qquad b \neq 0$$

1 is the identity element for multiplication. Any number multiplied by 1 is equal to itself.

1.3G Definition: Division for Rationals

Let a and b represent any rational numbers provided $b \neq 0$.

$$\frac{a}{b} = q \qquad \text{if} \qquad q \cdot b = a$$

There are two important ideas that follow from these definitions.

I. Suppose that $b = 1$. The definition of division becomes $\dfrac{a}{1} = q$ which implies that $a = 1 \cdot q$, but $1 \cdot q = q$ since 1 is the identity element for multiplication. Therefore, $q = a$ which means that $\dfrac{a}{1} = a$.

This means that anything divided by 1 is equal to itself. One is the identity element for both multiplication and division.

Example 24 □ $\quad\dfrac{x}{1} = x\quad,\quad \dfrac{\dfrac{a}{b}}{1} = \dfrac{a}{b}$ □

II. Now suppose that $b = a$ in $\dfrac{a}{b} = q$. We have $\dfrac{a}{a} = q$ which implies $a \cdot q = a$, but, since 1 is the identity element for multiplication, $a \cdot q = a$ means $q = 1$. Therefore, $\dfrac{a}{a} = 1$ provided that $a \neq 0$. In other words anything, except zero, divided by itself is one.

Example 25 □ $\quad\dfrac{4}{4} = 1,$

1, 1

$$\dfrac{-4}{-4} = \underline{\qquad}\quad,\quad \dfrac{x}{x} = \underline{\qquad}\quad,$$

1, 1, $\dfrac{3}{4}$

$$\dfrac{\dfrac{a}{b}}{\dfrac{a}{b}} = \underline{\qquad}\quad,\quad \dfrac{\dfrac{3}{4}}{\dfrac{3}{4}} = \underline{\qquad}\quad,\quad \dfrac{\dfrac{3}{4}}{1} = \underline{\qquad\qquad}$$ □

To change the form of a fraction, that is, to reduce the fraction to lower terms or build it to higher terms, use the following definition.

1.3H   Definition: Equivalent Fractions

The rational number $\dfrac{a}{b}$ is equivalent to the rational number $\dfrac{c}{d}$ if

$$\dfrac{a}{b} = 1 \cdot \dfrac{c}{d}\qquad b \neq 0,\ d \neq 0$$

> But multiplying by 1 doesn't change anything.

> That's the point! However, if we write 1 as $\dfrac{2}{2}, \dfrac{3}{3}, \dfrac{-4}{-4},$ things will look different.

When two rational numbers are equal but are written in different forms, they are said to be **equivalent.**

The multiplication property of 1 can be used to write fractions that are equivalent to a given fraction.

Example 26 □ Write a fraction equivalent to $-\dfrac{3a}{5}$ with a denominator of 15.

$$-\frac{3a}{5} = \frac{?}{15}$$

$$-\frac{3a}{5} = -\frac{3a}{5} \cdot 1 \qquad 5 \cdot \underline{\hspace{2cm}} = 15$$

Therefore, we write one as \underline{\hspace{2cm}}

$$-\frac{3a}{5} = -\frac{3a}{5} \cdot \frac{3}{3}$$

$$-\frac{3a}{5} = \underline{\hspace{2cm}}$$

$\dfrac{3}{3}$

3

$-\dfrac{9a}{15}$

That's why
$-\dfrac{3a}{5} \cdot 1 = -\dfrac{9a}{15}.$

Now you're catching on.

□

Reducing Fractions

To find equivalent fractions, multiply by 1. A fraction with no common factor except 1 in the numerator and denominator is called a **fraction in lowest terms**.

1.3I   Rule: To Reduce Fractions to Lowest Terms

1. Write the numerator and denominator in factored form.
2. Remove the common factors from the numerator and denominator or divide numerator and denominator by the same number. Continue the process until there is no common factor except 1 in the numerator and denominator.

Example 27 □ Reduce $\dfrac{-4x}{-12}$ to lowest terms.

$$\frac{-4x}{-12} = \frac{-4 \cdot x}{-4 \cdot 3}$$

1

$$= \underline{\hspace{2cm}} \cdot \frac{x}{3}$$

$$= \frac{x}{3}$$

□

Division of Rational Numbers

To carry out the division of rational numbers we need another definition.

Recall that the additive inverse of a number is what you would add to the number to get zero (the additive identity) for an answer. The symbol for the additive inverse of a is an a with a negative sign in front of it. And the additive inverse of $-a$ is a, hence,

$$a + (-a) = 0$$

Similarly, there is a multiplicative inverse for every number except zero. The symbol for the multiplicative inverse of a number a is $\frac{1}{a}$.

1.3J   Definition: Multiplicative Inverses

Two rational numbers are multiplicative inverses of each other if their product is one.

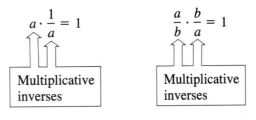

Reciprocal

Multiplicative inverses are sometimes called **reciprocals**.

Just as the sum of a number and its additive inverse gives the additive identity element zero, the product of a number and its multiplicative inverse gives the multiplicative identity element one.

How to Divide Rational Numbers

At this point we have defined division of rational numbers by saying $\frac{a}{b} = q$ if q exists so that $b \cdot q = a$, but we didn't tell you how to find q.

Here's how we can find a quotient $R \div S$ or $\frac{R}{S}$ when R and S are both rational numbers. Let $R = \frac{a}{b}$, $S = \frac{c}{d}$ when a, b, c, d are all integers, b, c, $d \neq 0$.

Then the quotient of two rational numbers can be written

$$\frac{a}{b} \div \frac{c}{d} = \frac{\dfrac{a}{b}}{\dfrac{c}{d}}$$

$$= \frac{\dfrac{a}{b}}{\dfrac{c}{d}} \cdot 1$$
Any number can be multiplied by 1 without changing its value

$$= \frac{\dfrac{a}{b}}{\dfrac{c}{d}} \cdot \frac{\dfrac{d}{c}}{\dfrac{d}{c}}$$
Any number divided by itself is 1. Therefore the expression in the box is 1

$$= \frac{\dfrac{a}{b} \cdot \dfrac{d}{c}}{\dfrac{c}{d} \cdot \dfrac{d}{c}}$$
Above the line is the product of two fractions; below the line is also the product of two fractions

$$= \frac{\dfrac{a}{b} \cdot \dfrac{d}{c}}{1}$$
The denominator is 1

$$= \frac{a}{b} \cdot \frac{d}{c}$$
Any number divided by 1 is itself

Division Using the Multiplicative Inverse

1.3K   Rule for Division of Fractions

To divide by a number, multiply by its multiplicative inverse.

$$\frac{a}{b} \div \frac{c}{d} = \frac{a}{b} \cdot \frac{d}{c}$$

To divide by a fraction, invert the divisor and multiply.

An example from arithmetic illustrates this. Six divided by two is the same as six times one half.

$$6 \div 2 = 6 \cdot \frac{1}{2}$$

Example 28 □

$$\frac{3}{4} \div \frac{1}{2} = \frac{3}{4} \cdot \frac{2}{1}$$

⟸ This is a multiplication

This is a division

$$= \frac{3}{\underset{2}{4}} \cdot \frac{\overset{1}{2}}{1}$$

$$= \frac{3}{2}$$ □

Example 29 □

$$\frac{5a}{4} \div \frac{15ab}{2} = \frac{5a}{4} \cdot \frac{2}{15ab}$$

$$= \frac{\overset{1}{5a}}{\underset{2}{4}} \cdot \frac{\overset{1}{2}}{\underset{3}{15ab}}$$

$$= \frac{1}{6b}$$ □

Since there is a close connection between multiplication and division, the same rules for signs apply to division as to multiplication.

$$(-2)(+3) = -6 \qquad \text{therefore} \qquad \frac{-6}{-2} = +3$$

$$(+2)(-3) = -6 \qquad \text{therefore} \qquad \frac{-6}{+2} = -3$$

$$(-2)(-3) = +6 \qquad \text{therefore} \qquad \frac{+6}{-2} = -3$$

1.3L Rule: Division of Signed Numbers

When two numbers with like signs are divided, the quotient is positive. When two numbers with unlike signs are divided, the quotient is negative.

$$\frac{+}{+} \qquad \text{and} \qquad \frac{-}{-} \quad \Rightarrow \quad +$$

$$\frac{+}{-} \qquad \text{and} \qquad \frac{-}{+} \quad \Rightarrow \quad -$$

Example 30 □

$$\frac{-36}{4} = \underline{\hspace{1cm}} \qquad \frac{48}{-6} = \underline{\hspace{1cm}} \qquad \frac{-56}{-8} = \underline{\hspace{1cm}}$$ □

−9, −8, 7

What Happens When You Multiply Two Fractions

Multiply.

$$\frac{8}{9} \cdot \frac{15}{16} = \frac{8(15)}{9(16)}$$

First use the definition of multiplication to write the product as a simple fraction.

$$= \frac{8 \cdot 15}{9 \cdot 16}$$

Factor the numerator and denominator.

$$= \frac{8 \cdot 3 \cdot 5}{3 \cdot 3 \cdot 8 \cdot 2}$$

Why these factors? You could have factored $8 = 2 \cdot 4$ or $16 = 4 \cdot 4$.

We might have tried that, but, since we've had experience, we identified the largest factors common to the numerator and denominator.

Use the commutative property of multiplication to rearrange the denominator.

$$= \frac{8 \cdot 3 \cdot 5}{8 \cdot 3 \cdot 3 \cdot 2}$$

Now rewrite the product as a multiplication of separate fractions.

$$= \frac{8}{8} \cdot \frac{3}{3} \cdot \frac{5}{3 \cdot 2}$$

Identify multiplication by 1.

$$= \frac{8}{8} \cdot \frac{3}{3} \cdot \frac{5}{3 \cdot 2}$$

Multiply by 1.

$$= \frac{5}{3 \cdot 2}$$

Write the numerator and denominator in simplest form.

$$= \frac{5}{6}$$

Why can't I just cancel?

The term **cancel** means do the first five steps above. In practice we do skip writing many of these steps. Most errors in cancelling come when we fail to recognize that what we really are doing is multiplying by one.

Cancelling is only a shorter way to divide out. Consider the previous multiplication

$$\frac{8}{9} \cdot \frac{15}{16} = \frac{\overset{1}{\cancel{8}} \cdot \overset{5}{\cancel{15}}}{\underset{3}{\cancel{9}} \cdot \underset{2}{\cancel{16}}}$$

Here we say that $8 \div 8 = 1$ and $16 \div 8 = 2$. Then we write the results after we divide out the common factor of 8.

Similarly we say that $9 \div 3 = 3$ and $15 \div 3 = 5$, and write the result. This leaves

$$= \frac{1 \cdot 5}{3 \cdot 2}$$

$$= \frac{5}{6}$$

Sometimes this process is called **cancelling.** We prefer to call this process **dividing out like factors.** It is equivalent to identifying multiplications by 1. You can divide out any factor above the line with the same factor anywhere below the line.

Example 31 □ Multiply $\dfrac{9x}{y} \cdot \dfrac{7y}{15}$.

$$\frac{9x}{y} \cdot \frac{7y}{15} = \frac{\overset{3}{\cancel{9}}x}{\underset{1}{\cancel{y}}} \cdot \frac{7\overset{1}{\cancel{y}}}{\underset{5}{\cancel{15}}}$$

$$= \frac{21x}{5}$$

□

Can you divide out any factor above the line with the same factor below the line?

Only if it's a pure multiplication problem. You can't cancel addition or subtraction.

$\dfrac{4 \cdot 3}{4 \cdot 5}$ can be reduced.

$\dfrac{4 + 3}{4 \cdot 5}$ **CANNOT** be reduced.

Evaluating Rational Expressions

Frequently algebraic expressions involve more than one operation, and the order in which the operations are done can change the result. The value of $5 + 3 \cdot 2$ depends on whether you do the addition or multiplication first.

1.3M   Rules of Order

Here is a brief reminder of the procedure you should use when evaluating expressions and the rules of order you should follow.

1. Replace all variables with their assigned value.
2. Simplify all expressions within symbols of grouping, such as parentheses.
3. Evaluate expressions with exponents.
4. Multiply and divide, working from left to right in order.
5. Add and subtract, working from left to right in order.

Example 32 □ Evaluate $(-28 + 8) \div 5 \cdot (-3)^2 - 4$.

$$(-28 + 8) \div 5 \cdot (-3)^2 - 4 =$$

-20

$$\underline{} \div 5 \cdot (-3)^2 - 4 = \qquad \text{Parentheses first}$$

9

$$-20 \div 5 \cdot \underline{} - 4 = \qquad \text{Evaluate exponents}$$

$$-4 \cdot 9 - 4 = \qquad \begin{array}{l}\text{Multiply and divide from}\\ \text{left to right in order}\end{array}$$

$$-36 - 4 = \qquad \begin{array}{l}\text{Multiplication before}\\ \text{subtraction}\end{array}$$

-4, -40

$$-36 + \underline{} = \underline{} \qquad \text{Add signed numbers} \qquad \square$$

Example 33 □ Evaluate $2 - \dfrac{a - b}{2}$ if $a = 2$ and $b = -4$.

$$2 - \frac{a - b}{2} =$$

$$2 - \frac{2 - (-4)}{2} = 2 - \frac{2 + 4}{2}$$

$$= 2 - \frac{(6)}{2}$$

$$= 2 - 3$$

Remember that fraction bars are symbols of grouping. That's why we add the 2 + 4 before we divide by 2.

-1

$$= \underline{} \qquad \qquad \square$$

Example 34 □ Evaluate $\dfrac{2a^2}{3} - \dfrac{(a-b)^3}{2}$ if $a = -2$ and $b = -4$.

$$\frac{2a^2}{3} - \frac{(a-b)^3}{2} = \frac{2(-2)^2}{3} - \frac{[-2-(-4)]^3}{2}$$

$$= \frac{2(4)}{3} - \frac{(2)^3}{2}$$

$$= \frac{8}{3} - \frac{8}{2}$$

$$= \frac{8}{3} - 4$$

$$= -\frac{4}{3}$$

□

Problem Set 1.3

Determine the multiplicative inverse of the following.

1. 4

2. -8

3. $\dfrac{1}{3}$

4. $-\dfrac{8}{9}$

5. $-\dfrac{12}{7}$

Determine the reciprocal of the following.

6. 6

7. $-\dfrac{1}{3}$

8. a

9. $-b$

10. 0

Evaluate the following.

11. $-9 \div 1$

12. $75 \div (-3)$

13. $(-0.4)(-0.3)(20)$

14. $(0.6)(-0.3)(0.2)$

15. $(-1.8)(30)$

16. $(-2.8)(-0.04)$

17. $60 \div (-2)$

18. $-8 \div (-1)$

19. $28 \div 0.4$

20. $0.28 \div (-0.4)$

21. $(0.2)(-0.3)(-0.4)$

22. $(1.5)(0.20)$

23. $(-1.2)(-0.60)$

24. $(-0.04)(-0.12)$

25. $-0.64 \div (-0.08)$

26. $(-7.2) \div (0.09)$

27. $\dfrac{(-8)(-6)}{-3}$

28. $\dfrac{12(-4)}{|-6|}$

29. $-11 + 3 - 8$

30. $\dfrac{(-8)(-12)}{-24}$

31. $\dfrac{(-24)(-6)}{|-12|}$

32. $\dfrac{15(-9)}{-45}$

33. $\dfrac{|-18|(-15)}{-9}$

34. $|-9| - (-3) + (-4)$

35. $(-15) - |-5| - (-3)$

36. $|-20| - |-10| - (10)$

37. $|-12| + |(-6) - 4| - |-10|$

38. $|10| - (-10) + |-8| - (-8)$

39. $(-12) + (-7) - |(-9) - 5|$

40. $-|-18 - (7 - 12)|$

41. $|-12| - |(-3) - (-8)|$

42. $(-15) + (-6) - |7 - 9|$

43. $|(-10) + (10)| - |6 - (-6)|$

44. $-|-15 + (6 - 9)| + (-8) - |-8 + 6|$

45. $-\dfrac{2}{3} \div \dfrac{8}{9}$

46. $-\dfrac{6}{5} \div \dfrac{-12}{-15}$

47. $\left(-4\dfrac{2}{3}\right) \div \left(-3\dfrac{1}{9}\right)$

48. $\left(3\dfrac{3}{4}\right) \div \left(4\dfrac{1}{6}\right)$

49. $(-1.44) \div (1.2)$

50. $(-6) \div (-0.03)$

51. $-5\dfrac{1}{3} \div \left(4\dfrac{4}{5}\right)$

52. $\left(-10\dfrac{2}{3}\right) \div \left(-6\dfrac{2}{5}\right)$

53. $\dfrac{-16y}{-15} \cdot \left(-\dfrac{14y}{28}\right)$

54. $-\dfrac{36x^2}{24} \cdot \left(\dfrac{-10x}{-9}\right)$

Looks like I'll need my calculator for these.

Perhaps, but if you look again I think you'll see that most of these can be done mentally.

55. $-\dfrac{24x}{13} \div \left(-\dfrac{15}{26x^2}\right)$

56. $\dfrac{-36a^2}{-35} \div \left(-\dfrac{12}{49a}\right)$

57. $-\dfrac{18y^2}{25} \div \left(-\dfrac{24}{35y}\right)$

58. $-\dfrac{42x^3}{55} \div \left(\dfrac{-72}{-77x}\right)$

59. $-\dfrac{1.72}{9} \cdot \left(-\dfrac{135}{4}\right)$

60. $-\dfrac{0.084}{125} \cdot \left(\dfrac{-625x}{-14}\right)$

61. $-\dfrac{72x}{75} \div \left(-\dfrac{18}{1.5x}\right)$

62. $-\dfrac{1.24m^2}{34} \div \left(-\dfrac{155}{85m}\right)$

63. $\dfrac{8.8x}{9} \cdot \dfrac{27}{1.32}$

64. $\dfrac{0.096x}{135} \cdot \dfrac{45x}{0.16}$

65. $\dfrac{7.2x}{35} \div \dfrac{2.4}{105x}$

66. $\dfrac{1.55x}{93} \div \dfrac{5.5}{66x}$

Evaluate the following.

67. $10 - (-1) + (-3) \cdot 4$

68. $11 - 3[12 - 3(5 - 2)]$

69. $8 - 2 \cdot (1 - 3)^3$

70. $-8 - 2 \cdot (-4) + 3(2 - 5)^2$

71. $(-1.2)^2 - (0.6)^2 \cdot 2$

72. $7 + 34 - \dfrac{2 - 10^2}{1 - 5}$

Evaluate the following if $a = -2$, $b = 3$, $c = 0$, $x = -\dfrac{2}{3}$, $y = \dfrac{1}{2}$.

73. $\dfrac{1}{2}a^2 - xb^3 + c$

74. $\dfrac{2}{3} \cdot b^3 \cdot c^4 - \dfrac{1}{2}y \cdot a^2 b$

75. $8 - 2(a^3 + bxy + b^3)$

76. $9x - 4y + 3(a^5yc - xb^2)$

77. $6x - 8y^2 + 4(ay - bx)$

78. $\dfrac{2}{3}b^2a - c(3xy^3 + 4a^2b) + 3xa^3y^2$

Evaluate the following for the indicated values.

79. $\dfrac{1}{3}(9a - 4bx + c^2)$ if $a = -2$, $b = -\dfrac{1}{4}$, $c = 3$, $x = 5$

80. $\dfrac{3}{4}x - 15ax - xy^3$ if $a = \dfrac{1}{6}$, $x = -4$, $y = -2$

81. $6(a + b + c) - 12abc - 8(b - c)$ if $a = -1$, $b = \dfrac{1}{2}$, $c = -\dfrac{1}{2}$.

82. $\dfrac{2}{3}a(5 - x^2 + y) - \dfrac{3}{4}y^2(axy)$ if $a = -\dfrac{1}{2}$, $x = -3$, $y = -2$

83. $9ab + a(b + x) - \dfrac{32}{9}c^2bx$ if $a = -\dfrac{2}{3}$, $b = -2$, $c = \dfrac{3}{4}$, $x = -1$

84. $24ax - \dfrac{1}{5}bc + 20ab - 9cx$ if $a = \dfrac{1}{4}$, $b = -5$, $c = -1$, $x = \dfrac{2}{3}$

Addition and Subtraction of Rational Numbers

The problem with adding rational numbers is what to do with the denominators. The distributive property provides the answer.

Example 35 □ Add the following.

$$\frac{2}{7} + \frac{3}{7} = 2 \cdot \frac{1}{7} + 3 \cdot \frac{1}{7} \qquad \text{To divide by 7 multiply by } \frac{1}{7}$$

$$= (2 + 3)\frac{1}{7} \qquad \text{Distributive property}$$

$$= 5 \cdot \frac{1}{7} \qquad \text{Addition of integers}$$

$$= \frac{5}{7} \qquad \text{Multiplication of rationals} \qquad \qquad \square$$

> In short, if two fractions have a common denominator, add the numerators and place the sum over the common denominator.

> Right! We can't add two fractions until we get a common denominator.

Addition and Subtraction with Unlike Denominators

Before adding fractions whose denominators are not alike we must express both fractions as fractions with a common denominator. The easiest denominator to use is the number that contains all the factors of each denominator called the least common denominator (LCD). Frequently, the least common denominator can be found by inspection. When the LCD is not obvious, here is how to find it.

1.4A   Rule: To Find the Least Common Denominator

1. Completely factor each denominator.
2. Write each factor that appears in any denominator.
3. Build the LCD by raising each factor to the highest power it has in any single denominator.

Example 36 □ Find the least common denominator for $\dfrac{5}{24a}$ and $\dfrac{7}{30a^2}$.

Step 1. Completely factor each denominator:

$$24a = 2 \cdot 2 \cdot 2 \cdot 3 \cdot a = 2^3 \cdot 3^1 \cdot a$$
$$30a^2 = 2 \cdot 3 \cdot 5 \cdot a \cdot a = 2^1 \cdot 3^1 \cdot 5^1 \cdot a^2$$

Step 2. Write each factor that appears in any denominator:

$$2 \cdot 3 \cdot 5 \cdot a$$

Step 3. Raise each factor to the highest power it has in any single denominator:

$$LCD = 2^3 \cdot 3^1 \cdot 5^1 \cdot a^2$$
$$= 8 \cdot 3 \cdot 5 \cdot a^2$$
$$= 120a^2 \qquad\qquad □$$

Notice that the least common denominator is the smallest number that can be divided by each of the denominators.

Example 37 □ Add.

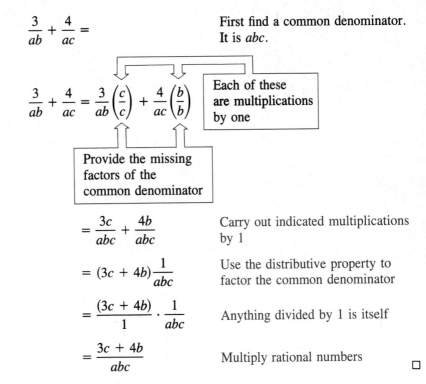

$$\dfrac{3}{ab} + \dfrac{4}{ac} = \qquad\qquad \text{First find a common denominator.}$$
It is abc.

$$\dfrac{3}{ab} + \dfrac{4}{ac} = \dfrac{3}{ab}\left(\dfrac{c}{c}\right) + \dfrac{4}{ac}\left(\dfrac{b}{b}\right)$$

Each of these are multiplications by one

Provide the missing factors of the common denominator

$$= \dfrac{3c}{abc} + \dfrac{4b}{abc} \qquad\qquad \text{Carry out indicated multiplications by 1}$$

$$= (3c + 4b)\dfrac{1}{abc} \qquad\qquad \text{Use the distributive property to factor the common denominator}$$

$$= \dfrac{(3c + 4b)}{1} \cdot \dfrac{1}{abc} \qquad\qquad \text{Anything divided by 1 is itself}$$

$$= \dfrac{3c + 4b}{abc} \qquad\qquad \text{Multiply rational numbers} \qquad □$$

1.4B Rule: Addition of Rational Numbers

In general, if b and d are not equal to zero, we can find the sum: $\dfrac{a}{b} + \dfrac{c}{d}$.

$$\frac{a}{b} + \frac{c}{d} = \frac{a}{b}\left(\frac{d}{d}\right) + \frac{c}{d}\left(\frac{b}{b}\right)$$
Multiply by 1 to get common denominator

$$= \frac{ad}{bd} + \frac{cb}{bd}$$
Rewrite each indicated product as a single fraction with a common denominator

$$= \left(\frac{ad + cb}{1}\right)\frac{1}{bd}$$
Distributive property

$$= \frac{ad + cb}{bd}$$
Rewrite the product indicated on the line above as a single fraction

Therefore:

$$\frac{a}{b} + \frac{c}{d} = \frac{ad + bc}{bd} \qquad \text{if } b, d \neq 0$$

Example 38 □ Add $\dfrac{7b}{15a} + -\dfrac{5a}{6b}$.

A common denominator is _____

$30ab$

$\dfrac{2b}{2b}, \dfrac{5a}{5a}$

$$\frac{7b}{15a} + -\frac{5a}{6b} = \underline{\quad\quad} \cdot \frac{7b}{15a} + \underline{\quad\quad} \cdot -\frac{5a}{6b}$$

$$= \frac{14b^2}{30ab} + -\frac{25a^2}{30ab}$$

$$= \underline{\quad\quad}$$

$\dfrac{14b^2 - 25a^2}{30ab}$

Remember, we multiply each of the original fractions by some form of 1 to convert the original denominator to the LCD.

□

Subtraction of Rational Numbers

The rule for subtraction of rationals is the same as for subtraction of integers. To subtract, add the additive inverse.

How do I find the additive inverse of $\dfrac{a}{b}$?

Ask what would you add to $\dfrac{a}{b}$ to get zero for an answer.

1.4C   Definition: Additive Inverse of a Rational Number

For every rational number $\frac{a}{b}$ there exists an additive inverse $-\frac{a}{b}$ so that

$$\frac{a}{b} + \left(\frac{-a}{b}\right) = 0 \qquad \text{if } b \neq 0$$

Example 39 □ Write the additive inverse of each of the following.

$$\frac{x}{y} \Rightarrow -\frac{x}{y} \qquad \frac{2}{3} \Rightarrow \underline{\hspace{1.5cm}} \qquad -\frac{2}{3} \Rightarrow \underline{\hspace{1.5cm}} \qquad \qquad □$$

▶ **Pointers for Better Style**

1. Notice that the above example uses $\Rightarrow$ and not an $=$ sign. That's because, except for zero, a number and its additive inverse are definitely not equal. The equal sign does not indicate merely where to write the answer. The equal sign indicates that the expressions on both sides of it represent the same number, or are in fact equal.

2. Notice that the above example has no negative signs in the denominators. The additive inverse of $-\frac{3}{4}$ is $\frac{3}{4}$. But, to prove they are additive inverses, we must find a common denominator and add. Therefore we prefer to write fractions in standard form. Standard form means the negative sign, if any, goes in the numerator.

What does $-\frac{1}{2}$ mean?

The negative sign isn't in either the numerator or the denominator.

$-\frac{1}{2}$ is the negative of $\frac{1}{2}$.

$-\frac{1}{2}$ in standard form is written $\frac{-1}{2}$.

Standard Form of a Fraction

In general, if a and b represent positive numbers, the standard form of a fraction is: $\frac{a}{b}$ for positive fractions $-\frac{a}{b}$ for negative fractions

What's the difference between rational numbers and fractions?

A rational number is the quotient of two integers. All rational numbers are fractions. But some fractions have numerators or denominators that are not integers, for example, $\frac{4}{\pi}, \frac{\sqrt{2}}{3}$.

1.4D Rule: Subtraction of Rational Numbers

$$\frac{a}{b} - \frac{c}{d} = \frac{a}{b} + \left(-\frac{c}{d}\right)$$

To subtract a rational number, add its additive inverse.

Example 40 □ Subtract.

$$\frac{1}{4} - \frac{3}{4} = \frac{1}{4} + \left(\frac{-3}{4}\right) \qquad \text{Write as addition}$$

$$= \frac{1 + (-3)}{4} \qquad \text{Indicate addition of the numerators over the common denominator}$$

$$= -\frac{2}{4} \qquad \text{Add the numerators}$$

$$= -\frac{1}{2} \qquad \text{Reduce} \qquad \qquad □$$

Example 41 □ Subtract.

$\dfrac{2}{b}$

$$\frac{3}{a} - \left(\frac{-2}{b}\right) = \frac{3}{a} + \underline{\qquad} \qquad \text{Write as addition}$$

$\dfrac{2a}{ab}$

$$= \frac{3b}{ab} + \underline{\qquad} \qquad \text{Common denominator}$$

$\dfrac{2a + 3b}{ab}$

$$= \underline{\qquad} \qquad \text{Add numerators} \qquad □$$

Problem Set 1.4

Perform the indicated operations. Reduce all answers to lowest terms and write fractions in standard form.

1. $-\dfrac{17a}{12} + \left(\dfrac{-8}{15}\right)$

2. $-\dfrac{13x}{15} + \left(\dfrac{-9}{10}\right)$

3. $-\dfrac{11xy}{18} - \left(\dfrac{-7}{12}\right)$

4. $-\dfrac{7ax}{12} - \dfrac{5}{9}$

5. $-\dfrac{ax^2}{8} + \dfrac{8ax^2}{14}$

6. $-\dfrac{3ax^3}{4} + \left(\dfrac{-13ax^3}{18}\right)$

7. $\dfrac{3x}{8a} - \left(\dfrac{-7x}{12a}\right)$

8. $\dfrac{5x}{9ab} - \left(\dfrac{-7x}{15ab}\right)$

9. $-\dfrac{5by}{a} + \dfrac{13by}{c}$

10. $\dfrac{3a}{8x} + \left(\dfrac{-7a}{12y}\right)$

11. $-\dfrac{7}{ax} - \dfrac{9}{bx}$

12. $-\dfrac{11}{xy} - \left(\dfrac{-13}{ay}\right)$

13. $-\dfrac{12x}{a^2y} - \left(\dfrac{-7}{a^2x}\right)$

14. $\dfrac{15x^2}{a^2y} + \left(\dfrac{-6}{a^2x}\right)$

15. $-\dfrac{19a}{12by} - \left(\dfrac{-17b}{15ay}\right)$

16. $-\dfrac{7a}{18bx} - \left(\dfrac{-13b}{12ax}\right)$

17. $-\dfrac{5x^2}{6aby} + \dfrac{9ay}{14bx^2}$

18. $\dfrac{5cx}{8ab} + \left(\dfrac{-17b}{20a^2x}\right)$

19. $\dfrac{7b}{18axy^2} - \left(\dfrac{-7a}{15bx^2y}\right)$

20. $-\dfrac{8ac}{9bx^2y} - \left(\dfrac{-10bc}{21axy^2}\right)$

21. $-\dfrac{12a}{bx^2} - \dfrac{6b}{ax^2}$

22. $\dfrac{12x^2}{b^2x} + \left(-\dfrac{9}{b^2y}\right)$

23. $-\dfrac{11x}{8by} - \left(-\dfrac{13c}{12ay}\right)$

24. $-\dfrac{5b}{12ay} - \left(-\dfrac{17a}{18bx}\right)$

25. $\dfrac{4a}{9x} - \dfrac{9a}{8y}$

26. $-\dfrac{7x}{8ay} - \dfrac{5y}{12ax}$

27. $-\dfrac{3ax}{8by} - \left(-\dfrac{11ay}{12bx}\right)$

28. $-\dfrac{23x}{18ay} - \left(-\dfrac{19y}{15ax}\right)$

29. $-\dfrac{7y^2}{6abx} + \dfrac{11ax}{14by^2}$

30. $\dfrac{5ab}{12xy} + \left(-\dfrac{13cx}{20ay^2}\right)$

31. $\dfrac{11a}{12bx^2y} - \left(-\dfrac{11b}{18axy^2}\right)$

32. $-\dfrac{9ab}{8cx^3y} - \left(-\dfrac{11bc}{12bx^2y^2}\right)$

33. $\dfrac{11by}{9ax^2} + \dfrac{7ay}{21bx^2}$

34. $\dfrac{3ay}{8bc} + \left(-\dfrac{9x}{20bc^2}\right)$

35. $\dfrac{11ba}{15cxy^2} - \left(-\dfrac{11ab}{18cx^2y}\right)$

36. $-\dfrac{10ax}{9cxy^3} - \left(-\dfrac{8by}{21cx^2y^2}\right)$

■ **1.5**   **Problem Solving**

This text makes a careful effort to help you develop the skills needed to apply mathematics to practical situations. Throughout the text we will help you:

—To develop translation skills to express ideas that are imprecisely expressed in English in precise mathematical terms.

—To critically analyze a problem so you can identify both important and irrelevant bits of information in a problem.

—To build a knowledge base of formulas that you can apply to problems in this book and throughout life.

—To visualize or draw mental pictures based on the information given in a problem.

Now we will start building translation visualization and formula skills by reviewing some of the most useful formulas dealing with perimeter, area, and volume.

Many geometric problems can be visualized as combinations of circles, rectangles, and triangles. In this section we will develop the formulas for perimeter, area, and volume of these common shapes. Also, to aid with the analysis of practical problems, we will practice visualizing how these shapes can be combined. These formulas have wide applications in business and science.

Perimeter

The perimeter of a geometric figure is the distance around its outside. You might think of perimeter as the length of the fence that would be needed to enclose a figure.

1.5A Perimeter of a Rectangle

The perimeter of a rectangle is:

$$P = 2L + 2W \qquad \text{where } L = \text{length}; w = \text{width}$$

W

$L \qquad\qquad\qquad\qquad L$

W

The perimeter or distance around this rectangle is:

$$W + L + W + L \text{ or } 2L + 2W$$

Example 42 □ A group of students on the way to the beach to play volleyball want to mark the outside of the court with a rope. If the court is twenty feet wide by forty feet deep, how long a rope will they need?

Since a volleyball court is a rectangle, we are looking for the perimeter of a rectangle.

$$P = 2L + 2W$$
$$= 2(20 \text{ ft}) + 2(40 \text{ ft})$$
$$= 40 \text{ ft} + 80 \text{ ft}$$
$$= 120 \text{ ft of rope is needed} \qquad □$$

Area

The area of a rectangle is given by the formula:

■ Figure 1.1

$$A = L \cdot W$$

W gives the number of rows each containing L square units

L gives the number of square units in one row

Since area is the product of two lengths, it is always measured in square units.

measure of length	abbreviation	measure of area	abbreviation
inches or in	in	square inches or in^2	in^2 or sq in
feet or ft	ft	square feet or ft^2	ft^2 or sq ft
centimeters or cm	cm	square centimeters or cm^2	cm^2 or sq cm
meters or m	m	square meters or m^2	m^2 or sq m

Example 43 ☐ Find the area of a rectangular patio that is twelve by ten feet. Use the formula for area of a rectangle.

$$A = L \cdot W$$
$$= 10 \text{ ft} \cdot 12 \text{ ft}$$
$$= 120 \text{ ft}^2$$

You substituted the smaller value for the length.

It doesn't matter which side you call the length or the width.

☐

Notice that when we multiplied ft · ft we got ft^2 which is an appropriate unit to measure area.

Example 44 ☐ Find the area of a square silicon computer chip that is eight-tenths of a centimeter on a side.

Since a square is just a rectangle with length equal to width, we can use the formula for the area of a rectangle.

$$A = L \cdot W$$
$$= 0.8 \text{ cm} \cdot 0.8 \text{ cm}$$
$$= 0.64 \text{ cm}^2$$

When numbers start with a decimal point we usually write a leading zero to make the decimal point more visible.

☐

Other Area Formulas

The area of a right triangle is obtained by visualizing it as half a rectangle.

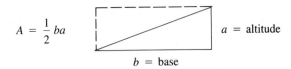

$$A = \frac{1}{2} ba$$

a = altitude

b = base

Base and altitude are the equivalent of width and length for a triangle. If the altitude or height of a triangle is defined as a perpendicular line from the base to the opposite vertex, the formula for area of a triangle applies to all triangles.

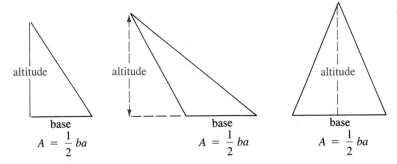

altitude

base

$$A = \frac{1}{2} ba$$

altitude

base

$$A = \frac{1}{2} ba$$

altitude

base

$$A = \frac{1}{2} ba$$

Example 45 □ Find the area of a triangular-shaped lawn that is ten feet along the base. The perpendicular distance from the base to the vertex of the triangle is four yards.

We can use the formula for area of a triangle with base = 10 feet and altitude = 4 yards.

What's feet times yards?

You can't simply multiply; both sides have to be in the same units.

To get both base and altitude in the same units, we will convert 4 yards to 12 feet.

$$A = \frac{1}{2} bh$$

$$= \frac{1}{2}(12 \text{ ft})(10 \text{ ft})$$

$$= \frac{1}{2}(120 \text{ ft}^2)$$

$$= 60 \text{ square ft is the area of the lawn} \qquad \square$$

Circles

A circle is defined as the set of points at a given distance (called the **radius**) from a fixed point (called the **center of the circle**).

■ Figure 1.4

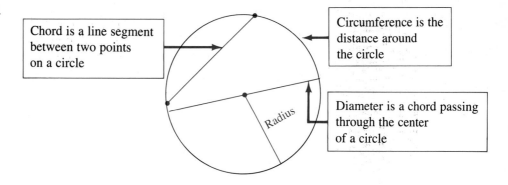

Chord is a line segment between two points on a circle

Circumference is the distance around the circle

Diameter is a chord passing through the center of a circle

Radius

Formulas for a Circle

$$\text{diameter:} \quad d = 2r$$
$$\text{circumference:} \quad c = \pi d$$
$$= 2\pi r$$
$$\text{Area:} \quad A = \pi r^2$$

We can combine the formulas for the areas of a circle and of a rectangle to find the surface area of a right circular cylinder.

■ Figure 1.5

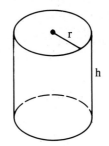

■ Figure 1.6

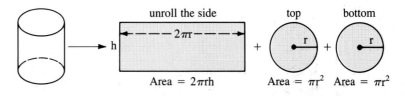

unroll the side

top   bottom

$2\pi r$

h

r   r

Area $= 2\pi rh$ Area $= \pi r^2$ Area $= \pi r^2$

To find the surface area of this figure, we will take it apart into component surfaces. Then add all the surface areas to find the surface area of the entire cylinder.

The surface area of a right circular cylinder is given by:

$$A = 2\pi rh + 2\pi r^2$$

Example 46 □ Find the surface area of a gasoline tank made in the form of a right circular cylinder with a diameter of 8 ft and a height of 12 ft.

■ Figure 1.7

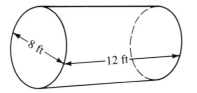

The diameter is 8 ft, but the radius is 4 ft.

Total area = area of side + 2 times area of one end

$$
\begin{aligned}
\text{Area} &= 2\pi rh + 2(\pi r^2) \\
&= 2\pi(4 \text{ ft})(12 \text{ ft}) + 2[\pi(4 \text{ ft})^2] \\
&= 96\pi \text{ ft}^2 + 2\pi \cdot 16 \text{ ft}^2 \\
&= 96\pi \text{ ft}^2 + 32\pi \text{ ft}^2 \\
&= 128\pi \text{ ft}^2 \qquad\qquad\qquad □
\end{aligned}
$$

Volume

The methods used to calculate volume are an extension of the methods used to calculate area. Area can be visualized as covering a surface with square stamps. Volume can be visualized as filling a container with cubes.

■ Figure 1.8

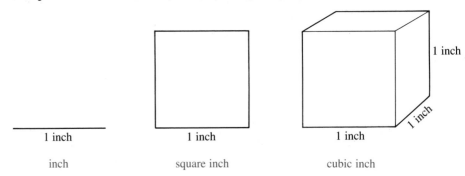

The volume of a rectangular solid is the number of cubes of a specific size that will fill it.

Example 47 □ Find the volume of a rectangular box 5 cm long by 3 cm wide by 2 cm high.

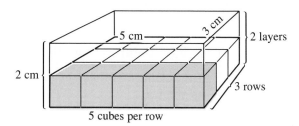

Think of this problem as filling the box with cubes 1 cm on a side:

$$1 \text{ cubic centimeter or } 1 \text{ cm}^3$$

The number of cubes needed to fill the box is:

$$5 = \text{number of cubes along the side in one row}$$

$$\underline{\times\ 3} = \text{number of rows}$$

$$15 = \text{number of cubes in one layer}$$

$$\underline{\times\ 2} = \text{number of layers}$$

$$30 = \text{number of cubes to fill the box} \qquad\qquad □$$

1.5B Formula for the Volume of a Rectangular Solid

The basic formula for volume is:

$$V = L \cdot W \cdot H \qquad \text{where } L = \text{Length}$$
$$W = \text{Width}$$
$$H = \text{Height}$$

Notice that $L \cdot W = $ Area of the base. The volume formula can be rewritten by substituting area for $L \cdot W$.

$$V = L \cdot W \cdot H \qquad\qquad \text{becomes}$$
$$V = \text{Area of the base} \cdot \text{height}$$

Volume of a Cylinder

The idea of volume as area times height applies also to the volume of a right circular cylinder.

1.5C   Volume of a Right Circular Cylinder

The volume of a right circular cylinder is the area of its base times its height.

$$V = \pi r^2 h$$

Example 48 □ Find the volume of a gasoline tank that has a diameter of 6 ft and a height of 15 ft.

■ Figure 1.10

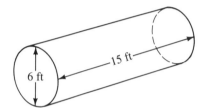

$$V = \pi r^2 h$$
$$= \pi (3 \text{ ft})^2 (15 \text{ ft})$$
$$= \pi 9 \text{ ft}^2 \cdot 15 \text{ ft}$$
$$= 135\pi \text{ ft}^3$$

You keep putting the height on the side.

The height is the distance perpendicular to the base. It doesn't matter which way you turn it.

Chapter 1 ■ Real Numbers and Their Properties

This tank has a volume of 54π cubic ft.
Replacing π with 3.14, the volume is:

$$V = 54(3.14)$$
$$= 169.56 \text{ cubic ft}$$

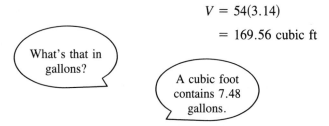

In gallons that would be approximately:

$$\text{Volume in gallons} = 169.56 \, \cancel{ft^3} \cdot 7.48 \, \frac{\text{gallons}}{ft^3}$$
$$= 1268.30 \text{ gallons} \qquad \square$$

Notice how the units ft³ cancelled in the calculation above.

PROBLEM SET 1.5

Solve the following problems.

A dollar bill measures $2\frac{5}{8}$ inches by $6\frac{1}{8}$ inches.

1. What is the area of a dollar bill?

2. What is the perimeter of a dollar bill?

The top of an office desk measures 5 feet by 34 inches.

3. What is the area of this desk top in square feet?

4. What is the area of this desk top in square inches?

5. What is the perimeter of this desk top in feet?

The diameter of a round kitchen table is 45 inches.

6. How many square feet of area is available for serving?

7. To the nearest tenth of an inch, what is the circumference of this table top?

The picture area on a portable television screen measures 15.8 inches by 12.5 inches.

8. Find the area of the television screen in square inches.

9. What is the area of the television screen in square feet?

NOTE: When an approximation for π was needed the authors used 3.14 in this section. If you use the calculator key for π your answers may be slightly larger than those given at the end of the book.

The outside measurement of a side by side refrigerator is 66 inches high, 36 inches wide, and 27 inches deep.

10. What is the volume in cubic feet of the outside of this refrigerator?

11. If the freezing compartment of this side by side refrigerator measures 53 inches by 14 inches by 17 inches, what is its volume in cubic feet?

12. If the cooling compartment of this refrigerator measures 53 inches by 19 inches by 18 inches, what is its volume in cubic feet?

13. What is the total inside capacity of the freezing compartment and cooling compartment in this refrigerator?

A triangular sail on a sailboat has a base of 15 feet and height of 20 feet.

14. How much area does one side of this sail expose to the wind?

The walls of a garage that measures 24 feet by 22 feet are to be painted.

15. If the walls are 8 feet high, how much paint is needed to paint the four walls if one gallon of paint will cover 425 square feet of surface?

A 52-inch ceiling fan is to be used to circulate the air in a family room.

16. What is the area covered when the fan is in motion? (52-inch fan means it has a diameter of 52 inches)

A pizza pan measures 15 inches in diameter.

17. What is the circumference of the 15-inch pan?

A pizza pan measures 12 inches in diameter.

18. What is the circumference of the 12-inch pan?

19. How many more square inches of pizza do you get in a 15-inch diameter pan than in a 12-inch diameter pan?

A typical motor oil can in the shape of a cylinder has a diameter of 4 inches and height of $5\frac{1}{2}$ inches.

20. What is the volume in cubic inches of the motor oil can?

21. What is the surface area in square inches of the oil can?

The stove top of a particular stove measures 36 inches by 18 inches.

22. How many square inches of stove top is available?

A circular lawn sprinkler covers a circle with a radius of 18 feet.

23. How many square feet of grass will be watered by the sprinkler?

24. What is the circumference of the circle covered by the sprinkler?

There are 72 parking spaces in a parking lot. Each space is rectangular in shape and measures 8 feet by 16 feet.

25. How much space is available for parking cars?

A coffee mug has a diameter of 3 inches and depth of $3\frac{1}{4}$ inches.

26. What is the maximum volume of coffee this cup will hold if it is completely filled?

27. What is the outer surface area of the coffee mug? (Remember, the mug is open at the top.)

An outside door on a house measures 3 feet by 80 inches.

28. What is the perimeter of the door in feet?

29. What is the area of the door in square inches?

30. What is the area of the door in square feet?

31. If the door is a solid wooden door and is $1\frac{7}{8}$ inches thick, what is its volume in cubic inches?

A countertop in a kitchen measures 6 feet by 2 feet.

32. What is the area in square feet of the countertop?

33. What is the area in square inches of the countertop?

34. What is the perimeter in feet of the countertop?

A queen-size mattress measures 81 inches by 62 inches by 8 inches.

35. What is the volume in cubic inches of the mattress?

36. What is the volume in cubic feet of the mattress?

A family room, which measures 22 feet by 15 feet, is to be carpeted.

37. Find the cost of carpeting in this family room if carpet costs $16.50 per square yard.

38. A quarter round molding is to be installed in the family room after the carpeting is completed. If there is one 3-foot door opening and an 8-foot door opening in the room, how much quarter round is to be purchased? Assume an additional two feet are purchased for waste due to cuttings.

A bedroom measures 15 feet by 12 feet with the ceiling 8 feet high.

39. How many square feet of wall space are available in the bedroom?

40. This bedroom has one window that measures 70 inches by 36 inches, a door and casing that measures 82 inches by 34 inches, and a closet door that measures 6 feet by 82 inches. How much plastered wall space is available to be papered?

41. If a single roll of wallpaper covers 35 square feet, how many single rolls of wallpaper must be purchased to paper this bedroom?

42. Afterwards it was decided to include a border at the ceiling of this bedroom. How much border is needed if one additional foot of border is purchased for waste?

A television station can televise programs for a distance of 90 miles in all directions.

43. What area in square miles does it cover?

44. What is the circumference of the area covered by the television station?

The bedroom door in a house measures 79 inches by 30 inches and is $1\frac{1}{2}$ inches thick.

45. What is the area in square inches of one side of the door?

46. What is the area in square feet side of one side of the door?

47. What is the volume in cubic inches of the door?

48. What is the volume in cubic feet of the door?

The coffee table in a home measures 54 inches by 2 feet.

49. What is the area in square feet of the top of the table?

A television screen measures 15.8 inches by 20 inches.

50. What is the perimeter of the television screen?

A student's textbook measures 11 inches by $8\frac{1}{2}$ inches by $1\frac{1}{4}$ inches.

51. What is the volume of the textbook?

At an R.V. park, a lot owner wants to cement two areas, one measures 13 feet by 50 feet and the other area 15 feet by 18 feet.

52. What is the total cost of cementing both areas if the cost of cementing is $18.00 per square yard?

A house and garage are to be built on a lot that is 64 feet wide and 120 feet deep.

53. What is the area of the lot?

Chapter 1 ■ Real Numbers and Their Properties

1. **Commutative property for addition and multiplication.**
   a. $a + b = b + a$
   b. $\quad a \cdot b = b \cdot a$
2. **Associative property for addition and multiplication.**
   a. $(a + b) + c = a + (b + c)$
   b. $\quad (a \cdot b) \cdot c = a \cdot (b \cdot c)$

3. **Multiplication by zero.**

   $a \cdot 0 = 0 \cdot a = 0$

4. **Distributive property.**
   a. $a(b + c) = a \cdot b + a \cdot c$
   b. $(b + c)a = b \cdot a + c \cdot a$

1. **Multiplication of signed numbers.**
   a. If two numbers have like signs, their product is positive.
   b. If two numbers have unlike signs, their product is negative.

2. **Multiplication of powers with like bases.**

   $a^x \cdot a^y = a^{x+y}$

3. The **set of rational numbers** is the set of all numbers that can be written as a ratio of two integers $\dfrac{a}{b}$ where $b \neq 0$.

4. The **product of two rational numbers** $\dfrac{a}{b}$ and $\dfrac{c}{d}$ is $\dfrac{a \cdot c}{b \cdot d}$.

5. The **identity element** for multiplication is one.

6. To **reduce fractions to lowest terms,** remove all common factors from the numerator and denominator.

7. Two numbers are **multiplicative inverses** of each other if their product is one.

   $\dfrac{a}{b} \cdot \dfrac{b}{a} = 1$

8. To **divide** by a number, multiply by its multiplicative inverse.

   $\dfrac{a}{b} \div \dfrac{c}{d} = \dfrac{a}{b} \cdot \dfrac{d}{c}$

9. **Division of signed numbers.**
   a. When you divide two numbers with like signs, their quotient is positive.
   b. When you divide two numbers with unlike signs, their quotient is negative.

54. If the garage measures 24 feet by 20 feet and the house measures 50 feet by 37 fe
open area remains on the lot?

A cinder block wall is to built across the back of the lot (64 feet) and along each side of th
of 105 feet from the back wall toward the front of the lot.

55. How many lineal feet of block are required?

56. If the block wall is to be 6 feet high, how many square feet of block wall will be

57. If each block along with the mortar measures 18 inches by 8 inches, how many
needed to complete the wall?

New turf is to be put on a football field.

58. If the area measures 100 yards by 50 yards, how many square feet of turf are ne

---

Chapter 1 □ Key Ideas

1.1
1. A **set** is a collection of objects.
2. Set A is **equal** to set B if both sets contain the same elements.
3. A **variable** is a symbol used to represent any element in a set of numbers.
4. The set of **natural numbers** is the set of counting numbers.
5. The set of **whole numbers** is the set of natural numbers combined with th
6. The **integers** are the set of natural numbers combined with zero and the neg
   of natural numbers.
7. **Closure**—A set is said to be closed under an operation if performing the ope
   two members of the set always yields a member of the set.
8. Zero is the **identity element** for addition.
9. Two numbers are **additive inverses** of each other if their sum is zero, $a +$
10. **Absolute value:**

$$|a| = a \qquad \text{if } a \geq 0$$
$$|a| = -a \qquad \text{if } a < 0$$

11. **Addition of signed numbers:**
    a. Numbers with **like signs:** Add their absolute value, keeping the sign of
    b. Numbers with **unlike signs:** Find the difference between their magn
       value), and use the sign of the number with the larger magnitude (absc
12. **Subtraction of signed numbers:**
    To subtract add the additive inverse.

$$a - b = a + (-b)$$

10. **Evaluating expressions.**
    a. Replace all variables with current value.
    b. Simplify all expressions within symbols of grouping.
    c. Evaluate expressions with exponents.
    d. Perform all multiplications and divisions, working from left to right.
    e. Perform all additions and subtractions, working from left to right.

1.4 1. To find the **least common denominator**
    a. Completely factor each denominator.
    b. Write each factor that appears in any denominator.
    c. Build the LCD by raising each factor to the highest power it has in any single denominator.

2. **Addition of rational numbers.**

$$\frac{a}{b} + \frac{c}{d} = \frac{ad + bc}{bd} \qquad \text{if } b, d \neq 0$$

3. The **additive inverse** of a rational number $\frac{a}{b}$ is $-\frac{a}{b}$.

4. **Subtraction of rational numbers.**

$$\frac{a}{b} - \frac{c}{d} = \frac{a}{b} + \left(-\frac{c}{d}\right)$$

$$= \frac{ad - bc}{bd}$$

Chapter 1 Review Test

Evaluate the following.

1. $|-35|$

2. $|0|$

3. $8 + (-5)$

4. $8 - 5$

5. $|-16| - (-5)$

6. $(-12) - |-6| - (-3)$

7. $5 - |(-6) - 3| - (-4)$

8. $|-11 - 12| - |8 - (-12)|$

9. $(-15) + (-12) - |(-19) + 14|$

10. $|7 - (10 - 7) + (6 - 11)|$

11. $-|18 - (4 - 6)|$

Name the property for each statement.

12. $6 + (7 + y) = (6 + 7) + y$ _____

13. $a \cdot (x \cdot 6) = a \cdot (6 \cdot x)$ _____

14. $a \cdot 0 = 0$ _____

Using only the property indicated, write an expression equivalent to the given expression.

15. Distributive property $\qquad\qquad\qquad p(2 + q) =$ _____

16. Identity element for multiplication $\qquad 1(x + 0) =$ _____

17. Identity element for addition $\qquad\quad 1(x + 0) =$ _____

18. Commutative property of multiplication $\quad 1(x + 0) =$ _____

19. Commutative property of addition $\qquad 1(x + 0) =$ _____

Evaluate the following.

20. $-\dfrac{3}{4}[5(-2) + 5 - 3^3]$ $\qquad$ **21.** $\dfrac{5(-4) + 4(-2)}{8 + (-1)}$ $\qquad$ **22.** $-35\left(-\dfrac{3}{5}\right) + 2^3 - 32 \div 4$

Evaluate the following for specified values.

23. $4a^2 - 7b + \dfrac{4}{5}c$ $\qquad$ if $a = -2, b = 3, c = -5$

24. $\left(2ab - \dfrac{a}{b}\right) + \dfrac{1}{3}(b^2 - a)$ $\qquad$ if $a = -8, b = 2$

25. $\dfrac{[a + (-a) \div b] + b}{ab \div a + 2}$ $\qquad$ if $a = 6, b = 3$

Perform the indicated operations.

26. $(-1.08) \div (-0.9) \cdot (-0.05)$ $\qquad$ **27.** $\dfrac{(-12)(-15)}{|-10|}$ $\qquad$ **28.** $-\dfrac{44x^2}{27} \cdot \dfrac{45y}{33}$

29. $\dfrac{35x^4}{65y} \cdot \dfrac{39z^3}{x^4}$ $\qquad$ **30.** $-10n \div \left(-\dfrac{5}{2n}\right)$ $\qquad$ **31.** $-15n^2 \cdot \left(-\dfrac{8n}{25}\right)$

32. $-\dfrac{36x}{11} \div \left(-\dfrac{27}{22x^3}\right)$ $\qquad$ **33.** $-\dfrac{72x}{45} \cdot \left(\dfrac{-1.5x^2}{-5.4}\right)$ $\qquad$ **34.** $\dfrac{4x}{9a} + \left(-\dfrac{5x}{12a}\right)$

35. $-\dfrac{11a}{8bx} - \dfrac{7}{12by}$ $\qquad$ **36.** $-\dfrac{5ab}{6x^2y} - \left(\dfrac{-8ab}{15xy^2}\right)$

37. A local restaurant, using an eight-inch pie tin, served a pie that was one inch thick. The pie was cut into five equal-sized pieces. How many cubic inches are in one piece of pie?

38. The transition road from one freeway to another is in the form of a circle. If the radius of this transition is 120 yards, how far to the nearest yard does a person drive when going from one freeway to the other?

39. A high school athlete was required to run five laps around the perimeter of the football field. If the field measures 160 feet by 120 yards, how far did the athlete run? Give your answer in yards.

40. A cylindrical water pitcher that measures six inches in diameter and ten inches high was filled to capacity. How many cubic inches of water does the pitcher hold? If there are 231 cubic inches in a gallon, how many gallons or what part of a gallon does the pitcher hold?

41. A family room that measures 15 feet by 22 feet has walls that are 8 feet high. On one wall there is a sliding glass door 7 feet by 8 feet and on another wall there is a window $3\frac{1}{2}$ feet by 4 feet. If a gallon of paint covers 475 square feet, how many gallons of paint are needed to paint the walls?

42. A circular sprinkler with a radius of 25 feet is set on a rectangular lot that measures 50 feet by 70 feet.
    a. How much of the lot is watered during one setting of the sprinkler?
    b. How much of the lot is not watered during one setting of the sprinkler?

2

First-Degree Equations

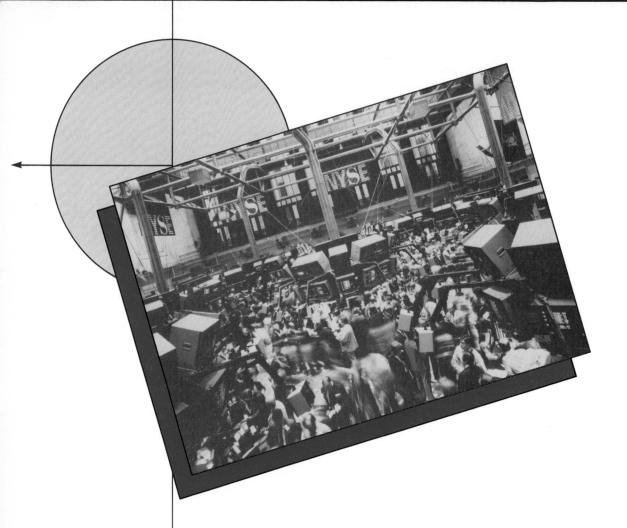

Contents

Τhis chapter will apply the basic ideas of mathematics to the solution of equations. After listing the basic properties of equality, we will solve first-degree equations.

The second section, using skills you learned in the first section, generalizes to solve equations for the value of a specified variable. This skill is particularly useful in applying general formulas to specific situations. When you have mastered this skill you will be able to solve any first-degree equation.

Many important relationships in business, for example programs that buy and sell stock, are inequalitites. Therefore, this chapter will cover the rules for the solution of inequalities. It will also cover the concept of absolute value and show you how to manipulate equations involving absolute value.

Finally it will help you develop one of the basic but frequently overlooked skills of problem solving.

■ 2.1 Solving First-Degree Equations

Properties of Equality

Earliest symbols for equality did not look like our equal sign. They tended to convey ideas like *yields, produces,* or *together.* An Egyptian hieroglyphic was used to express ideas like *2 and 3 result 5.* Our use of the equal sign has evolved to a more precise meaning, generally, along the lines of *is the same as* or *is identical to.*

When we use an equal sign between two quantities, we mean all four of the following.

2.1A Properties of Equality

For any real numbers a, b, c, there is a

1. Reflexive Property:
   $a = a$
2. Symmetric Property:
   If $a = b$ then $b = a$
3. Transitive Property:
   If $a = b$ and $b = c$, then $a = c$
4. Substitution Property:
   If $a = b$, then a may be substituted for b in any statement using b

---

Reflexive Property of Equality

John is as old as himself.

Symmetric Property of Equality

If John is as old as Mary, then Mary is as old as John.

Transitive Property of Equality

If John is as old as Mary and Mary is as old as Sue, then John is as old as Sue.

Substitution

If John is as old as Mary and Mary is 21, then John is 21.

Conditional Equation

An equation is made up of a left member and a right member joined by an equal sign. $4x + 3 = 15$ is an equation. It is a **conditional equation** because it is true only on the condition that you replace x with 3.

2.1B Definition: Solution Set

The **solution set** of an equation is the set of values that makes the equation true.

In this chapter, solution sets of equations will have one member. Later chapters will deal with equations that have more than one solution. The solution set of the equation $x + 6 = 10$ is 4, because replacing x with 4 makes the equation true.

2.1C Definition: Identity

An **identity** is an equation that is true when the variable is replaced by any member of the replacement set.

$x + x = 2x$ is an identity because it is true no matter which real number is used to replace x.

2.1D Definition: Equivalent Equations

Equivalent equations are equations that have the same solution set.

$2x + 5 = 11$ and $2x = 6$ are equivalent equations because $x = 3$ is the solution to both equations. To solve an equation, we keep writing equivalent equations until we get one where the solution is obvious.

To write equivalent equations we use the following properties:

2.1E Additive Property of Equations

If the same quantity is added to both sides of a true equation, the resulting equation is also true.

In symbols,   if $a = b$

then $a + c = b + c$

Since subtraction is addition of the additive inverse, we could work very well without a subtraction property of equality. But, for convenience we define the following:

2.1F Addition and Subtraction Property of Equivalent Equations

If the same quantity is added to or subtracted from both sides of an equation, the new equation is equivalent to the original equation.

Example 1 □ Solve $16 = a - 4$ for a.

$$16 = a - 4$$
$$16 + 4 = a - 4 + 4 \qquad \text{Add 4 to both sides}$$
$$\underline{\qquad} = \underline{\qquad} \qquad \text{Simplify}$$

20, a

In this example we have a good illustration of the ways in which an answer may be written.

$$20 = a \qquad \text{can be written} \qquad a = 20$$

The statement that $a = 20$ also means $20 = a$ is an example of the symmetric property of equality.   □

Example 2 □ Solve $2(5z + 2) = 4(2z + 5) + z$ for z.

$$2(5z + 2) = 4(2z + 5) + z \qquad \text{Use the distributive property}$$
$$\text{to eliminate the parentheses}$$
$$10z + 4 = 8z + 20 + z$$
$$10z + 4 = 9z + 20 \qquad \text{Combine like terms}$$
$$10z - 9z + 4 = 9z - 9z + 20 \qquad \text{Subtract } 9z \text{ from both sides}$$
$$z + 4 = 20 \qquad \text{Combine like terms}$$

$-4, -4$

$$z + 4 \underline{\hspace{1cm}} = 20 \underline{\hspace{1cm}} \qquad \text{Subtract 4 from both sides}$$

16

$$z = \underline{\hspace{1cm}} \qquad\qquad\qquad\qquad □$$

2.1G   Multiplication Property of Equations

If both sides of a true equation are multiplied by the same quantity, the resulting equation is also true.

In symbols, if $a = b$

then $a \cdot c = b \cdot c$

Since division is multiplication by the multiplicative inverse, the multiplication property of equality allows us to write a very useful rule.

2.1H   Multiplication and Division Property of Equivalent Equations

If both sides of an equation are multiplied or divided by a nonzero quantity, the new equation is equivalent to the original equation.

Example 3 □ Solve $3x = 5$ for x

$$3x = 5 \qquad x \text{ is multiplied by 3}$$

$$\frac{3x}{3} = \frac{5}{3} \qquad \text{Dividing both sides by 3}$$

$$x = \frac{5}{3}$$

Dividing by 3 is the same as multiplication by the multiplicative inverse of 3 which is $\frac{1}{3}$.

□

I understand that you can't divide by zero. Why can't you multiply both sides by zero?

If you multiply both sides of an equation by zero, then any statement of inequality can be made into a true equation. $3 \neq 5$ but $3 \cdot 0 = 5 \cdot 0$.

Example 4 □ Solve $3x - 4 = 6 - 2x$.

$$3x - 4 = 6 - 2x$$

$$3x + 2x - 4 = 6 - 2x + 2x \qquad \text{Add } 2x \text{ to both sides}$$

$$5x - 4 = 6$$

$$5x - 4 + 4 = 6 + 4 \qquad \text{Add 4 to both sides}$$

$$5x = 10$$

$$\frac{5x}{5} = \frac{10}{5} \qquad \text{Divide both sides by 5}$$

$$x = 2 \qquad\qquad □$$

Problem Set 2.1A

Solve for the variable. Check your solutions.

1. $2x + 3 = 9$

2. $2 - 5a = -1$

3. $5z = 3z + 6$

4. $4x = x - 12$

5. $3x + 6 = x + 16$

6. $6t - 7 = 2t + 9$

7. $2x + 4 = 6 - x$

8. $8 - 2x = 3x - 7$

9. $6x + 7 = 2x - 5$

10. $-4x + 5 = -7x + 9$

11. $8 - 3t = 6t - 10$

12. $4 - 5t + 8 = 6 - 3t$

13. $9 - 2x = 4x - 15$

14. $7x - 8 = 12 - 3x$

15. $3t - 7 + 2t = 11 - 4t$

16. $2y + 12 - 5y = 3y + 15$

17. $x - 13 - 5x = 8 + 2x$

18. $6 - 2x + 5 = 7x - 7$

19. $8 - 3x + 4 = 5x - 8$

20. $2 + 4x - 9 = 5 + 2x - 7$

21. $9 - 3x + 2 = 5x - 4 - 2x$

22. $6x - 4 + 3x = 6 + 5x + 8$

23. $-7x + 6 + 5x = 12 - 9x + 9$

24. $5x - 8 - 11x = 5 + 2x + 9$

When fractions are involved in an equation, it's a good idea to eliminate the denominator in the first step.

Example 5 □ Solve $\dfrac{7y}{5} - y = \dfrac{1}{2}$.

$$\frac{7y}{5} - y = \frac{1}{2}$$

$$10\left[\frac{7y}{5} - y\right] = \frac{1}{2} \cdot 10 \qquad \text{Multiply both sides by the common denominator}$$

$$10 \cdot \frac{7y}{5} - 10y = \frac{1}{2} \cdot 10 \qquad \text{Distributive property}$$

14y, 5

$$\underline{\hspace{2cm}} - 10y = \underline{\hspace{1.5cm}} \qquad \text{Simplify}$$

$$4y = 5 \qquad \text{Combine like terms}$$

1.25

$$y = \underline{\hspace{1.5cm}} \qquad \text{Divide both sides by 4} \qquad \square$$

That didn't come out even.

In many applications of mathematics the values are not integers. We just tend to prefer integer or simple fractions in books, so that the arithmetic doesn't get in the way of the idea we are illustrating.

The following is an example of how we can sometimes clear the decimals from an equation. However it still may be necessary to use your calculator.

Example 6 □ Solve for x.

$$0.3x - 0.23 = 5.77$$

$$30x - 23 = 577 \qquad \text{Multiply both sides by 100}$$

$$30x = 600 \qquad \text{Add 23 to both sides}$$

$$x = 20 \qquad \text{Divide both sides by 30} \qquad \square$$

Why did you multiply both sides by 100 in step 2?

To eliminate all decimals. Find the largest number of decimal places in any one term; then multiply by that power of 10 to eliminate the decimal within the equation.

First-Degree Equations

Equations with only a single variable raised to the first power are called **first-degree, or linear, equations.**

▶ **Pointers to Solve First-Degree Equations**

1. Use the distributive property to remove parentheses.
2. Clear any fractions by multiplying both sides of the equation by the least common denominator.
3. Simplify by combining any like terms.
4. Use addition or subtraction to get all terms with the specified variable on one side and all other terms on the other side.
5. Combine like terms.
6. Use multiplication or division to make the coefficient of the variable 1.

Must I check or test the solution for all equations?

It's a good practice to make sure that no mistake was made. However, some kinds of equations should always be checked. Those equations will be covered in a later chapter.

Problem Set 2.1B

Solve the following equations.

1. $\dfrac{4x - 3}{5} = 5$

2. $3 - \dfrac{4x}{3} = 2$

3. $\dfrac{6y}{7} - 5 = 3$

4. $\dfrac{8s - 5s}{2} = 12$

5. $-3 = \dfrac{5x - 7x}{4}$

6. $\dfrac{5x + 9}{6} = -1$

7. $\dfrac{5x}{3} - \dfrac{3x}{4} = 3$

8. $\dfrac{2y}{5} - \dfrac{y}{3} = -2$

9. $\dfrac{4s}{7} - \dfrac{2s}{3} = \dfrac{2}{5}$

10. $\dfrac{5t}{12} - \dfrac{7t}{4} = \dfrac{8}{3}$

11. $4 - \dfrac{2x}{6} = 3$

12. $5 - \dfrac{2x}{3} = -1$

13. $\dfrac{5x}{3} - \dfrac{3x}{2} = 1$

14. $\dfrac{6x}{5} - \dfrac{8x}{3} = -\dfrac{11}{15}$

15. $\dfrac{6x + 12x}{4} = \dfrac{3}{2}$

16. $\dfrac{4t}{3} - \dfrac{6t}{5} = -\dfrac{1}{5}$

17. $\dfrac{9y}{2} + \dfrac{6y}{3} = -\dfrac{13}{3}$

18. $\dfrac{6s}{5} - \dfrac{9s}{3} = -\dfrac{3}{2}$

19. $8x - 5 = 5(2 + x)$

20. $2x = 9 - (x + 3)$

21. $5 - (2x + 1) = 3x$

22. $6(2x - 3) = -(x - 8)$

23. $6(x - 5) + 24 = 2(x + 1) - 4$

24. $3(2x + 1) - x = 3(x - 2) + 1$

25. $5(x - 4) - 2x = 3(x - 7) + 1$

26. $5x + 3(x + 4) + 5 = (2x + 11) + 3x$

27. $6x - 3(4x - 3) = 1 - (8x - 7)$

28. $4x + (8x - 7) + 6 = 3(4x - 5) + 8x + 8$

29. $3(x - 4) = 2x - (12 - x)$

30. $6 - (4 - x) + 3x = 2(2x + 1)$

31. $(x - 6) - (7x + 4) - (5 - 4x) = 5$

32. $-(x - 2) - (4x - 3) = 8 + 3(x - 1)$

33. $5(a - 3) - 3(a - 2) = 6(a + 2) - 3$

34. $-3(t - 5) - (2t - 3) = -1 + (3t + 12)$

35. $5x - 2(x - 3) - 4 = 2x - (1 - 4x)$

36. $-6(x - 2) - (4 - 3x) - 2 = 2(x - 2) - x$

37. $-8(x + 2) + (5 - 2x) + 15 = 5 - (3x + 7) - 5x$

38. $3(2x - 5) - 4(6x - 15) + 12 = 2x - (2x - 6)$

39. $2(3x + 7) - (3x + 5) - 4x = x + (3 - 2x)$

40. $-(2x - 3) - (4x - 3) = 2 + 2(x - 4)$

41. $3(x + 3) - (2 - 3x) + 7 = -7 - (6x - 5)$

42. $6(x - 2) + 3(2x + 4) - 9 = 2(3x + 4) - 6x + 3$

43. $0.05x + 0.15 = 0.3$

44. $-0.25x - 0.6 = 1.15$

45. $1.25x - 0.06 = 2(0.5x - 1.28)$

46. $4 - 0.06x = -3(0.6 - 0.08x)$

47. $0.8x + 4 = 3(0.4 + 0.1x) - 0.2$

48. $1.32x - 0.16 = 3(0.41x + 0.12) + 0.02$ **49.** $2 - 0.05x = -5(0.8 - 0.02x)$

50. $1.52x + 2 = 6(0.25x + 0.3)$

51. $\dfrac{x - 2}{5} + \dfrac{x}{6} = \dfrac{1}{3}$ **52.** $\dfrac{x - 3}{5} + \dfrac{x}{4} = \dfrac{3}{10}$ **53.** $\dfrac{x + 2}{4} - \dfrac{1}{2} = \dfrac{x}{5}$

54. $\dfrac{x - 2}{5} + \dfrac{x}{4} = \dfrac{x}{3} + \dfrac{3}{10}$ **55.** $\dfrac{x + 3}{3} + \dfrac{x}{4} = \dfrac{2x}{5} + \dfrac{9}{20}$ **56.** $\dfrac{x - 4}{6} + \dfrac{1}{5} = \dfrac{2x}{5}$

57. $\dfrac{2x - 3}{7} + \dfrac{1}{2} = \dfrac{x}{2} + \dfrac{2}{7}$ **58.** $\dfrac{x - 2}{4} - \dfrac{x}{5} = \dfrac{2x}{5} + \dfrac{1}{5}$ **59.** $\dfrac{3x - 1}{5} - \dfrac{2}{3} = \dfrac{x}{5} - \dfrac{1}{15}$

■ 2.2   Solving Equations for a Specified Variable

When working with formulas we don't try to remember all possible forms of the formula. Instead, we memorize one form and solve the equation for the variable that is needed. Consider the formula for distance.

$$D = R \cdot T$$

where D = Distance traveled

R = Rate or speed

T = Time

If you wish to know the formula for the time to travel a certain distance you could memorize another formula. However, it's more practical to solve the first equation for T.

$$D = R \cdot T \text{ is readily solved}$$

to yield

$$T = \frac{D}{R}$$

or $R = \dfrac{D}{T}$

To solve any formula for a specific variable use the same procedures you used in solving first-degree equations. The only difference is that more than one letter will remain in the solved equation.

Example 7 □ Solve $p = 4q(s + t)$ for q.

$$p = 4q(s + t)$$

$4(s + t)$

What factors are multiplying q? _____

$4(s + t)$

Therefore, we will divide both sides of the equation by _____ .

$$\frac{p}{4(s + t)} = \frac{4q(s + t)}{4(s + t)}$$

$$\frac{p}{4(s + t)} = \frac{\overset{1}{4}q(s \overset{1}{\cancel{+} t})}{\underset{1}{\cancel{4}}(s \underset{1}{\cancel{+} t})}$$

$$\frac{p}{4(s + t)} = q$$

□

Example 8 □ Solve $y = mx + b$ for x.

$$y = mx + b \qquad \text{b is added to mx}$$

$$y - b = mx + b - b \qquad \text{Remove the b by subtraction}$$

mx

$$y - b = \text{_____} \qquad \text{x is multiplied by m}$$

$$\frac{y - b}{m} = \frac{mx}{m} \qquad \text{Divide both sides by m}$$

$$\frac{y - b}{m} = x \qquad \text{Careful! The entire side is divided by m}$$

or $\qquad x = \dfrac{y - b}{m}$

□

Example 9 □ Find the formula for converting from the Celsius scale to the Fahrenheit scale.

The formula for conversion from the Fahrenheit temperature scale to the Celsius scale is

$$C = \frac{5}{9}(F - 32)$$

Solve the equation for F.

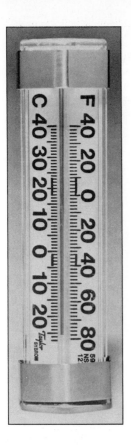

Since $\frac{5}{9}$ is multiplying the quantity in the parentheses with F, we will remove it first by multiplying both sides by $\frac{9}{5}$.

$$C = \frac{5}{9}(F - 32)$$

$$\frac{9}{5} \cdot C = \frac{9}{5} \cdot \frac{5}{9}(F - 32) \qquad \text{Multiply by } \frac{9}{5}$$

$$\frac{9}{5} \cdot C = F - 32 \qquad\qquad \text{32 is subtracted from F}$$

$$\frac{9}{5} \cdot C + 32 = F - 32 + 32 \qquad \text{Add 32 to both sides}$$

$$\frac{9}{5} \cdot C + 32 = F$$

Water boils at 100°C. If we put this in our formula and evaluate it, we will find the Fahrenheit temperature of boiling water.

$$\frac{9}{5} \cdot C + 32 = F$$

$$\frac{9}{5}(100) + 32 = F$$

$$\underline{\qquad\qquad} + 32 = F$$

$$212 = F$$

180

212°

Water boils at _____ Fahrenheit. ☐

Example 10 ☐ Solve $A = Lb + Lc$ for L.

This time L appears in two places on the right side. We can use the distributive property to change a sum into a product. This process is called **factoring.** We will factor L from the right side.

$$A = Lb + Lc$$

$$A = L(b + c) \qquad \text{Use distributive property to factor } L$$

$$\frac{A}{b + c} = \frac{L(b + c)}{(b + c)} \qquad \text{Divide both sides by } (b + c)$$

$$\frac{A}{b + c} = L$$ ☐

Example 11 □ Solve the lensmaker's equation $\dfrac{1}{f} = (n-1)\left(\dfrac{1}{R_1} - \dfrac{1}{R_2}\right)$ for R_2, which is the radius of curvature of the second lens required to produce a desired focal length f.

$$\frac{1}{f} = (n-1)\left(\frac{1}{R_1} - \frac{1}{R_2}\right)$$

$$\frac{1}{f} = (n-1)\frac{1}{R_1} - (n-1)\frac{1}{R_2}$$

Use the distributive property to get the expression involving R_2 out of the parentheses

$$R_1R_2f\frac{1}{f} = \frac{(n-1)}{R_1}R_1R_2f - \frac{(n-1)}{R_2}R_1R_2f$$

Multiply both sides by the common denominator

$$R_1R_2 = (n-1)R_2f - (n-1)R_1f$$

Simplify

Now R_2 is on both sides.

Move all terms with R_2 to the left side by subtracting $(n-1)R_2f$ from both sides.

$$R_1R_2 - (n-1)R_2f = -(n-1)R_1f$$

$$R_2[R_1 - (n-1)f] = -(n-1)R_1f$$

Factor R_2

$$R_2 = \frac{-(n-1)R_1f}{[R_1 - (n-1)f]}$$

Divide by $[R_1 - (n-1)f]$

$$R_2 = \frac{(1-n)R_1f}{[R_1 - (n-1)f]}$$

Rewrite:
$-1(n-1) = 1 - n$ □

Problem Set 2.2

Solve for the indicated variable.

1. $y = kx$ for x (Direct variation)

2. $F = ma$ for a (Force formula)

3. $A = lw$ for w (Area of a rectangle)

4. $A = lw$ for l (Area of a rectangle)

5. $v = lwh$ for h (Volume of a rectangular solid)

6. $v = lwh$ for w (Volume of a rectangular solid)

7. $A = \frac{1}{2}bh$ for h (Area of a triangle)

8. $A = \frac{1}{2}bh$ for b (Area of a triangle)

9. $V = \frac{1}{3}\pi r^2 h$ for h (Volume of a cone)

10. $V = \pi r^2 h$ for h (Volume of a cylinder)

11. $P = 2l + 2w$ for w (Perimeter of a rectangle)

12. $P = 2l + 2w$ for l (Perimeter of a rectangle)

13. $P = a + b + c$ for c (Perimeter of a triangle)

14. $P = a + b + c$ for b (Perimeter of a triangle)

15. $d = rt$ for r (Distance formula)

16. $c = 2\pi r$ for r (Circumference of a circle)

17. $A = \pi r^2$ for π (Area of a circle)

18. $I = prt$ for t (Simple interest formula)

19. $I = prt$ for p (Simple interest formula)

20. $d = rt$ for t (Distance formula)

21. $y = mx + b$ for m (Linear equation)

22. $y = mx + b$ for x (Linear equation)

23. $y = mx + b$ for b (Linear equation)

24. $y = \frac{3}{4}x + 3$ for x (Linear equation)

25. $F = \frac{9}{5}C + 32$ for C   (Celsius-Fahrenheit)

26. $F = G\frac{m_1 m_2}{r^2}$ for m (Gravitational force)

27. $E = \frac{1}{2}mv^2$ for m (Kinetic energy)

28. $V = V_0 + at$ for a (Velocity formula)

29. $y = V_0 t + \frac{1}{2}at^2$ for a (Velocity formula)

30. $\dfrac{1}{R} = \dfrac{1}{R_1} + \dfrac{1}{R_2}$ for R (Electric circuit)

31. $A = \dfrac{h}{2}(B + b)$ for h (Area of a trapezoid)

32. $A = \dfrac{h}{2}(B + b)$ for B (Area of a trapezoid)

33. $s = a + (n - 1)d$ for d (Arithmetic progression)

34. $s = a + (n - 1)d$ for n (Arithmetic progression)

35. $s = a + (n - 1)d$ for a (Arithmetic progression)

36. $S = \dfrac{n}{2}(a + s)$ for s (Sum of an arithmetic progression)

37. $S = \dfrac{n}{2}(a + s)$ for n (Sum of an arithmetic progression)

38. $S = \dfrac{n}{2}[2a + (n - 1)d]$ for a (Sum of an arithmetic progression)

39. $T = Q(V_2 - V_1)$ for V_1 (Thrust by a turbo jet)

40. $R = 101.3V\left(\dfrac{T - D}{w}\right)$ for D (Rate of climb of a jet)

41. $a = \dfrac{q}{w}(T - D - F)$ for D (Acceleration of an aircraft during take-off)

42. $Q = nC(T_2 - T_1)$ for T_1 (Heat to change mass of body from time T_1 to T_2)

43. $S = \left[\dfrac{V_1 - V_2}{2}\right]T$ for V_1 (Uniform acceleration)

44. $S = \left[\dfrac{V_1 - V_2}{2}\right]T$ for T (Uniform acceleration)

45. $Q = mCT_1 - mCT_2$ for C (Specific heat capacity)

46. $S = C + rC$ for C (Selling price)

First-Degree Inequalities

Pension plans and large investors, like insurance companies, make decisions to buy and sell stocks using computer programs. After analyzing many factors, including the sales price of the stock, interest rates, and inflation rates, these programs use inequalities to make a buy or sell, a decision which is then executed electronically before the market price can change.

Inequalities

$3 < 5$ is an example of an inequality. It is read "3 is less than 5." Inequalities using numerals are either true or false. When variables are used, the truth of the inequality depends upon the value of the variable. $x < 5$ is true for all values of x less than 5. Some values of x that make this inequality true are 4, 4.99, 0, −2000. There are obviously many more replacements for x that make the inequality true. One of the values that makes $x < 5$ false is 5, because 5 is not less than 5. One way to represent the replacements of x that make an inequality true is with a number line.

■ Figure 2.1

Notice these things about this graph.

1. The line itself has arrows on both ends. These arrows mean that the line extends forever in both directions.
2. The heavy portion of the line represents the values of x that make the inequality true. It has an arrow only on the left side, which means that the shaded portion and all numbers to the left make the inequality true. Therefore, the graph includes 4, −8, and −101.
3. The point 5 has an open circle around it. This indicates that 5 is not included. If 5 was included in the replacement set, then the circle would be solid.
4. It is possible to test if the graph is shaded correctly by substituting values from the shaded region in the inequality. These values should make the inequality true. Values from the unshaded region should make the inequality false.

$$x < 5$$

Value to be tested	Region	Substitution	Truth	
0	shaded	$0 < 5$	true	Notice that the larger number is always on the right on a number line.
−6	shaded	$−6 < 5$	_____	
6	unshaded	$6 < 5$	false	Just like reading from left to right.
5	unshaded	$5 < 5$	false	
7	unshaded	$7 < 5$	_____	

true

false

□

There are only three possible relationships that can exist between two numbers. The trichotomy axiom states these relationships.

2.3A Trichotomy Axiom

For any real numbers *a* and *b*, exactly one of the following is true:

$$a < b \qquad a = b \qquad a > b$$

| Less than | Equal to | Greater than |

Axiom

What's an axiom?

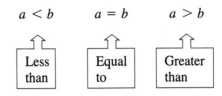

An **axiom** is a statement that we accept as true without proof. While trichotomy looks very reasonable, it's impossible to prove. Therefore, trichotomy is an axiom.

Just as there is a transitive property of equality, there is also a transitive property of inequality.

2.3B Transitive Property of Inequality

For any real numbers *a*, *b*, and *c*,

if $a < b$ and $b < c$, then $a < c$.

$5 < 7$ and $7 < 10$, therefore $5 < 10$ is an example of the transitive property of inequality.

John is younger than his mother who is younger than her mother. Therefore, John is younger than his grandmother.

         Chapter 2 ∎ First-Degree Equations

Here is a summary of the symbols we can use to compare two numbers.

Symbol	How Read
$a = b$	a is equal to b
$a > b$	a is greater than b
$a < b$	a is less than b
$a \leq b$	a is less than or equal to b
$a \geq b$	a is greater than or equal to b
$a \neq b$	a is not equal to b

Here are some typical inequalities and their graphs.

■ Figure 2.2

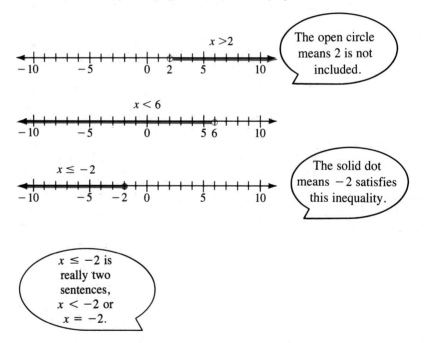

Solving First-Degree Inequalities in One Variable

The rules for solving inequalities are very similar to the rules for solving equations. The basic rule remains: **What you do to one side of the inequality you must do to the other.**

2.3C Addition and Subtraction Rule for Inequalities

Any number may be added to or subtracted from both sides of an inequality and the direction of the inequality will remain the same. If $a < b$, then $a + c < b + c$ and $a - c < b - c$.

Example 12 □ Solve $2x - 7 \leq x + 5$ for x.

$$2x - 7 \leq x + 5$$
$$2x \leq x + 12 \qquad \text{Add 7 to both sides}$$
$$x \leq 12 \qquad \text{Subtract } x \text{ from both sides} \qquad □$$

Things are almost as easy for multiplication and division.

Example 13 □ Consider the true inequality $3 < 5$.

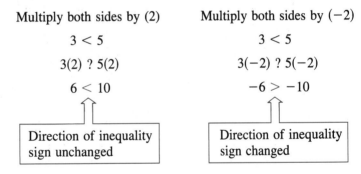

Multiply both sides by (2)	Multiply both sides by (-2)
$3 < 5$	$3 < 5$
$3(2) \; ? \; 5(2)$	$3(-2) \; ? \; 5(-2)$
$6 < 10$	$-6 > -10$

Direction of inequality sign unchanged

Direction of inequality sign changed □

Example 14 □ Consider $10 > -4$.

Divide both sides by (2)	Divide both sides by (-2)
$10 > -4$	$10 > -4$
$\dfrac{10}{2} \; ? \; -\dfrac{4}{2}$	$\dfrac{10}{-2} \; ? \; \dfrac{-4}{-2}$
$5 > -2$	$-5 < +2$

Direction of inequality signed unchanged

Direction of inequality sign changed □

Notice whenever an inequality is multiplied or divided by a positive number, the direction of the inequality remains the same. However, multiplying or dividing an inequality by a negative number reverses the direction of the inequality symbol.

2.3D Multiplication and Division Rule for Inequalities

1. If both sides of a true inequality are multiplied or divided by the same positive number, n, another true inequality in the same direction is produced.

$$\text{If} \qquad a < b,$$

$$\text{then} \qquad a \cdot (n) < b \cdot (n)$$

$$\text{and} \qquad \frac{a}{n} < \frac{b}{n}$$

where n is a positive number.

2. If both sides of a true inequality are multiplied or divided by the same negative number, another true inequality in the opposite direction is produced.

$$\text{If} \qquad a < b,$$

$$\text{then} \qquad a(-n) > b(-n)$$

$$\text{and} \qquad \frac{a}{(-n)} > \frac{b}{(-n)}$$

where $(-n)$ is a negative number.

Example 15 □ Solve $-2x < 4$.

$$-2x < 4$$

$$\frac{-2x}{-2} > \frac{4}{-2} \qquad \text{Divide both sides by } -2 \text{ and change the direction of the inequality sign}$$

−2

$$x > \underline{\hspace{1cm}}$$

Notice the direction of the inequality sign changes as soon as both sides are divided by a negative number.

□

Example 16 □ Solve $\dfrac{x}{-3} > 4$.

$$\frac{x}{-3} > 4$$

$$(-3)\frac{x}{-3} \underline{\hspace{2cm}} 4(-3) \quad \Longleftarrow \boxed{\text{Multiply by } -3; \text{ change direction of inequality}}$$

$$x < -12 \qquad \qquad \square$$

$<$

Example 17 □ Solve $2x + 3 \geq 7$.

$$2x + 3 \geq 7$$

$$2x \geq 4 \qquad \text{Subtract 3}$$

$$x \geq 2 \qquad \text{Divide by 2} \qquad \square$$

Example 18 □ Solve $6 - 5x > 3x + 14$.

$$6 - 5x > 3x + 14$$

$$6 > 8x + 14 \qquad \text{Add } 5x$$

$$-8 > 8x \qquad \text{Subtract 14}$$

$$\frac{-8}{8} > \frac{8x}{8} \qquad \text{Divide by } +8$$

$$-1 > x$$

What would happen if I subtracted $3x$ first?

Watch.

$$6 - 5x > 3x + 14$$

$$6 - 8x > 14 \qquad \text{Subtract } 3x$$

$$-8x > 8 \qquad \text{Subtract 6}$$

$$\frac{-8x}{-8} < \frac{8}{-8} \qquad \text{Divide by } -8$$

$$x < -1$$

$x < -1$ is the same thing as $-1 > x$.

$\square$

Chapter 2 ■ First-Degree Equations

Example 19 □ Solve and graph $\dfrac{1-x}{5} \le \dfrac{1}{2}$.

$$\dfrac{1-x}{5} \le \dfrac{1}{2}$$

$$\dfrac{10(1-x)}{5} \le \dfrac{1}{2} \cdot 10 \qquad \text{Multiply both sides by } +10$$

$$2(1-x) \le 5 \qquad \text{Simplify}$$

$$2 - 2x \le 5 \qquad \text{Distributive property}$$

$$-2x \le 3 \qquad \text{Subtract } +2$$

$$x \ge -\dfrac{3}{2} \qquad \text{Divide by } -2$$

> Note the change in direction of the inequality in the last step.

■ Figure 2.3

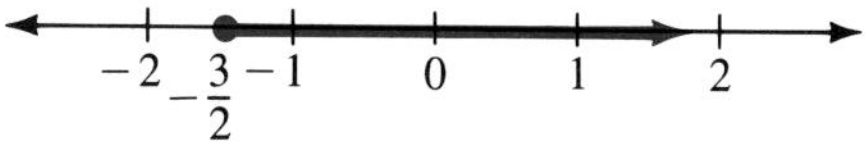

Example 20 □ Solve $3(2x - 4) \ge 4\left(3x + \dfrac{5}{3}\right)$.

$$3(2x - 4) \ge 4\left(3x + \dfrac{5}{3}\right)$$

$$6x - 12 \ge 12x + \dfrac{20}{3} \qquad \text{Distributive property}$$

$$-12 \ge 6x + \dfrac{20}{3} \qquad \text{Subtract } 6x$$

$$-36 \ge 18x + 20 \qquad \text{Multiply by } +3$$

$$-56 \ge 18x \qquad \text{Add } -20$$

$$-\dfrac{28}{9} \ge x \qquad \text{Divide by } +18 \text{ and reduce}$$

$$x \le -\dfrac{28}{9} \qquad \text{Rewrite with } x \text{ first}$$

> Notice the direction of the inequality did change.

> There is no symmetric property of inequality.

Solve and graph the following inequalities.

1. $x - 4 < 2$

2. $x - 8 < -3$

3. $3x \geq 12$

4. $5x > 20$

5. $-6x > -18$

6. $-4x < 12$

7. $x + 3 \leq 8$

8. $2x + 6 \geq 16$

9. $-3x + 7 < 9$

10. $-6x > 2x + 12$

11. $9x - 5 < 7x + 9$

12. $11x + 4 < 8x - 5$

13. $-5x + 6 \leq 3x - 10$

14. $-3x - 8 \leq 4x + 6$

15. $8x - 4 > 6x + 8$

16. $-5x - 6 \leq 2x + 4$

17. $-4(3x - 7) > 4$

18. $-6(3x - 5) \geq 6$

19. $-3(y + 8) \leq 5y + 4$

20. $-5(y - 3) < 6y - 7$

21. $-7(2x - 4) \leq 3x - 6$

22. $5(6 - 3y) > -6y + 9$

23. $-2y + 7 < 4(2y - 2)$

24. $3y - 4 \geq 2(4y - 7)$

25. $-8(3x + 4) > -6x - 20$

26. $-4(y + 7) < 6y + 2$

27. $6(1 - y) + 4 \geq 2y - (5y - 1)$

28. $4(2 - y) - 6 > 3 - (5 - 4y)$

29. $-6(2n - 4) \leq 5(3n + 6)$

30. $4(5 - 3n) - 3(2n - 4) - 2 \geq 0$

31. $3(2 - y) + 5 \geq -3y - (4y - 3)$

32. $4(1 - 3y) - 7 \leq 5y - 2(y - 6)$

33. $-6(2n - 3) + 4 \leq 2(3 - n) - 2n$

34. $3(2 - 3x) - 6 \geq -2x + 2(3 - 2x)$

35. $-5(x + 2) - 3 - 4(2 - 3x) < 0$

36. $3(2y - 6) - 4(2 - y) - 4 > 0$

37. $\dfrac{x}{4} \leq 12$

38. $\dfrac{x}{5} < -15$

39. $-\dfrac{x}{4} > 6$

40. $-\dfrac{x}{3} > 18$

41. $-\dfrac{3x}{2} \geq 15$

42. $-\dfrac{2}{3}x \geq 12$

43. $-\dfrac{3x}{2} \leq 6$

44. $\dfrac{4}{3}x > -8$

45. $-\dfrac{3}{4}y \geq -3$

46. $-\dfrac{2}{5}x < -12$

47. $\dfrac{5}{2}x < -15$

48. $-\dfrac{12}{5}x \geq -3$

49. $\dfrac{2-x}{4} \leq \dfrac{1}{3}$

50. $\dfrac{5-x}{7} \leq \dfrac{2}{3}$

51. $\dfrac{x-2}{3} + \dfrac{x+2}{2} > 1$

52. $\dfrac{x+4}{5} - \dfrac{x+2}{3} > \dfrac{1}{2}$

53. $\dfrac{2n+3}{8} - \dfrac{n+2}{5} \leq -1$

54. $\dfrac{4-x}{5} \leq \dfrac{1}{2}$

55. $\dfrac{6-5x}{3} \geq \dfrac{1}{2}$

56. $\dfrac{x-3}{2} + \dfrac{x+3}{3} \geq 1$

57. $\dfrac{x+3}{4} - \dfrac{x+2}{3} \leq \dfrac{1}{2}$

58. $\dfrac{3n-4}{6} - \dfrac{2n-3}{5} < \dfrac{1}{3}$

59. $\dfrac{5x-3}{4} - \dfrac{2x+3}{3} < -1$

60. There are four properties of equalities given in Definition 2.1A. Which of these four properties hold for inequality?

Compound Inequalities

Frequently two inequalities are combined in one statement. You probably prefer room temperatures greater than 70°F but less than 80°F. We can express this idea with the compound inequality.

$$70° < T < 80°.$$

You can read this statement from left to right "70° is less than T and T is less than 80°." You can also read it starting in the middle: "T is greater than 70° and less than 80°."

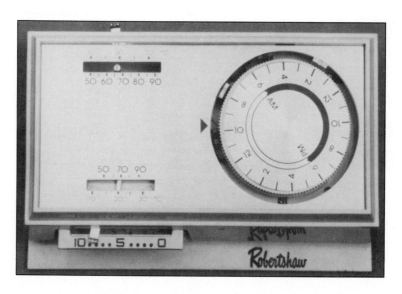

Notice both readings use the word *and*. This compound inequality is really two separate inequalitites.

$$70° < T \qquad \text{AND} \qquad T < 80°$$

Here is how we graph this.

■ Figure 2.4

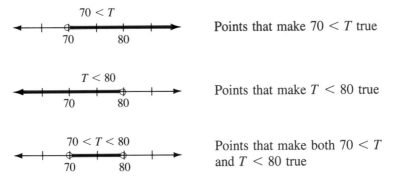

Points that make $70 < T$ true

Points that make $T < 80$ true

Points that make both $70 < T$ and $T < 80$ true

Some values of T that make the inequality $70 < T < 80$ true are 72, 75, and 79.99. 90 is greater than 70 but it is not less than 80. Therefore, 90 is not included on the graph.

Compound inequalitites using *and* are called **conjunctions.**

Here are some other conjunctions and their graphs.

■ Figure 2.5

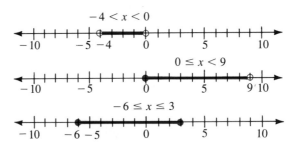

Conjunctions make sense only if they satisfy the transitive law of inequality.

$7 < x < 3$ is meaningless because $7 < 3$ is false.

Compound inequalities using *or* are called **disjunctions.** $x < -1$ or $x > 2$ is a disjunction.

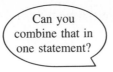

Can you combine that in one statement?

Sorry. There is no compact way to write a disjunction. We need two inequalities.

To graph a disjunction just draw both inequalities on the same line.

Example 21 □ Graph $x < -1$ or $x > 2$.

■ Figure 2.6

□

Example 22 □ Graph $x \geq -1$ or $x > 3$.
Since the graphs overlap, we'll draw each separately, and then combine them.

■ Figure 2.7

$x \geq -1$

$x > 3$

$x \geq -1$ or $x > 3$

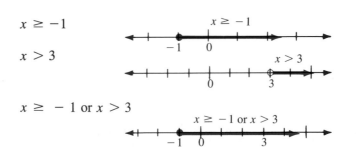

□

Example 23 □ Solve the conjunction $2x + 1 \geq -5$ and $3x + 2 < 8$, then graph on a number line.
Solve both inequalities.

$$2x + 1 \geq -5 \qquad \text{and} \qquad 3x + 2 < 8$$

$$2x \geq -6 \qquad \text{and} \qquad 3x < 6$$

$$x \geq -3 \qquad \text{and} \qquad x < 2$$

■ Figure 2.8

□

Example 24 □ Solve the disjunction $3x - 1 < -7$ or $2x + 3 \geq 5$, then graph on a number line.

$$3x - 1 < -7 \qquad \text{or} \qquad 2x + 3 \geq 5$$
$$3x < -6 \qquad \text{or} \qquad 2x \geq 2$$
$$x < -2 \qquad \text{or} \qquad x \geq 1$$

■ Figure 2.9

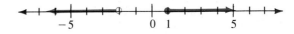

□

Problem Set 2.3B

Graph the following compound inequalities.

1. $-6 \leq x < 2$

2. $-3 < x \leq 3$

3. $0 < x \leq 4$

4. $-5 \leq x < 0$

5. $-2 \leq x < 3$

6. $-1 < x < 5$

7. $-6 \leq x < -1$

8. $-1 \leq x \leq 4$

9. $x \geq 1$ or $x > 3$

10. $x \geq -5$ or $x > -4$

11. $x < -2$ or $x \geq 2$

12. $x \leq 3$ or $x > 5$

13. $x \leq -1$ or $x < 4$

14. $x < 0$ or $x > 3$

15. $x \leq -5$ or $x \geq 3$

16. $x \leq -3$ or $x > 2$

17. $x > -2$ or $x > 2$

18. $x \leq -4$ or $x < 0$

Solve and graph the following inequalities.

19. $x - 3 < 2$ and $x - 3 > -2$

20. $\dfrac{x + 2}{3} \leq -1$ or $x + 2 \geq 3$

21. $2x - 3 \leq 0$ and $3x + 7 > 0$

22. $4x + 7 \leq 5$ or $5x - 4 > 6$

23. $3x - 2 < -2$ and $2x + 4 > 6$

24. $6x + 7 > 1$ or $3x - 4 > 5$

25. $2x + 4 \leq 6$ and $4x - 5 \geq -1$

26. $4x - 8 < 0$ or $5x - 12 > 8$

27. $2x + 6 \leq -2$ or $-3x + 5 < 8$

28. $3x - 4 \leq 2$ and $-7x + 5 < -7$

29. $6x + 8 \leq 2$ and $-4x + 5 < -11$

30. $-2x + 4 > 10$ or $-3x + 4 \leq 1$

Absolute values form interesting compound statements.

Example 25 □ Graph $|x| = 4$.

Since $|+4| = 4$ and $|-4| = 4$, x can be either $+4$ or -4. Hence, both $+4$ and -4 are graphed.

■ Figure 2.10.

□

Example 26 □ Graph $|x| < 4$.

In Example 25 we saw that x could be either -4 or $+4$. Since we have inequality, let's try some values of x to find the solution to this inequality.

Try	$x = 3$	$|3| < 4$	True
	$x = -3$	$|-3| < 4$	True
	$x = 5$	$|5| < 4$	False
	$x = -5$	$|-5| < 4$	False

Hence, we see that the values between -4 and $+4$ satisfy the inequality. Therefore this statement is equivalent to the conjunction $-4 < x < +4$.

■ Figure 2.11

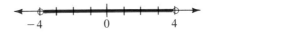

□

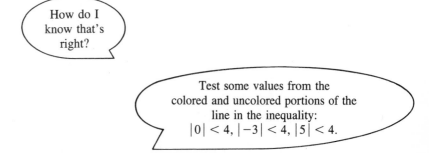

How do I know that's right?

Test some values from the colored and uncolored portions of the line in the inequality: $|0| < 4$, $|-3| < 4$, $|5| < 4$.

Example 27 □ Graph $|x| \geq 3$.

The solution is all numbers that are three or more units from zero.

■ Figure 2.12

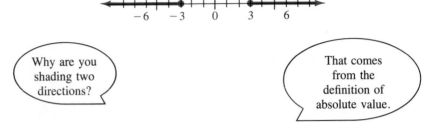

Why are you shading two directions?

That comes from the definition of absolute value.

Recall the definition of absolute value

If $a \geq 0$ then $|a| = a$

If $a < 0$ then $|a| = -a$

In the case of $|x| \geq 3$

If x is positive then $|x| \geq 3$ is the same as $+(x) \geq 3$

If x is negative then $|x| \geq 3$ is the same as $-(x) \geq 3$

So the values of x that satisfy $|x| \geq 3$ are those that satisfy either of these equations:

$$x \geq 3 \qquad \text{or} \qquad -(x) \geq 3 \qquad \text{These are equivalent}$$
$$x \leq -3 \qquad \text{inequalities}$$

Therefore, this statement is equivalent to the disjunction $x \leq -3$ or $x \geq +3$.

Drawing the Graph

■ Figure 2.13

$x \leq -3$ 　　　　　　　 $x \geq +3$

□

How do I know if I'm dealing with a conjunction or a disjunction?

It depends on whether the expression inside the absolute value sign is less than, equal to, or greater than the constant on the other side.

If a statement involving absolute values can be written with the expression inside the absolute value signs on one side of the equation, then the following rules apply:

Solutions of Equations Involving $|ax + b|$

Since a number inside an absolute value sign may be either positive or negative, there are two cases.

Positive Case	Negative Case
$ax + b \geq 0$	$ax + b \leq 0$

For a conjunction $|ax + b| \leq c$ is equivalent to

$$-(ax + b) \leq c \text{ and } +(ax + b) \leq c$$

Which may be written as

$$ax + b \geq -c \text{ and } ax + b \leq c$$

Writing $-c$ on the left

$$-c \leq ax + b \text{ and } ax + b \leq c$$

or more compactly

$$-c \leq ax + b \leq c$$

For a disjunction (*or* statement) $|ax + b| \geq c$ is equivalent to

$$-(ax + b) \geq c \text{ or } (ax + b) \geq c$$

equivalently

$$ax + b \leq -c \text{ or } ax + b \geq c$$

2.4A   Rules for Absolute Value Inequalities

If c is a positive constant

$|ax + b| \leq c$ is equivalent to the conjunction $-c \leq ax + b \leq +c$

$|ax + b| \geq c$ is equivalent to the disjunction $ax + b \leq -c$ or $ax + b \geq +c$

Don't try to memorize things like Rule 2.4A. Rules like "port is on the left and starboard is on the right" can be hard to keep straight. Instead of relying on your memory, set up a very simple example of the relation and see if you have the direction correct.

Relation	Simple Example
$\lvert ax + b \rvert > c$	$\lvert x \rvert > 2$

You know that numbers like -3 and $+3$ both satisfy the simple example. Therefore, this must be the disjunction

$$x < -2 \text{ or } x > +2$$

and the rule for the relationship $\lvert ax + b \rvert > c$ must be

$$\lvert ax + b \rvert > c \text{ implies } (ax + b) < -c \text{ or } (ax + b) > +c.$$

Relation	Simple Example
$\lvert ax + b \rvert < c$	$\lvert x \rvert < 2$

Numbers like 0, $-\dfrac{1}{2}$, and 0.3 satisfy this simple example. Numbers like $+5$ and -5 do not.

Because $\lvert x \rvert < 2$ implies $x > -2$ and $x < +2$ or equivalently $-2 < x < 2$, the rule for the relationship $\lvert ax + b \rvert < c$ must be

$$\lvert ax + b \rvert < c \text{ implies } (ax + b) > -c \text{ and } (ax + b) < +c$$

or equivalently $-c < ax + b < c$.

■ Figure 2.14

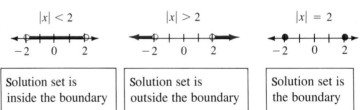

$\lvert x \rvert < 2$	$\lvert x \rvert > 2$	$\lvert x \rvert = 2$
Solution set is inside the boundary	Solution set is outside the boundary	Solution set is the boundary

Example 28 □ Solve and graph $\lvert 2x - 6 \rvert > 4$.

$$\lvert 2x - 6 \rvert > 4 \qquad \text{is equivalent to the inequalities}$$

$$2x - 6 < -4 \qquad \text{or} \qquad 2x - 6 > +4$$

> The *greater than* sign implies a disjunction.

Solving each

$$2x < 2 \qquad\qquad 2x > 10$$

$$x < 1 \qquad\qquad x > 5$$

Graphing

■ Figure 2.15

An Alternate Method to Graph Inequalities

Look again at Example 28. This time it will be done in a different way. Solve and graph $|2x - 6| > 4$.

Step 1: Replace the inequality with an equal sign in order to establish boundary points.

$$|2x - 6| = 4$$

Step 2: Solve the absolute value equality.

$$
\begin{array}{ccc}
2x - 6 = 4 & \text{or} & 2x - 6 = -4 \\
2x = 10 & & 2x = 2 \\
x = 5 & & x = 1
\end{array}
$$

Step 3: Graph the boundary points found in Step 2. Use an open circle to exclude the boundary point(s). Use a closed circle to include the boundary point(s).

■ Figure 2.16

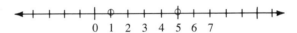

Step 4: The graph is now separated into sections. Test one value from each section. If it makes the statement true, graph all points in that section. If it makes the statement false, don't graph that section.

Test $x = 0$

$|2(0) - 6| \overset{?}{>} 4$

$\quad 6 > 4 \qquad$ True, so graph

Test $x = 2$

$|2(2) - 6| \overset{?}{>} 4$

$\quad 2 > 4 \qquad$ False, don't graph

Test $x = 6$

$|2(6) - 6| \overset{?}{>} 4$

$\quad 6 > 4 \qquad$ True, so graph

■ Figure 2.17

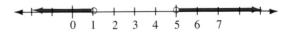

Step 5: Write the correct symbol description to correspond to your graph.

$$x < 1 \qquad \text{or} \qquad x > 5$$

Example 29 □ Solve and graph $\left| \dfrac{x}{3} - 1 \right| \leq \dfrac{1}{2}$.

$\left| \dfrac{x}{3} - 1 \right| \leq \dfrac{1}{2}$ is equivalent to the inequalities

$$\dfrac{x}{3} - 1 \geq -\dfrac{1}{2} \qquad \text{and} \qquad \dfrac{x}{3} - 1 \leq +\dfrac{1}{2}$$

Solving $\qquad \dfrac{x}{3} \geq +\dfrac{1}{2} \qquad\qquad\qquad \dfrac{x}{3} \leq \dfrac{3}{2}$

$$x \geq \dfrac{3}{2} \qquad\qquad\qquad\qquad x \leq \dfrac{9}{2}$$

The *less than* sign implies a conjunction.

■ Figure 2.18

$$\begin{array}{cc} & \\ 0 & 5 \end{array}$$

□

An alternate solution:

$\left| \dfrac{x}{3} - 1 \right| \leq \dfrac{1}{2}$ is equivalent to the conjunction $-\dfrac{1}{2} \leq \dfrac{x}{3} - 1 \leq \dfrac{1}{2}$

$\dfrac{1}{2} \leq \dfrac{x}{3} \leq \dfrac{3}{2} \qquad$ Add 1 to each part

$\dfrac{3}{2} \leq x \leq \dfrac{9}{2} \qquad$ Multiply each part by 3

Problem Set 2.4

Solve the following equations for the variable.

1. $|x| = 2$

2. $|x| = 5$

3. $|a| = 1$

4. $|x| < 2$

5. $|x| < 5$

6. $|y| \geq 3$

7. $|x| = 0$

8. $|x| < 0$

9. $|x| \leq 0$

10. $|x| \geq 0$

11. $|x| < -5$

12. $|x| > -5$

13. $|x| = -5$

14. $|x| = 4$

15. $|x| > 2$

16. $|x| \leq 6$

17. $|x| \geq 4$

18. $|x| > -2$

19. $|x + 1| \geq 2$

20. $|x + 2| \leq 5$

21. $|2x - 1| \geq 7$

22. $|3x + 2| > 8$

23. $|5m + 2| = 9$

24. $|4y - 3| = 4$

25. $|5a - 1| + 2 = 6$

26. $|4b + 3| - 1 = 6$

27. $|x - 1| \leq 5$

28. $|x + 2| < 3$

29. $|2x - 1| > 5$

30. $|3x - 1| \geq 8$

31. $|y + 1| - 1 \leq 4$

32. $|y - 3| + 2 \leq 3$

33. $|4x - 3| - 2 > 3$

34. $|4a - 3| + 5 \leq 6$

35. $\left| \dfrac{b}{2} + 2 \right| = 7$

36. $\left| \dfrac{y}{3} - 4 \right| = 6$

37. $|3x - 6| + 2 \geq 10$

38. $|4x + 5| - 3 > 6$

39. $|2x - 5| + 3 \leq 4$

40. $|4x - 5| - 7 < -4$

41. $\left| \dfrac{b}{3} - b \right| = 3$

42. $\left| \dfrac{x}{4} + 5 \right| = 6$

43. $\left| \dfrac{a}{2} + 5 \right| + 4 = 7$

44. $\left| \dfrac{b}{5} - 3 \right| - 2 = 3$

45. $\left| \dfrac{x}{4} - 2 \right| \leq \dfrac{1}{2}$

46. $\left| \dfrac{x}{3} + 4 \right| < 8$

47. $\left| \dfrac{x}{6} - \dfrac{2}{3} \right| > \dfrac{2}{5}$

48. $\left| \dfrac{x}{5} + \dfrac{3}{4} \right| > \dfrac{2}{3}$

49. $\left| \dfrac{x}{3} + 2 \right| \geq 2$

50. $\left| \dfrac{y}{4} - 3 \right| \geq 6$

51. $\left| \dfrac{x}{6} - \dfrac{1}{2} \right| < \dfrac{1}{3}$

52. $\left| \dfrac{x}{5} - \dfrac{1}{6} \right| \leq \dfrac{1}{2}$

53. $\left| \dfrac{x}{4} + \dfrac{2}{3} \right| > \dfrac{5}{6}$

54. $\left| \dfrac{x}{7} - \dfrac{1}{2} \right| > \dfrac{3}{14}$

55. $\left| \dfrac{2x}{3} - \dfrac{1}{6} \right| \leq \dfrac{1}{2}$

56. $\left| \dfrac{3}{4}x + \dfrac{2}{3} \right| < \dfrac{5}{6}$

One basic skill in problem solving is to find the value of a single unit. If you know the value of one unit, you can calculate the value of any number of units desired. In this chapter we will focus on finding the value of one unit.

Example 30 □ If three pencils cost 30 cents, how much do eight pencils cost?

> The first question to answer is "What does one pencil cost?"

> That's 30 cents divided by 3 pencils, or 10 cents per pencil.

Now that you know that the cost per pencil is 10 cents, you can determine the price of any quantity of pencils by multiplying the price of one pencil times the number desired.

Eight pencils cost:

$$8 \text{ pencils} \cdot 10 \frac{\text{cents}}{\text{pencil}} = 80 \text{ cents} \qquad \qquad \square$$

In the line above we divided out the word *pencils*. Units such as pencils, cents, miles, or hours are what scientists refer to as the dimensions of a quantity. Dimensions tell you what kind of thing you are dealing with. Miles are different from gallons. Both miles and gallons are different from miles per gallon.

Dimensions can also help in finding the value per basic measure of a quantity. Consider how you calculate miles per hour when driving. If you travel 100 miles in two hours, you compute your average speed by dividing miles driven by hours spent.

$$\frac{100 \text{ miles}}{2 \text{ hours}} = 50 \frac{\text{miles}}{\text{hr}}$$

Notice the units of measure provide the formula for calculating the miles per unit of time. Miles per hour means divide miles driven by hours used to find the miles driven in one hour. The same idea applies to many quantities. For example, miles per gallon is determined by dividing miles driven by gallons of gas used.

Example 31 □ If a plane travelled 315 miles in one hour and 45 minutes, how far will it travel in two hours?

First we'll find how far the plane will travel in one hour.

The speed of the plane is measured in miles per hour; therefore to find how far the plane goes in one hour, divide the miles travelled by the time used.

$$\text{speed in } \frac{\text{miles}}{\text{hour}} = \frac{315 \text{ miles}}{1.75 \text{ hours}}$$

$$= 180 \frac{\text{miles}}{\text{hour}}$$

Why 1.75 hours?

The units must be consistent. We converted 45 minutes to 0.75 hours.

Now that we know the airplane traveled 180 miles in one hour, we can find out how far it travels in two hours.

$$\text{Distance in two hours} = 180 \frac{\text{miles}}{\text{hour}} \cdot 2 \text{ hours}$$

$$= 360 \text{ miles} \qquad \Box$$

Notice that you can multiply and divide units of measure just as you do numbers.

Yes, and the units in your final answer should be the correct units for the quantity being measured. In this case distance has a dimension of miles.

Example 32 □ If there are 20 oranges in an 8-pound bag, how many oranges are there in a 10-pound bag?

In this case, if we know the number of oranges per pound, we can find the number of oranges in any specified number of pounds.

$$\frac{\text{oranges}}{\text{pound}} = \frac{20 \text{ oranges}}{8 \text{ pounds}}$$

$$= 2.5 \frac{\text{oranges}}{\text{pound}}$$

Knowing there are 2.5 oranges per pound, we can answer how many oranges in 10 pounds.

$$\text{oranges in 10 pounds} = \frac{2.5 \text{ oranges}}{\text{pound}} \cdot 10 \text{ pounds}$$

$$= 25 \text{ oranges} \qquad \qquad □$$

The concept of how much per single unit of measure is helpful in other problems as well. The next example applies this concept to finding the true airspeed of an airplane.

Example 33 □ Airspeed indicators in airplanes work by measuring the air pressure in a small tube below the wing, called a pitot tube. The faster the airplane flies, the higher the pressure as the pitot tube is forced through the air. Because air gets thinner as altitude increases, the speed that is indicated by the airspeed indicator is not the true airspeed. Pilots can estimate their true airspeed by adding a correction of 2% of the indicated airspeed for each 1000 feet of altitude for the indicated airspeed.

If an airplane flying at 8000 feet indicates a speed of 200 knots, what is the true airspeed of the plane?

The discussion told us the percent of error in 1000 feet of altitude. We can compute the percent of error in 8000 feet altitude.

$$\text{\% error in 8 thousand feet} = \frac{2\% \text{ error}}{1 \text{ thousand feet}} \cdot 8 \text{ thousand feet}$$

$$= 16\% \text{ error}$$

Now that we know the percent of error at 8000 feet, we can add a correction of 16% of the indicated airspeed to the indicated airspeed to find the true airspeed.

$$16\% \text{ of } 200 \text{ knots} = 200 \text{ knots} \cdot 0.16$$

$$= 32 \text{ knots}$$

$$\text{True airspeed} = \text{indicated airspeed} + \text{correction}$$

$$= 200 \text{ knots} + 32 \text{ knots}$$

$$= 232 \text{ knots} \qquad \qquad □$$

Problem Set 2.5

Solve the following problems.

1. If three pencils cost 18 cents, how much does one pencil cost?

2. If three pencils cost 18 cents, how much do 12 pencils cost?

3. If three pounds of coffee cost $9.90, how much is coffee per pound?

4. If three pounds of coffee cost $9.90, how much do eight pounds cost?

5. If approximately 260 cups of coffee can be made from two pounds of coffee beans, how many cups can be made from one pound of coffee beans?

6. If approximately 260 cups of coffee can be made from two pounds of coffee beans, how many cups can be made from nine pounds of this kind of coffee beans?

7. If approximately 125 cups of coffee can be made from one pound of coffee beans, and a pound of coffee beans costs $3.25, how much does one cup of this coffee cost?

8. If approximately 140 cups of coffee can be made from one pound of a certain brand of coffee bean, and the coffee costs $7.40 for two pounds of coffee, what is the cost per cup of coffee?

9. If approximately 420 cups of coffee can be made from three pounds of coffee beans, and if two pounds of coffee beans cost $7.20, how much is coffee per cup?

10. If approximately 270 cups of coffee can be made from two pounds of coffee beans, and the coffee costs $10.80 for three pounds, how much does it cost to make 75 cups of coffee?

11. If approximately 435 cups of coffee can be made from three pounds of coffee beans, and these coffee beans cost $8.40 for two pounds, how many cups of coffee can be made for $1.60?

12. If a high school athlete can run approximately 105 feet in three seconds, how far can the athlete run in one second?

13. If a second athlete can run approximately eight feet in one-fourth of a second, how far can the athlete run in one second?

14. If a third athlete can run approximately 100 feet in three seconds, how far can the athlete run in five seconds?

15. If a certain athlete can run 43 meters in four seconds, how long will it take to run 100 meters?

16. A vehicle travels at the rate of 40 feet in $\frac{1}{2}$ second. At this same speed, how many feet will it travel in ten seconds?

17. A light airplane flies 44 feet in $\frac{1}{4}$ second. How many feet will it travel in ten seconds?

18. If an airplane can fly 33 feet in $\frac{1}{8}$ second, how fast does it fly in miles per hour? (1 mile equals 5280 feet.)

Chapter 2 □ Key Ideas

2.1 **1.** Properties of equality.
    a. Reflexive property
        $a = a$
    b. Symmetric property
        If $a = b$ then $b = a$
    c. Transitive property
        If $a = b$ and $b = c$, then $a = c$
    d. Substitution property
        If $a = b$, then a may be substituted for b in any statement using b
2. A **solution set** is a set of values that make an equation true.
3. An **identity** is an equation that is true for all values of a replacement set.
4. **Equivalent equations** are equations that have the same solution set.
5. **Addition property of equations**—For any real numbers a, b, and c:
    If $a = b$, then $a + c = b + c$
6. **Multiplication property of equations**—For any real numbers a, b, and c:
    If $a = b$, then $a \cdot c = b \cdot c$

2.2 To solve an equation for a **specified variable** apply the addition property of equations and the multiplication property of equations.

2.3 **1. Trichotomy Axiom**
    For any real numbers a and b:
    $a = b$, or $a < b$ or $a > b$
2. The **transitive property** of inequality:
    For any real numbers a, b, and c:
    If $a < b$, and $b < c$ then $a < c$
3. Addition and **subtraction** rule for inequalities:
    If $a < b$
    Then $a + c < b + c$
    And $a - c < b - c$
4. Multiplication and **division** rule for inequalities:
    a. If $a < b$ and $n > 0$ **b.** If $a < b$ and $n < 0$
        Then $a \cdot n < b \cdot n$ Then $a \cdot n > b \cdot n$
        And $\dfrac{a}{n} < \dfrac{b}{n}$ And $\dfrac{a}{n} < \dfrac{b}{n}$

5. Compound inequalities are of two types:
   a. A **conjunction**
      $a \leq x$ and $x \leq b$
      Then $a \leq x \leq b$
   b. A **disjunction**
      $x \leq a$ or $x \geq b$

2.4 Inequalities involving absolute value
   If c is a positive constant:
   $|ax + b| \leq c$ is equivalent to the conjunction $-c \leq ax + b \leq c$
   $|ax + b| \geq c$ is equivalent to the disjunction $ax + b \leq -c$ or $ax + b > c$

Chapter 2   Review Test

Solve the following equations. **(2.1)**

1. $\dfrac{3x - 5}{4} = 6$

2. $\dfrac{8y}{7} - 2 = 3$

3. $\dfrac{3a}{4} - \dfrac{7a}{9} = \dfrac{1}{6}$

4. $6 - (3x + 2) = 13$

5. $3(2y - 4) = y - (3y - 12)$

6. $2.8 - 0.05x = 2(0.6x - 1.1)$

7. $\dfrac{x - 3}{4} - \dfrac{7}{8} = \dfrac{3x}{4}$

8. $\dfrac{x + 2}{4} + \dfrac{5}{6} = \dfrac{2x}{3} - \dfrac{1}{3}$

Solve for the indicated variable. **(2.2)**

9. $d = rt$ for t (Distance formula)

10. $P = 2l + 2w$ for w (Perimeter of a rectangle)

11. $P = s - rs$ for s (Profit formula)

12. $A = \dfrac{h}{2}(B + b)$ for b (Area of a trapezoid)

13. $y = -\dfrac{3}{4}x + 2$ for x (Linear equation)

14. $\dfrac{PV}{T} = k$ for T (Gas law)

Solve the following inequalities. **(2.3)**

15. $-8x \geq 40$

16. $10y - 2 < 4 - 8y$

17. $4(2a - 3) > 8$

18. $-6(x + 2) \leq 9$

19. $5(3 - 2n) - 4(3n - 10) \geq 0$

20. $\dfrac{x - 3}{8} - \dfrac{2x + 3}{6} < -\dfrac{2}{3}$

Solve and graph the following inequalities. **(2.3)**

21. $3x - 4 > 5$

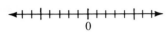

22. $-4(3x - 4) \geq \dfrac{1}{2}(16 - 8x)$

Graph the compound inequality. **(2.3)**

23. $-2 \leq x < 5$

24. $x < -3$ or $x \geq 2$

Solve and graph the following. **(2.4)**

25. $|x| = 6$

26. $|4x - 3| \geq 5$

27. $|3y - 2| < 7$

Solve each of the following.

28. $|5n - 4| = 6$

29. $\left| \dfrac{x}{2} - 3 \right| + 4 = 9$

30. $|2x - 5| + 2 < 6$

31. $\left| \dfrac{x}{4} + \dfrac{1}{2} \right| \geq \dfrac{2}{3}$

32. If 28 servings of homemade ice cream can be made by a 6-quart mixer, how many servings can be made by a 9-quart mixer? **(2.5)**

33. If it takes 16 seconds for a car to travel 1400 ft., how long will it take the same car to travel 1000 ft? (Assume the speed is uniform. Give your answer to the nearest tenth of a second.)

3 Linear Equations, Graphs, and Inequalities

Contents

Preview

Graphs are an important method of communication in business, science and social studies. Graphs help you visualize a relationship between two variables. This chapter will deal with relationships where graphs are straight lines. Chapters 4, 10, 11, 12, and 13 will expand the study of graphs and include graphs that are not straight lines. A key idea to understand is the concept of **slope.** The alternate forms of the equation of a line are restatements of the definition of slope.

■ 3.1   Graphing First-Degree Equations

Graphing Ordered Pairs

Two ways to represent the ordered pairs that satisfy an equation are tables and sets of ordered pairs. Both of these methods are very accurate. However, for a large set of ordered pairs a more convenient method is needed to represent the ordered pairs and to show any pattern that may exist. A graph is the most commonly-used method.

x-axis
y-axis
Origin

    Figure 3.1 is a **rectangular graph.** The horizontal line, called the ***x*-axis,** is placed at right angles to the vertical line, called the ***y*-axis.**

    The point where the two lines cross is called the **origin.**

    By agreement, the *x*-axis is marked with positive numbers to the right of the origin and negative numbers to the left of the origin, with zero at the origin. The *y*-axis is marked off with positive numbers going up from the origin and negative numbers going down from the origin. Zero is at the origin.

■ Figure 3.1

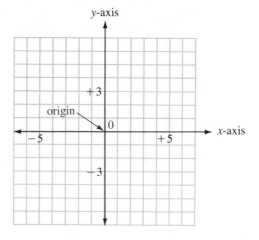

    We use points to represent each ordered pair. The first component of the ordered pair tells how far right or left in the direction of the *x*-axis to place the point. If the first component is a positive number, the point will be to the right of origin along the *x*-axis, and, if it is a negative number, the point will be to the left of origin along the *x*-axis. The second component of an ordered pair tells how far up or down on the *y*-axis to place the point. If the second component is a positive number, the point will be above the *x*-axis, and, if

                                             Chapter 3 ■ Linear Equations, Graphs, and Inequalities

it is a negative number, it will be below the *x*-axis. When the two components of an ordered pair are used to represent a point on a graph, they are referred to as the **coordinates** of the point.

Example 1 □ Give the coordinates of each point labeled on the graph. □

■ Figure 3.2

$A(4, 2)$
$B(2, 4)$
$C(-1, 4)$
$D(-5, 2)$
$E(-4, -3)$
$F(-1, -1)$
$G(1, -1)$
$H(5, 0)$
$I(0, -5)$

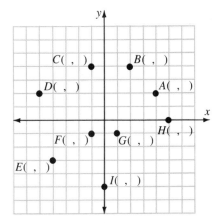

Graphing First-Degree Equations

An equation that gives the value of one variable as determined by the value assigned to another variable raised to the first power only is a **first-degree equation.**

$y = 3x + 2$ and $y = -\dfrac{2}{3}x - 7$ are two different first-degree equations.

To draw a graph of a first-degree equation we evaluate the equation for several values of *x*.

Example 2 □ Graph $y = 3x + 2$.

	Equation		Ordered Pair
x	$3x$	$+ 2 = y$	(x, y)
1	$3(1)$	$+ 2 = 5$	$(1, 5)$
0	$3(0)$	$+ 2 = 2$	$(0, 2)$
-1	$3(-1)$	$+ 2 = -1$	$(-1, -1)$
-3	$3(-3)$	$+ 2 = -7$	$(-3, -7)$

How did you get those values for *x*?

I could have picked any values, so I chose ones that were easy to work with.

Then we plot the points that correspond to the ordered pairs.

■ Figure 3.3

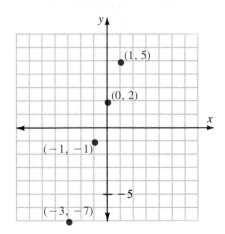

Hey, these points all lie on a straight line! Can I draw a line through them?

Yes, but read the following to see what a solid line means.

Drawing a solid line implies that the coordinates of every point on the line satisfy the equation. Notice that the line also passes through the point $(-2, -4)$.

■ Figure 3.4

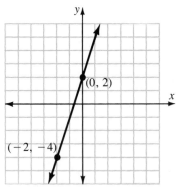

Test the point $(-2, -4)$

$$y = 3x + 2$$
$$-4 \stackrel{?}{=} 3(-2) + 2$$
$$-4 \stackrel{?}{=} -6 + 2$$
$$-4 = -4$$

It works.

The point $(-2, -4)$ is one of the points that satisfy the equation. You may wish to try the coordinates of a few other points that are on the line of the function. You will find that the coordinates of any point on the line satisfy the equation $y = 3x + 2$. You will also find that the coordinates of any point that is not on the line will not satisfy the equation.

Chapter 3 ■ Linear Equations, Graphs, and Inequalities

Notice that the equation $y = 3x + 2$ says that the value of y is three times the value of x plus two. The points along the line are the set of points whose second coordinate is three times the first coordinate plus two. The line is therefore said to be the graph of the equation $y = 3x + 2$. ☐

Example 3 ☐ Graph $y = -\dfrac{2}{3}x - 1$.

	Equation			Ordered Pair
x	$-\dfrac{2}{3}x$	$-1 = y$		(x, y)
0	$-\dfrac{2}{3}(0)$	$-1 = -1$		$(0, -1)$
3	$-\dfrac{2}{3}(3)$	$-1 = -3$		$(3, -3)$
6	$-\dfrac{2}{3}(6)$	$-1 = -5$		(___, ___)
-3	$-\dfrac{2}{3}(-3) - 1 = 1$			(___, ___)

6, −5

−3, 1

■ Figure 3.5

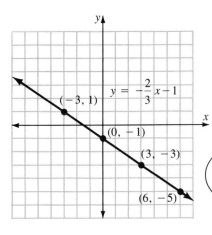

How come all the x values are multiples of 3?

You can pick any value of x you wish. Therefore, you might as well pick values of x that make it easy to calculate y. When the denominator of the fraction is 3, multiples of 3 are easy. ☐

Two points are enough to draw a line, but it's a good practice to use at least three points so that you can tell if you make a mistake.

Are all graphs straight lines?

No, only the graphs of first-degree equations are straight lines. That's why they are sometimes called **linear** equations.

Problem Set 3.1A

Complete the following tables and make a set of ordered pairs for the following equations. Sketch the graphs.

1. $y = 3x - 4$

x	$3x - 4$	(x, y)
-2		
0		
1		
3		
4		

2. $y = -\dfrac{1}{2}x - 3$

x	$-\dfrac{1}{2}x - 3$	(x, y)
-4		
-2		
0		
2		
-6		

3. $2x + 3y = 6$ (Solve for y first)

x	$-\dfrac{2}{3}x + 2$	(x, y)
-6		
-3		
0		
3		
6		

4. $3x - 2y = 6$ (Solve for y first)

x		(x, y)
-4		
-2		
0		
2		
6		

Make a table for ordered pairs and sketch the graph for each of the following equations.

5. $y = 2x - 4$

6. $y = -3x - 1$

7. $y = \dfrac{3}{4}x - 2$

8. $y = \dfrac{2}{3}x + 2$

9. $y = -\dfrac{2}{3}x + 3$

10. $y = -\dfrac{3}{4}x - 2$

11. $y = \dfrac{3}{2}x - 1$

12. $y = \dfrac{5}{3}x + 2$

13. $y = -\dfrac{3}{2}x - 1$

14. $y = -\dfrac{5}{3}x - 2$

15. $y = -\dfrac{5}{6}x + 2$

16. $y = \dfrac{3}{5}x - 3$

17. $5x - 2y = 6$

18. $2x - 4y = -12$

19. $3x + 3y = -6$

20. $3x + 2y = 4$

21. $2x + 3y = 6$

22. $3x - 4y = 8$

23. $4x - 3y = 9$

24. $5x - 6y = -12$

25. $6x - 5y = -10$

Slope

Think for a minute about the minimum information required to locate a line. Two points will do it. You could also locate a line if you knew one point on the line and the proper angle to draw the line through that point. **The inclination of a line is measured by its slope.**

Example 4 □

■ Figure 3.6

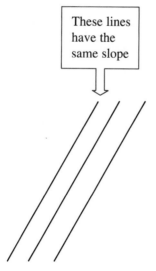

These lines have the same slope

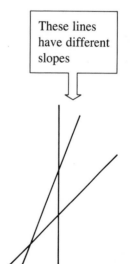

These lines have different slopes

□

Example 5 □ Consider the equation $y = 2x$.

A table of values
for the equation is:

x	$y = 2x$
0	0
1	2
2	4
3	6
4	8

Using the table of values,
we can sketch the graph of $y = 2x$.

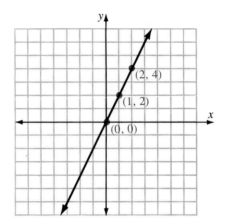

2, 4

1, 2

2, 6

2, 4

2, 8

3, 6

2

4

6

2

As x varies from 1 to 2, y varies from _____ to _____. The
change in x is _____ and the corresponding change in y is _____.

As x varies from 1 to 3, y varies from _____ to _____. The
change in x is _____ and the corresponding change in y is _____.

As x varies from 1 to 4, y varies from _____ to _____. The
change in x is _____ and the corresponding change in y is _____.

If x increases by 1, y increases by _____.

If x increases by 2, y increases by _____.

If x increases by 3, y increases by _____.

The change in y is always _____ times the change in x.   □

The steepness or inclination of a line is dependent upon its slope.

3.1 Definition: Slope of a Line

The slope of a line between two points with coordinates (x_1, y_1) and
(x_2, y_2) is the ratio of the change in the y-coordinates of the two points
to the change in the x-coordinates of the points.

$$\text{Slope} = \frac{\text{change in } y}{\text{change in } x} = \frac{\text{rise}}{\text{run}} = \frac{y_2 - y_1}{x_2 - x_1}$$

> What do the little 1's and 2's next to the *x*'s and *y*'s mean?

> The little numbers are called subscripts; y_2 is read "*y* sub 2." When we want to use the same variable letter for two different quantities, we use subscripts to show that they are different.

> Then y_1 and y_2 are different quantities?

> Yes. In the above definition we are talking about two points, each with an *x*- and a *y*-coordinate; so we say that the first point has coordinates (x_1, y_1) and the second point has coordinates (x_2, y_2).

The letter *m* is used to represent the slope.

■ Figure 3.8

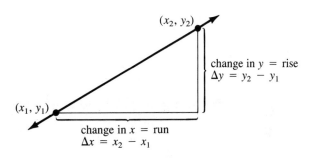

change in y = rise
$\Delta y = y_2 - y_1$

change in x = run
$\Delta x = x_2 - x_1$

$$m = \frac{\text{change in } y}{\text{change in } x} = \frac{\Delta y}{\Delta x}$$

$$= \frac{\text{rise}}{\text{run}}$$

$$m = \frac{y_2 - y_1}{x_2 - x_1}$$

> Δ is used as shorthand for "change in," thus "Δx" means "change in *x*."

Example 6 □ Find the slope of the line below.

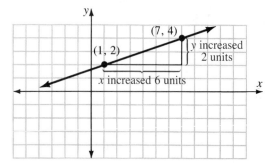

To find the change in y, count squares or notice that the y-coordinate of the second point (4) minus the y-coordinate of the first point (2) is $4 - 2 = 2$ units.

The change in x is $7 - 1 = 6$ units.

$$\text{Slope} = m = \frac{\Delta y}{\Delta x} = \frac{y_2 - y_1}{x_2 - x_1} = \frac{4 - 2}{7 - 1} = \frac{2}{6} = \frac{1}{3}$$

A slope of $\frac{1}{3}$ tells us that the change in the value of y will be $\frac{1}{3}$ the change in the value of x.

If x is increased 6 units, y will increase $\frac{1}{3} \cdot (6) = 2$ units.   □

Example 7 □ Find the slope of the line below.

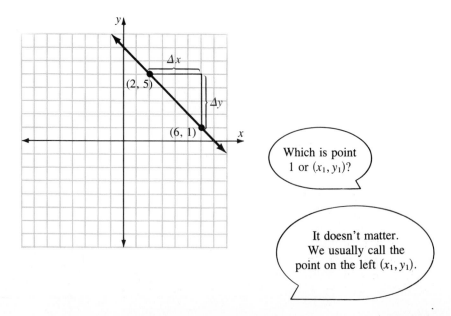

Chapter 3 ■ Linear Equations, Graphs, and Inequalities

Notice this time that the run (Δx) is positive, but the rise (Δy) is negative. This is a line with a negative slope.

$$m = \frac{\Delta y}{\Delta x} = \frac{y_2 - y_1}{x_2 - x_1}$$

$$= \frac{\overset{y_2}{\overset{\downarrow}{1}} - \overset{y_1}{5}}{\underset{6}{6} - \underset{x_1}{2}}$$

$$= -\frac{4}{4}$$

$$= -1$$

Try it the other way around; call (6, 1) point 1. Then

$$m = \frac{y_2 - y_1}{x_2 - x_1} = \frac{5 - 1}{2 - 6} = \frac{+4}{-4} = -1 \qquad \square$$

Example 8 $\square$ Complete the following table showing the slope between selected pairs of points on the line below.

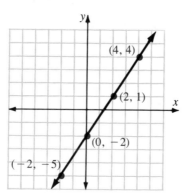

Point 2	Point 1	Slope
$(4, 4)$	$(2, 1)$	$\dfrac{4 - 1}{4 - 2} = \dfrac{3}{2}$
$(2, 1)$	$(4, 4)$	$\dfrac{1 - 4}{2 - 4} = \dfrac{-3}{-2} = \dfrac{3}{2}$
$(4, 4)$	$(0, -2)$	$\dfrac{4 - (-2)}{4 - 0} = \dfrac{6}{4} = \dfrac{3}{2}$
$(2, 1)$	$(-2, -5)$	$\dfrac{1 - (-5)}{2 - (-2)}$, $\dfrac{6}{4}$, $\dfrac{3}{2}$
$(-2, -5)$	$(0, -2)$	$\dfrac{-5 - (-2)}{-2 - 0}$, $\dfrac{-3}{-2}$, $\dfrac{3}{2}$

Hey! The slope between any two points on a straight line is always the same.

Yup.

Lines with positive slope rise in the direction you read.

Lines with negative slope fall to the right.

■ Figure 3.12
■ Figure 3.13

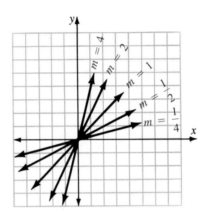

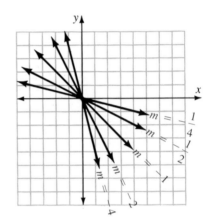

Problem Set 3.1B

Sketch a line through each pair of points and find the slope of the line through the points.

1. $(2, 6), (-3, 4)$

2. $(-6, -4), (4, 3)$

3. $(4, -6), (2, 3)$

4. $(-4, 6), (3, 2)$

5. $(-4, -2), (3, -4)$

6. $(-5, 4), (4, -6)$

7. $(-7, 2), (-3, 1)$

8. $(8, 4), (2, -2)$

9. $(6, 7), (-2, -4)$

10. $(-1, 2), (4, 6)$

11. $(-4, 5), (6, -7)$

12. $(-2, -3), (1, 4)$

13. $(5, 6), (-2, -4)$

14. $(3, -4), (-5, 2)$

15. $(3, 1), (-6, 4)$

16. $(-1, 2), (-5, 3)$

17. $(-2, -5), (-6, -8)$

18. $(-1, 3), (3, -1)$

Chapter 3 ■ Linear Equations, Graphs, and Inequalities

Slope-Intercept Form

Every straight line (except a vertical one) crosses the y-axis somewhere. Therefore, in general, the straight line crosses the y-axis at $(0, b)$. The y-coordinate, b, is called the **y-intercept.**

y-intercept

■ Figure 3.14

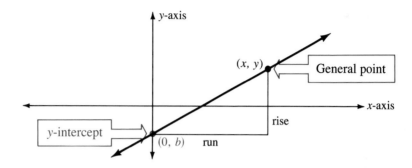

The slope of any straight line is

$$m = \frac{y_2 - y_1}{x_2 - x_1}$$

A general point on a line can be represented by the ordered pair (x, y). The slope of a line passing through the general point (x, y) with a y-intercept of b is

$$m = \frac{y - b}{x - 0} \qquad \text{Using the points } (0, b) \text{ and } (x, y)$$

$$m = \frac{y - b}{x}$$

$$mx = y - b \qquad \text{Multiplying both sides by } x$$

$$mx + b = y \qquad \text{Adding } b \text{ to both sides}$$

$$y = mx + b \qquad \text{Symmetric property of equality}$$

3.2A   Definition: Slope-Intercept Form of the Equation of a Line

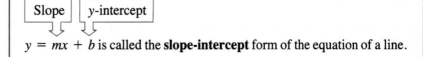

$y = mx + b$ is called the **slope-intercept** form of the equation of a line.

The slope-intercept form is one of the most useful forms of the equation of a straight line because we can graph the line very easily from it. In the equation $y = 3x - 4$, the slope is 3 and the y-intercept is -4.

$$y = 3x - 4$$

| Slope | y-intercept |

Example 9 ☐ Give m, b, and the coordinates of the y-intercept for each of the equations below.

Equation	m	b	Coordinates of y-intercept
$y = 2x + 3$	2	3	$(0, 3)$
$y = 2x - 3$	2	-3	_____
$y = 5x + 4$	5	_____	_____
$y = -5x + 4$	_____	_____	_____
$y = \frac{2}{3}x + 3$	_____	_____	_____
$y = -4x$	_____	_____	_____
$y = x$	_____	_____	_____

(Margin answers)

		$(0, -3)$
4		$(0, 4)$
-5	4	$(0, 4)$
$\frac{2}{3}$	3	$(0, 3)$
-4	0	$(0, 0)$
1	0	$(0, 0)$

☐

Example 10 ☐ Find the equation of the line with slope $\frac{2}{3}$ and y-intercept 4.

$$y = mx + b$$

$$y = \underline{\quad\quad} \; x + \underline{\quad\quad}$$

Just substitute the proper values for m and b.

(Margin answer: $\frac{2}{3}$, 4)

☐

Problem Set 3.2A

Find the equation of the line with the given slope and y-intercept.

1. Slope 2, y-intercept 4

2. Slope 3, y-intercept 7

3. Slope -5, y-intercept 6

4. Slope -6, y-intercept 8

5. Slope $\frac{3}{4}$, y-intercept 4

6. Slope $\frac{5}{6}$, y-intercept -3

7. Slope $-\frac{4}{7}$, y-intercept -5

8. Slope $-\frac{3}{5}$, y-intercept 2

9. Slope 4, y-intercept 0

10. Slope -3, y-intercept 0

11. Slope 3, y-intercept 5

12. Slope 3, y-intercept -5

13. Slope 4, y-intercept -2 **14.** Slope 3, y-intercept -3 **15.** Slope -2, y-intercept 3

16. Slope -5, y-intercept 1   **17.** Slope $\frac{2}{3}$, y-intercept -1 **18.** Slope $\frac{1}{6}$, y-intercept 6

19. Slope $-\frac{3}{2}$, y-intercept -3 **20.** Slope $-\frac{6}{5}$, y-intercept 3

21. Slope $\frac{3}{7}$, y-intercept 0   **22.** Slope $-\frac{6}{7}$, y-intercept -2

23. Slope $\frac{6}{5}$, y-intercept -3 **24.** Slope $-\frac{5}{4}$, y-intercept 5

25. Slope -2, passing through $(0, 3)$ **26.** Slope $\frac{1}{2}$, passing through $(0, -1)$
   Hint: Since $(0, 3)$ is on the y-axis the y-intercept is 3.

27. Slope $-\frac{2}{3}$, passing through $(0, 0)$ **28.** Slope $\frac{5}{6}$, passing through the origin

29. Slope -3, passing through $(0, 2)$ **30.** Slope -4, passing through $(0, -2)$

31. Slope $-\frac{6}{5}$, passing through the origin   **32.** Slope $-\frac{7}{5}$, passing through $(0, 0)$

Graphing Lines by the Slope-Intercept Method

We can use the slope-intercept form of an equation to sketch its graph.

Example 11 □ Sketch a graph of $y = \frac{4}{3}x - 2$.

-2

$\dfrac{4}{3}$

The y-intercept in this equation is _____ .

The slope of this equation is $\dfrac{\Delta y}{\Delta x} =$ _____ .

■ Figure 3.15

Step 1.
Plot the y-intercept

Step 2.
Starting from the y-intercept point, $(0, -2)$, draw Δx and Δy to locate a second point on the line

Step 3.
Draw the line through these points

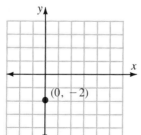

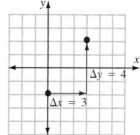

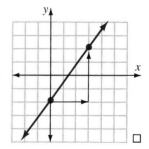

□

Example 12 □ Sketch a graph of $3x + 4y = 8$.

First rewrite the equation in slope-intercept form by solving for y. Then graph using the slope and y-intercept.

$$3x + 4y = 8$$

$$4y = -3x + 8$$

$$\frac{4}{4}y = -\frac{3}{4}x + \frac{8}{4}$$

$$y = -\frac{3}{4}x + 2$$

□

■ Figure 3.16

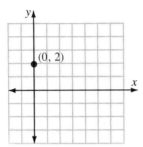 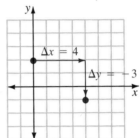 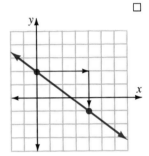

Problem Set 3.2B

Graph the following using the slope-intercept method.

1. $y = \frac{2}{3}x + 2$

2. $y = -\frac{2}{3}x + 2$

3. $y = -\frac{2}{3}x - 2$

4. $y = \frac{2}{3}x - 2$

5. $y = \frac{3}{5}x - 2$

6. $y = -\frac{3}{5}x + 2$

7. $y = -\frac{5}{2}x - 3$

8. $y = \frac{5}{2}x + 3$

9. $-2x + 3y = 6$

10. $-3x + 2y = 6$

11. $-3x + 2y = -6$

12. $y = \frac{2}{3}x$

13. $y = -3x$

14. $y = x$

15. $4x - 3y = 12$

16. $3x - y = 6$

17. $3x - 4y = 12$

18. $4x - 5y = 10$

19. $-3x + 4y = 12$

20. $-4x + 3y = 12$

21. $7x - 4y = 8$

22. $-4x + 7y = 14$

23. $y = -\frac{2}{3}x$

24. $y = -\frac{3}{2}x$

Chapter 3 ■ Linear Equations, Graphs, and Inequalities

Equation of a Line, Given the Slope and a Point

There are several methods of writing the equation of a line, but upon examination they are all restatements of the definition of slope.

We have learned that the slope of a straight line is the same between any two points on the line. The ordered pair (x, y) is a general point on the line. The slope of a line passing through the general point (x, y) and the fixed point (x_1, y_1) is:

■ Figure 3.17

$$m = \frac{y - y_1}{x - x_1}$$

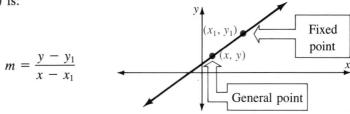

Example 13 □ Use the slope formula to find the equation of a line through $(-3, 4)$ with a slope of $\frac{2}{5}$.

The known point on the line $(-3, 4)$ corresponds to (x_1, y_1). The slope is $\frac{2}{5}$.

$$m = \frac{y - y_1}{x - x_1} \qquad \text{Definition of slope from above}$$

$$\frac{2}{5} = \frac{y - 4}{x - (-3)} \qquad \text{Substitute values for } m, x_1, y_1$$

$$\frac{2}{5} = \frac{y - 4}{x + 3} \qquad \text{Simplify}$$

$$\frac{2}{5}(x + 3) = y - 4 \qquad \text{Multiply both sides by } x + 3$$

$$\frac{2}{5}x + \frac{6}{5} = y - 4$$

$$\frac{2}{5}x + \frac{6}{5} + 4 = y \qquad \text{Add 4 to both sides}$$

$$\frac{2}{5}x + \frac{6}{5} + 4\left(\frac{5}{5}\right) = y \qquad \text{Common denominator}$$

$$\frac{2}{5}x + \frac{26}{5} = y \qquad \text{Simplify}$$

$$y = \frac{2}{5}x + \frac{26}{5} \qquad \text{Symmetric property}$$

This is the equation of a line through $(-3, 4)$ with a slope of $\frac{2}{5}$. □

One way to reduce the amount of equation solving for each problem is to solve the problem once in general terms.

The definition of the slope, m, between a general point on a line (x, y) and a specific point on the line (x_1, y_1) is

$$m = \frac{y - y_1}{x - x_1}$$

$$\frac{y - y_1}{x - x_1} = m \qquad \text{Symmetric property of equality}$$

$$(x - x_1)\frac{y - y_1}{x - x_1} = m(x - x_1) \qquad \text{Multiply both sides by } (x - x_1)$$

$$y - y_1 = m(x - x_1) \qquad \text{Reduce}$$

3.2B Definition: Point-Slope Form of the Equation of a Line

$y - y_1 = m(x - x_1)$ is called the **point-slope** form of the equation of a line.

To use the point-slope form of the equation of a line, we must know the slope of the line and the coordinates of one point on the line.

Example 14 □ Find the equation of a line with slope -2 and passing through the point $(-4, 3)$.

$y - y_1 = m(x - x_1)$ The point-slope form of the equation of a line

$y - 3 = -2\left[x - (-4)\right]$ Substitute values in the formula

$y - 3 = -2(x + 4)$ Simplify the right side

Where did you get values for (x_1, y_1)?

The problem gives a point on the line: $(-4, 3)$. This tells us $x_1 = -4$, and $y_1 = 3$.

$y - 3 = -2x - 8$ Distributive property

$y = -2x - 5$ Add 3 to both sides   □

Chapter 3 ■ Linear Equations, Graphs, and Inequalities

Example 15 □ Find the equation of a line with slope $\frac{4}{3}$ and passing through the point $(6, -2)$.

$$y - y_1 = m(x - x_1)$$ Point-slope form of equation of a line

$$y - (-2) = \frac{4}{3}(x - 6)$$ Substitute values in the formula

$$y + 2 = \frac{4}{3}x - 8$$ Simplify

$$y = \frac{4}{3}x - 10$$ Combine like terms □

Problem Set 3.2C

Find the equation of a line with the given slope and passing through the given point. Use the point-slope method.

1. Slope $= 2$, passing through $(3, 2)$

2. Slope $= \frac{2}{3}$, passing through $(3, 2)$

3. Slope $= \frac{3}{5}$, passing through $(-2, 4)$

4. Slope $= -\frac{3}{2}$, passing through $(4, -2)$

5. Slope $= -\frac{4}{3}$, passing through $(6, -7)$

6. Slope $= -\frac{3}{7}$, passing through $(-1, 5)$

7. Slope $= 0$, passing through $(4, -7)$

8. Slope $= \frac{6}{5}$, passing through $(-2, -5)$

9. Slope $= \frac{4}{5}$, passing through $(-4, 2)$

10. Slope $= -\frac{4}{5}$, passing through $(-3, 5)$

11. Slope $= -3$, passing through $(3, -6)$

12. Slope $= \frac{1}{3}$, passing through $(1, 0)$

13. Slope $= -\frac{5}{4}$, passing through $(0, -4)$

14. Slope $= \frac{5}{4}$, passing through $(-4, 0)$

15. Slope $= -\frac{7}{6}$, passing through $(5, -1)$

16. Slope $= 0$, passing through $(7, -4)$

Equation of a Line Through Two Points

If we know only two points on the line, we must find the slope before we can write the equation of the line.

Example 16 ☐ Find the equation of a line through the points $(5, 6)$ and $(-5, 0)$. Find the slope of the line through the two points.

$$m = \frac{y_2 - y_1}{x_2 - x_1} = \frac{6 - 0}{5 - (-5)} = \frac{6}{10} = \frac{3}{5}$$

We can now use either point $(5, 6)$ or $(-5, 0)$ in the point-slope form of the equation of a line.

Substituting $(x_1, y_1) = (5, 6)$ and $m = \frac{3}{5}$ in the point-slope form of a line

$$y - y_1 = m(x - x_1)$$

becomes

$$y - 6 = \frac{3}{5}(x - 5)$$

$$y - 6 = \frac{3}{5}x - 3 \qquad \text{Distributive property}$$

$$y = \frac{3}{5}x + 3 \longleftarrow \boxed{\text{This is the equation of a line through the points } (5, 6) \text{ and } (-5, 0).}$$

Test the other point in the equation to see that it fits.

$$y = \frac{3}{5}x + 3$$

$$0 \overset{?}{=} \frac{3}{5}(-5) + 3$$

$$0 \overset{?}{=} -3 + 3 \qquad \qquad \text{It checks.}$$

$$0 = 0$$

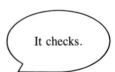

☐

Problem Set 3.2D

Find the equation of the line passing through these given points.

1. $(0, 2), (4, -1)$ **2.** $(0, -4), (-5, 2)$ **3.** $(0, 6), (-5, -6)$

4. $(-2, 8), (6, 7)$ **5.** $(-4, -3), (-1, 7)$ **6.** $(-6, 3), (4, -5)$

7. $(2, 3), (1, 4)$ **8.** $(2, 3), (-1, -4)$ **9.** $(2, 3), (4, 1)$

Chapter 3 ■ Linear Equations, Graphs, and Inequalities

10. $(0, -6)$, $(3, -2)$ **11.** $(0, 4)$, $(-5, 2)$ **12.** $(-1, 0)$, $(5, 2)$

13. $(-3, 0)$, $(3, -5)$ **14.** $(-3, 8)$, $(6, 8)$ **15.** $(-6, -4)$, $(-6, 5)$

16. $(2, -3)$, $(-5, 1)$ **17.** $(-1, -3)$, $(6, 2)$ **18.** $(-1, 3)$, $(4, 7)$

■ 3.3   Graphing Lines by the Intercepts Method

The previous methods for graphing linear equations will allow you to graph any line. However, when the equation of the line is not solved for y explicitly, the method known as the intercepts method is frequently a convenient way to graph the line.

Example 17 □ Sketch the graph of $3x - 4y = 8$ using the x- and y-intercepts.

One way to graph $3x - 4y = 8$ is by finding its x- and y-intercepts, that is, the points where the line crosses each axis. We have observed that when the line crosses the y-axis, the x-coordinate is 0.

Substituting $x = 0$ in the equation yields

$$3(0) - 4y = 8$$

$$-4y = 8$$

$$y = -2 \qquad (0, -2) \text{ are the coordinates}$$
$$\text{of the } y\text{-intercept}$$

Where the line crosses the x-axis, the y-coordinate is 0. We find the x-intercept by substituting $y = 0$ in the equation.

$$3x - 4(0) = 8$$

$$3x = 8$$

$$x = \frac{8}{3} \qquad \left(\frac{8}{3}, 0\right) \text{ are the coordinates}$$
$$\text{of the } x\text{-intercept}$$

$$\frac{8}{3} = 2\frac{2}{3}$$

■ Figure 3.18

Step 1.
Plot the intercepts

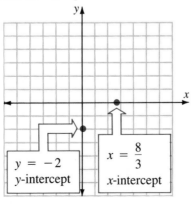

$y = -2$
y-intercept

$x = \dfrac{8}{3}$
x-intercept

Step 2.
Draw a line connecting the intercepts

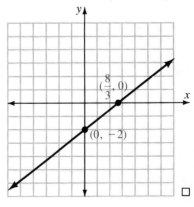

$\left(\dfrac{8}{3}, 0\right)$

$(0, -2)$

□

Example 18 □ Sketch the graph of $3x - y = 3$ by the intercepts method.

x

To find the y-intercept we substitute _____ = 0

Substituting in the equation yields

$$3(0) - y = 3$$

$$-y = 3$$

$$y = \text{_____} \text{ is the y-intercept}$$

−3

y

To find the x-intercept we substitute _____ = 0

$$3x - (0) = 3$$

$$3x = 3$$

1

$$x = \text{_____} \text{ is the x-intercept}$$

■ Figure 3.19

Step 1.
Plot the intercepts

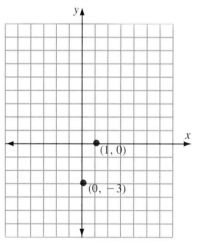

Step 2.
Draw a line connecting the intercepts

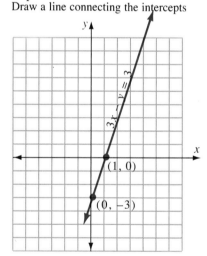

When graphing lines there are two special cases where one variable appears to be missing. □

Case 1

Example 19 □ Sketch the graph of $y = -2$. Since there is no x-term in this equation, the coefficient of the x-term is considered to be zero. The equation can be written $0x + y = -2$.

Chapter 3 ■ Linear Equations, Graphs, and Inequalities

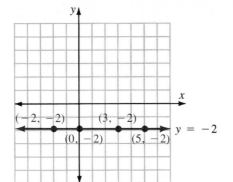

Make a table		Ordered pair
x	y	(x, y)
−2	−2	(−2, −2)
0	−2	(0, −2)
3	___	(_____)
5	___	(_____)

−2, (3, −2)

−2, (5, −2)

In the equation $y = -2$, what is the change in y when x changes from −2

0, 0, 0

to 0? ———— from 0 to 3? ———— from any point to another? ————

Using the coordinates $(0, -2)$ and $(5, -2)$, we can calculate the slope of $y = -2$.

0

$$m = \frac{y_2 - y_1}{x_2 - x_1} = \frac{-2 - (-2)}{5 - 0} = \frac{0}{5} = \text{————}$$

$y = -2$ means all points that are 2 down from the x-axis. □

3.3A   Graph of $y = b$

In general, the graph of $y = b$ is a horizontal line parallel to the x-axis with zero slope. It crosses the y-axis at $y = b$.

Case 2

Example 20 □ Sketch the graph of $x = -3$.

Since there is no y-term in the equation $x = -3$, the coefficient of the y-term may be considered to be zero. The equation becomes $x + 0y = -3$.

In the equation $x + 0y = -3$, the only value that can be assigned to x is −3. Notice, however, you may assign any value to y and the equation will be true as long as x is −3.

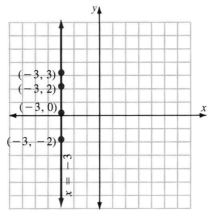

Make a table		Ordered pair
x	y	(x, y)
−3	−2	(−3, −2)
−3	0	(−3, 0)
___	2	(_____)
___	3	(_____)

−3, (−3, 2)

−3, (−3, 3)

The graph of $x = -3$ is a vertical line that crosses the x-axis at $x = -3$.

$x = -3$ means all points that are -3 units from the y-axis. For the equation $x = -3$, x does not change. By definition, slope $= \dfrac{\text{change in } y}{\text{change in } x}$.

Since the change in x is zero, the denominator of our fraction will always be zero. But division by zero is undefined. Therefore, the slope of the vertical line $x = 2$ is undefined. ◻

3.3B Graph of $x = a$

In general, the graph of $x = a$ is a vertical line parallel to the y-axis. It crosses the x-axis at $x = a$. Its slope is undefined.

Problem Set 3.3A

Sketch the following graphs using the intercepts method.

1. $2x + 3y = 6$	**2.** $3x + 2y = 6$	**3.** $3x - 2y = 6$
4. $6x - 3y = 6$	**5.** $2x - 5y = 10$	**6.** $-3x + 4y = 12$
7. $-2x + 2y = 8$	**8.** $2x + y = 10$	**9.** $3x - 2y = 8$
10. $-3x + 2y = 8$	**11.** $3x - 4y = -4$	**12.** $-3x + 4y = -4$
13. $3x - 5y = 10$	**14.** $3x + 5y = -10$	**15.** $3x + 7y = -14$
16. $4x - 7y = -14$	**17.** $3x - y = 9$	**18.** $x + y = 7$
19. $x - y = 7$	**20.** $2x + y = 7$	**21.** $3x - y = 5$
22. $-4x + y = 3$	**23.** $-5x - y = 4$	**24.** $x = 3$
25. $y = 4$	**26.** $y = 0$	**27.** $x = 0$
28. $x = -3$	**29.** $x = -2$	**30.** $y = -2$
31. $x = 5$	**32.** $y = -5$	

Chapter 3 ■ Linear Equations, Graphs, and Inequalities

Parallel and Perpendicular Lines

Example 21 □ Sketch $y = -3x - 3$ and $y = -3x + 2$.

■ Figure 3.22

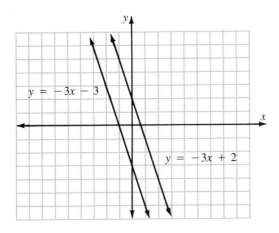

Slope

y-intercept

Direction

Where they cross the
y-axis

Parallel

What is alike in the two equations? _____

What is different in the two equations? _____

What is alike in the two graphs? _____

What is different in the two graphs? _____

□

Example 21 illustrates the fact that lines with the same slope are **parallel.**

Problem Set 3.3B

Sketch the graphs of each pair of lines on the same coordinate system.

1. $y = \dfrac{2}{3}x + 5$ and $y = \dfrac{2}{3}x - 5$

2. $y = -\dfrac{2}{3}x + 5$ and $y = -\dfrac{2}{3}x - 5$

3. $y = 3x - 3$ and $y = 3x$

4. $y = -2x + 6$ and $y = -2x + 1$

5. $y = \dfrac{3}{5}x$ and $y = \dfrac{3}{5}x + 3$

6. $y = -\dfrac{3}{5}x + 2$ and $y = -\dfrac{3}{5}x - 2$

7. $y = -3x - 3$ and $y = -3x + 1$

8. $y = \dfrac{1}{3}x - 6$ and $y = \dfrac{1}{3}x + 2$

It is possible to sketch lines parallel to a given line using the slope-intercept method of sketching graphs.

Example 22 □ Sketch a line parallel to the line $y = \frac{3}{4}x - 4$ with a y-intercept 5.

To sketch the line $y = \frac{3}{4}x - 4$, we see that the slope is $\frac{3}{4}$ and the y-intercept is -4.

■ Figure 3.23

Locate the point $(0, -4)$, the y-intercept. The line has a rise of 3 and a run of 4. Connect the points.

To sketch the line parallel to the given line with y-intercept 5, locate the point $(0, 5)$. This line also has a rise of 3 and a run of 4. Connect the points.

Since the lines have the same slope, they are parallel.

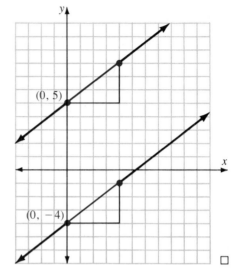

Problem Set 3.3C

Sketch a second line parallel to the given line. The second line should cross the y-axis as specified.

1. Parallel to $y = 3x - 4$, passing through $(0, 3)$

2. Parallel to $y = -3x$, passing through $(0, -4)$

3. Parallel to $y = \frac{2}{5}x + 3$, y-intercept of -1

4. Parallel to $y = \frac{2}{5}x + 3$, crossing the y-axis at the origin

5. Parallel to $y = -\frac{3}{5}x + 4$, passing through $(0, -2)$

6. Parallel to $y = \frac{3}{5}x - 3$, passing through $(0, 4)$

7. Parallel to $y = \frac{7}{4}x + 2$, y-intercept of 3

8. Parallel to $y = -\frac{7}{4}x - 3$, crossing the y-axis at $y = 4$

Chapter 3 ■ Linear Equations, Graphs, and Inequalities

It is possible to write the equation of a line parallel to another line passing through any point we specify.

Example 23 □ Write the equation of a line passing through the point $(-5, 4)$ and parallel to the line $y = \frac{2}{5}x - 4$.

Since the desired line is to be parallel to $y = \frac{2}{5}x - 4$ we know its slope must be $\frac{2}{5}$.

Therefore this is really a problem of writing the equation of a line with slope $\frac{2}{5}$ passing through the point $(-5, 4)$.

We will use one of the methods we used earlier to do this.

$$y = mx + b \qquad \text{General equation of a line}$$

Since we know values for x, y and m, we can substitute them to find the value of b.

$$4 = \frac{2}{5}(-5) + b$$

$$4 = -2 + b$$

$$4 + 2 = b$$

$$6 = b$$

Therefore the equation we want is

$$y = \frac{2}{5}x + 6$$

It is parallel to $y = \frac{2}{5}x - 4$ because it has the same slope.

We can establish that $y = \frac{2}{5}x + 6$ contains the point $(-5, 4)$ by substituting these values in the equation.

$$y = \frac{2}{5}x + 6$$

$$4 \stackrel{?}{=} \frac{2}{5}(-5) + 6 \qquad \text{Substituting}$$

$$4 \stackrel{?}{=} -2 + 6 \qquad \qquad \text{It checks.}$$

$$4 = 4 \qquad \qquad \qquad \qquad \qquad \square$$

In higher math classes it is possible to prove the following theorem.

3.3C Theorem: Perpendicular Lines

Two lines are perpendicular if their slopes are negative reciprocals, that is, a line with slope m_1 is perpendicular to line 2 with slope m_2

if $$m_2 = -\frac{1}{m_1} \qquad \text{OR} \qquad m_1 m_2 = -1$$

Example 24 □ Use Theorem 3.3C to sketch a line perpendicular to $y = \frac{3}{2}x - 4$ with y-intercept 3.

To sketch the given line, locate the point $(0, -4)$, the y-intercept, on the graph; then use the slope to find the direction.

Theorem 3.3C states that the slope of the line perpendicular to the given line is given by

$$m_2 = -\frac{1}{m_1}$$

The desired slope is

■ Figure 3.24

$$m_2 = -\frac{1}{\dfrac{3}{2}}$$

$$m_2 = -\frac{2}{3}$$

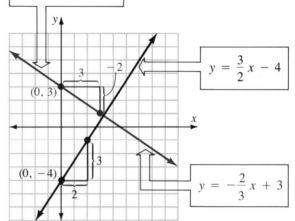

With y-intercept 3 and slope $-\dfrac{2}{3}$ we can sketch the graph. □

Chapter 3 ■ Linear Equations, Graphs, and Inequalities

Use the method of Example 24 to sketch a line perpendicular to the given line with the specified intercept.

1. Perpendicular to $y = 3x - 2$, y-intercept 2

2. Perpendicular to $y = -3x$, passing through $(0, -3)$

3. Perpendicular to $y = \frac{2}{3}x - 3$, passing through $(0, 4)$

4. Perpendicular to $y = -\frac{3}{5}x + 2$, crossing the y-axis at the origin

5. Perpendicular to $y = -4x + 2$, passing through $(0, 6)$

6. Perpendicular to $y = 4x - 2$, passing through $(0, -3)$

7. Perpendicular to $y = \frac{4}{3}x + 3$, crossing the y-axis at $y = -4$

8. Perpendicular to $y = \frac{6}{5}x - 2$, y-intercept 3

We can also use this theorem to write the equation of a line perpendicular to a given line passing through any desired point.

Example 25 □ Find the equation of a line perpendicular to $y = \frac{2}{3}x + 4$ passing through the point $(-4, 4)$.

Since the slope of the desired line m_2 is perpendicular to a line with slope $\frac{2}{3}$, the desired slope is

$$m_2 = -\frac{1}{m_1}$$

$$m_2 = -\frac{1}{\frac{2}{3}}$$

$$= -\frac{3}{2}$$

Now write the equation of a line passing through $(-4, 4)$ with slope $-\frac{3}{2}$.

Using the point-slope form of a line

$$y - y_1 = m(x - x_1)$$

$$y - (4) = -\frac{3}{2}(x - (-4))$$

$$y - 4 = -\frac{3}{2}(x + 4)$$

$$y - 4 = -\frac{3}{2}x - 6$$

$$y = -\frac{3}{2}x - 2$$

Plotting both lines,

■ Figure 3.25

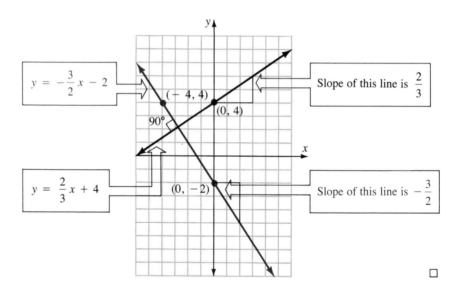

Example 26 □ Find the equation of a line through $(-1, 3)$ and perpendicular to the line $3x + 5y = -10$. Sketch the lines.

We must find the slope of the given equation before we can find the desired slope of the required line.

Hence, solve the given equation for y.

$$3x + 5y = -10$$

$$5y = -3x - 10$$

$$y = -\frac{3}{5}x - 2$$

Since the slope of the desired line m_2 is perpendicular to the line with slope $-\dfrac{3}{5}$ the desired slope is

$$m_2 = -\frac{1}{m_1}$$

$$= -\frac{1}{-\dfrac{3}{5}}$$

$$= \frac{5}{3}$$

Now we write the equation of a line passing through $(-1, 3)$ with slope $\dfrac{5}{3}$.

Using the point slope form of a line

$$y - y_1 = m(x - x_1)$$

$$y - (3) = \frac{5}{3}(x - (-1))$$

$$y - 3 = \frac{5}{3}(x + 1)$$

$$y - 3 = \frac{5}{3}x + \frac{5}{3}$$

$$y = \frac{5}{3}x + \frac{5}{3} + 3$$

$$y = \frac{5}{3}x + \frac{5}{3} + \frac{9}{3}$$

$$y = \frac{5}{3}x + \frac{14}{3}$$

Plotting both lines,

■ Figure 3.26

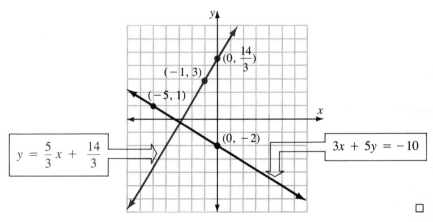

$$y = \frac{5}{3}x + \frac{14}{3}$$

$$3x + 5y = -10$$

a) Find the equation of a line which is parallel to each given line and contains the given point.

b) Find the equation of the line which is perpendicular to each given line and contains the given point.

1. $y = \dfrac{2}{3}x + 5,$ $\quad (-6, 4)$

2. $y = \dfrac{-2}{5}x - 6,$ $\quad (-2, -3)$

3. $x - 2y = 3,$ $\quad (2, -3)$

4. $3x + 2y = 4,$ $\quad (4, 5)$

5. $y = -\dfrac{2}{3}x - 5,$ $\quad (4, -6)$

6. $y = \dfrac{2}{5}x + 6,$ $\quad (-3, -2)$

7. $2x - 3y = 4,$ $\quad (1, 4)$

8. $2x - 5y = 4,$ $\quad (-3, 5)$

Find the equation of the line with the following conditions.

9. Parallel to $2x - 3y = 6$, crossing the x-axis at -2

10. Perpendicular to $2x - 3y = 9$, with x-intercept -2

11. Perpendicular to $3x + 4y = 12$, passing through $(0, 3)$

12. Parallel to $3x + 4y = 8$, crossing the y-axis at 3

13. Parallel to $3x - 4y = 8$, passing through $(-2, 0)$

14. Perpendicular to $4x - 3y = 6$, passing through $(-3, 0)$

15. Perpendicular to $5x + 2y = 4$, crossing x-axis at -3

16. Parallel to $3x + 5y = 5$, with x-intercept 4

17. Parallel to $x = 2$, passing through $(-3, 4)$

18. Perpendicular to $x = -3$, crossing the y-axis at 3

19. Perpendicular to $y = 4$, with x-intercept $(-5, 0)$

20. Parallel to $y = -2$, passing through $(-4, 5)$

21. Parallel to $y = -2$, crossing the y-axis at -2

22. Perpendicular to $y = -3$, with x-intercept -2

23. Perpendicular to $x = -6$, passing through $(-5, 4)$

24. Parallel to $x = 5$ passing through $(2, -5)$

Just as equations in two variables have an infinite number of ordered pairs that make them true, so do inequalities in two variables.

An inequality in two variables divides the graph into two regions. All points whose coordinates make the inequality true are in the shaded region, and all points that make the inequality false are in the unshaded region.

Example 27 □ Graph $y \leq 3x + 4$.

First graph the line $y = 3x + 4$.

Every point on the line satisfies the inequality because the inequality includes the case $y = 3x + 4$.

Every point below the line $y = 3x + 4$ satisfies the inequality $y < 3x + 4$. For each of these points with coordinates (x, y), the y-coordinate is less than three times the x-coordinate plus four.

■ Figure 3.27

Consider the point $(1, 2)$. It is below the line $y = 3x + 4$. Its y-coordinate is 2, which is less than 3 times its x-coordinate, 1 plus 4.

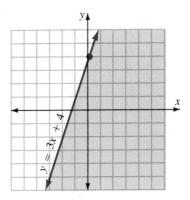

You can think of an equation as a boundary or fence between two portions of the graph. The only question is who owns the fence. If the inequality includes an equal sign, the line of the equation is included in the graph. If the inequality does not contain an equal sign, represent the line with a dashed line to show it is not included in the shaded area.

> How can you test if the graph is shaded correctly?

> Try a few points on either side of the line.

	Value	Region	Substitution	Inequality	Truth
			$y \leq 3x + 4$		
	$(-4, 0)$	unshaded	$0 \leq 3(-4) + 4$	$0 \leq -8$	False
False	$(-2, 5)$	unshaded	$5 \leq 3(-2) + 4$	$5 \leq -2$	———
True	$(0, 0)$	shaded	$0 \leq 3(0) + 4$	$0 \leq 4$	———
True	$(2, 2)$	shaded	$2 \leq 3(2) + 4$	$2 \leq 10$	———

The shaded portion of the graph represents the points that make the inequality true. □

Example 28 □ Graph $y > -\dfrac{1}{2}x - 1$.

Step 1.

■ Figure 3.28

Draw the line $y = -\dfrac{1}{2}x - 1$.

Since the inequality does not contain an equal sign, use a dashed line for the graph of the line.

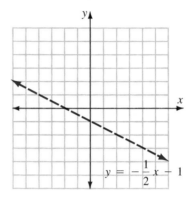

Step 2.

■ Figure 3.29

Try $(0, 0)$ in the inequality.

$$y > -\frac{1}{2}x - 1$$

$$0 > -\frac{1}{2}(0) - 1$$

$$0 > -1 \qquad \text{True!}$$

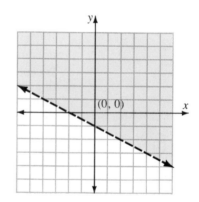

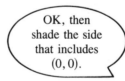

OK, then shade the side that includes $(0, 0)$.

□

Example 29 □ Graph $y < -2x + 3$.

■ Figure 3.30

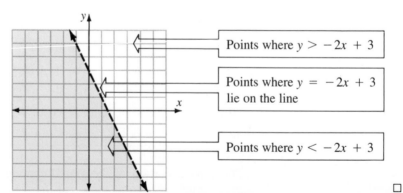

Points where $y > -2x + 3$

Points where $y = -2x + 3$ lie on the line

Points where $y < -2x + 3$

□

How do I know which side of the line to shade?

First draw the line; then pick any point not on the line. (0, 0) is the easiest. If it satisfies the inequality, shade that side.

Example 30 □ Graph $x \geq -4$.
Graph $x = -4$.

■ Figure 3.31

Step 1.

Since the inequality includes an equal sign, draw a solid line.

Step 2.

Test $(0, 0)$ in the inequality.

$$x \geq -4$$

$$0 \geq -4 \qquad \text{is true.}$$

Therefore, shade the side that contains $(0, 0)$.

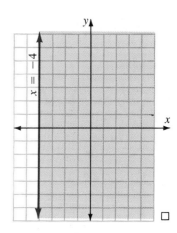

Example 31 □ Graph $y < 2$.

■ Figure 3.32

Step 1.

Graph $y = 2$.
Since no equal sign is included in the inequality, use a dashed line.

Step 2.

Try $(0, 0)$ in the inequality.

$$y < 2$$

$$0 < 2 \qquad \text{is true.}$$

Therefore, shade the side that does contain $(0, 0)$.

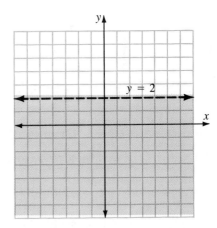

What if $(0, 0)$ falls on the line?

No big thing. Just pick another point and use it to test.

□

Graph the following linear inequalities.

1. $y < x$

2. $y < x - 2$

3. $y \geq x - 2$

4. $y > x + 2$

5. $y \leq x + 2$

6. $y \leq x$

7. $y > 0$

8. $y \geq -3$

9. $y > 2x + 2$

10. $y \geq 3x + 2$

11. $y \leq 2x - 4$

12. $y < 3x - 3$

13. $y \leq \frac{1}{2}x$

14. $y < \frac{1}{2}x + 1$

15. $y < -\frac{1}{2}x + 1$

16. $y \leq -\frac{1}{2}x + 1$

17. $y > -3x + 5$

18. $y \geq -4x - 1$

19. $x \geq -2$

20. $x < 3$

21. $y \leq -\frac{3}{4}x + 2$

22. $y < -\frac{2}{3}x - 1$

23. $y > 4 - \frac{2}{3}x$

24. $y \geq 3 - \frac{3}{4}x$

25. $y \leq -2x + 3$

26. $y \geq 2x - 3$

27. $y \geq 3x - 6$

28. $y < -3x + 2$

29. $y \geq \frac{2}{3}x + 1$

30. $y > \frac{2}{3}x$

31. $y < -\frac{2}{3}x + 1$

32. $y \leq -\frac{2}{3}x + 2$

33. $x < -2$

34. $y \geq 4$

35. $x \leq 0$

36. $y \leq 0$

37. $y > 4x + 1$

38. $y < 4x + 3$

39. $y \leq -4x + 2$

40. $y \geq -4x - 3$

41. $y < \frac{3}{5}x$

42. $y > \frac{3}{5}x + 2$

43. $y \geq -\frac{3}{5}x - 2$

44. $y \leq -\frac{3}{5}x + 3$

45. $y > 5 + \frac{5}{2}x$

46. $y < -2 + \frac{5}{2}x$

47. $y \leq -3 - \frac{5}{2}x$

48. $y \geq 2 - \frac{5}{2}x$

Graphs of Absolute Value Relations

It is possible to graph equations and inequalities involving absolute values.

Example 32 □ Graph $y = |x|$.
Before you graph $y = |x|$,
build a table of values.

x	$y = |x|$
0	0
1	1
2	2
3	3
-1	$+1$
-2	$+2$
-3	$+3$

Plot the points of the table on a rectangular graph.

■ Figure 3.33

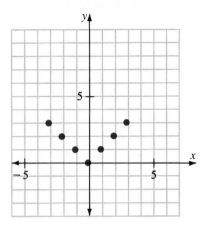

Since absolute value is defined for all real numbers, connect the dots.

■ Figure 3.34

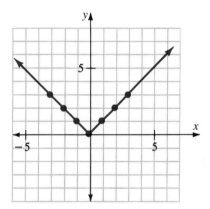

Do all absolute value graphs look like this one?

For first-degree variables they have a V shape. The slope of the lines and the y-intercept may change.

☐

■ Figure 3.35

Example 33 □ Graph $y = |2x|$.
Make a table.

x	$y = |2x|$
0	0
1	2
2	4
−2	4
−1	2

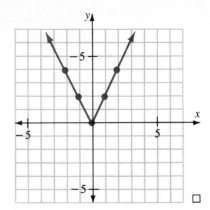

■ Figure 3.36

Example 34 □ Graph $y \geq |x| - 3$.
Make a table of values.

x	$y = |x| - 3$
0	−3
2	−1
3	0
−3	0
−2	−1

Shade the area above the line since $(0, 0)$ satisfies the inequality.

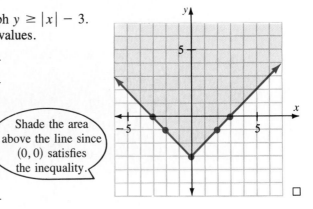

■ Figure 3.37

Example 35 □ Graph $y = |2x| - 3$.
Make a table.

x	$y = |2x| - 3$
0	−3
1	−1
2	1
−2	1
−1	−1

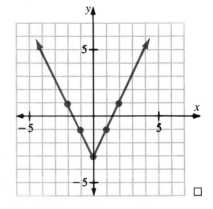

Notice from Examples 32 and 33 that changing the coefficient of x in an absolute value graph changes the slope of the characteristic V. A constant term added or subtracted outside the absolute sign raises or lowers the V on the graph, as illustrated by Examples 33 and 35. Examples 32 and 34 also illustrate the effect of adding a constant outside the absolute value sign.

Let's look at what happens when a constant is added inside the absolute value sign.

Example 36 □ Graph $y = |x + 3| - 2$.

Make a table.

x	$y = |x + 3| - 2$
2	3
−1	0
−2	−1
−3	−2
−4	−1
−5	0
−6	1

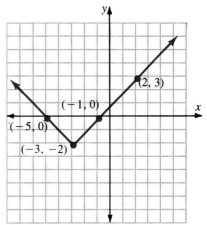

Notice that the minimum value for y occurs when the quantity inside the absolute value sign is equal to zero. In this case $x = -3$ makes $|x + 3| = 0$. Therefore, the x-coordinate of the vertex is -3. The y-coordinate of the vertex is found by substituting $x = -3$ in the equation.

$$y = |-3 + 3| - 2$$
$$y = -2$$

The coordinates of the vertex are $(-3, -2)$. □

Example 37 □ Graph $y = |x - 3| - 2$.

Make a table.

x	$y = |x - 3| - 2$
6	1
4	−1
3	−2
2	−1
0	+1
1	0
5	0

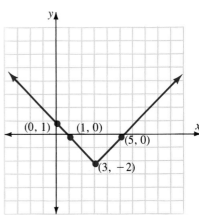

This graph is shifted 3 units to the right.

□

Example 38 □ Graph $y \leq |2x - 7| + 1$.

Examine what makes the expression inside the absolute value signs equal zero.

$$2x - 7 = 0$$
$$2x = 7$$
$$x = \frac{7}{2}$$

> Notice the minimum of this graph is 1 when $2x - 7 = 0$.

> This is the y-coordinate of the point of the V.

$y \leq |2x - 7| + 1$ can now be treated as the union of two inequalities.

First Case	AND	**Second Case**

First Case

When $x > \frac{7}{2}$, $(2x - 7)$ represents a positive number and $y \leq |2x - 7| + 1$ becomes

$$y \leq +(2x - 7) + 1$$
$$y \leq +2x - 6$$

Second Case

When $x < \frac{7}{2}$, $(2x - 7)$ represents a negative number and $y \leq |2x - 7| + 1$ becomes

$$y \leq -(2x - 7) + 1$$
$$y \leq -2x + 8$$

Graphing the left and right inequalities separately,

■ Figure 3.40

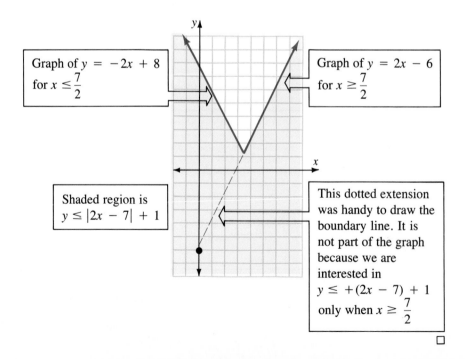

Graph of $y = -2x + 8$ for $x \leq \frac{7}{2}$

Graph of $y = 2x - 6$ for $x \geq \frac{7}{2}$

Shaded region is $y \leq |2x - 7| + 1$

This dotted extension was handy to draw the boundary line. It is not part of the graph because we are interested in $y \leq +(2x - 7) + 1$ only when $x \geq \frac{7}{2}$

□

Chapter 3 ■ Linear Equations, Graphs, and Inequalities

▶ Pointer for Graphing Absolute Value Relations the Easy Way

All first-degree absolute value relations have a characteristic V shape for $y = |mx + a| + b.$

$a \longrightarrow$ moves the V right or left

$x = -\dfrac{a}{m} \longrightarrow$ gives the x-coordinate of the vertex of the V

$y = b \longrightarrow$ gives the y-coordinate of the vertex of the V

To sketch the graph, plot the vertex at $\left(-\dfrac{a}{m}, b\right)$. Set up a table of values, choosing some x values less than $-\dfrac{a}{m}$ and some x values greater than $-\dfrac{a}{m}.$

$$\text{Graph} \qquad y = |3x - 5| - 4$$

$$\text{Set} \qquad 3x - 5 = 0$$

$$\text{Then} \qquad 3x = 5$$

$$x = \dfrac{5}{3}$$

The coordinates of the vertex are $\left(\dfrac{5}{3}, -4\right)$. Plot this point first. Next make a table of values

■ Figure 3.41

		x	y
x values less than $\dfrac{5}{3}$	$\left\{\begin{array}{l}\\\\\end{array}\right.$	−1	4
		0	1
		$\dfrac{5}{3}$	−4
x values greater than $\dfrac{5}{3}$	$\left\{\begin{array}{l}\\\\\end{array}\right.$	2	−3
		3	0

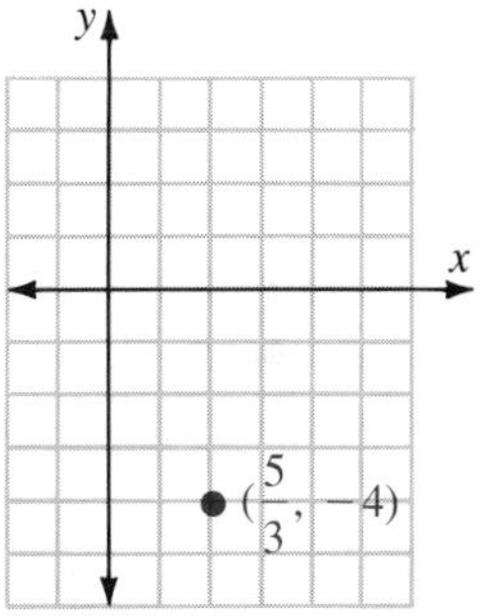

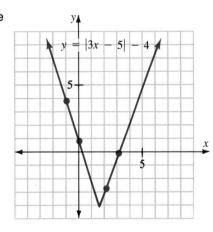

■ Figure 3.42

Now plot the points from the table and finish the graph.

Graph the following absolute value relations.

1. $y = |x| + 1$

2. $y = |3x|$

3. $y = |3x| - 5$

4. $y = |x + 2|$

5. $y = |x - 2|$

6. $y = |x + 3| - 4$

7. $y = |x - 3| - 4$

8. $y = |3x - 2|$

9. $y = |3x + 2|$

10. $y = |2x - 3| + 4$

11. $y = |2x + 3| - 4$

12. $y = |x - 3| + 4$

13. $y = |x + 1| - 5$

14. $y = |x + 5| - 6$

15. $y = |x + 2| - 6$

16. $y = |4x - 1| + 2$

17. $y = |4x + 1| - 2$

18. $y = |3x - 4| + 2$

19. $y = |3x + 4| - 2$

20. $y = |3x - 1| - 3$

21. $y = |3x + 1| + 3$

22. $y > |x| + 2$

23. $y < |x - 3| + 4$

24. $y \geq |x + 2| - 3$

25. $y \geq |3x - 2| + 1$

26. $y \geq |2x + 5| - 4$

27. $y > |4x - 3| + 2$

28. $y \leq |5x + 2| + 3$

29. $y > |3x - 4| + 1$

30. $y < |3x - 4| - 1$

31. $y \leq |3x - 2| - 3$

32. $y \geq |3x - 2| + 4$

33. $y < |5x + 3| - 2$

34. $y > |3x + 5| + 4$

35. $y \geq |3x - 5| - 4$

36. $y \geq |5x - 3| + 2$

37. $y \leq |4x + 3| - 5$

38. $y = \left| \dfrac{x}{2} - 2 \right| + 1$

39. $y = \left| \dfrac{x}{2} - 3 \right|$

40. $y < \left| \dfrac{x}{3} + 2 \right| - 1$

41. $y \geq \left| \dfrac{2}{3}x - 4 \right|$

42. $y \leq \left| \dfrac{2x}{3} + 4 \right| - 1$

Chapter 3 □ Key Ideas

3.1   **1.** Terms associated with graphs

         x-axis—a horizontal line

         y-axis—a vertical line

         origin—point where the x- and y-axes cross

2. The **first-degree equation** is an equation that gives the value of one variable as determined by the value assigned to another variable raised to the first power only.

3. **Slope of a line**

$$\text{Slope} = \frac{\text{change in } y}{\text{change in } x} = \frac{\text{rise}}{\text{run}} = \frac{y_2 - y_1}{x_2 - x_1}$$

3.2 1. **Slope-intercept** form of the equation of a line

$$y = mx + b \qquad \text{where } m = \text{slope}, b = y\text{-intercept}$$

2. Graphing lines by **slope-intercept** method
   a. Locate the y-intercept on the graph
   b. From the y-intercept, determine the run and rise to a second point. Connect the points.

3. **Point-slope** form of the equation of a line

$$y - y_1 = m(x - x_1)$$

3.3 1. Graphs of lines can be sketched using the **intercepts method**
   a. Set $x = 0$ to determine the y-intercept
   b. Set $y = 0$ to determine the x-intercept
   c. Draw a line through the two points

2. **Graph of $y = b$**
   A line parallel to the x-axis crossing the y-axis at $y = b$.

3. **Graph of $x = a$**
   A line parallel to the y-axis crossing the x-axis at $x = a$.

4. **Parallel lines** are lines with the same slope.

5. **Perpendicular lines** are lines whose slopes are negative reciprocals of each other.

$$m_2 = -\frac{1}{m_1}$$

3.4 1. **Graphing linear inequalities**
   a. Sketch the line itself
      If the inequality includes an equal sign ($\leq$ or $\geq$), use a solid line for the graph of the line.
      If the inequality is an inequality only ($<$ or $>$), use a dashed line for the graph of the line.
   b. Use a test point to determine which part of the graph to shade
   c. Shade the region that makes the inequality true

2. **Graphing absolute value equations**
   Build a table of values and connect the points.

3. **Graphing absolute value inequalities**
   a. Build a table of values
   b. Determine if lines are solid or dashed
   c. Connect the points
   d. Determine the region to be shaded

Make a table of values and sketch the following graphs. **(3.1)**

1. $y = 2x + 1$

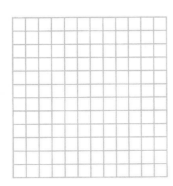

2. $3x - 4y = -8$

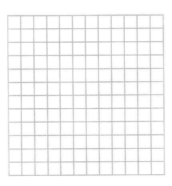

3. Find the slope of the line through the two points $(8, -9)$ and $(-4, -5)$. **(3.1)**

4. Find the equation of the line with slope -6 and y-intercept 3. **(3.2)**

5. Find the equation of the line with slope $\dfrac{2}{5}$ passing through $(0, -4)$. **(3.2)**

6. Use the point-slope method to find the equation of the line with slope 4 passing through $(-6, 7)$. **(3.2)**

7. Use the point-slope method to find the equation of the line with slope $-\dfrac{4}{3}$, passing through $(4, -2)$. **(3.2)**

8. Find the equation of the line through the points $(-1, -6)$ and $(-5, 3)$. **(3.2)**

9. Find the equation of the line through the points $(0, -3)$ and $(-4, 6)$. **(3.2)**

10. Find the value of b in the slope and y-intercept form of the equation of the line, then write the equation of the line with slope $\dfrac{2}{5}$, passing through the point $(3, -2)$. **(3.2)**

11. Write the equation of the line with slope $-\dfrac{1}{2}$, passing through the point $(-7, 3)$. **(3.2)**

12. Find the equation of the line parallel to the line $2x - 5y = 6$, passing through the point $(-1, 2)$. **(3.3)**

13. Find the equation of the line perpendicular to the line $-5x + 3y = 6$, passing through the point $(-3, -4)$. **(3.3)**

Sketch the following graphs using the slope and *y*-intercept method. **(3.2)**

14. $y = -x + 2$

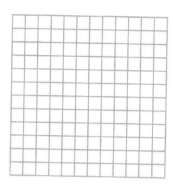

15. $3x - 5y = 15$

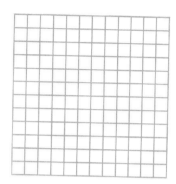

Sketch the following graphs using the intercept method. **(3.3)**

16. $3x + 4y = 12$

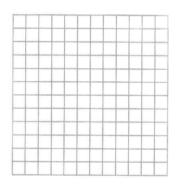

17. $2x - y = 6$

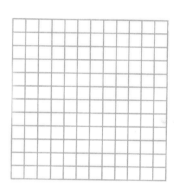

18. Sketch a line parallel to the given line with the specified *y*-intercept. **(3.3)**

$$y = -\frac{2}{7}x + 3 \qquad \text{\textit{y}-intercept } -2$$

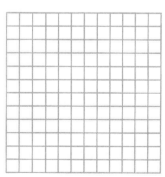

19. Sketch a line perpendicular to the given line with the specified y-intercept. **(3.3)**

$$y = \frac{3}{4}x + 2 \qquad \text{passing through } (0, -3)$$

Graph the following linear equations or inequalities. **(3.4)**

20. $y \geq -5x + 4$

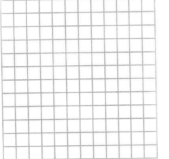

21. $x < -3$

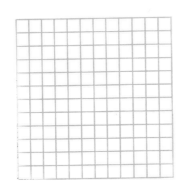

22. $y = |x + 2| - 3$

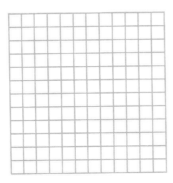

23. $y = |3x - 5| + 2$

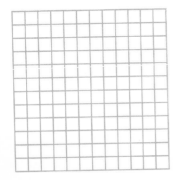

24. $y \leq |x - 3| - 4$

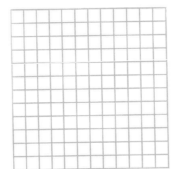

25. $y > |2x + 5| + 2$

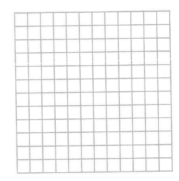

Chapter 3 ■ Linear Equations, Graphs, and Inequalities

4 System of Equations

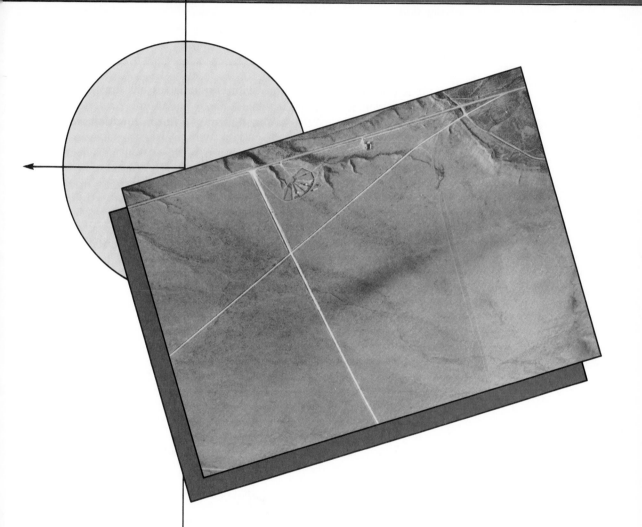

Contents

This chapter will show how to find the point of intersection, if it exists, for any two straight lines. This point of intersection is referred to as **the solution of a system of equations.**

The ability to solve systems of equations allows us to use more than one variable in application problems. The last two sections of this chapter will show you how two or more variables can be used to extend your problem-solving skills.

■ 4.1   Solving Systems of Equations

Standard Form of a Linear Equation

The graphs of first-degree equations with two variables are straight lines; therefore, first-degree equations are called **linear equations.**

$Ax + By = C$ is called the **standard form of a linear equation,** where A, B, and C are constants.

That means that once A, B, and C are given, we have a specific equation of a line.

When we work with two or more equations at once, we refer to them as a **system of equations.**

Example 1 □ Consider the system

$$4x + y = 1$$

$$x - 2y = 7$$

Plot both equations on the same graph:

■ Figure 4.1

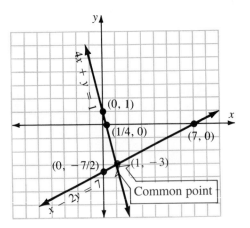

In Fig. 4-1 notice the point where the two lines cross. Point $(1, -3)$ is on both lines. This point satisfies either equation. The point that is common to both lines is called the **simultaneous solution** to the system of equations. Point $(1, -3)$ is the simultaneous solution in Example 1.

Simultaneous Solution

Test the solution by substituting $(1, -3)$ in both equations.

$$4x + y = 1 \qquad\qquad x - 2y = 7$$
$$4(1) + (-3) \overset{?}{=} 1 \qquad 1 - 2(-3) \overset{?}{=} 7$$
$$4 - 3 \overset{?}{=} 1 \qquad\qquad 1 + 6 \overset{?}{=} 7$$
$$1 = 1 \qquad\qquad\qquad 7 = 7$$

$(1, -3)$ checks in both equations.

□

Graphic methods of solving equations work, but they are time-consuming. Also, it is hard to get precise answers with graphic methods. As you might expect, mathematicians have easier ways of finding the simultaneous solution to two equations. One method rests on the **addition property of equality.**

Consider the following pair of equations.

Example 2 □ Find the solution to the system of equations

■ Figure 4.2

(1) $2x - y = 4$

(2) $4x + y = 2$

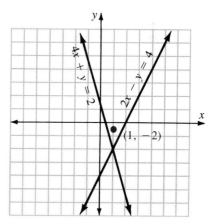

Graphing tells us that $(1, -2)$ is the simultaneous solution.

Now let's see how to get this result without graphing.

To solve a system of equations with two variables, we need a technique to eliminate one variable from one equation of the system.

The addition property of equality says that you can add the same thing to both sides of an equation. We could use this property to add 2 to both sides of the first equation.

$$2x - y = 4$$
$$+ \qquad\qquad 2 = 2$$
$$\overline{2x - y + 2 = 4 + 2}$$

This result is true but not very useful

However, if you take the second equation of the system quite literally, it says

$$4x + y = 2 \qquad \text{which says that } 4x + y \text{ is the same as 2.}$$

If we add 2 to equation 1 in the form of the second equation, we get a very useful result.

This is the same as adding 2 to both sides of equation 1	$2x - y = 4$	First equation of the system
	$+\ 4x + y = 2$	Second equation of the system
	$6x \quad\ = 6$	This result is useful because it has only one variable
	$x = 1$	

Now substitute $x = 1$ into either equation to find the value for y. Using the second equation

$$4x + y = 2$$
$$4(1) + y = 2$$
$$4 + y = 2$$
$$y = -2$$

Again, we find that $(1, -2)$ is the simultaneous solution to both equations. □

Example 3 □ Find the simultaneous solution to the system

$$y + 4x = 8$$
$$5x - y = 1$$

First rewrite the system so the x's and y's are lined up in columns.

$$4x + y = 8$$
$$\underline{5x - y = 1}$$
$$9x \quad\ = 9 \qquad\qquad \text{Adding both equations}$$
$$x = \underline{\qquad} \qquad\qquad \text{Solving for } x$$

Next substitute $x = 1$ in either equation to find y.

$$4x + y = 8$$
$$4(\underline{\qquad}) + y = 8$$
$$4 + y = 8$$
$$y = 4$$

Check the solution $(1, 4)$ in both equations.

$$4x + y = 8 \qquad\qquad\qquad 5x - y = 1$$
$$4(1) + (4) \stackrel{?}{=} 8 \qquad 5(\underline{\qquad}) - (\underline{\qquad}) \stackrel{?}{=} 1$$
$$4 + 4 \stackrel{?}{=} 8 \qquad\qquad\qquad 5 - 4 \stackrel{?}{=} 1$$
$$8 = 8 \qquad\qquad\qquad\qquad 1 = 1$$

$(1, 4)$ satisfies both equations. □

In both of the examples on the previous page it was possible to eliminate one variable by simply adding the two equations. Sometimes we have to manipulate one or both equations before we add.

Example 4 □ Solve the system

$$4x + y = 5$$
$$3x - 4y = -1$$

We cannot solve the system by simply adding the two equations. However, if we multiply both sides of the first equation by 4, the y terms will be eliminated when we add because $4y + (-4y) = 0$.

$$4x + y = 5 \quad \boxed{\text{Multiply by 4}} \Rightarrow 16x + 4y = 20$$

$$3x - 4y = -1 \quad \boxed{\text{No change}} \Rightarrow \underline{3x - 4y = -1}$$

$$\boxed{\text{Add}} \Rightarrow 19x \qquad = 19$$
$$x = 1$$

Substitute $x = 1$ into one of the original equations.

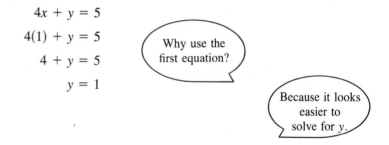

$$4x + y = 5$$
$$4(1) + y = 5$$
$$4 + y = 5$$
$$y = 1$$

Why use the first equation?

Because it looks easier to solve for y.

1, 1

(___ , ___) is the simultaneous solution.

Check the solution in both original equations.

$$4x + y = 5 \qquad\qquad 3x - 4y = -1$$
$$4(1) + (1) \stackrel{?}{=} 5 \qquad\qquad 3(1) - 4(1) \stackrel{?}{=} -1$$
$$4 + 1 \stackrel{?}{=} 5 \qquad\qquad 3 - 4 \stackrel{?}{=} -1$$
$$5 = 5 \qquad\qquad -1 = -1$$

$(1, 1)$ satisfies both equations. □

Example 5 □ Solve the system

$$3x + 4y = 20$$
$$2x - 5y = -2$$

In this system, multiplying only one equation by an integer will not eliminate a variable. However, if we multiply the first equation by 2 and the second equation by -3, the x terms will be eliminated when we add because $6x + (-6x) = 0$.

$-6x + 15y = 6$

$3x + 4y = 20$ ⟶ Multiply by 2 ⟶ $6x + \quad 8y = 40$

$2x - 5y = -2$ ⟶ Multiply by -3 ⟶ $\underline{\quad} x + \underline{\quad} y = \underline{\quad}$

Add ⟶ $23y = 46$

$y = 2$

Substituting $y = 2$ into one of the original equations,

$$3x + 4y = 20$$
$$3x + 4(2) = 20$$
$$3x + 8 = 20$$
$$3x = 12$$
$$x = 4$$

$(4, 2)$

The solution is (____ , ____).

Check the solution by substituting in both original equations.

$3x + 4y = 20$ $\qquad$ $2x - 5y = -2$

$3(4) + 4(2) \stackrel{?}{=} 20$ $\qquad$ $2(4) - 5(2) \stackrel{?}{=} -2$

$12 + 8 \stackrel{?}{=} 20$ $\qquad$ $8 - 10 \stackrel{?}{=} -2$

$20 = 20$ $\qquad$ $-2 = -2$

$(4, 2)$ checks. □

Dependent Equations

Sometimes the addition method yields puzzling results.

Example 6 □ Solve the system

$$3x + 4y = 8$$
$$-6x - 8y = -16$$

We can eliminate the x terms if we multiply the first equation by

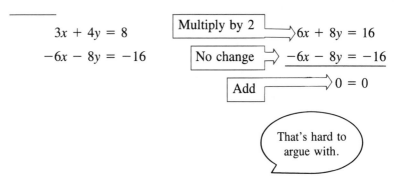

$$3x + 4y = 8$$
$$-6x - 8y = -16$$

Multiply by 2 $\longrightarrow$ $6x + 8y = 16$

No change $\Longrightarrow$ $-6x - 8y = -16$

Add $\longrightarrow$ $0 = 0$

That's hard to argue with.

The reason for this result is that the second original equation is just the first equation multiplied by -2.

First Equation　　　　　　　　　　　　　　**Second Equation**

$$3x + 4y = 8$$　　Multiply by -2 $\Longrightarrow$ $-6x - 8y = -16y$ 　□

Dependent Equations

We really don't have two independent equations. We have two equivalent equations. This case is called **dependent equations.**

■ Figure 4.3

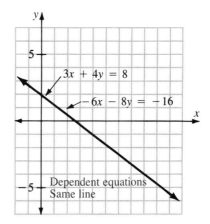

In English,
independent
means
separate.

It's the same in math. Independent equations represent separate lines. Independent lines may or may not intersect.

Inconsistent Equations

Here is another case you should be aware of.

Example 7 □ Solve the system

$$x - 2y = 2$$
$$2x - 4y = 8$$

We can eliminate the x terms if we multiply the first equation by

−2 _____

$$x - 2y = 2 \quad \boxed{\text{Multiply by } -2} \Rightarrow -2x + 4y = -4$$

$$2x - 4y = 8 \quad \boxed{\text{No change}} \Rightarrow \underline{\quad 2x - 4y = 8 \quad}$$

That's obviously false.

0

$$\boxed{\text{Add}} \Rightarrow \underline{\quad\quad} = 4$$

Inconsistent Equations

A graph of both equations shows the problem. Since the two lines are parallel, they do not intersect. There is no common solution. Equations whose graphs are parallel lines are called **inconsistent.** □

■ Figure 4.4

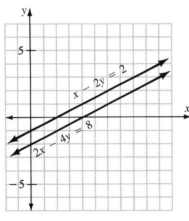

If inconsistent equations are parallel, what are consistent equations?

Two equations whose graphs intersect.

Why do inconsistent equations yield false results?

Read below.

▶ **Pointer for Better Understanding**

Recall from the definition of a variable that the letter must stand for the same value each time it is used in an equation.
Now consider two intersecting lines.

$$x + y = 5$$
$$x - y = 3$$

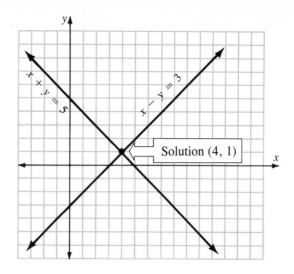

There is only one point on the graph where x in the first equation represents the same number as x in the second equation. It is the point where the lines intersect, $(4, 1)$.

We really have two distinct equations.

$$x_1 + y_1 = 5$$

and

$$x_2 - y_2 = 3$$

If we add these equations we get

$$x_1 + x_2 + y_1 - y_2 = 8$$

which leads nowhere. However, if there is a point (x, y) where the two lines intersect

then $x_1 = x$ and $x_2 = x$ ⎫   at the point of
and $y_1 = y$ and $y_2 = y$ ⎭   intersection only

If we substitute x for x_1 and x_2 and also substitute y for y_1 and y_2,

$$x_1 + x_2 + y_1 - y_2 = 8$$

becomes $$x + x + y - y = 8$$

$$2x = 8$$

$$x = 4$$

which is very useful. We can now solve either equation for the y-coordinate of the point of intersection. BUT if there is no point of intersection, our entire process is based on a false assumption. Hence the contradictory result when you try to add two equations which represent parallel lines.

Here are two dependent equations.

$$3x_1 + 4y_1 = 8$$
$$-6x_2 - 8y_2 = -16$$

These are two representations of the same line

For every point on the line above, $x_1 = x_2$ and $y_1 = y_2$. Therefore, using the addition method of solving equations yields the tautology $0 = 0$. That's because any simultaneous solution that satisfies the first equation also satisfies the second.

> A tautology is a statement that is always true.

Here is a summary of the cases that can occur with a system of two equations.

Typical System of Equations	$2x + 3y = -13$ $4x - y = -5$	$x - 2y = 3$ $x - 2y = 5$	$x - 2y = 3$ $2x - 4y = 6$
Algebraic Result	$x = -2$ $y = -3$	$0 = -2$	$0 = 0$
Typical Graph			
Descriptive Name	Intersecting lines Consistent equations	Parallel lines Inconsistent equations	Same line Dependent equations
Number of Solutions	One	None	Since both equations are actually the same line, any pair that satisfies one equation satisfies the other.

It is possible to find a meaningful solution only for the case with intersecting lines.

Use the addition method to find the simultaneous solution, if one exists, for the following equations.

1. $x + y = 8$
$x - y = 6$

2. $2x + y = 2$
$4x - y = -8$

3. $3x - y = 6$
$-6x + 2y = -12$

4. $4x - 3y = 10$
$-9x + 15y = -6$

5. $3y + 2x = 1$
$3x + 2y = -6$

6. $3x - 2y = 4$
$-6x + 4y = -8$

7. $3x = 8 - 2y$
$4x + 3y = 13$

8. $2x + 5y = 11$
$8y = 16 - 3x$

9. $4y + 3x = 7$
$x = 2y + 9$

10. $3x = 5y + 10$
$15y = 9x - 20$

11. $5x + 2y = 3$
$3y = 2x + 14$

12. $-4y = 3x + 8$
$5x = 8y - 17$

13. $x + y = 0$
$6x + 4y = -1$

14. $y = -3x - 2$
$6x = -3y + 4$

15. $6y = 2x - 3$
$3x = -10y + 14$

16. $3x - 7y = 3$
$x + 2y = -8$

17. $3x - 2y = 7$
$3x + 10y = 19$

18. $7y = 9x - 3$
$2y = -x - 8$

19. $4x - 9y = -1$
$6x + 6y = 5$

20. $3x = 8y + 6$
$9x + 4y = 4$

21. $3y = -5x - 6$
$7x = 2y + 35$

22. $4x + 9y = -9$
$3x = 6y - 11$

23. $3x = 8y + 6$
$12y = -5x - 9$

24. $10x = 9y - 5$
$5x + 6y = 1$

25. $\dfrac{1}{3}x - \dfrac{1}{4}y = -2$

$2x + \dfrac{1}{2}y = -4$

26. $\dfrac{3}{4}x - \dfrac{2}{3}y = 10$

$\dfrac{1}{2}x + \dfrac{1}{3}y = 2$

27. $\dfrac{1}{5}x + \dfrac{2}{3}y = 3$

$\dfrac{4}{5}x - \dfrac{1}{3}y = 2$

28. $0.2x - 0.1y = 0.9$
$2.2x + 0.5y = 1.9$

29. $1.3x + 0.5y = -0.3$
$-0.9x + 0.4y = 1.7$

30. $1.4x + 0.9y = 1.3$
$1.0x - 1.2y = -0.3$

Solving Systems of Equations by Substitution

It is possible to solve any system of equations by addition, but for some systems the substitution method is more efficient.

Example 8 □ Solve the system

$$y = 2x - 4$$

$$x + 3y = 16$$

In the system above, the first equation neatly specifies the value of y in terms of x. If we substitute this value of y in the second equation, we will have an equation in one variable that we can solve for x.

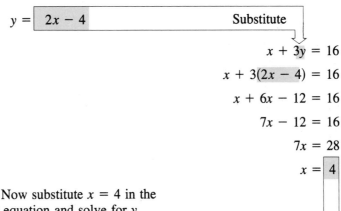

First Equation

$$y = \boxed{2x - 4}$$

Second Equation

Substitute

$$x + 3y = 16$$

$$x + 3(2x - 4) = 16$$

$$x + 6x - 12 = 16$$

$$7x - 12 = 16$$

$$7x = 28$$

$$x = \boxed{4}$$

Now substitute $x = 4$ in the first equation and solve for y.

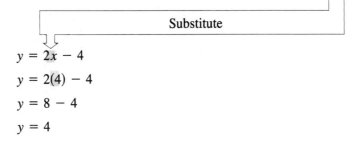

Substitute

$$y = 2x - 4$$

$$y = 2(4) - 4$$

$$y = 8 - 4$$

$$y = 4$$

The simultaneous solution is $(4, 4)$.

$$y = 2x - 4 \qquad\qquad x + 3y = 16$$

$$(4) \overset{?}{=} 2(4) - 4 \qquad\qquad (4) + 3(4) \overset{?}{=} 16$$

$$4 \overset{?}{=} 8 - 4 \qquad\qquad 4 + 12 \overset{?}{=} 16$$

$$4 = 4 \qquad\qquad\qquad 16 = 16 \qquad\qquad □$$

Example 9 □ Solve the system

$$2x + 3y = -4$$
$$x - 5y = -2$$

Look at the second equation. The coefficient of the x term is 1; therefore, it's easy to solve this equation for x in terms of y. Then substitute this solution for x into the first equation and solve for y.

First Equation   **Second Equation**

$$x - 5y = -2$$
$$x = \boxed{5y - 2}$$

Substitute

$$2x + 3y = -4$$

$5y - 2$

$$2(\quad) + 3y = -4$$
$$10y - 4 + 3y = -4$$
$$13y - 4 = -4$$
$$13y = 0$$
$$y = \boxed{0} \quad \text{Substitute}$$

$$x = 5y - 2$$
$$x = 5(0) - 2$$
$$x = -2$$

The solution is $(-2, 0)$. Check the solution in the original equations.

$$2x + 3y = -4 \qquad\qquad x - 5y = -2$$
$$2(-2) + 3(0) \stackrel{?}{=} -4 \qquad (-2) - 5(0) \stackrel{?}{=} -2$$
$$-4 + 0 \stackrel{?}{=} -4 \qquad\qquad -2 - 0 \stackrel{?}{=} -2$$
$$-4 = -4 \qquad\qquad\qquad -2 = -2$$

It checks.   □

The substitution method generally works best when the coefficient of one of the variables is 1 in either equation.

Use the substitution method to find the simultaneous solution, if one exists, for the following equations.

1. $2x + y = 1$
   $y = 3 - x$

2. $2x + 3y = -8$
   $y = 3x + 12$

3. $2x - 3y = 1$
   $x - 3y = 5$

4. $2x = 2 - 5y$
   $2y = 2 - x$

5. $2y = 6 - x$
   $x + 2y = 6$

6. $2y = 1 - 3x$
   $-2x = y + 1$

7. $3y - x = 7$
   $-5y = 3x + 7$

8. $8y = -x - 1$
   $2x = 1 - 10y$

9. $3x = y + 6$
   $2y = 6x - 10$

10. $2x = y - 8$
   $3x = 4y - 12$

11. $2x + y = \dfrac{1}{2}$
   $8x = y + 7$

12. $3y = x - 3$
   $2y + 3x = -13$

13. $x + 3y = -3$
   $2x - 3y = 21$

14. $4x = -y + 7$
   $6x - 4y = -17$

15. $3x - 5y = 2$
   $9x = 15y + 5$

16. $x + 5y = 2$
   $3x = 10y + 11$

17. $2x + 3y = -18$
   $x - 5y = 4$

18. $3x + y = 5$
   $2y = -6x + 10$

19. $3x + y = 2$
   $9x + 2y = 2$

20. $6x = -y - 3$
   $5y = 8x + 4$

21. $x - 3y = 5$
   $6y = 2x - 10$

22. $9x + 5 = -4y$
   $3x - y = 3$

23. $x - 8 = 5y$
   $3x + 2y = -10$

24. $5x - y = -3$
   $2y = 10x + 5$

25. $x + 2y = 5$
   $3y + 2x = 4$

26. $5x = 8 - y$
   $y = 9 - 5x$

27. $2x + y = 7$
   $3x = 4y + 5$

28. $2x + 5y = 5$
   $4y = -3x - 3$

29. $x - 5y = 4$
   $15y = 3x + 12$

30. $2x = 3y - 7$
   $6x + 4y = 5$

31. $2x + y = 5$
$x = 2y + 20$

32. $5x + 2y = -3$
$10x = 3y + 8$

33. $3x = 4y - 2$
$2x - y = -3$

34. $5x - 2y = 8$
$4y = 10x + 16$

35. $4x + 2y = -6$
$8x + 3y = -15$

36. $2y = 4x - 8$
$2x = y + 4$

37. $3x + 9y = 15$
$3y = 6x + 12$

38. $4x + 8y = -20$
$3y = 8x - 17$

39. $3x + 2y = 2$
$-6x + y = -14$

40. $-x + 3y = 8$
$6y = 2x + 7$

41. $\frac{1}{3}x + \frac{3}{4}y = -1$
$\frac{5}{6}x + 2y = -3$

42. $x - \frac{3}{4}y = 5$
$2y = -\frac{1}{2}x - 7$

43. $x - 2y = 11$
$7x + 3y = 9$

44. $3x + 4y = 8$
$-5x + 3y = 6$

45. $x - 2y = 7$
$4y = 2x - 14$

46. $2x + 3y = 3$
$8x = 9y - 2$

47. $7y = 11 - x$
$3x - 7 = 5y$

48. $2x - 3y = 7$
$4x = 6y + 11$

49. $5x = y - 6$
$10x - 2y = 9$

50. $-2x + y = -6$
$\frac{1}{2}x + 3y = 8$

51. $\frac{1}{3}x = \frac{1}{5}y + 1$
$\frac{5}{6}x + y = 10$

52. $\frac{1}{3}x + \frac{1}{2}y = -4$
$\frac{1}{5}x - \frac{1}{4}y = 2$

53. $\frac{1}{3}x + y = 2$
$\frac{1}{6}x = \frac{1}{4}y - 2$

54. $\frac{1}{3}x + \frac{1}{4}y = \frac{1}{12}$
$\frac{2}{3}x = -\frac{1}{2}y + \frac{1}{6}$

55. $\frac{1}{3}x + 3y = -2$
$4y = \frac{1}{2}x + 3$

56. $4x + 3y = -1$
$9y = 8x + 12$

57. $0.3x = 1.0y - 1.2$
$1.5x - 5y = -6$

58. $3x + y = -3$
$-5y = 6x + 3$

59. $3x + y = -6$
$\frac{1}{2}x = \frac{1}{3}y - 4$

60. $0.2y = 1.3x + 1.4$
$6.5x - y = -7$

The ability to solve equations in two variables makes it easier to solve many applications problems.

Example 10 □ The difference of two numbers is 6. The sum of the numbers is 20. Find the numbers.

Let x = larger number

     y = smaller number

$x - y = 6$ The difference of the numbers is 6.

$\underline{x + y = 20}$ The sum of the numbers is 20.

$2x = 26$ Add the two equations to eliminate y.

13          $x =$ _____

To find the other number, substitute in either equation.

$$x - y = 6$$
$$(13) - y = 6$$
$$-y = -7$$

7          $y =$ _____

The numbers are 13 and 7.
Check.

$$
\begin{array}{cc}
x - y = 6 & x + y = 20 \\
13 - 7 \stackrel{?}{=} 6 & 13 + 7 \stackrel{?}{=} 20 \\
6 = 6 & 20 = 20
\end{array}
$$

         □

▶ **Pointers for Problem Solving**

Be specific when you choose a variable.

     This time we are going to stumble into a common error, and then we will tell you how to avoid it.

Example 11 □ The difference in two people's ages is 15 years. If four times the younger person's age equals the older person's age, what are the ages?

Let x = first person

     y = second person.

$x - y = 15$ The difference in ages is 15.

$4x = y$ Four times the younger is the older.

Substituting the second equation into the first:

$$x - y = 15$$

becomes

$$x - 4x = 15$$

$$-3x = 15$$

$$x = -5$$

Wait a minute, you can't have a negative age!

This problem happened because we were sloppy about the meaning of x.

When we said $x - y = 15$, we implied that x represented the older's age. When we said $4x = y$, we implied that x represented the younger's age. X can't be both the younger's and the older's age. If we had specified,

$$x = \text{age of the younger person}$$

$$y = \text{age of the older person}$$

this problem would have worked out correctly.

Sometimes you can be quite confident just by using a relationship given in the problem. In this case we subtract the younger's age from the older's age.

How do I know if it's $x - y = 15$, or $y - x = 15$?

Another almost certain way to get the relationship is to pick any two numbers that have the correct relationship. Establish the relationship; then substitute the variables in the simple relationship you have established.

Relationship: The difference of two ages is 15 years.

Numbers that could satisfy the relationship: 25 and 10.

But these aren't the answers!

At this point it doesn't matter. All we need are two numbers with a difference of 15.

Establish the relationship $\qquad$ $25 - 10 = 15$

Substitute the variables $\qquad$ $25 - 10 = 15$

Older age	Younger. age

$$y \ - \ x = 15$$

Let's apply the same technique to the second equation from the problem.
Relationship: Four times the younger's age equals the older's age.
Numbers that could satisfy the relationship: $4 \cdot 10 = 40$.
   Substitute the variables. Since 10 is obviously younger than 40, substitute x for 10 and y for 40.

$$4 \cdot 10 = 40 \qquad \text{becomes}$$

$$4 \cdot x = y$$

Now we can solve the two equations.

$$y - x = 15$$

$$4x = y$$

Substituting $\qquad$ $4x - x = 15$

$$3x = 15$$

$$x = 5 \qquad \text{Age of the younger person}$$

Using this information in the second equation

$$4x = y$$

$$4(5) = y$$

$$20 = y \qquad \text{Age of the older person} \qquad \square$$

Problem Set 4.2A

Solve the following problems by using simultaneous solutions. State in words exactly what your variables represent. Set up two equations and solve them.

1. The sum of two numbers is 20. If one number is four more than the other, find the numbers.

2. The difference of two numbers is 36. If one number is twice the other, find the numbers.

3. One number is 15 less than another. If the larger is six times the smaller, find the numbers.

4. The difference of two numbers is 4. If twice the smaller plus three times the larger is 52, find the numbers.

5. The sum of two integers is -4. If two times the second minus three times the first is 42, find the integers.

Coin Problems

Coin problems are typical of a whole class of problems. Mixture problems in chemistry, atomic particle problems in physics, and counting problems in probability are essentially coin problems. Most of us are more familiar with coins than chemistry, physics and probability, so we'll use coins to learn the principles.

4.2A   First Principle

The value of any number of identical coins is equal to the value of one coin times the number of coins.

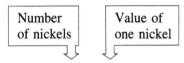

Two nickels are worth 2 ·   5 cents.
12 nickels are worth 12 ·   5 cents.
When we work with variables, it's common to write the product with the value of one coin written first.

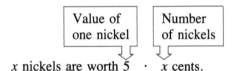

x nickels are worth 5 · x cents.

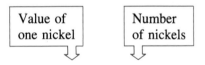

5, y y nickels are worth _____ _____ cents.

$5(x + 3)$ $x + 3$ nickels are worth _____ cents.

$10n$ n dimes are worth _____ cents.

$10 \cdot 2n$ $2n$ dimes are worth _____ cents.

4.2B   Second Principle

The total value of any mixture is equal to the sum of the values of the parts.

Example 12 □ A bank contains three times as many nickels as dimes. The total value of the nickels and dimes in the bank is $1.25. Find the number of each kind of coin in the bank.

| Value of nickels | + | Value of dimes | = | Total value of nickels and dimes in the bank |

The problem tells us there are three times as many nickels as dimes in the bank. If we let x = number of nickels

y = number of dimes

The value of the nickels is _____.

5x

10y

The value of the dimes is _____.

Now, we can write an equation from the diagram.

| Value of all the nickels | + | Value of all the dimes | = | Total value |

$$5x \quad + \quad 10y \quad = \quad 125$$

To avoid confusion you can express the value of all coins in pennies rather than in dollars.

We have one equation but we have two unknowns! To solve a system of equations, we need as many independent equations as we have unknowns. We get a separate equation from the sentence in the problem which says "A bank contains three times as many nickels as dimes." That equation is

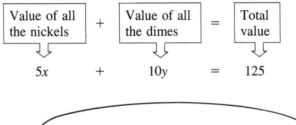

| Number of nickels | = | 3 | · | Number of dimes |

$$x \quad = \quad 3y$$

Substituting in the first equation

$$5x + 10y = 125$$

$$5\,(3y) + 10y = 125$$

$$15y + 10y = 125$$

$$25y = 125$$

$$y = 5$$

There are 5 dimes in the bank.
Our second equation says

$$x = 3y$$

$$x = 3(5) \qquad \text{Substituting the value of } y$$

$$x = 15$$

There are 15 nickels and 5 dimes in the bank.
It's always a good idea to test your solution against the original problem.

Value of 15 nickels	$= 5 \cdot 15 =$	75 cents	
+ Value of 5 dimes	$= 10 \cdot 5 =$	50 cents	
Total value of coins		125 cents	

This matches the statement in the original problem that the total money in the bank is \$1.25. □

▶ Pointers for Problem Solving

Be clear about what the variable represents.

You can usually get a five-year-old child to trade you two dimes for five pennies. When we choose variables for word problems we want to be careful to avoid the child's mistake which is confusing number and value.

In the following set of problems, be very careful to distinguish between the number of coins and the value of the coins.

Is it all right to say x = nickels?

You should specify x = number of nickels or x = value of all the nickels.

Assuming that you had 35 cents in nickels, you should be clear whether

$$x = 5 \qquad \text{the value of one nickel,}$$

$$x = 7 \qquad \text{the number of nickels, or}$$

$$x = 35 \qquad \text{the value of all the nickels.}$$

It's usually best to let x represent the number of coins.

Then $\qquad x$ = number of nickels

$5x$ = value of nickels

Why not let x = 5 cents?

5 cents is a constant. There is no point in assigning it to an unknown x.

Solve the following problems by using simultaneous solutions. State in words exactly what your variables represent. Set up two equations and solve them.

1. The parents of the players in a little league baseball team sold candy, ice cream, and soft drinks. At the end of a particular game, the coin box contained twice as many dimes as quarters. If the value of these coins was $8.10, how many of each were there?

2. Teryle's bank contains 60 coins, all nickels and dimes. If the total amount of money in the bank is $5.40, how many of each coin are in the bank?

3. At a football game student tickets were $1.50 each and adult tickets were $4.00 each. If 1800 people were in attendance and $4050 was collected, how many students and how many adults attended the game?

4. 3000 tickets were sold for a concert. General admission tickets sold for $6 each and reserved seats sold for $11 each. If there were 600 more general admission tickets sold than reserved, how many of each were sold?

5. A parking meter brought in $7.70. The collection of coins consisted of nickels and dimes only. If there were one third as many nickels as dimes, how many of each type were in the meter?

6. Toan made a collection at a pay phone. The phone contained $25.30 made up of dimes and quarters. If there were 154 coins in the phone, how many of each type were in the phone?

7. One of Gary's kiddy rides contained dimes and quarters worth $55.10. If there were 12 fewer quarters than dimes, how many of each type of coin were in the kiddy ride.

8. Sharon gave her son Matthew $14.00 to buy some 22 cents stamps and 25 cent stamps at the post office. If Matthew picked up 17 more 25 cent stamps than 22 cent stamps and got 35 cents change, how many of each did he purchase?

Business Applications

Many students study algebra because it is required as part of a business major. One reason it is required is that the methods of coin problems are also useful in dealing with business problems. Investments and product pricing are illustrations of how coin problems can be applied in business.

First let's look at how interest is paid on an investment. A 9% rate of return on an investment means that each year 9% interest is paid for the use of the investor's money. Therefore, a person who invests $3000 for one year at 9% could expect to collect.

$$\$3000 \cdot 9\% =$$

$$\overset{30}{\cancel{3000}} \cdot \frac{9}{\cancel{100}} = \$270$$

Recall
$$9\% = \frac{9}{100}$$

That is, each year the investor would collect an income of $270 as payment for the use of the money.

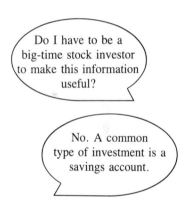

Do I have to be a big-time stock investor to make this information useful?

No. A common type of investment is a savings account.

Example 13 □ Find the income from an investment of $4000 at 8% interest for one year.

$$\$4000 \cdot 8\% = \text{income for one year}$$

$$\overset{40}{\cancel{4000}} \cdot \frac{8}{\cancel{100}} = \$320$$

The investor will receive $320 per year return on the investment of $4000.   □

Example 14 □ Give the yearly income from the following investments.

180

$2000 at 9% returns _____

55

$500 at 11% returns _____

220

$2200 at 10% returns _____

$\dfrac{7}{100}x$

$x at 7% returns _____

$\dfrac{9}{100}y$

$y at 9% returns _____   □

► Pointers for Problem Solving

Model Making

A model is a representation of a relationship. No model can be exactly the object it models or it would be the real thing and not a model. However, all models have some relationships in common with what they model. Consider model airplanes. A small child can make a model of an airplane by nailing two boards together to represent a wing and a fuselage. The model may not fly or have any movable parts but it is a start toward the idea of an airplane. At least the idea of wing and fuselage are represented.

We can use boxes and words to make models of the relationships in word problems. The main relationship we usually model is a statement of equality or inequality. Consider the following problem:

Example 15 □ An apartment owner was charged $205 to have an apartment painted. When he asked the painter "Why so much?" he was told, "The paint alone cost $40 and my helper's hourly rate is half my hourly rate." With this information can the apartment owner figure out how much the painter made?

We can use boxes to model the relationships.

First consider where the $205 went:

Model I

| $205 paid by owner | = | $40 for paint | + | Money paid painter | + | Money paid helper |

Next consider the relationship between the pay of the painter and his helper.

Model II

| Painter's pay | = 2 · | Helper's pay |

Now if we choose variables, we can translate the models into equations.
Let x = painter's pay
$\quad y$ = helper's pay
Writing the first model as an equation:

Model I

| $205 paid by owner | = | $40 for paint | + | Money paid to painter | + | Money paid to helper |

becomes:
Equation I: $205 = $40 + x + y

Writing the second model as an equation:

Model II

| Painter's pay | = 2 · | Helper's pay |

becomes
Equation II: $x = 2 \cdot y$

What's the difference between equations and models?

Equations can be thought of as a form of models. Equations are just a little more abstract.

One test of the value of a model is how well it can be used to yield additional information. By manipulating the model in the form of equations we can find the information we are seeking. How much did the painter make?
Equation I: $205 = 40 + x + y$
Equation II: $x = 2y$

$$205 = 40 + 2y + y \qquad \text{Substituting II into I}$$

$$165 = 3y$$

$$55 = y$$

The helper made $55. Therefore, the painter made:

$$y = 2y$$

$$= 2(55)$$

$$= \$110 \qquad \text{Painter's pay} \qquad \square$$

Example 16 □ Roberta Sanchez, a skillful marketing representative, received a bonus of $4000 at the end of the year for her outstanding sales record. She plans to invest part in a risky private bond paying 12% interest and the remainder in a safe savings account paying 6% interest. If she needs a total interest income of $450 a year to pay her house insurance, how much should she invest at each rate to meet her income goal without risking all of her money?

First build a model for each major idea in the problem.

Model I

| Part invested at 12% | + | Part invested at 6% | = | Total investment |

Model II

| Income from 12% investment | + | Income from 6% investment | = | Total Income |

Select letters to make equations from the models.
Let x = part invested at 12%
y = part invested at 6%
Make an equation for each diagram.

Model I

| Part invested at 12% | + | Part invested at 6% | = | Total investment |

Equation I

$$x \quad + \quad y \quad = \quad 4000$$

Total investment was $4000

Model II

| Income from 12% investment | + | Income from 6% investment | = | Total Income |

Equation II

$$\frac{12}{100} \cdot x \quad + \quad \frac{6}{100} \cdot y \quad = \quad 450$$

Income from 12% investment plus income from 6% investment totals 450 dollars.

Multiply both sides of the equation by 100 to eliminate the 100 in the denominator.

Multiply the first equation by -6 to eliminate the y terms.

$$x + y = 4000 \qquad \boxed{\text{Multiply by } -6} \Rightarrow -6x - 6y = -24000$$

$$\frac{12}{100}x + \frac{6}{100}y = 450 \qquad \boxed{\text{Multiply by } 100} \Rightarrow \underline{12x + 6y = 45000}$$

$$\boxed{\text{Add}} \Rightarrow 6x = 21000$$

$$x = 3500$$

Substitute $x = 3500$ in the first equation.

$$x + y = 4000$$
$$(3500) + y = 4000$$
$$y = 500$$

$x = \$3500$ invested at 12%. $y = \$500$ invested at 6%. □

Mixture problems also use the same reasoning as coin problems.

Example 17 □ A gourmet coffee shop found that sales of their special blend of coffee at $7.95 a pound were poor. The manager decided that she needed a new blend that would sell for $5.95 a pound to increase sales. To allow for profit she wanted her cost per pound of the new blend to be $4.00 How many pounds of beans costing the shop $7.00 per pound should be mixed with beans costing $2.00 per pound in order to get 30 pounds of coffee costing the shop $4.00 per pound?

Let x = number of lbs of $7.00 per lb beans
y = number of 1bs of $2.00 per lb beans

How do I write the equations?

Look for things that are equal and make models.

One model is

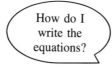

Total pounds of coffee in the mixture	=	Total weight of final mixture

Refining this model

| Lbs of 7.00/lb coffee | $+$ | Lbs of 2.00/lb coffee | $= 30$ lbs |

Rewriting the model as an equation

$$x + y = 30 \qquad \text{First Equation}$$

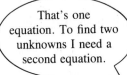

That's one equation. To find two unknowns I need a second equation.

You can get a second equation if you think about the money involved.

A model for the costs is

| Cost of all 7.00/lb coffee in the blend | $+$ | Cost of all 2.00/lb coffee in the blend | $=$ | Total cost of the resulting blend of coffee |

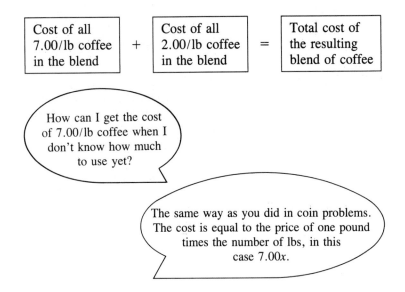

How can I get the cost of 7.00/lb coffee when I don't know how much to use yet?

The same way as you did in coin problems. The cost is equal to the price of one pound times the number of lbs, in this case 7.00x.

Writing the model as an equation

$$7.00x \qquad + \qquad 2.00y \qquad = \qquad 4.00\,(30) \qquad \text{Second Equation}$$

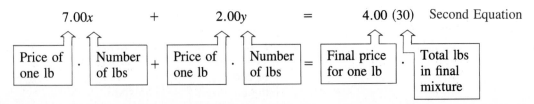

| Price of one lb | $\cdot$ | Number of lbs | $+$ | Price of one lb | $\cdot$ | Number of lbs | $=$ | Final price for one lb | $\cdot$ | Total lbs in final mixture |

Now we have a system of equations with two unknowns.

$$x + y = 30$$

First equation representing weight in = weight out

$$7x + 2y = 120$$

Second equation without the zeros in the prices, also
$$4 \cdot 30 = 120$$

$$-2x - 2y = -60 \qquad \text{First equation times } -2$$

$$\underline{7x + 2y = 120}$$

$$5x \quad\quad = 60 \qquad \text{Adding}$$

$$x = 12$$

Substituting the value for x in the first equation

$$x + y = 30$$

becomes

$$12 + y = 30$$

$$y = 18$$

There should be 12 lbs of 7.00/lb coffee and 18 lbs of 2.00/lb coffee.

We can insure that we have the right answer by testing our solution against the original statement of the problem.

12 lbs at $7.00/lb = $84.00

$$\underline{18 \text{ lbs at } \$2.00/\text{lb} = \$36.00}$$

30 total pounds $120 total dollars

$$\frac{120 \text{ total price}}{30 \text{ total lbs}} = \$4.00 \text{ cost per lb of mixture}$$

☐

Pointers for Problem Solving

Organization

Frequently a problem contains so much information that it's difficult to keep track of what is known and what is unknown. A chart can help.

Consider the previous example which we have repeated for the reader's convenience:

☐ A gourmet coffee shop found that sales of their special blend of coffee selling at $7.95 a pound were poor. The manager decided that she needed a new blend that would sell for $5.95 a pound to increase sales. To allow for profit she wanted her cost per pound of the new blend to be $4.00. How many pounds of beans costing the shop $7.00 per pound should be mixed with beans costing $2.00 per pound in order to get 30 pounds of coffee costing the shop $4.00 per pound?

There are three kinds of coffee involved here:

  a) coffee with a cost price of $2.00/lb
  b) coffee with a cost price of $7.00/lb
  c) mixture with a cost price of $4.00/lb

> What about the $7.95/lb and the $5.95/lb coffee?

> That's the sales price. The critical information here is the cost price.

To make an effective chart we must define at least one relationship in this case:

$$\boxed{\text{Price per pound}} \cdot \boxed{\text{Number of pounds}} = \boxed{\text{Cost of coffee}}$$

This relationship is the top line on the chart. The left side of the chart lists the different elements involved. Fill in as many blanks on the chart as possible.

	price/lb	number of lbs	= cost
2.00 coffee	2.00		
7.00 coffee	7.00		
final mixture	4.00	30	

> Now what? Most of the chart is blank.

> Pick variables to represent the unknowns.

We let x = number of lbs of 2.00/lb coffee
  y = number of lbs of 7.00/lb coffee
Using these variables we can fill in the second column of the chart.
  Using the relationship and the first two columns of the chart, we can fill in the third column of the chart.

	price/lb	number of lbs	= cost
2.00 coffee	2	x	$2x$
7.00 coffee	7	y	$7y$
final mixture	4	30	120

We can use the information on the chart to write equations from models

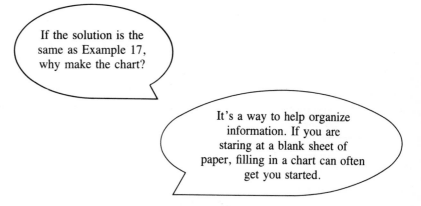

| Weight of final mixture | = | Weight of 2.00/lb coffee | + | Weight of 7.00/lb coffee |

30 = x + y

| Cost of final mixture | = | Cost of 2.00/lb coffee | + | Cost of 7.00/lb coffee |

$120 = $2x$ + $7y$

Now that the two equations are defined, the solution is the same as in Example 17.

If the solution is the same as Example 17, why make the chart?

It's a way to help organize information. If you are staring at a blank sheet of paper, filling in a chart can often get you started.

Charts won't solve every problem and certainly aren't needed every time. However, they are one more way to approach a problem. □

Solve the following problems by using simultaneous equations. State in words exactly what your variables represent. Set up the equations and solve them.

1. Yolanda spent $30.60 for walnuts and pecans. The walnuts cost $2.80 per pound and the pecans cost $3.20 per pound. If she has a total of 10 pounds of nuts, how many pounds of each does she have?

2. The Stay Slim Ice Cream Company wishes to blend some Irish Toffee with Brazilian Chocolate ice cream for a new flavor. If the Irish Toffee cones sell for 80 cents and the Brazilian Chocolate cones sell for 60 cents, how should the ice cream be blended to produce 900 cones costing 65 cents each?

3. The Kind to Animals Pet Shop has two grades of dog food. One grade sells for 72 cents a pound and the other sells for 52 cents a pound. How many pounds of each must be used to produce 600 pounds of mixture selling for 60 cents a pound?

4. The Thrifty Coffee Shop blends coffee worth $3.00 a pound with coffee worth $4.25 per pound for a blend worth $3.50. The weight of the $4.25 per pound coffee was $\frac{2}{3}$ of the weight of the $3.00 per pound coffee. How many pounds of each was used to make a mixture with a total value of $140?

5. Two stocks yield 8% annually and 11% annually. If an investor invests $25,000 and derives $2510 from the two investments, how much is invested at each rate?

6. A real estate firm invested $50,000 in second mortgages. Part of the money was invested at 12% and the remainder at 10%. If three times as much was invested at 12% as at 10%, how much was invested at each rate?

7. A total of $16,200 was invested in two accounts. One account paid 8% interest and the other paid 10% interest. How much was invested at each rate if the annual yield was the same for each investment?

8. An investor invested her savings in two enterprises. She invested $6000 more at 12% than at 9%. If her total annual income from both investments was $4500, how much was invested at each rate?

9. The Noble Oil Company bond pays 12% annual interest and the Uing Oil Company bond pays 10%. If you have $10,000 invested, how much is invested in each company to receive an income of $1,120 annually?

10. For a party the hostess purchased party mix nuts for $5.70 a pound and some cereal mix for $2.20 a pound. If she wanted 5 pounds of the mixture worth $3.60 a pound, how many pounds of each did she purchase?

11. A confectioner has $4.00 per pound candy and $6.00 per pound candy. If he has a total of 16 pounds of mixture worth $4.75 a pound, how many pounds of each were used to make the mixture?

Systems of Three Linear Equations

A system of two linear equations provides a powerful tool for solving problems. To solve a system of equations you need as many distinct equations as you have unknowns. As long as this requirement is met, it is possible to solve systems with any number of unknowns. We will start with systems of three equations in three unknowns.

Example 18 □ Solve the system.

(1) $\quad x + 2y + z = 2$

(2) $\quad x + y - z = 6$

(3) $\quad x - y + 2z = -7$

> Notice we have labeled the equations (1), (2), (3), so it's easier to tell which equations we're talking about.

The first step is to eliminate the same variable from each of two pairs of equations. We usually use addition to do this. In the following example we will eliminate the variable x.

$$
\begin{array}{rl}
(1) & x + 2y + z = 2 \\
-1 \cdot (2) \quad | & -x - y + z = -6 \\
\hline
(4) & y + 2z = -4 \\
\end{array}
$$

> $-1 \cdot (2) \;|$ indicates that equation 2 was multiplied by -1.

$$
\begin{array}{rl}
(2) & x + y - z = 6 \\
-1 \cdot (3) \quad | & -x + y - 2z = 7 \\
\hline
(5) & 2y - 3z = 13 \\
\end{array}
$$

Now we have two equations in two unknowns. Use the methods of the previous sections to find y and z.

$$
\begin{array}{lll}
(4) & y + 2z = -4 \;\Rightarrow\; -2 \cdot (4) \;| & -2y - 4z = 8 \\
(5) & 2y - 3z = 13 \;\Rightarrow & 2y - 3z = 13 \\
& & \hline \\
& & -7z = 21 \\
& & z = -3 \\
\end{array}
$$

Now that z is known, we can use either (4) or (5) to find y.

Using (4)

$$y + 2z = -4$$

$$y + 2(-3) = -4$$

$$y = -4 + 6$$

$$y = 2$$

Now that y and z are known, substitute these values in any of the original equations to find x.

Using (1)

$$x + 2y + z = 2$$
$$x + 2(2) + (-3) = 2$$
$$x + 1 = 2$$
$$x = 1$$

Since we used equation (1) to find x we know that $x = 1$, $y = 2$, and $z = -3$ satisfy (1), but we still must test our solution in equations (2) and (3).

(2)

$$x + y - z = 6$$
$$(1) + (2) - (-3) = 6$$
$$6 = 6 \qquad \text{OK.}$$

(3)

$$x - y + 2z = -7$$
$$1 - (2) + 2(-3) = -7$$
$$-1 - 6 = -7$$
$$-7 = -7 \qquad \text{OK.} \qquad \square$$

Example 19 □ Solve the system

(1) $\qquad x + y + 3z = 2$

(2) $\qquad x + z = 3$

(3) $\qquad 3x + y = -2$

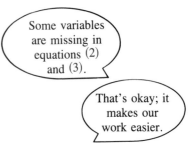

Some variables are missing in equations (2) and (3).

That's okay; it makes our work easier.

Let's eliminate z first using equations (1) and (2).

(1) $\quad x + y + 3z = 2 \Rightarrow \qquad\qquad x + y + 3z = 2$

(2) $\quad x \qquad + z = 3 \Rightarrow \underline{-3\cdot(2)} \qquad \underline{-3x \qquad - 3z = -9}$

(4) $\qquad\qquad -2x + y = -7$

Since (4) has x and y terms, we'll next work with equations (3) and (4).

(3) $\quad 3x + y = -2 \Rightarrow \qquad\qquad 3x + y = -2$

(4) $\quad -2x + y = -7 \Rightarrow \underline{-1(4)} \qquad \underline{2x - y = 7}$

$$5x = 5$$
$$x = 1$$

Next use (3) to find y.

(3)

$$3x + y = -2$$
$$3(1) + y = -2$$
$$y = -2 - 3$$
$$y = -5$$

Now use (2) to find z.

$$\begin{aligned} x + z &= 3 \\ (1) + z &= 3 \\ z &= 2 \end{aligned}$$

(2)

Equation (1) is the only equation left to test our solution.

$$\begin{aligned} x + y + 3z &= 2 \\ (1) + (-5) + 3(2) &= 2 \\ -4 + 6 &= 2 \\ 2 &= 2 \qquad \text{OK.} \end{aligned}$$

(1)

□

Example 20 □ Solve the system

(1) $\qquad 2x - y + 3z = 1$

(2) $\qquad 4x + 7y - z = 7$

(3) $\qquad x + 4y - 2z = 3$

Let's use (2) and (3) to eliminate z.

(2) $\quad 4x + 7y - z = 7 \Rightarrow \quad -2 \cdot (2) \mid \quad -8x - 14y + 2z = -14$

(3) $\quad x + 4y - 2z = 3 \Rightarrow \qquad\qquad\qquad\underline{x + 4y - 2z = \quad 3}$

$\qquad\qquad\qquad\qquad\qquad\qquad$ (4) $\quad -7x - 10y \qquad = -11$

Now eliminate z from (1) and (2).

(1) $\quad 2x - y + 3z = 1 \Rightarrow \qquad\qquad\qquad 2x - y + 3z = \quad 1$

(2) $\quad 4x + 7y - z = 7 \Rightarrow \quad 3 \cdot (2) \mid \quad \underline{12x + 21y - 3z = 21}$

$\qquad\qquad\qquad\qquad\qquad\qquad$ (5) $\quad 14x + 20y \qquad = 22$

Now use (4) and (5) to eliminate y.

(4) $\quad -7x - 10y = -11 \Rightarrow \quad 2 \cdot (4) \mid \quad -14x - 20y = -22$

(5) $\quad 14x + 20y = \quad 22 \Rightarrow \qquad\qquad\qquad \underline{14x + 20y = \quad 22}$

$\qquad\qquad\qquad\qquad\qquad\qquad\qquad\qquad 0 + 0 \quad = \quad 0$ □

What happened?

Same thing as with a system of two equations. There are an infinite number of solutions to this system. Read on.

We graph an equation with three variables as a plane in space.

■ Figure 4.7

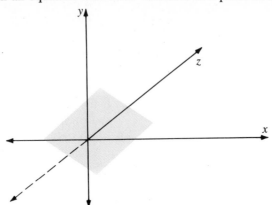

Some of the possibilities for the graph of a system of three equations are

■ Figure 4.8
■ Figure 4.9

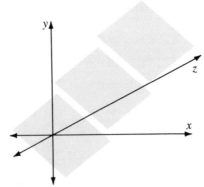

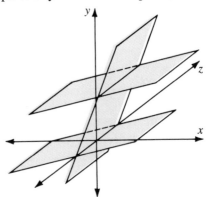

Three parallel planes.
No solutions.

Two parallel planes.
No solutions.

■ Figure 4.10

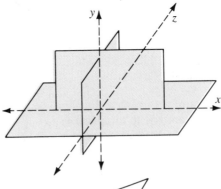

If three planes meet at one point like the corner of a room, there is one solution for the system.

■ Figure 4.11

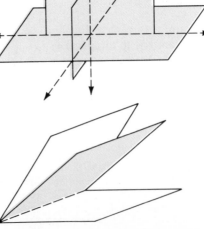

If planes meet in a line like the pages of a book, there is an infinite number of solutions.

Chapter 4 ■ System of Equations

Dependent Systems

We call systems that yield true but meaningless statements like $0 = 0$ **dependent systems.**

Inconsistent Systems

We call systems that yield obviously false statements like $0 = 7$ **inconsistent systems.**

It is possible to solve systems of equations with four or more unknowns using the same techniques. As you can imagine, the process is tedious and requires great care. A later chapter will show you how to use the determinant of a matrix to solve large systems of equations efficiently, or, better yet, use a computer to solve them for you,

The next example will show how systems of three equations can be used to solve problems.

Example 21 □ In a triangle the first angle is $2°$ more than three times the second. The third angle is $8°$ more than the second. Find the angles.

Let x = Degrees in first angle

y = Degrees in second angle

z = Degrees in third angle

Translating the written statement into algebra,

The first angle is $2°$ more than three times the second

$$x \qquad = 2° \qquad + \qquad 3y$$

The third angle is $8°$ more than the second

$$z \qquad = 8° \qquad + \qquad y$$

Wait a minute. That's only two equations. You have three unknowns.

Add your knowledge that the sum of the angles of a triangle is $180°$. Therefore $x + y + z = 180°$.

Now we can solve the system.

(1) $\qquad x = 2° + 3y \qquad \Rightarrow \quad x - 3y \qquad = \quad 2°$

(2) $\qquad z = 8° + y \qquad \qquad -y + z = \quad 8°$

(3) $\qquad x + y + z = 180 \qquad x + \ y + z = \ 180°$

Working with equations (1) and (3) to eliminate x

(1) $x - 3y = 2°$ $\Rightarrow$ $\qquad\qquad x - 3y = 2°$

(3) $x + y + z = 180°$ $\Rightarrow$ $-1 \cdot (3)\big|$ $\underline{-x - y - z = -180°}$

(4) $\qquad\qquad\qquad\qquad\qquad\qquad\qquad -4y - z = -178°$

Now using (4) and (2)   Now use (1) to find x

(2) $\quad -y + z = \quad 8°$ $\qquad x = 2° + 3y$

(4) $\quad \underline{-4y - z = -178°}$ $\qquad\quad = 2° + 3(34°)$

$\qquad -5y \quad = -170°$ $\qquad\quad = 2° + 102°$

$\qquad\quad y \quad = \quad 34°$ $\qquad\quad = 104°$

Use (2) to find z

$$z = 8° + y$$
$$z = 8° + 34°$$
$$z = 42°$$

Now checking in equation (3)

$$x \quad + \quad y \quad + \quad z \quad = 180°$$
$$(104°) + (34°) + (42°) = 180°\qquad \square$$

Problem Set 4.3

Find the simultaneous solution, if one exists, for the following equations.

1. $x - 2y + z = 20$
   $2x + y - 3z = -15$
   $3x + 2y - z = -4$

2. $2x - y + z = -7$
   $3x + 2y + z = -2$
   $x - 2y + z = -8$

3. $x + y - z = -3$
   $2x - 2y + z = 16$
   $2x + y - 2z = -2$

4. $x + y - z = 0$
   $x + 4y + z = 4$
   $2x - 3y - 2z = 10$

5. $-2x - 2y + z = -6$
   $x + 2y - z = 5$
   $x - 2y + 3z = -7$

6. $x - y + 3z = -1$
   $2x - y + 4z = 0$
   $3x - y + 5z = 1$

7. $x + y + 2z = 1$
   $3x + y + 3z = 2$
   $2x \quad + z = 2$

8. $x + 2y + z = 1$
   $x - y + z = 1$
   $2x + 2y = 2$

9. $-x + y + 2z = -3$
   $2x + y + 3z = 4$
   $x + 2y + z = -3$

10.
$$x \qquad + z = 0$$
$$x + y \qquad = 1$$
$$2x + y + z = 2$$

11. $-x + 2y + 3z = -6$
$$x - 2y + 3z = -6$$
$$-x - 5y + 4z = -43$$

12. $-\dfrac{x}{3} - 10y + \dfrac{z}{6} = 0$
$$2x - 20y + \dfrac{z}{3} = \dfrac{4}{3}$$
$$-40x - 40y + z = 1$$

13. $3x + y - z = 4$
$$5x - 2y + z = 6$$
$$6x + 2y - z = 5$$

14. $x + 2y - z = -4$
$$-2x - 3y + z = 5$$
$$2x - 5y + 2z = 4$$

15. $x + 2y + z = 0$
$$-x + 4y - 3z = 9$$
$$2x - 6y + 2z = -5$$

16. $2x - 3y + 4z = 13$
$$3x + 2y - 2z = -8$$
$$5x - 4y + 2z = 14$$

17. $3x + y + 3z = -2$
$$6x - 3y - 4z = -4$$
$$2y - 3z = 7$$

18. $x - y + 2z = 3$
$$2x + y - 3z = 6$$
$$2x - 2y + 4z = 6$$

19. $2x + 4y + 3z = 5$
$$4x + 8y - 5z = 10$$
$$-3x + 7z = -12$$

20. $3x + 6y - 2z = -10$
$$2x - 3y + 5z = 17$$
$$5x + 9y + 3z = 3$$

21. $\dfrac{x}{3} + y - \dfrac{z}{2} = 1$
$$-2x - 6y + 3z = -6$$
$$\dfrac{x}{4} - \dfrac{y}{3} + 2z = \dfrac{1}{2}$$

22. $2x + \dfrac{y}{2} - \dfrac{z}{3} = 4$
$$4x - \dfrac{y}{4} + 2z = -5$$
$$6x + 2y - \dfrac{2z}{3} = 13$$

23. $\dfrac{x}{2} + 2y - 3z = -6$
$$\dfrac{x}{3} + 4y + \dfrac{z}{2} = 5$$
$$\dfrac{2}{3}x + 6y - \dfrac{z}{2} = 4$$

24. $\dfrac{x}{2} - \dfrac{y}{3} - \dfrac{z}{6} = \dfrac{1}{6}$
$$\dfrac{x}{4} + \dfrac{y}{2} - z = -1$$
$$-3x + 2y + z = -1$$

Specify what each variable represents. Set up three equations and solve them.

25. The sum of three numbers is 37. The second is eight less than twice the first. Twice the third is 28 more than the second minus the first. Find the numbers.

26. Rashelle sold magazines for her school. On Monday, Wednesday, and Friday she sold $106 worth. On Monday she sold $8 less than on Wednesday. On Friday she sold $30 less than twice the amount sold on Monday. How much did she sell each day?

27. The Serve Yourself Supermarket sells three kinds of ice cream. The ice milk brand sells for $1.75 a carton. The half cream variety sells for $2.25 a carton and all cream variety sells for $3.25 a carton. The total receipts from the sale of 88 boxes in one day was $204. Three boxes of ice milk were sold for every two boxes of all cream that was sold. How many boxes of each were sold?

Note to Students: At this point many teachers like to cover an additional method for solving systems of equations called **Cramer's Rule.** Sections 15.1 and 15.2 in this book develop this technique.

4.1 **1. Standard form of a linear equation**

$$Ax + By = C \qquad \text{where } A, B, \text{ and } C \text{ are constants}$$

2. Simultaneous solution of a system of two equations
The point that is common to both lines on a graph.

3. To use the addition method to solve systems of equations
  a. Write one equation under the other so the x's and y's (variables) are lined up in columns.
  b. If necessary, multiply one or both equations by some constant so the coefficients of one of the variables have the same numerical value but opposite sign.
  c. Add the equations so one variable is eliminated.
  d. Solve for the value of the remaining variable.
  e. Substitute this value in one of the equations and solve for the other variable.

4. Dependent equations
Two equivalent equations, whose graphs are the same line. There are an infinite number of solutions that satisfy both equations.

5. Inconsistent equations
Two equations whose graphs are parallel lines. Since parallels never intersect they have no common order pairs or solution.

6. To use the substitution method to solve systems of equations
  a. solve one of the equations for one of the variables;
  b. substitute this expression for that variable in the other equation and solve for the remaining variable;
  c. substitute this value into one of the original equations and solve for the other variable.

4.2 Simultaneous equations have applications in various fields. Some applications are:
  a. number problems
  b. age problems
  c. coin problems
  d. business problems
  e. mixture problems

4.3 **1. Possible simultaneous solutions of a system of three equations**
  a. Consistent equations—intersection of three planes—a point.
  b. Dependent systems—Systems that yield true but meaningless statements like $0 = 0$—Planes meet in a line, hence an infinite number of solutions.
  c. Inconsistent systems—Systems that yield obviously false statements like $0 = 7$—Planes that are parallel to each other, hence no solution.

  2. Simultaneous systems of three equations allow us to extend applications in various fields.

Use the substitution method to find the simultaneous solution for the following. **(4.1)**

1. $2x - 5y = -7$

$3x + 2y = 18$

2. $2x + 3y = \quad 5$

$4x - 5y = -12$

3. $\frac{1}{2}x + 2y = -5$

$\frac{1}{3}x - \frac{1}{2}y = 4$

Use the substitution method to find the simultaneous solution for the following. **(4.1)**

4. $3x + \quad y = \quad 5$

$x - 2y = 11$

5. $x - 6y = -4$

$2x + 3y = -3$

6. $x - 6y = 8$

$\frac{1}{2}x = 3y + 4$

Use either method to find the simultaneous solution for the following. **(4.1)**

7. $\frac{1}{2}x - \quad y = \quad 8$

$\frac{1}{3}x + 2y = -8$

8. $3x - 2y = -5$

$y - 13 = -2x$

9. $3x + \quad y = \frac{3}{2}$

$2x - 3y = \frac{17}{6}$

Set up two equations and solve. **(4.2)**

10. The sum of two numbers is 54. If twice the smaller is 6 more than the larger, find the numbers.

11. The sum of $3800 was invested, part at 8% and part at 10%. If the yearly income from the 10% investment was $92 more than the income from the 8% investment, how much was invested at each rate?

12. A confectioner sells English walnuts for $3.00 a pound and pecans for $4.50 a pound. How many pounds of each must be mixed to obtain a mixture of 25 pounds selling for $3.60 per pound?

Find the simultaneous solution for the following equations. **(4.3)**

13. $-x + 2y + 4z = \quad 21$

$4x + 3y - \quad z = -2$

$3x + 4y + 2z = \quad 13$

14. $\frac{1}{6}x + \frac{1}{3}y + \quad z = \quad 1$

$\frac{1}{2}x + \frac{1}{4}y \qquad = \quad z$

$\frac{1}{2}x + \quad y - \frac{1}{3}z = -7$

Set up three equations and solve. **(4.3)**

15. A child's bank contains $4.25 consisting of nickels, dimes, and quarters. The bank has 4 more dimes than nickels and twice as many nickels as quarters. How many of each type of coin is in the bank?

5 Polynomials

Contents

Preview

In the last two chapters all the variables were raised to the first power. The first section of this chapter will develop the rules of operations for quantities that have integer exponents. The second section will propose one useful application of the laws of exponents, scientific notation. The next section will define some well-behaved expressions, called **polynomials,** that are sums of terms with whole number exponents. Then it will examine multiplication and division of polynomials.

In the final section of this chapter, a technique for rapid division of some polynomials, called **synthetic division,** is presented. This technique is quite useful in finding roots of higher order polynomials and differential calculus.

■ 5.1　Operations with Integer Exponents

Throughout your study of mathematics, you will see the laws of exponents applied to many number systems. However, to help you develop a feel for the laws of exponents, we will start by using exponents that are natural numbers and bases that are real numbers.

Recall the following from Chapter 1.

5.1A　Multiplication of Powers with Like Bases

When you multiply monomials with the same base, add the exponents.

$$a^x \cdot a^y = a^{x+y}$$

Example 1 □ Multiply $2x^3 \cdot 4x^2$.

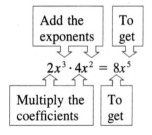

$$2x^3 \cdot 4x^2 = 8x^5$$

□

Zero as an Exponent

The multiplication law of powers says that to multiply two powers of the same base you keep the base and add the exponents. Using this law, even though we don't know what a^0 means yet

$$a^3 \cdot a^0 = a^{3+0}$$

$$= a^3$$

Since one is the identity element for multiplication,

$$a^3 \cdot 1 = a^3$$

Now, since both expressions $a^3 \cdot 1$ and $a^3 \cdot a^0$ are equal to a^3, they are equal to each other

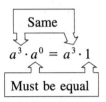

Same

$$a^3 \cdot a^0 = a^3 \cdot 1$$

Must be equal

Therefore, the following definition seems reasonable.

5.1B   Definition: Zero Exponent

Any number except zero raised to the zero power is one.

$$a^0 = 1 \qquad x^0 = 1$$

$$(-3)^0 = 1 \qquad (2x)^0 = 1$$

Problem Set 5.1A

Use the multiplication law of powers to find the monomial products.

1. $x^2 \cdot x$

2. $y^2 \cdot (-3y^5)$

3. $-4a \cdot 5a^3$

4. $(-5x^4)(-6x^3)$

5. $-x^2 y \cdot 5x^3 y^4$

6. $(-3x^4 y)(-4xy)$

7. $4a^0 b \cdot 5ab^2$

8. $-a^2 \cdot b^2 \cdot 3a^4 \cdot b^2$

9. $-8a \cdot 9a^2 x^3$

10. $a^0 \cdot 5ab^2$

11. $-x^0 \cdot y^0 \cdot 7xa^2$

12. $(-3x^0)(-4x^0 y^2)$

13. $-a^0 \cdot x^2 \cdot 2a^2 \cdot x^3$

14. $-6a^2 \cdot 7a^3 \cdot b^0$

15. $3a^0 b^3 \cdot (-ba^3 \cdot b^0)$

16. $-4x^3 y^0 \cdot (-6x^0 y^0)$

17. $-a^0 \cdot (-5ab^3)$

18. $(-a^3) \cdot (-b^3)(2a^0)(-3b^2)$

19. $-x^3 y \cdot (-4x^0 y^4)$

20. $(-3x^0 y) \cdot (-x^2 y^3)$

21. $8x^0 y^2 \cdot (-4xy^0)$

22. $-x^2 y \cdot (3x^0 y)$

23. $-x^0 \cdot (-y^0) \cdot (8x^2 y^0)$

24. $(-6a) \cdot (8a^0 y^2) \cdot (-x^0 y^3)$

Now let's examine division of powers.

To divide different powers of the same base, first write out what the numerator and denominator mean.

$$\frac{a^5}{a^3} \quad \text{means} \quad \frac{\overbrace{a \cdot a \cdot a \cdot a \cdot a}^{5 \ a's}}{\underbrace{a \cdot a \cdot a}_{3 \ a's}}$$

Dividing out common factors yields

$$\frac{a^5}{a^3} = \frac{\overset{1}{\cancel{a}} \cdot \overset{1}{\cancel{a}} \cdot \overset{1}{\cancel{a}} \cdot a \cdot a}{\underset{1}{\cancel{a}} \cdot \underset{1}{\cancel{a}} \cdot \underset{1}{\cancel{a}}} = a^2$$

5.1C Division of Powers

To find the quotient of two powers of the same base, keep the base and subtract the exponents.

$$\frac{a^x}{a^y} = a^{x-y}$$

for any $a \neq 0$.

Example 2 □ Divide $\dfrac{p^7}{p^3}$.

Notice, the base, p, remains unchanged.

Subtract exponents

To divide powers $\Rightarrow \dfrac{p^7}{p^3}$ $= p^{7-3}$ $= p^4$

□

Example 3 □ Divide $\dfrac{a^{5+x}}{a^{1-2x}}$.

What do I do about the variables in the exponent?

$$\frac{a^{5+x}}{a^{1-2x}} = a^{(5+x)-(1-2x)}$$

$$= a^{4+3x}$$

Follow the definition of division. To divide, subtract the exponents.

□

But what if the exponent of the denominator is larger than the exponent of the numerator?

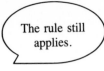

The rule still applies.

Let's try a simple case using the definition of exponents to divide.

$$\frac{a^3}{a^5} = \frac{\overset{1}{\cancel{a}} \cdot \overset{1}{\cancel{a}} \cdot \overset{1}{\cancel{a}}}{\underset{1}{\cancel{a}} \cdot \underset{1}{\cancel{a}} \cdot \underset{1}{\cancel{a}} \cdot a \cdot a} = \frac{1}{a \cdot a} = \frac{1}{a^2}$$

But if the law for division of powers is followed,

$$\frac{a^3}{a^5} = a^{3-5} = a^{-2}$$

Therefore, we will define $a^{-2} = \dfrac{1}{a^2}$.

5.1D Definition: Negative Exponents

Any power in the numerator with a negative exponent can be written in the denominator with a positive exponent.

$$a^{-x} = \frac{1}{a^x}$$

for any $a \neq 0$.

The law for multiplication of powers still holds for negative exponents.

Powers multiplied	Exponents added

$$a^5 \cdot a^{-2} \quad = \quad a^{5+(-2)} \quad = a^3$$

Using the definition of negative exponents yields the same result.

$$a^5 \cdot a^{-2} = a^5 \cdot \frac{1}{a^2} \qquad \text{Definition of negative exponents}$$

$$= a^{5-2} \qquad \text{Division of powers}$$

$$= a^3$$

$\dfrac{1}{x^5}$

$\dfrac{1}{x^3 y^5}$

Example 4 □ Write the following with positive exponents.

a. $p^{-3} = \dfrac{1}{p^3}$ **b.** $x^{-5} = $ _____

c. $p^{-2} \cdot q^3 = \dfrac{1}{p^2} \cdot q^3 = \dfrac{q^3}{p^2}$ **d.** $x^{-3} \cdot y^{-5} = \dfrac{1}{x^3} \cdot \dfrac{1}{y^5} = $ _____

e. $(2x)^{-4} = \dfrac{1}{(2x)^4} = \dfrac{1}{16x^4}$ **f.** $2x^{-4} = \dfrac{2}{x^4}$

□

How come the "2" in the last example stayed in the numerator?

Because the exponent acts only on the symbol it follows. If you want it to act on more than one symbol, you must enclose them in parentheses.

Denominators with Negative Exponents

Negative exponents can also appear in the denominator.

$$\dfrac{1}{x^{-2}} = \dfrac{1}{\dfrac{1}{x^2}}$$ Definition of negative exponents

$$= 1 \div \dfrac{1}{x^2}$$

$$= 1 \cdot \dfrac{x^2}{1}$$ Definition of division

$$= x^2$$

5.1E   Negative Exponents in the Denominator

Any power in the denominator with a negative exponent can be written with a positive exponent in the numerator.

$$\dfrac{1}{x^{-n}} = x^n$$

where $x \neq 0$.

Example 5 □ Write the following without negative exponents.

$\dfrac{1}{x^3}, x^3, \dfrac{y^3}{x^2}$

a. $x^{-3} = $ _____

b. $\dfrac{1}{x^{-3}} = $ _____

c. $\dfrac{x^{-2}}{y^{-3}} = $ _____

$x^4 y^3, \dfrac{2}{x^3 y^2}, 5x^2 y^2$

d. $\dfrac{x^4}{y^{-3}} = $ _____

e. $\dfrac{2x^{-3}}{y^2} = $ _____

f. $\dfrac{5}{x^{-2} \cdot y^{-2}} = $ _____

□

Powers Raised to Powers

$(a^5)^3$ means $(a^5)(a^5)(a^5) = a^{5+5+5} = a^{15}$

also $5 \cdot 3 = 15$

These observations lead to the following property.

5.1F Power Raised to a Power

To raise a power to a power, keep the base, multiply the exponents.

$$(a^x)^y = a^{x \cdot y}$$

Example 6 □ Simplify the following.

Power of a power	Multiply exponents

a^{25}

y^{10}

$(x^4)^3 \quad = \quad x^{4 \cdot 3} \quad = x^{12}$

$(a^5)^5 \quad = \quad a^{5 \cdot 5} \quad = $ _____

$(y^5)^2 \quad = $ _____

□

Power of a Product

To see what happens when a product like (ab^2) is raised to a power, use the basic definition of exponents as repeated multiplication.

$$(ab^2)^3 \qquad \text{means} \qquad (ab^2)(ab^2)(ab^2)$$

Therefore,

$$(ab^2)^3 = (ab^2)(ab^2)(ab^2)$$

Since multiplication is commutative,

$$= a \cdot a \cdot a \cdot b^2 \cdot b^2 \cdot b^2$$
$$= a^3 \cdot b^6$$

Notice that this is the same result as if we had multiplied each of the exponents.

$$(ab^2)^3 = (a^{1 \cdot 3})(b^{2 \cdot 3}) = a^3 \cdot b^6$$

5.1G Power of a Product

$$(a^x \cdot b^y)^z = a^{x \cdot z} \cdot b^{y \cdot z}$$

Example 7 ☐ Simplify each of the following.

$p^9 q^3$

a. $(x^2 y^4)^2 = x^{2 \cdot 2} \cdot y^{4 \cdot 2}$ **b.** $(p^3 q)^3 = $ _____

$\qquad\qquad\quad = x^4 y^8$ ☐

Example 8 ☐ Simplify $(2a^2 b^{-3})^{-2}$.

$(2a^2 b^{-3})^{-2} = 2^{-2}(a^2)^{-2}(b^{-3})^{-2}$	Power of a product
$= 2^{-2} a^{-4} b^6$	Power of a power
$= \dfrac{1}{2^2} \cdot \dfrac{1}{a^4} \cdot \dfrac{b^6}{1}$	Negative exponent
$= \dfrac{1}{4} \cdot \dfrac{1}{a^4} \cdot \dfrac{b^6}{1}$	Evaluate 2^2
$= \dfrac{b^6}{4a^4}$	☐

Power of Quotients

Quotients may also be raised to a power.

$$\left(\frac{a}{b}\right)^2 \qquad \text{means} \qquad \frac{a}{b} \cdot \frac{a}{b} = \frac{a^2}{b^2}$$

and

$$\left(\frac{x^2}{y^3}\right)^2 \qquad \text{means} \qquad \frac{x^2}{y^3} \cdot \frac{x^2}{y^3} = \frac{x^4}{y^6}$$

5.1H   Power of a Quotient

$$\left(\frac{a^x}{b^y}\right)^z = \frac{a^{xz}}{b^{yz}}$$

where $b \neq 0$.

Example 9 □ Simplify the following.

a. $\left(\dfrac{a^3}{b^4}\right)^2 = \dfrac{a^{3 \cdot 2}}{b^{4 \cdot 2}} = \dfrac{a^6}{b^8}$

$\dfrac{9a^{10}}{16b^6}$

b. $\left(-\dfrac{3a^5}{4b^3}\right)^2 = \dfrac{(-3)^2 a^{10}}{(4)^2 b^6} = $ _____

c. $\left(\dfrac{2a^4}{3b^3}\right)^{-3} = \dfrac{2^{-3} a^{-12}}{3^{-3} b^{-9}}$

$\qquad = \dfrac{3^3 b^9}{2^3 a^{12}}$

$\dfrac{27b^9}{8a^{12}}$

$\qquad = $ _____

d. $\left(-\dfrac{2a^3 b^{-2}}{4c^{-4}}\right)^{-2} = \dfrac{(-2)^{-2} a^{-6} b^4}{4^{-2} c^8}$

$\qquad = \dfrac{4^2 b^4}{(-2)^2 a^6 c^8}$

$\qquad = \dfrac{16b^4}{4a^6 c^8}$

$\dfrac{4b^4}{a^6 c^8}$

$\qquad = $ _____   □

Simplify the following. Write the result without negative exponents.

1. $(5x^2)(-4x^3)$

2. $(-3x^4)(-6x^3)$

3. $\dfrac{12x^4}{3x^3}$

4. $-\dfrac{18x^7}{6x^3}$

5. $\dfrac{a^{2n}}{a^n}$

6. $\dfrac{a^{3x}b}{ab^y}$

7. $\dfrac{x^{n-1}}{x^{1-n}}$

8. $\dfrac{-9x^{2n-1}}{-3x^{5-2n}}$

9. $-\dfrac{15x^5y^3}{3x^4y^4}$

10. $\dfrac{24a^3b^5}{-15a^5b^3}$

11. $\dfrac{-8x^{-2}y^{-3}}{-6x^{-4}y^0}$

12. $\dfrac{-18x^{-5}a}{-12x^{-4}y^0}$

13. $\dfrac{x^{3n}}{x^{2n}}$

14. $\dfrac{a^{-n}}{a^{2n}}$

15. $\dfrac{x^{1-2n}}{x^{1-3n}}$

16. $\dfrac{-12x^{3n-2}}{-4x^{4-3n}}$

17. $\dfrac{-24x^{8-3n}}{6x^{2n-3}}$

18. $\dfrac{36x^{4n-3}}{-4x^{9-5n}}$

19. $\dfrac{-18x^4y^3}{3x^2y^6}$

20. $\dfrac{-15a^{-2}x^0}{-12a^0x^{-3}}$

21. $\dfrac{36x^{-3}y^{-4}}{-15x^0y^{-5}}$

22. $\dfrac{-12x^0y^{-3}}{-15x^{-2}y^4}$

23. $\dfrac{-15a^{-2}x^0}{-12a^0x^{-3}}$

24. $\dfrac{-24b^3y^0}{-28b^{-3}y^{-2}}$

25. $(-2t^{-3}s^{-4})(3t^{-4}s^4)$

26. $(-4a^{-3}b^{-5})(-5ab^{-3})$

27. $(a^2b^{-2})^2$

28. $(p^{-3}q^2s^{-2})^3$

29. $(x^{-2}y^3)^{-3}$

30. $(a^{-5}b^2c^0)^{-4}$

31. $(-2x^2y^{-3})^2(3x^{-3}y)^3$

32. $(-2x^{-3}y^2)^3(-3x^{-2}y^{-3})^2$

33. $(-4s^3t^{-3})\cdot(-8s^0t^3)$

34. $(3x^0y^{-2})^3\cdot(4x^{-3}y^4)^2$

35. $(x^5y^{-3}z^0)^{-4}$

36. $(a^3y^{-3}w^{-4})^{-2}$

37. $(-3x^2y^{-3}z^0)(-4x^{-2}y^3z^3)^3$

38. $(-4x^0y^{-4}a^5)^{-2}\cdot(-3x^{-3}y^3a^0)^3$

39. $(-2a^2b^{-3}c^0)^2\cdot(3a^{-3}bc^{-4})^{-3}$

40. $(-4a^0b^3x^{-2})^2\cdot(-3a^{-2}b^{-1}x^0)^{-1}$

41. $\dfrac{(3x^{-2}y)^3}{(-6xy^{-3})^2}$

42. $\dfrac{(-4x^{-3}y^0)^3}{(-8x^{-4}y^{-2})^2}$

43. $-16\left(\dfrac{x^{-3}y^0}{4x^{-4}y^2}\right)^2$

44. $-\dfrac{27}{16}\left(-\dfrac{2x^{-5}y^{-2}}{3x^{-4}y^{-3}}\right)^3$

45. $\left(\dfrac{a^{-2}b^3}{2a^0b^{-4}}\right)^{-2}$

46. $\left(\dfrac{a^{-1}b^{-1}c^2}{a^2b^3c^{-3}}\right)^{-3}$

47. $\left(-\dfrac{y^7}{2x^7}\right)^2$

48. $\left(-\dfrac{y^7}{2x^7}\right)^{-2}$

49. $\dfrac{(4x^{-3}y^{-2})^3}{(-3x^2y^{-3})^2}$

50. $\dfrac{(-3a^{-2}b^{-3}c^0)^{-2}}{(-2a^4b^2c^{-2})^3}$

51. $\left[-\dfrac{x^3y^{-3}w^0}{2x^4y^{-4}w}\right]^2$

52. $\left[\dfrac{a^{-2}b^3c^0}{a^0b^{-3}c^4}\right]^{-2}$

53. $\left[-\dfrac{a^8}{2b^{-3}}\right]^{-3}$

54. $\left[-\dfrac{8x^3y^{-6}z^3}{3x^4y^{-5}z^{-4}}\right]^0$

55. $\left[\dfrac{-2a^{-3}b^{-4}}{-3a^0b^{-2}}\right]^{-2}$

56. $\left[\dfrac{a^{-2}b^0c^{-3}}{a^4b^{-3}c^{-4}}\right]^{-3}$

57. $\dfrac{-16(a-b)^{-2}(x+y)}{-20(x+y)^{-4}(a-b)^{-3}}$

58. $\dfrac{18(2a-b)^0(x+3y)^{-4}}{-24(x+3y)^{-2}(2a-b)^{-3}}$

59. $\dfrac{-20(a+b)^{-3}(x-2y)^2}{-18(x-2y)^{-3}(a+b)^0}$

60. $\dfrac{24(3a-b)^{-2}(x+4y)^{-5}}{-15(x+4y)^3(3a-b)^{-6}}$

Evaluate for $a=-2$, $b=3$, $c=0$, $x=-\dfrac{1}{2}$, $y=\dfrac{2}{3}$.

61. a^2bx

62. ab^3x

63. $(a^3bx)^2$

64. $(b^3xy^2)^3$

65. $\dfrac{1}{2}ab+\dfrac{a}{3y}$

66. $-\dfrac{1}{4}by^2-\dfrac{a}{b}$

67. $8x^4-\dfrac{1}{8}ab$

68. $-27y^4-\dfrac{1}{3}a^2x$

69. $\dfrac{y\cdot b}{a}-\dfrac{a+b}{x}$

70. $\dfrac{ax}{b}+\dfrac{b-a+c}{y}$

71. $\dfrac{a+x^{-1}}{by}-\dfrac{y^{-1}x^{-1}}{a+b}$

72. $\dfrac{b^2-a^2+c^2}{(xy)^{-1}}+\dfrac{ca^{-2}b^3}{x+y^{-1}}$

73. $\left(\dfrac{a^2x}{b^2y}\right)^{-2}\cdot\left(\dfrac{b^2y^2}{b^{-1}a^4y^3}\right)^{-1}$

74. $\left(\dfrac{a^{-4}b^2y}{x^2}\right)^{-3}\cdot\left(\dfrac{b^{-3}a^5}{x^{-2}y^2}\right)^{-1}$

If $a=\dfrac{2}{3}$, $b=0$, $c=-3$, $x=-\dfrac{1}{2}$, $y=2$

75. $a^2y^2c^3$

76. $(abc)^2\cdot xy$

77. $(a^2c^3y)^2$

78. $a^{-2}x^2y^3$

79. $-9a^3xc^2$

80. $(acx^2y^3)^2$

81. $(a^2c^3x^3y)^{-2}$

82. $(3a^3c^2x^2)^{-3}$

83. $\dfrac{ab}{c}-\dfrac{y}{x}$

84. $\dfrac{ac}{xy}+\dfrac{3a}{y}$

85. $\dfrac{a-c^{-1}}{y}+\dfrac{y^{-2}+x^2}{ac}$

86. $\dfrac{x^2+b^2+y^{-2}}{(ac)^2}+\dfrac{abc}{xy}$

87. $\dfrac{x^2y^4}{a^3c^2}\cdot\dfrac{x^{-4}y^{-2}}{ac^{-2}}$

88. $\left[\dfrac{a^{-3}y^2c^{-3}}{x^{-2}}\right]^{-1}\cdot\left[\dfrac{c^3a}{a^{-2}x^{-2}}\right]^{-2}$

The laws of exponents are used to express very big or very small numbers. The mass of an electron is 0.00000000000000000000000000091 grams. The distance from earth to the sun is 92900000 miles.

Neither of these figures is convenient to work with. However, the properties of exponents and the way the decimal moves when you multiply or divide by a power of ten can simplify things.

$$1 = 1 = 10^0$$
$$10 = 10 = 10^1$$
$$100 = 10 \cdot 10 = 10^2$$
$$1000 = 10 \cdot 10 \cdot 10 = 10^3$$

| 4 zeros | 4 tens | Exponent 4 |

$$10{,}000 = 10 \cdot 10 \cdot 10 \cdot 10 = 10^4$$

Notice that the number of zeros in the power of ten is equal to the number of times ten is used as a factor. The exponent also gives us the number of times ten is used as a factor.

To multiply by a power of ten, move the decimal point as many places to the right as there are zeros in the power of ten.

Example 10 □ Multiply 5.14 by the following powers of ten.

$$5.14 \times 10 = 51.4$$
$$5.14 \times 100 = 514$$
$$5.14 \times 1000 = \underline{\hspace{1cm}}$$
$$5.14 \times 10{,}000 = \underline{\hspace{1cm}} \qquad □$$

5140

51,400

The exponent on the base (ten) is the same as the number of zeros in the power of ten. When multiplying by a power of ten written with exponents, move the decimal point to the right the number of places indicated by the exponent. Add zeros if necessary.

Example 11 □ Multiply 1.23 using these powers of ten with exponents.

$$1.23 \times 10^1 = 12.3$$
$$1.23 \times 10^2 = 123$$
$$1.23 \times 10^3 = \underline{\hspace{1cm}}$$
$$1.23 \times 10^4 = \underline{\hspace{1cm}} \qquad □$$

1230

12,300

Move the decimal point the proper number of places to complete the multiplications.

1. 3.654×100 **2.** 5.6×10 **3.** 8.62×10^3 **4.** 5.646×10^2

5. 8.265×10^1 **6.** 2.3×10^2 **7.** $3.86 \times 10,000$ **8.** 1.4×1000

9. 5.6×10^4 **10.** 2.3×10^3 **11.** 4×100 **12.** 6×1000

13. 5.9×10^4 **14.** 7.216×10^2 **15.** 8.0×10 **16.** 3.4×10^3

17. 3.869×1000 **18.** 9.46×10^3 **19.** 1.08×10^1 **20.** 3.006×10^2

21. 7.10×10 **22.** 4.93×10^4 **23.** $2.901 \times 10,000$ **24.** 1.8×10^5

Supply the missing powers of ten so each statement is true.

25. $5.6 \times$ _____ $= 560$ **26.** $2.34 \times$ _____ $= 23.4$

27. $8.65 \times$ _____ $= 86,500$ **28.** $9.674 \times$ _____ $= 96.74$

29. $3.2 \times$ _____ $= 3200$ **30.** $1.0 \times$ _____ $= 100$

31. $5 \times$ _____ $= 5000$ **32.** $6.74 \times$ _____ $= 67,400$

33. $3.26 \times$ _____ $= 326$ **34.** $6.5 \times$ _____ $= 6500$

35. $6.86 \times$ _____ $= 68.6$ **36.** $9 \times$ _____ $= 900,000$

37. $4.09 \times$ _____ $= 40.9$ **38.** $6.57 \times$ _____ $= 6.57$

39. $5.923 \times$ _____ $= 59230$ **40.** $1.093 \times$ _____ $= 109.3$

41. $3.704 \times$ _____ $= 3704$ **42.** $2.009 \times$ _____ $= 20.09$

43. $8 \times$ _____ $= 80,000$ **44.** $7.6 \times$ _____ $= 76000$

45. $2.0 \times$ _____ $= 20$ **46.** $3.101 \times$ _____ $= 310.1$

47. $6 \times$ _____ $= 6000$ **48.** $9.260 \times$ _____ $= 92.6$

5.2A Scientific Notation

To write a number in **scientific notation,** write it as a number with a magnitude between one and ten multiplied by a power of ten.

Example 12 □ Study these numbers as they are written in scientific notation.

Number	Scientific Notation
65	6.5×10^1
4768	4.768×10^3
$-32,900$	-3.29×10^4
5,600,000	5.6×10^6

□

In Example 12 the magnitude of each number is greater than one, and all exponents are positive. However, scientific notation can also be used to represent numbers with magnitudes less than one. To express numbers with magnitudes between zero and one in scientific notation use a negative exponent.

When a number is multiplied by a power of ten with a positive exponent, the decimal moves to the right. Multiplying by a power of ten with a negative exponent results in moving the decimal to the left the number of places the exponent indicates.

Example 13 □ Write as a decimal numeral without using exponents.

a. $6.53 \times 10^{-2} = 0.0653$

b. $-5.86 \times 10^{-1} = -0.586$

c. $8.36 \times 10^{-3} = 0.00836$

−0.013

d. $-1.3 \times 10^{-2} = $ _____

□

5.2B Rule: Conversion from Scientific to Decimal Notation

To convert numbers from scientific notation to decimal notation, move the decimal point the number of places indicated by the exponent.

Positive exponents: Move the decimal right.

Negative exponents: Move the decimal left.

Example 14 □ Write as a decimal numeral.

    a. $2.81 \times 10^4 = 28,100$

    b. $4.12 \times 10^{-3} = 0.00412$

−750   **c.** $-7.5 \times 10^2 = $ _____   □

5.2C   Rule: Conversion from Decimal to Scientific Notation

To convert decimals to scientific notation:

1. Move the decimal point so that it is immediately after the first nonzero digit.
2. Multiply the number by the power of ten that would move the decimal to its original position.

Example 15 □ Write the following in scientific notation.

    a. $4210 = 4.21 \times 10^3$

    b. $0.0673 = 6.73 \times 10^{-2}$

    c. $3.75 = 3.75 \times 10^0$

    d. $-0.24 = -2.4 \times 10^{-1}$

10^2 **e.** $754.5 = 7.545 \times$ _____

10^{-4} **f.** $0.00035 = 3.5 \times$ _____

10^0 **g.** $-8.5 = -8.5 \times$ _____

10^2 **h.** $312 = 3.12 \times$ _____   □

In chemistry, physics, biology, and other scientific fields, scientific notation is frequently used to simplify expressions containing large or small numbers.

Example 16 □ Use scientific notation to find the value of 0.00024×0.0000013.

Write in scientific notation.

$$0.00024 \times 0.0000013 = 2.4 \times 10^{-4} \times 1.3 \times 10^{-6}$$
$$= 2.4 \times 1.3 \times 10^{-4} \times 10^{-6}$$
$$= 3.12 \times 10^{-10}$$
$$= 0.000000000312$$

It is easier to see the size of the number in scientific notation.

□

          Chapter 5 ■ Polynomials

Example 17 □ Use scientific notation to find the value of $\dfrac{0.0042}{12000}$. Write in scientific notation.

$$\frac{0.0042}{12000} = \frac{4.2 \times 10^{-3}}{1.2 \times 10^{4}}$$

$$= \frac{4.2}{1.2} \times \frac{10^{-3}}{10^{4}} \qquad 1.2\overline{)4.2\,0}$$

$$\begin{array}{r} 3.5 \\ 1.2\overline{)4.2\,0} \\ \underline{3\;6} \\ 6\;0 \end{array}$$

$$= 3.5 \times 10^{-3-4} \qquad \text{Subtract the exponents}$$

$$= 3.5 \times 10^{-7} \qquad\qquad\qquad \square$$

Example 18 □ A single B-1 bomber has an estimated cost of $290 million. The population of the United States is about 220 million people. Use scientific notation to determine how much each man, woman, and child in the United States would have to pay to buy a fleet of 100 B-1 bombers.

$$\frac{(\text{Cost of one bomber})(\text{Number of bombers})}{(\text{Number of people})} = \frac{2.9 \times 10^{8} \times 10^{2}}{2.2 \times 10^{8}}$$

$$= \frac{2.9}{2.2} \times 10^{2}$$

$$= 1.32 \times 10^{2}$$

$$= \$132 \qquad\qquad \square$$

Problem Set 5.2B

Convert the following numbers in scientific notation to decimal numbers.

1. 6.7×10^{3} **2.** 7.56×10^{2} **3.** 1.8×10^{-3} **4.** 2.1×10^{-1}

5. 4.768×10^{1} **6.** 9.4×10^{4} **7.** -5×10^{-5} **8.** -8×10^{2}

9. 2.1×10^{4} **10.** -5.768×10^{2} **11.** 3.769×10^{-3} **12.** 7.698×10^{-4}

13. 3.86×10^{-4} **14.** 5.6×10^{-3} **15.** 4.924×10^{2} **16.** 6.092×10^{4}

17. 2.395×10^{4} **18.** 7.9021×10^{5} **19.** 1.7403×10^{3} **20.** 8.9002×10^{-3}

21. 9.0092×10^{-5} **22.** 5.810×10^{2} **23.** 2.9089×10^{-4} **24.** 7.308×10^{-6}

Convert the following decimal numbers to numbers in scientific notation.

25. 0.654   **26.** 0.0862   **27.** 856   **28.** 4900

29. 0.879   **30.** 0.00486   **31.** 859,000   **32.** 24,000,000,000

33. 0.0000065   **34.** 0.0000876   **35.** -391 **36.** 90,000,560

37. $-4,000,000$ **38.** 100,000,000   **39.** 0.0000001   **40.** -0.0001

41. 92.8   **42.** 301.5   **43.** 0.516   **44.** 0.00168

45. -8649 **46.** -1612 **47.** 0.0000531   **48.** 0.00000637

49. 3,000,456   **50.** 8,005,602   **51.** 1,000,000   **52.** 5,000,002

53. 0.00005   **54.** 0.00000078   **55.** -0.000721 **56.** -0.000030045

Use scientific notation to simplify the following.

57. 0.0008×0.0000065 **58.** $380,000 \times 0.00084$ **59.** $18,700,000 \times 24,000$

60. $48,000 \times 0.000015$ **61.** 0.0025×0.00000018 **62.** 2500×0.000000018

63. $\dfrac{350,000}{2500}$ **64.** $\dfrac{0.000065}{2500}$ **65.** $\dfrac{0.00217}{350}$ **66.** $\dfrac{88,400}{2,600,000}$

67. 0.0006×0.00000072 **68.** $0.000008 \times 0.0000000075$ **69.** $4,900,000 \times 78,200,000$

70. $3,200,000 \times 89,300,000$ **71.** $820,000 \times 0.000000015$ **72.** $68,400,000 \times 0.0000055$

73. $\dfrac{7600}{2,500,000}$ **74.** $\dfrac{2,520,000}{3500}$ **75.** $\dfrac{0.00000143}{550}$ **76.** $\dfrac{1870}{8,500,000}$

77. The federal standard for clean air says that the maximum permissible level of pollution from particulate matter over our cities is 2.6×10^{-3} grams of particles per cubic meter. An area of a city that is 2 miles wide by 5 miles long has about 8×10^{9} cubic meters of air in the first 1000 feet above ground level. Estimate the total weight of the particle pollution in this amount of air at the maximum permissible level of pollution.

78. Astronomers measure large distances in light years. One light year is the distance travelled by a beam of light in one year. The speed of light is 300,000,000 meters per second. In a year there are 3.15×10^{7} seconds. Use scientific notation to calculate the distance in 5 light years.

79. A single F-18 fighter jet cost $29 million in 1984. At that time, the annual tuition, books, room and board for a student at a private 4-year college was about $9000. Estimate how many students could receive 4-year degrees for the price of one F-18 fighter.

       Chapter 5 ■ Polynomials

Recall the following definition from Chapter 1. A variable is a symbol that can represent any element in a set of numbers.

Factors

Product

Numbers or variables that are multiplied are called **factors.** The result of multiplication is called a **product.**

$$3 \cdot 4 \cdot b = \quad 12b$$

Factors   Product

Terms

Sum

Numbers or variables that are added are called **terms.** The result of an addition is called a **sum.**

$$3 + 4 = \quad 7$$

Terms   Sum

These are polynomials	These are *not* polynomials
$7, a, 3b, x^3, x^2 + y^2$	$x^{-3}, \sqrt{x}$
$\dfrac{1}{3}$, and $2x^2 + x + 1$	$\dfrac{1}{x}, \dfrac{1}{x^2 + 1}$

5.3A   Definition: Polynomial

A **polynomial** is a sum of one or more terms, each term consisting of a constant or a product of a constant and one or more variables. The variables may be raised to any whole number power.

A **polynomial in one variable** is a polynomial whose terms contain only one variable. The variable may be raised to different whole number powers in the separate terms.

Why bother? It looks like polynomials are almost any expression.

Not quite. Notice the definition of polynomials requires that the exponent of variables be whole numbers. This avoids division by zero. Polynomials also avoid problems with roots of negative numbers.

5.3B Definition: Monomial

A polynomial with only one term is called a **monomial.** Some typical monomials are $3, -17, \frac{3}{4}, a, -a, 3a^2, -7a^3, \pi, \sqrt{3}$.

Coefficient

In the monomial $3a$ there are two factors, 3 and a. A numerical factor is referred to as the numerical **coefficient** of the term. The numerical coefficient of $3a$ is 3.

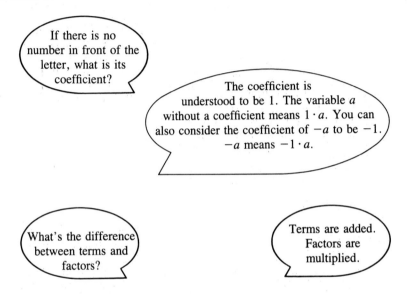

If there is no number in front of the letter, what is its coefficient?

The coefficient is understood to be 1. The variable a without a coefficient means $1 \cdot a$. You can also consider the coefficient of $-a$ to be -1. $-a$ means $-1 \cdot a$.

What's the difference between terms and factors?

Terms are added. Factors are multiplied.

5.3C Definition: Like Terms

Terms having exactly the same variable raised to exactly the same power are called **like terms.** The numerical coefficients of like terms do *not* have to be alike.

$3ax^2$ and $-2ax^2$ are like terms.
$3ax^2$ and $3ax$ are not like terms because the powers of the variables are different.

5.3D Definition: Binomial

A **binomial** is a polynomial with exactly two terms. Some typical binomials are

$$a + 2 \qquad a + b \qquad x^2 + 6 \qquad a^2 - b^2$$

5.3E Definition: Trinomial

A **trinomial** is a polynomial with exactly three terms. Some typical trinomials are

$$x^2 + 2x + 1 \qquad\qquad ab^2 - ab - a^2$$

A polynomial may involve more than one variable.

$$4x^3y + 7x^2y^2 - 2xy^3 \qquad \text{is a polynomial in two variables.}$$

5.3F The Distributive Property

$$a(b + c) = a \cdot b + a \cdot c \qquad \text{and} \qquad (b + c)a = b \cdot a + c \cdot a$$

where a, b, and c stand for any algebraic expression.

Since a, b, and c may stand for any algebraic expression, we can use the distributive property to multiply polynomials. Let's start with a monomial times a polynomial.

Example 19 □ Multiply $6a^2(4a^2 - a + 2)$.

$$6a^2(4a^2 - a + 2) = 6a^2 \cdot 4a^2 - 6a^2 \cdot a + 6a^2 \cdot 2$$
$$= 24a^4 - 6a^3 + 12a^2 \qquad\qquad \square$$

Next we'll multiply two binomials.

Example 20 □ Multiply $(x + 3)(2x + 4)$.
Think of $(x + 3)$ as Z.
Then

$$(x + 3)(2x + 4) = Z(2x + 4)$$

and

$$Z \cdot (2x + 4) = Z \cdot 2x + Z \cdot 4 \qquad\qquad \text{Using the distributive property}$$

$$= (x + 3)2x + (x + 3) \cdot 4 \qquad \text{Restoring } (x + 3) \text{ for } Z$$

Now distribute the $2x$ and the 4

$$= x(2x) + 3(2x) + x(4) + 3(4)$$
$$= 2x^2 + 6x + 4x + 12$$
$$= 2x^2 + 10x + 12 \qquad\qquad \text{Collecting like terms} \square$$

Example 21 □ Multiply $(a - 5)(a - 4)$.

Think of the first factor as a single quantity.

Let

$$(a - 5) = Z$$

Then

$$(a - 5)(a - 4)$$

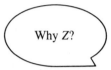
Why Z?

Any symbol will do. The idea is to emphasize that $(a - 5)$ is a single quantity.

becomes

$$Z \cdot (a - 4) = Za + Z(-4)$$

How come the subtraction sign on the left became an addition sign on the right?

You caught us, we skipped a step. $Z(a - 4) = Z[a + (-4)]$. If we write it as an addition problem, the next step will be easier.

Now restoring Z with $a - 5$

$$(a - 5)(a - 4) = (a - 5)(a) + (a - 5)(-4)$$

Applying the distributive property

$$= a(a) - 5(a) + a(-4) - 5(-4)$$
$$= a^2 - 5a - 4a + 20$$
$$= a^2 - 9a + 20$$

Notice: The effect of this line is to multiply every term of the first binomial by every term of the second □

This same process can be done vertically in a method similar to multiplication in arithmetic. To multiply two polynomials, *multiply every term of the first polynomial by every term of the second polynomial,* and then collect the like terms.

$$a - 5$$
$$\underline{a - 4}$$
$$a^2 - 5a \qquad\qquad \text{Product of } a(a - 5)$$
$$\underline{\quad -4a + 20} \qquad \text{Product of } -4(a - 5)$$
$$a^2 - 9a + 20 \qquad \text{Collect like terms}$$

Why write the $-4a$ under $-5a$?

Writing the like terms in a column makes it easier to collect like terms.

Example 22 □ Use the distributive property to multiply $(2x + y)(3x - y)$.

Keep sign with term

2x + y, 2x + y

$$(2x + y)(3x - y) = (\underline{\qquad})3x + (\underline{\qquad})(-y)$$
$$= 6x^2 + 3xy + (-2xy) + (-y^2)$$

$6x^2 + xy - y^2$

$$= \underline{\qquad} \qquad \text{Collect like terms} \qquad □$$

Example 23 □ Use the vertical method to multiply.
$(2a^2 + 3a - 4)(5a^2 - 6a - 7)$

$$2a^2 + \ 3a - 4$$
$$\underline{5a^2 - \ 6a - 7}$$
$$10a^4 + 15a^3 - 20a^2 \qquad\qquad \text{Product of } 5a^2(2a^2 + 3a - 4)$$
$$\quad - 12a^3 - 18a^2 + 24a \qquad\quad \text{Product of } -6a(2a^2 + 3a - 4)$$
$$\underline{\qquad\quad - 14a^2 - 21a + 28} \qquad \text{Product of } -7(2a^2 + 3a - 4)$$
$$10a^4 + \ \ 3a^3 - 52a^2 + \ \ 3a + 28 \qquad \text{Collect like terms} \qquad □$$

Problem Set 5.3A

Find the following products.

1. $8ax^2 \cdot (-7ay)$ **2.** $(9a^2b)(3a^2x)$ **3.** $(3x^3y)(-x^2y^2)$ **4.** $(-ab^2)(-2ab)$

5. $(-8a^3b^2)(4a^2x^2)$ **6.** $(-x^2y)(-3ax^3y^2)$ **7.** $(-5x^4y^3)(-a^3x^2)$ **8.** $(7a^2b^2)(-2a^3x^3)$

Use the distributive property to find the following products.

9. $6x(m^2 - 5m)$ **10.** $-2y(3y^2 + 4y)$ **11.** $2a^2(a^2 + 8a)$

12. $-2xy(6x^2 - 4y^2)$ **13.** $(x - 2y)(x + 3y)$ **14.** $(4t - 7)(3t - 2)$

15. $(3a - 2b)(6a - 2b)$ **16.** $(8a - 7x)(4a + 3x)$ **17.** $-4xy^2(5x - 7y^2)$

18. $-6a^2b^3(4a^3 - 5b^2)$ **19.** $(3a^3x^2)(-9a^4 - 4x^3)$ **20.** $9x^4y^3(8x^3 - 3y^5)$

21. $(5x - y)(3x - 4y)$ **22.** $(3x - 2y)(4x + y)$ **23.** $(6a + 5b)(4a + 7b)$

24. $(7t - 5s)(8t + 3s)$

Line up the following problems vertically, then multiply.

25. $(5a - 6)(3a + 4)$ **26.** $(3a - 4b)(a + 3b)$

27. $(x - 3)(x^2 - 2x + 1)$ **28.** $(x + 2)(x^2 + 5x - 6)$

29. $(2x - 1)(x^2 + 5x - 3)$ **30.** $(3x + 4)(x^2 - 6x + 5)$

31. $(2x - 3)(5x^2 + 7x - 9)$ **32.** $(5y - 2)(3y^2 - 4y + 7)$

33. $(4y - 2)(3y^2 + 6y - 7)$ **34.** $(5a - 2)(4a^2 - 3a + 5)$

35. $(3x + 7)(2x^2 - 7x + 4)$ **36.** $(7t - 4)(5t^2 - 2t + 4)$

37. $(8y - 1)(3y^2 - 4y - 6)$ **38.** $(5x + 1)(6x^2 - 5x - 9)$

39. $(6a + 5)(4a^2 + 3a - 4)$ **40.** $(6y + 7)(3y^2 + 5y - 7)$

41. $(x^2 + 2x - 3)(x^2 + 4x - 5)$ **42.** $(x^2 + 3x - 4)(x^2 + 9x - 5)$

43. $(2a^2 - 6a + 3)(3a^2 + 7a - 7)$ **44.** $(5a^2 + 5a - 6)(3a^2 - 7a + 8)$

45. $(3a^2 - 2a + 4)(2a^2 + 5a - 7)$ **46.** $(5x^2 + 3x - 6)(3x^2 - 5x + 4)$

47. $(7x^2 - 3x + 4)(3x^2 + 4x - 1)$ **48.** $(6a^2 + 2a - 3)(5a^2 - 4a + 5)$

Special Products

Because the product of a binomial times a binomial occurs often, it is helpful to learn how to find it quickly.

Let's examine the product of $(2x + 3)(4x + 5)$. Notice that in this problem the first terms from each binomial are like terms, and the second terms from each binomial are like terms.

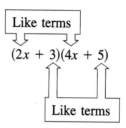

Example 24 □ Multiply $(2x + 3)(4x + 5)$ using the distributive property.

$$(2x + 3)(4x + 5) = (2x + 3) \cdot 4x + (2x + 3) \cdot 5$$

Applying the distributive property again,

$$= 2x(4x) + 3(4x) + (2x)(5) + 3(5)$$
$$12x + 10x$$
$$= 8x^2 + 22x + 15$$

Now let's multiply vertically and notice which factors produce each term of the result.

$$
\begin{array}{r}
2x + 3 \\
4x + 5 \\
\hline
8x^2 + 12x \\
+ 10x + 15 \\
\hline
8x^2 + 22x + 15
\end{array}
$$

$(2x)(4x)$ $(5)(3)$

Sum of $5(2x)$ and $(4x)(3)$

Finally let's write the result directly.
First, find the product of the first two terms of each binomial.

$$(2x + 3)(4x + 5) = 8x^2 +$$

Second, add the product of the two outer factors plus the product of the two inner factors.

$$(2x + 3)(4x + 5) = 8x^2 + 22x +$$

outer

inner

$12x$

$+ 10x$

$22x$

Third, add the product of the last terms in each binomial.

$$(2x + 3)(4x + 5) = 8x^2 + 22x + 15 \qquad \square$$

A device many people use to remember how to multiply two binomials is "FOIL."

FOIL

F–Product of the First terms	$(2x \quad)(4x \quad) \longrightarrow 8x^2$
O–Product of the Outside terms	$(2x \quad)(\quad 5) \longrightarrow + 10x$
I–Product of the Inside terms	$(\quad 3)(4x \quad) \longrightarrow + 12x$
L–Product of the Last terms	$(\quad 3)(\quad 5) \longrightarrow + 15$

$$(2x + 3)(4x + 5) = 8x^2 + 22x + 15$$

Example 25 $\square$ Multiply $(4x - 2)(3x + 1)$.

$12x^2$

The product of the first terms is _____ .

$4x$

The product of the outside terms is _____ .

$-6x$

The product of the inside terms is _____ .

$-2x$

The sum of the outside product plus the inside product is _____ .

-2

The product of the last terms is _____ .

$12x^2 - 2x - 2$

The product $(4x - 2)(3x + 1)$ is _____ . $\qquad \square$

Chapter 5 ■ Polynomials

Example 26 □ Multiply $(3x + y)(4x - y)$ directly using FOIL.

$12x^2, xy, (-y^2)$

$(3x + y)(4x - y) = \underline{\hspace{1.5cm}} + \underline{\hspace{1.5cm}} + \underline{\hspace{1.5cm}}$

| Think $(3x)(4x)$ | Think $3x(-y)$ $+ 4x(y)$ | Think $y(-y)$ |

$\quad\quad 12x^2 \quad + \quad xy \quad - \quad y^2$ □

Problem Set 5.3B

Use the FOIL method to multiply the following binomials.

1. $(x + 2)(x - 3)$ **2.** $(y + 2)(y + 4)$ **3.** $(x - 3)(x + 4)$

4. $(x - 3)(x - 4)$ **5.** $(x - 3)(2x - 1)$ **6.** $(3b + 1)(b - 4)$

7. $(4a - 1)(a - 4)$ **8.** $(2y + 3)(3y + 1)$ **9.** $(2x - 1)(4x - 3)$

10. $(8x - 7)(2x + 1)$ **11.** $(5x - 2)(6x - 7)$ **12.** $(8t + 9)(5t + 4)$

13. $(3x - 4)(2x + 1)$ **14.** $(8y + 2)(5y - 3)$ **15.** $(7a + 3)(3a + 5)$

16. $(2x + 7)(9x + 2)$ **17.** $(6y - 3)(7y - 5)$ **18.** $(5a - 9)(6a - 7)$

19. $(4x - 1)(5x + 7)$ **20.** $(9x - 2)(8x + 3)$ **21.** $(3x + 7)(4x + 9)$

22. $(7x + 9)(8x + 5)$ **23.** $(5x - 9)(3x - 8)$ **24.** $(8x - 7)(9x - 4)$

25. $(x - y)(2x + y)$ **26.** $(x + y)(3x - y)$ **27.** $(4x + 3y)(5x - 4y)$

28. $(5x - 6y)(3x + 4y)$ **29.** $(6r + 7s)(3r - 5s)$ **30.** $(2a + 3b)(3a + 4b)$

31. $(2a - 3b)(3a + 4b)$ **32.** $(2a + 3b)(3a - 4b)$ **33.** $(2a - 3b)(3a - 4b)$

34. $(2x - y)(5x + y)$ **35.** $(4x - 3y)(3x + 2y)$ **36.** $(2a - 3b)(4a - 5b)$

37. $(5a - 4b)(6a - 7b)$ **38.** $(2a + 3b)(4a + 6b)$ **39.** $(5a + 3b)(4a + 5b)$

40. $(6x - 7y)(3x + 2y)$ **41.** $(7x - 8y)(6x + 5y)$ **42.** $(9x + 4y)(8x - 5y)$

Two Special Products

Square of a Binomial

To find the **square of a binomial** we can use the FOIL method; however, let's study the product first using the vertical method.

Example 27 □ Find the square $(a + b)^2$.

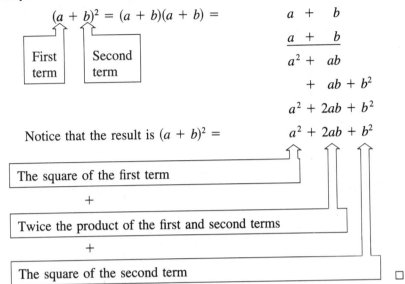

$$(a + b)^2 = (a + b)(a + b) =$$

First term

Second term

$$
\begin{array}{r}
a + b \\
a + b \\
\hline
a^2 + ab \\
+ \ ab + b^2 \\
\hline
a^2 + 2ab + b^2
\end{array}
$$

Notice that the result is $(a + b)^2 = \quad a^2 + 2ab + b^2$

The square of the first term

$+$

Twice the product of the first and second terms

$+$

The square of the second term
□

5.3G Square of a Binomial

The square of a binomial is the square of the first term plus twice the product of the two terms, plus the square of the second term.

$$(a + b)^2 = a^2 + 2ab + b^2$$

$$(a - b)^2 = a^2 - 2ab + b^2$$

Example 28 □ Find the square of $(5x - y)^2$.

$25x^2$

$-5xy$

$-10xy$

$+y^2$

$25x^2 - 10xy + y^2$

The square of the first term is ———.

The product of the two terms is ———.

Twice the product of the terms is ———.

The square of the second term is ———.

The product $(5x - y)^2 = $ ———.

Notice that squaring a binomial always gives us three terms.

The middle term of the result is twice the product of the first and second term of the binomial.

□

Chapter 5 ■ Polynomials

5.3H   Product of a Sum and Difference

A binomial product like $(a + b)(a - b)$ is called the **product of a sum and difference** of two terms because in the first binomial the terms are added and in the second binomial the same terms are subtracted. These are called **conjugate** factors.

$$(a + b)(a - b) = a^2 - b^2$$

Example 29 □ Use the vertical method to multiply $(a + b)(a - b)$.

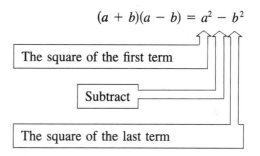

$$
\begin{array}{r}
a \ + \ b \\
a \ - \ b \\
\hline
a^2 + ab \\
-\,ab - b^2 \\
\hline
a^2 \quad - \quad b^2
\end{array}
$$

Notice that the middle term is zero because of the opposite signs of b.

The product of the sum and difference of two terms

$$(a + b)(a - b) = a^2 - b^2$$

The square of the first term

Subtract

The square of the last term

The result is the difference of two squares.

That is because the first terms of the binomials are the same and so are the last terms.

□

Example 30 □ Multiply $(3p^2 - 5)(3p^2 + 5)$.

$3p^2$ The first term is _____ .

$9p^4$ The square of the first term is _____ .

5 The second term is _____ .

25 The square of the second term is _____ .

$9p^4 - 25$ The difference of the squares is _____ .

$9p^4 - 25$ The product $(3p - 5)(3p + 5) =$ _____ .

□

Find the following products using the short method.

1. $(x + 3)^2$

2. $(x - 3)^2$

3. $(x + 3)(x - 3)$

4. $(2a + 1)^2$

5. $(2a - 1)^2$

6. $(2a - 1)(2a + 1)$

7. $(x + y)^2$

8. $(x - y)^2$

9. $(x - y)(x + y)$

10. $(a - 2b)^2$

11. $(2a + b)^2$

12. $(3r - s)(3r + s)$

13. $(a + 2b)^2$

14. $(2a - b)^2$

15. $(3a - 2)(3a + 2)$

16. $(4x - 5)(4x + 5)$

17. $(3x - 4)^2$

18. $(4x + 5)^2$

19. $(3x - 5)(3x + 5)$

20. $(2y - 3)(2y + 3)$

21. $(3r - 5)^2$

22. $(6s - 1)^2$

23. $(2x - 7)(2x + 7)$

24. $(3x - 7)(3x + 7)$

25. $(3x + 2y)^2$

26. $(3x + 2y)^2$

27. $(3x + 2y)(3x - 2y)$

28. $(4x - 5y)(4x + 5y)$

29. $(3a + 4b)(3a - 4b)$

30. $(2x + 5y)(2x - 5y)$

31. $(x + 7y)^2$

32. $(x^2 - 3)^2$

33. $(3x^2 - y)^2$

34. $(3x^2 + 2)(3x^2 - 2)$

35. $(4x^2 + 3y)^2$

36. $(4x^2 - 3y)(4x^2 + 3y)$

37. $(5x - 2y^2)(5x + 2y^2)$

38. $(4a - 3b^2)(4a + 3b^2)$

39. $(2x^2 - 3)^2$

40. $(3y^2 + 4)^2$

41. $(4x^2 - 3y^2)(4x^2 + 3y^2)$

42. $(5a^2 - 6b^2)(5a^2 + 6b^2)$

43. $(x^2 + 3y^2)^2$

44. $(a^2 - 5b^2)^2$

45. $(6r^2 - 7t^2)(6r^2 + 7t^2)$

46. $(7r - 6t^2)(7r + 6t^2)$

47. $(8a^2 - 9b)^2$

48. $(9a + 8b^2)^2$

■ **5.4**   ## Division of a Polynomial by a Polynomial

To divide a polynomial by a polynomial recall the definition of division.

5.4   Definition: Division of a Polynomial by a Polynomial

$$a \div b = a \cdot \frac{1}{b}$$

where $b \neq 0$.

Example 31 □ Divide $\dfrac{3a^3 + 18a^2 - 12a}{3a}$.

$$\dfrac{3a^3 + 18a^2 - 12a}{3a} = (3a^3 + 18a^2 - 12a) \cdot \dfrac{1}{3a} \qquad \text{Definition of division}$$

$$= \dfrac{3a^3}{3a} + \dfrac{18a^2}{3a} - \dfrac{12a}{3a} \qquad \text{Distributive property}$$

$a^2 + 6a - 4$

$$= \underline{\qquad\qquad} \qquad □$$

Are the results the same if you divide each term of the polynomial by $3a$ and add the results?

Yes, but be careful to divide every term by the divisor $3a$.

Example 32 □ Divide $\dfrac{5b^3 - b^2 + 5}{5b}$.

$\dfrac{1}{5b}$

$$\dfrac{5b^3 - b^2 + 5}{5b} = \underline{\qquad}(5b^3 - b^2 + 5)$$

$$= \dfrac{5b^3}{5b} - \dfrac{b^2}{5b} + \dfrac{5}{5b}$$

$$= b^2 - \dfrac{b}{5} + \dfrac{1}{b} \qquad □$$

When dividing we usually write the results with positive exponents.

Example 33 □ Divide $\dfrac{8a^3b^3 - 12a^2b^2 + 2ab}{-2ab^2}$.

$$\dfrac{8a^3b^3 - 12a^2b^2 + 2ab}{-2ab^2} = \left(\dfrac{8a^3b^3}{-2ab^2}\right) - \left(\dfrac{12a^2b^2}{-2ab^2}\right) + \left(\dfrac{2ab}{-2ab^2}\right)$$

Write all fractions in standard form
$$= \left(\dfrac{-8a^3b^3}{2ab^2}\right) - \left(\dfrac{-12a^2b^2}{2ab^2}\right) + \left(\dfrac{-2ab}{2ab^2}\right)$$

$$= -4a^2b - (-6a) + \left(\dfrac{-1}{b}\right)$$

$$= -4a^2b + 6a - \dfrac{1}{b} \qquad □$$

Divide. Write your answers with positive exponents.

1. $\dfrac{4x^3 + 12x^2}{x^2}$

2. $\dfrac{4x^3 + 12x^2}{4x^2}$

3. $\dfrac{4x^3 + 12x^2}{4x}$

4. $\dfrac{12a^2 - 6a + 6}{6a}$

5. $\dfrac{16x^2 - 10x + 4}{2x}$

6. $\dfrac{15y^3 + 25y^2 - 40y}{5y^2}$

7. $\dfrac{18a^2b^2 + 6a^2b}{2ab}$

8. $\dfrac{18a^2b^2 - 12a^2b}{6ab^2}$

9. $\dfrac{2x^4y^2 + 3x^3y + x^2}{x^2y}$

10. $\dfrac{6a^3b^2 - 8a^2b^3 + 4ab^4}{2ab^2}$

11. $\dfrac{7a^4b^2 - 9a^3b^3 - 8a^2b^4}{a^2b^2}$

12. $\dfrac{16a^3 - 12a^4 + 24a^5}{4a^3}$

13. $\dfrac{-6a^3b - 9a^3b^2 + 18ab}{-3ab}$

14. $\dfrac{9a^3b - 12ab^2 + 15ab}{3a^2b}$

15. $\dfrac{a^2b^3 - 5ab^2 - 10a^3b}{5ab^3}$

16. $\dfrac{x^4y - 7x^2 + 4x^3y^3}{x^2y}$

17. $\dfrac{3x^3y - 6x^4y + 9x^2y^2}{3x^2y^2}$

18. $\dfrac{4x^2y^3 - 12x^3y^2 + 16xy^4}{-4x^2y^2}$

19. $\dfrac{-6ay^2 + 36a^2y^3 - 42a^3y}{-6ay^2}$

20. $\dfrac{-7x^4 + 14x^5y^3 - 21xy^4}{7xy^3}$

21. $\dfrac{-5ab^4 + 15ab^2 - b^4}{-5a^2b^2}$

22. $\dfrac{-30ab - 15b^3 + 12a^4}{-3ab}$

23. $\dfrac{-6x^2y^2 + 9x^3y^3 - 12xy^4}{-6x^2y^3}$

24. $\dfrac{-18a^2b + 9a^4b^2 - 21a^3b^3}{-9a^2b^2}$

25. $\dfrac{3a^2b^3 - 6ab^3 + 8b}{2ab^2}$

26. $\dfrac{4x^2y^3 - 6xy^2 - 9x}{3x^2y}$

27. $\dfrac{x^3y - 3x^3y^2 + 5xy^2}{-x^2y}$

28. $\dfrac{-ab^4 + 2a^4b^5 - 3a^4b}{-a^2b^3}$

29. $\dfrac{12a^3b^4 - 18a^2b^3 - 3a^2b^2}{6a^2b^3}$

30. $\dfrac{8x^4y^3 - 12x^3y^2 - 2x^2y^2}{4x^3y^2}$

31. $\dfrac{-16x^3y^4 + 24x^2y^3 - 4xy}{-8x^2y^3}$

32. $\dfrac{24a^4b^5 - 12a^3b^2 + 3ab^2}{-6a^3b^3}$

33. $\dfrac{20ab^2 - 15a^2b^3 + a^3b}{5ab^2}$

34. $\dfrac{-28a^3b^4 - 14a^2b^3 + a^3b}{7a^2b^2}$

35. $\dfrac{14a^3b - 21a^3b^4 + 7a^3b^3}{-7a^2b^3}$

36. $\dfrac{xy + 10x^4y^2 - 15xy^3}{-5x^3y}$

37. $\dfrac{12x^2y^5 - 18x^4y^4 - 9x^2y}{6x^3y^4}$

38. $\dfrac{36a^2b^5 - 18a^2b^4 + 6ab^3}{9ab^4}$

39. $\dfrac{-48a^3b^5 + 36a^4b^4 - 9a^2b^2}{-12a^3b^4}$

40. $\dfrac{-24x^3y^4 + 32xy^4 - 6xy^4}{-8x^2y^3}$

41. $\dfrac{35x^2y^4 - 45x^2y^2 + 55x^3y}{15xy^2}$

42. $\dfrac{35a^2b^2 - 28a^3b^3 - 12ab}{14a^2b}$

43. $\dfrac{9a^3b - 36a^3b^2 + 15ab^2}{-18a^2b^2}$

44. $\dfrac{-45x^3y^6 + 25xy^3 - 10y}{-15x^2y^5}$

45. $\dfrac{30x^3y - 25x + 6}{10x^2y}$

46. $\dfrac{48ab^3 - 18ab^2 + 9ab}{12ab^2}$

47. $\dfrac{-96a^3b^4 + 12a^6b^6 - 18ab^5}{-24a^2b^4}$

48. $\dfrac{-60x^5y^5 - 10x^3y^6 + 18x^4y}{-20x^4y^5}$

Long Division of Polynomials

The **degree** of a polynomial in one variable is the highest power used in any of its variables.

$$5x^3 - 1 \qquad \text{is a 3rd-degree polynomial}$$

$$x^4 \qquad \text{is a 4th-degree polynomial}$$

$$x^2 + 2x + 1 \qquad \text{is a 2nd-degree polynomial}$$

When we write a polynomial, we usually order it so that the variable is written in decreasing powers.

$$2x - x^3 + 14 + 7x^2 \quad \text{is usually written} \quad -x^3 + 7x^2 + 2x + 14$$

$$x^2 + x^4 + 2 \qquad\qquad \text{is usually written} \quad x^4 + x^2 + 2$$

The degree of a polynomial in more than one variable is the highest sum of the powers in any one term.

$$5x^3y^2 \qquad\qquad\qquad \text{is a 5th-degree polynomial}$$

$$6p^4q^2 + 2p^3q - 7q \qquad \text{is a 6th-degree polynomial}$$

Polynomials with two variables are usually written with one variable in descending order and the other in increasing order.

$$x^5 + 5x^4y + 10x^3y^2 + 10x^2y^3 + 5xy^4 + y^5$$

To divide polynomials you must not only write them in order of decreasing powers but also supply zero coefficients for any missing powers.

Missing powers

$$x^4 + x^2 + 2 = x^4 + 0x^3 + x^2 + 0x^1 + 2$$

Supply zero coefficients

Notice how the exponents decrease: 4, 3, 2, 1.

The process for long division of polynomials is very similar to long division in arithmetic. It works because division can be thought of as repeated subtraction.

Example 34 □ $(2x^2 + 10x + 12) \div (x + 3)$.
To divide $(2x^2 + 10x + 12) \div (x + 3) = $?

Dividend   Divisor   Quotient

set it up as a long division problem.

? ← Quotient

Divisor → $x + 3 \overline{)2x^2 + 10x + 12}$ ← Dividend

Step 1.

Ask what you must multiply x by to get $2x^2$. In other words,

2x

$2x^2 \div x = $ _____

$$x + 3 \overline{)2x^2 + 10x + 12}$$

$2x^2 \div x = ?$

Step 2.

Write $2x$ above the $2x^2$.

$$\begin{array}{r} 2x \\ x + 3 \overline{)2x^2 + 10x + 12} \end{array}$$

Step 3.

Multiply this $2x$ by $x + 3$.

$$
\begin{array}{r}
2x \\
x + 3\overline{)2x^2 + 10x + 12} \\
2x^2 + 6x
\end{array}
$$

This is
$2x(x + 3)$

Step 4.

Subtract $2x^2 + 6x$ from $2x^2 + 10x + 12$.

$$
\begin{array}{r}
2x \\
x + 3\overline{)2x^2 + 10x + 12} \\
\underline{2x^2 + 6x} \\
4x + 12
\end{array}
$$

Step 5.

Ask what you must multiply x by to get $4x$. In other words

$4x \div x =$ _____

$$
\begin{array}{r}
2x \\
x + 3\overline{)2x^2 + 10x + 12} \\
\underline{2x^2 + 6x} \\
4x
\end{array}
$$

$4x \div x = ?$

Step 6.

Write the 4 as the next term in the answer.

$$
\begin{array}{r}
2x + 4 \\
x + 3\overline{)2x^2 + 10x + 12} \\
\underline{2x^2 + 6x} \\
4x + 12
\end{array}
$$

Step 7.

Multiply this 4 by $x + 3$.

$$
\begin{array}{r}
2x + 4 \\
x + 3\overline{)2x^2 + 10x + 12} \\
\underline{2x^2 + 6x} \\
4x + 12 \\
4x + 12
\end{array}
$$

This is
$4(x + 3)$

Step 8.

Subtract $4x + 12$ from $4x + 12$.

$$
\begin{array}{r}
2x + 4 \\
x + 3 \overline{)2x^2 + 10x + 12} \\
\underline{2x^2 + 6x } \\
4x + 12 \\
\underline{4x + 12} \\
0
\end{array}
$$

Our answer is $(2x^2 + 10x + 12) \div (x + 3) = 2x + 4$. $\quad\square$

Example 35 $\square$ Divide $(3n^2 - 14n + 8) \div (3n - 2)$.

Part 1.

$$
\begin{array}{r}
n \\
3n - 2 \overline{)3n^2 - 14n + 8} \\
\underline{3n^2 - 2n } \\

\end{array}
$$

$\boxed{3n^2 \div 3n = n}$

$\boxed{n(3n - 2)}$

$\boxed{\text{Subtract } 3n^2 - 2n \text{ from } 3n^2 - 14n + 8}$

$-12n + 8$

Subtraction is defined as addition of the additive inverse. To avoid errors it is better to *find the additive inverse and then add*.
The additive inverse of

$$3n^2 - 2n = -(3n^2 - 2n)$$
$$= -3n^2 + 2n$$

That is, to find the additive inverse of a polynomial, change the sign of every term.

$$
\begin{array}{r}
n \\
3n - 2 \overline{)3n^2 - 14n + 8} \\
+ + \\
\underline{3n^2 - 2n } \\

\end{array}
$$

$\boxed{\text{Change the signs}}$

$\boxed{\text{Add } 3n^2 - 14n \text{ and } -3n^2 + 2n}$

$-12n + 8$

(Notice how the changed signs are written above the old signs.)

Part 2.

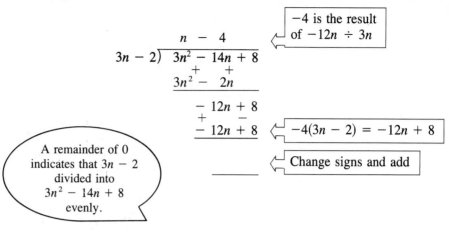

$$n \quad - \quad 4$$
$$3n - 2 \overline{)\ 3n^2 - 14n + 8}$$
$$\underset{+}{} \quad \underset{+}{}$$
$$3n^2 - 2n$$

-4 is the result of $-12n \div 3n$

$$-12n + 8$$
$$\underset{+}{} \quad \underset{-}{}$$
$$-12n + 8$$

$-4(3n - 2) = -12n + 8$

Change signs and add

A remainder of 0 indicates that $3n - 2$ divided into $3n^2 - 14n + 8$ evenly.

0

Therefore,

$$(3n^2 - 14n + 8) \div (3n - 2) = n - 4 \qquad \square$$

Example 36 $\square$ Divide $(x^3 + 1) \div (x + 1)$.

In this case the x^2 and x terms are missing in the dividend $x^3 + 1$. Therefore, supply them, and make their coefficients zero.

The complete dividend is $x^3 + 0x^2 + 0x + 1$.

$0x^2$ and $0x$ are still zero, but they keep the columns straight.

Part 1.

$$x^2$$
$$x + 1 \overline{)\ x^3 + 0x^2 + 0x + 1}$$
$$\underline{\overset{-}{x^3} \overset{-}{+} x^2}$$
$$- x^2 + 0x + 1 \quad \text{Add}$$

Result of $x^3 \div x$

Change signs

Part 2.

$$x^2 - x$$
$$x + 1 \overline{)\ x^3 + 0x^2 + 0x + 1}$$
$$\underline{\overset{-}{x^3} \overset{-}{+} x^2}$$
$$- x^2 + 0x + 1 \quad \text{Add}$$
$$\underline{\overset{+}{-} x^2 \overset{+}{-} x}$$
$$x + 1 \quad \text{Change signs and add}$$

Result of $-x^2 \div x$

Change signs

$-x(x + 1)$

Part 3.

$$\begin{array}{r} x^2 - x + 1 \quad\Longleftarrow \boxed{\text{Result of } x \div x} \\ x + 1 \overline{)\, x^3 + 0x^2 + 0x + 1} \\ \underline{\overset{-}{x^3} \overset{=}{+} x^2} \qquad\qquad \\ -\ x^2 + 0x + 1 \\ \underline{\overset{+}{\pm}\ x^2 \overset{+}{\pm}\ x} \qquad \\ x + 1 \\ \underline{\overset{-}{x} \overset{=}{+} 1} \Longleftarrow \boxed{1(x + 1)} \\ 0 \qquad \text{Change signs and add} \end{array}$$

Our answer is $x^2 - x + 1$.   □

In the preceding examples the remainders were zero. However, in many cases, the remainder is not zero.

Example 37 □ Divide $(7x + x^2 + 15) \div (x + 3)$.

$x^2 + 7x + 15$ First arrange the dividend in descending powers _____

$$\begin{array}{r} x\ + 4 \qquad\quad \\ x + 3 \overline{)\, x^2 + 7x + 15} \\ \underline{\overset{-}{x^2} \overset{=}{+} 3x} \qquad\quad \\ 4x + 15 \\ \underline{\overset{-}{4x} \overset{=}{+} 12} \\ 3 \Longleftarrow \boxed{\text{Remainder}} \end{array}$$

In arithmetic the remainder was frequently written as a fractional part of the divisor. The same procedure is used in algebra.

The answer is written $x + 4 + \dfrac{3}{x + 3}$.   □

Problem Set 5.4B

Divide the following.

1. $(x^2 + 7x + 10) \div (x + 2)$

2. $(x^2 - 7x + 12) \div (x - 3)$

3. $(2x^2 - x - 6) \div (x - 2)$

4. $(3x^2 + 5x - 12) \div (x + 3)$

5. $(2x^2 - 7x + 6) \div (x - 2)$

6. $(4x^2 - 17x - 42) \div (x - 6)$

7. $(3x^2 + 9x - 30) \div (x + 5)$

8. $(5x^2 - 8x - 21) \div (x - 3)$

9. $(6x^2 + 23x + 25) \div (2x + 5)$

10. $(15x^2 + 2x - 18) \div (3x + 4)$

11. $(42x^2 + 17x - 19) \div (7x - 3)$

12. $(12x^2 - 11x - 41) \div (3x + 4)$

13. $(3x^2 - 14x - 24) \div (x - 6)$

14. $(4x^2 + 7x - 15) \div (x + 3)$

15. $(5x^2 + 19x - 30) \div (x + 5)$

16. $(3x^2 - 23x + 30) \div (x - 6)$

17. $(4x^2 + 7x - 16) \div (4x - 5)$

18. $(3x^2 - 14x - 20) \div (3x + 4)$

19. $(3x^2 - 23x + 30) \div (3x - 5)$

20. $(5x^2 + 19x - 30) \div (5x - 6)$

21. $(4x^2 + 4x - 15) \div (2x - 3)$

22. $(6x^2 + 7x - 20) \div (3x - 4)$

23. $(15x^2 - 22x + 10) \div (5x - 4)$

24. $(20x^2 - 31x - 12) \div (4x - 3)$

25. $(x^3 - 3x^2 + 5x - 3) \div (x - 1)$

26. $(x^3 + 6x^2 + 5x - 12) \div (x + 3)$

27. $(x^3 + 6x^2 + 3x - 10) \div (x + 2)$

28. $(x^3 - 3x^2 + 8x - 12) \div (x - 2)$

29. $(2x^3 + 9x^2 - x - 15) \div (2x + 3)$

30. $(3x^3 + 2x^2 - 23x + 20) \div (3x - 4)$

31. $(2x^3 - 3x^2 - 13x + 12) \div (x - 3)$

32. $(3x^3 - 14x^2 + 7x + 4) \div (x - 4)$

33. $(4x^3 - 23x^2 + 19x - 18) \div (x - 5)$

34. $(3x^3 + 22x^2 + 19x - 27) \div (x + 6)$

35. $(4x^3 - 12x^2 + 17x - 14) \div (2x - 3)$

36. $(9x^3 + 3x^2 - 5x + 20) \div (3x + 5)$

37. $(x^3 + 2x^2 + 32) \div (x + 4)$

38. $(x^3 - 12x - 16) \div (x - 4)$

39. $(4x^4 - 11x^2 + 3x - 4) \div (2x - 3)$

40. $(7x^2 + 6x^3 + 4) \div (3x - 1)$

41. $(10x^3 - 4x^2 + 5x) \div (5x - 2)$

42. $(12x^3 - 10x^2 + 5) \div (4x + 2)$

43. $(4x^3 + x + 15) \div (2x + 3)$

44. $(9x^3 + 14x + 12) \div (3x + 2)$

45. $(8x^3 + 22x^2 - 9) \div (4x + 3)$

46. $(9x^3 + 3x^2 - 50) \div (3x - 5)$

47. $(15x^3 - 37x^2 + 18) \div (5x - 4)$

48. $(12x^3 + 25x^2 - 19) \div (3x + 4)$

There are times in college algebra, calculus, and differential equations where we have to do a lot of divisions by a binomial. The process of synthetic division is a quick way to divide. Unfortunately it works only if we are dividing by a binomial with the coefficient of the variable equal to one.

Let's do a long division problem and see how much writing we can cut out.

Example 38 □ Divide $(3x^3 + 10x^2 - 3x + 20) \div (x + 4)$.

$$
\begin{array}{r}
3x^2 - 2x + 5 \\
x + 4\overline{)3x^3 + 10x^2 - 3x + 20} \\
\underline{3x^3 + 12x^2} \\
-2x^2 - 3x \\
\underline{-2x^2 - 8x} \\
5x + 20 \\
5x + 20
\end{array}
$$

Notice the variables are always written in descending order $x^3 x^2 x^1 x^0$.

Writing the problem without the variables

$$
\begin{array}{r}
3 - 2 + 5 \\
+4\overline{)3 + 10 - 3 + 20} \\
\underline{3 + 12} \\
-2 - 3 \\
\underline{-2 - 8} \\
5 + 20 \\
5 + 20
\end{array}
$$

Notice the colored numbers are identical to the numbers above. Therefore we omit writing them.

Instead of bringing numbers down in each step, we subtract directly from the dividend.

$$
\begin{array}{r}
3 - 2 + 5 \\
+4\overline{)3 + 10 - 3 + 20} \\
\underline{+ 12 - 8 + 20} \\
-2 + 5 \quad 0
\end{array}
$$

We can avoid subtraction if we change the sign of the divisor. Then we can add each partial product instead of subtracting.

$$
\begin{array}{r}
3 - 2 + 5 \\
-4\overline{)3 + 10 - 3 + 20} \\
\underline{-12 + 8 - 20} \\
-2 + 5 + 0
\end{array}
$$

> The bottom line would be identical to the top line if we copied the 3 in the blank space on the bottom line.

Writing the problem without the top line and modifying the division bracket yields

$$
\begin{array}{r}
-4\,\big|\,3 + 10 - 3 + 20 \\
\underline{-12 + 8 - 20} \\
3 - 2 + 5
\end{array}
$$

Here's how to do it the quick way.
Divide $x + 4$ into $3x^3 + 10x^2 - 3x + 20$

Step 1. Write the opposite of the constant in the divisor.

$$-4\,\big|$$

Step 2. Write just the coefficients of the dividend. Use 0 for any missing terms.

$$-4\,\big|\,3 \quad + 10 \quad - 3 + 20$$

Step 3. Write the first coefficient below the line.

$$
\begin{array}{l}
-4\,\big|\,3 \quad + 10 \quad - 3 + 20 \\
\hline
3
\end{array}
$$

Step 4. Multiply 3 by the altered divisor (-4). Write the product (-12) in the next column.

$$
\begin{array}{l}
-4\,\big|\,3 \quad + 10 \quad - 3 + 20 \\
- 12 \\
\hline
3
\end{array}
$$

Step 5. Add $+10 + (-12)$. Repeat Step 4 using $(-2)(-4)$.

$$
\begin{array}{l}
-4\,\big|\,3 \quad + 10 \quad - 3 + 20 \\
- 12 \quad + 8 - 20 \\
\hline
3 \quad - 2 \quad + 5
\end{array}
$$

Step 6. Add $(-3) + 8$. Repeat Step 4 using $(5)(-4)$.

Step 7. Add $20 + (-20)$. If this is zero the divisor divides the dividend evenly.

$$
\begin{array}{l}
-4\,\big|\,3 \quad + 10 \quad - 3 + 20 \\
- 12 \quad + 8 - 20 \\
\hline
3 \quad - 2 \quad + 5 + 0
\end{array}
$$

Answer: $\quad 3x^2 - 2x + 5$

> This last number is the remainder

□

In the last example, how did you know the answer was $3x^2 - 2x + 5$ instead of $3x - 2 + \dfrac{5}{x + 4}$?

You must look at the original problem. The degree of the first term of the quotient will always be one less than the degree of the dividend.

Example 39 □ Divide $\dfrac{2x^3 + 4x^2 - 5}{x + 3}$.

$$
\begin{array}{r}
-3\,\lfloor\,2 + 4 + 0 - 5 \qquad \text{Note missing } x \text{ term} \\
\underline{-6 + 6 - 18} \\
2 - 2 + 6 - 23 \qquad \Leftarrow \text{Remainder} = -23
\end{array}
$$

Answer: $2x^2 - 2x + 6 + \dfrac{-23}{x + 3}$ □

Example 40 □ Divide $\dfrac{x^5 - 2x^3 + 4x + 24}{x + 2}$.

$$
\begin{array}{r}
-2\,\lfloor\,1 + 0 \;\; - 2 \;\; + 0 \;\; + 4 + 24 \\
\underline{-2 \;\; + 4 \;\; - 4 \;\; + 8 - 24} \\
1 - 2 \;\; + 2 \;\; - 4 \;\; + 12 \qquad 0
\end{array}
$$

Answer: $\qquad x^4 - 2x^3 + 2x^2 - 4x + 12$ □

Example 41 □ Divide $\dfrac{x^3 - 2}{x - 4}$.

$$
\begin{array}{r}
4\,\lfloor\,1 + 0 + 0 - 2 \\
\underline{+ 4 + 16 + 64} \\
1 + 4 + 16 + 62
\end{array}
$$

Answer: $\qquad x^2 + 4x + 16 + \dfrac{62}{x - 4}$ □

Chapter 5 ■ Polynomials

Problem Set 5.5

Use synthetic division to find the quotient and remainder.

1. $\dfrac{x^2 + x - 6}{x - 2}$

2. $\dfrac{x^2 - x - 6}{x - 3}$

3. $\dfrac{x^2 + 12x + 12}{x + 3}$

4. $\dfrac{7x^2 - 37x + 10}{x - 5}$

5. $\dfrac{x^3 + 4x^2 + x - 6}{x + 2}$

6. $\dfrac{2x^3 - 5x^2 + 7x - 4}{x - 1}$

7. $\dfrac{3x^3 + 4x^2 - 8}{x + 2}$

8. $\dfrac{4x^3 - 41x + 15}{x - 3}$

9. $\dfrac{4x^3 + x^2 + x + 3}{x + 1}$

10. $\dfrac{x^3 - x^2 - 2x + 3}{x - 2}$

11. $\dfrac{x^3 + 3x - 2}{x - 2}$

12. $\dfrac{x^3 - 5x - 2}{x + 2}$

13. $\dfrac{3x^3 + 10x^2 - 5x - 12}{x + 4}$

14. $\dfrac{2x^3 - 8x^2 + 7x - 5}{x - 3}$

15. $\dfrac{3x^3 - 7x^2 + 4x - 5}{x - 2}$

16. $\dfrac{2x^3 + 8x^2 + 3x + 3}{x + 3}$

17. $\dfrac{x^4 + 5x^3 - x^2 - 4x + 5}{x + 5}$

18. $\dfrac{x^4 + 5x^3 + 4x^2 - 2x - 8}{x + 4}$

19. $\dfrac{3x^4 - 12x^3 - x + 4}{x - 4}$

20. $\dfrac{2x^4 - 6x^3 + x^2 - 3x}{x - 3}$

21. $\dfrac{4x^4 + 20x^3 - x^2 - 5x + 3}{x + 5}$

22. $\dfrac{2x^4 - 10x^3 + 3x^2 - 15x - 4}{x - 5}$

23. $\dfrac{3x^4 + 8x^3 + 5x^2 - x - 6}{x + 2}$

24. $\dfrac{2x^4 + 4x^3 - 5x^2 + 4x - 3}{x + 3}$

25. $\dfrac{2x^4 - 8x^3 - 12x^2 + 9x + 7}{x - 5}$

26. $\dfrac{3x^4 - 11x^3 - 5x^2 + 6x - 5}{x - 4}$

27. $\dfrac{4x^4 - 5x^3 + 3x^2 - 4x - 3}{x - 1}$

28. $\dfrac{3x^4 + 8x^3 + 5x^2 - x - 8}{x + 2}$

29. $\dfrac{x^4 - 10x^2 - 2x - 1}{x + 3}$

30. $\dfrac{x^4 - 23x^2 - 12x + 8}{x - 5}$

31. $\dfrac{3x^4 + 5x^3 + 7x - 1}{x + 2}$

32. $\dfrac{2x^4 + 5x^3 + 8x - 6}{x + 3}$

33. $\dfrac{x^3 - 64}{x - 4}$

34. $\dfrac{x^3 + 27}{x + 3}$

35. $\dfrac{x^5 + 1}{x + 1}$

36. $\dfrac{x^5 - 1}{x - 1}$

37. $\dfrac{2x^5 + 2x^2 + 5}{x + 1}$

38. $\dfrac{3x^5 - 6x^3 + 3x^2 - 2}{x - 1}$

39. $\dfrac{x^6 - 8x^4 - 10x^2 + 6}{x - 3}$

40. $\dfrac{x^6 + 6x^3 + 24}{x + 2}$

41. $\dfrac{x^4 - 16}{x + 2}$

42. $\dfrac{x^3 - 8}{x - 2}$

43. $\dfrac{x^6 - 64}{x - 2}$

44. $\dfrac{x^6 - 64}{x + 2}$

45. $\dfrac{2x^6 - 3x^4 + 4x^3 - 2x^2 - 5x}{x - 1}$

46. $\dfrac{3x^6 + x^5 + 4x^3 - 3x^2 + 2}{x + 1}$

47. $\dfrac{x^6 - 4x^5 - 8x^3 - 10x + 1}{x - 3}$

48. $\dfrac{2x^6 - 9x^4 - 9x - 5}{x + 2}$

Chapter 5 □ Key Ideas

5.1

1. **Exponential form—a^n**
   a is the base and n is the exponent
2. **Multiplication of powers with like bases**
   $$a^x \cdot a^y = a^{x+y}$$
3. **Zero exponent**
   $$a^0 = 1, \qquad a \neq 0$$
4. **Division of powers**
   $$\frac{a^x}{a^y} = a^{x-y}$$
5. **Negative exponents in the numerator**
   $$a^{-x} = \frac{1}{a^x} \qquad \text{where } a \neq 0$$

6. **Negative exponents in the denominator**
   $$\frac{1}{x^{-n}} = x^n \qquad \text{where } x \neq 0$$
7. **Power raised to a power**
   $$(a^x)^y = a^{x \cdot y}$$
8. **Power of a product**
   $$(a^x \cdot b^y)^z = a^{x \cdot z} \cdot b^{y \cdot z}$$
9. **Power of quotients**
   $$\left(\frac{a^x}{b^y}\right)^z = \frac{a^{x \cdot z}}{b^{y \cdot z}}$$

5.2 Scientific Notation

To write a number in scientific notation:
1. Write a number as a number with a magnitude between one and ten multiplied by a power of ten.
2. To convert numbers from scientific notation to decimal notation, move the decimal point the number of places indicated by the exponent.
   Positive exponent: Move the decimal point right.
   Negative exponent: Move the decimal point left.
3. Scientific notation is useful when multiplying and/or dividing very large or very small numbers.

5.3
1. A **polynomial** is the sum of one or more terms, each term consisting of a constant or a product of a constant and one or more variables each raised to a whole number power.
2. A **polynomial in one variable** is a polynomial whose terms contain only one variable.
3. A **monomial** is a polynomial with only one term.
4. A **binomial** is a polynomial with exactly two terms.
5. A **trinomial** is a polynomial with exactly three terms.
6. The **distributive property** is used to multiply polynomials.
7. To **multiply two polynomials** you multiply every term of the first polynomial by every term of the second polynomial, and then collect like terms.
8. The **product of two binomials** can be remembered by FOIL.
9. The **square of a binomial** is given by:
$$(a + b)^2 = a^2 + 2ab + b^2$$
10. The **product of the sum and difference** of two terms is the square of the first term minus the square of the second term.
$$(a + b)(a - b) = a^2 - b^2$$
11. The sum of two terms and the difference of two terms are said to be **conjugates** of each other.

5.4
1. The **definition of division** is given by:
$$a \div b = a \cdot \frac{1}{b}$$
2. The definition of division is particularly useful in dividing a polynomial by a monomial.
3. The **degree** of a polynomial in one variable is the highest power used in any of its variables.
4. **To divide polynomials** using long division:
   a. Arrange two terms in the dividend and divisor in order of decreasing powers.
   b. Supply zero coefficients for any missing powers of the variable in the dividend.
   c. Divide as in arithmetic.

5.5 **Synthetic** division can be done if we are dividing by a binomial with coefficient of the variable equal to one.

Chapter 5   Review Test

Simplify the following. Write the result without negative exponents. **(5.1)**

1. $-a^0 \cdot b^0 \cdot (-8ab^2)$

2. $\left(-\dfrac{x^3}{2}\right)^{-2}$

3. $(-2a^2b^{-2})^3(-3a^0b^{-1})^2$

4. $-\dfrac{24x^8y^2}{3x^5y^{-2}}$

5. $\left(\dfrac{x^{-2}y^{-1}c^3}{x^{-1}y^3c^{-2}}\right)^{-3}$

6. $\dfrac{36(2x-1)^0(x+2y)^{-1}}{-9(2x-1)^{-2}(x+2y)^{-3}}$

Evaluate the following for $a = -1$, $b = -2$, $c = 0$, $x = -\dfrac{1}{2}$, $y = \dfrac{2}{3}$. **(5.1)**

7. ab^3x

8. $12x^2y - 9ay - \dfrac{bc}{x}$

9. $\dfrac{6a^3bxy}{a} + \dfrac{9b^2y^2}{4}$

10. $\dfrac{a^3x^{-2}}{b^3} \cdot \dfrac{3b^2x}{a^{-2}y^{-1}}$

Write the following using scientific notation. **(5.2)**

11. 564200   **12.** 0.00031   **13.** 0.04   **14.** 781.921

Write the following without using scientific notation. **(5.2)**

15. 3.0×10^4 **16.** -2.15×10^3 **17.** -4.071×10^{-3} **18.** 6.5×10^{-4}

Find the following products. **(5.3)**

19. $8y(a^3 - 3a)$

20. $-3xy^2(x^3 - y^3)$

21. $(4s - 8)(5s + 2)$

22. $(2x - 3y^2)^2$

23. $(6x - 7y^3)(6x + 7y^3)$

24. $(3x - 1)(x^2 - 2x + 4)$

25. $(a^2 - 2a + 1)(3a^2 + 4a - 5)$

26. $(x^3 - y^3)(x^3 + 2y^3)$

Divide. Write your answers with positive exponents.

27. $\dfrac{12x^3 - 18x^4}{3x^3}$

28. $\dfrac{20x^4 - 3x^3 + 10}{5x^3}$

29. $\dfrac{-6x^3y + 12xy^2 - 18x^2y^4}{-6x^2y^2}$

Use long division to find the following quotients. **(5.3)**

30. $(6x^2 + x - 12) \div (2x + 3)$

31. $(9x^3 - 3x^2 - 20x) \div (3x + 4)$

32. $(24x^2 + 20x^3 - 35x - 45) \div (5x + 6)$

33. $(9x + 8x^3 - 18x^2 + 6) \div (2x - 3)$

Use synthetic division to find the following quotients. **(5.4)**

34. $(x^2 - 6x - 8) \div (x - 3)$

35. $(x^3 + 21x + 11x^2 - 9) \div (x + 3)$

36. $(x^4 - x^2 + 3x) \div (x + 2)$

6 Factoring

Contents

T he last chapter showed how to multiply two polynomials. This chapter will describe the inverse process to multiplication. It is called **factoring.** When it is possible to factor a sum of terms we can express the sum as a product. This skill will be very useful in the next chapter when we add, subtract, multiply and divide algebraic fractions. The factoring skills you learn in this chapter will be used throughout your study of mathematics.

■ 6.1   Removing Monomial Factors, Factoring Trinomials

Monomial Factors

Factors

Algebraic expressions that are multiplied together are called **factors.**

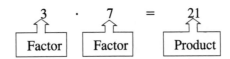

Factoring

The process of finding which factors were multiplied to give a product is called **factoring.**

The first type of factoring we will do is an application of the distributive property. The **distributive property** changes a product into a sum.

Distributive Property

$$a \cdot (b + c) = a \cdot b + a \cdot c$$

In factoring, we change a sum into a product.

$$a \cdot b + a \cdot c = a \cdot (b + c)$$

To factor a polynomial, first remove any monomial factors that are common to every term.

Example 1 □ Factor $ax + ay$.

$$ax + ay = a(x + y)$$

a is common to both terms

The sum of the terms $ax + ay$ was converted into a product of two factors, $a(x + y)$.

□

230

Example 2 □ Factor $3x + 6$.

To work this problem we think of 6 as $3 \cdot 2$.

$$3x + 6 = 3x + 3 \cdot 2$$
$$= 3(x + 2) \qquad \qquad □$$

Frequently, an expression will have more than one common factor in each term. In that case we remove all common factors.

Example 3 □ Factor $18a^3b^2 - 24a^2b + 30ab^2$.

The factor common to each term is $6ab$.

5b

$$18a^3b^2 - 24a^2b + 30ab^2 = 6ab(3a^2b - 4a + \underline{\qquad})$$

How do I find the number to put in the blank space?

Ask yourself, "By what do I multiply $6ab$ to get $30ab^2$ for an answer?"

□

Problem Set 6.1A

Use the distributive property to factor the following.

1. $2x + 4$

2. $3a + 3$

3. $6a - 10$

4. $8x - 12$

5. $5x^2 + 6x$

6. $4a^2 + 3a$

7. $4x^2 - 4x$

8. $3x^2 - 9x$

9. $y^3 - 2y^2 - 5y$

10. $a^3 + 3a^2 + a$

11. $8x^3 - 12x$

12. $15x^4 + 3x^3$

13. $12a^3 - 15a^2$

14. $18a^4 + 12a^3$

15. $21a^2 - 14a^5$

16. $24x^3 - 18x^5$

17. $9x^4 - 27x^3$

18. $6x^3 - 15x^5$

19. $x^3 - 2x^2 + 3x$

20. $3x^3 + 4x^2 - x$

21. $3x^3 - 6x^2 + 9x$

22. $2x^3 - 8x^2 + 6x$

23. $4a^3 - 2a^2 + 6a$

24. $5a^3 + 15a^2 - 10a$

25. $4x^3 - 2x^2 + 10x$

26. $6a^3 - 6a^2$

27. $7y^4 - 14y^2 - 21y$

28. $-5y^5 - 10y^3 + 15y$

29. $bx^2 + 3bx - 5b$

30. $ay^3 + 3ay^2 + 8ay$

31. $2ay^4 + 6ay^3 - 10ay^2$ **32.** $6ax^5 - 9ax^3 + 12ax$ **33.** $6x^3 + 9x^2 - 12x$

34. $10x^3 - 15x^2 + 20x$ **35.** $-8y^5 + 12y^4 - 6y^2$ **36.** $9y^6 - 3y^3 + 12y^2$

37. $ax^4 - ax^3 + 2ax^2$ **38.** $by^5 - 2by^4 - 4by^2$ **39.** $6a^2y^6 - 12a^2y^4 + 15a^2y^3$

40. $8b^2x^6 - 12b^2x^3 + 16b^2x^2$

Factoring Trinomials

In Chapter 5 we multiplied two binomials. Let's look carefully at the result.

Example 4 □ Study these binomials and what happens when they are multiplied.

a) $(x + 2)(x + 3) = x^2 + 5x + 6$
b) $(x + 4)(x + 5) = x^2 + 9x + 20$
c) $(x - 4)(x - 5) = x^2 - 9x + 20$

Notice the form of the binomials above. Each is $(x + \text{constant})$. Let's examine the results in Example 4a term by term.

The first term is always x^2. It results from the product of the first two terms in each binomial factor.

$$(x + \quad)(x + \quad) = x^2 +$$

If both binomials have a first term of x, the resulting product will start with x^2. The last term is the product of the constant terms in each binomial.

$$(\quad + 2)(\quad + 3) = \underline{\quad\quad} + \underline{\quad\quad} + 6$$

In general

$$(x + a)(x + b) = \underline{\quad\quad} + \underline{\quad\quad} + ab$$

The middle term is the sum of the product of the inside terms plus the product of the outside terms.

$$(x + 2)(x + 3) = \underline{\quad\quad} + 5x + \underline{\quad\quad}$$

$$2 \quad x$$
$$3x$$
$$5x$$

In general

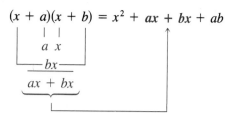

The process of determining which factors were multiplied to yield a trinomial is called **factoring.**

Example 5 □ Factor $x^2 + 7x + 10$.
First, write the trinomial followed by two empty parentheses.

$$x^2 + 7x + 10 = (\qquad)(\qquad)$$

The only way to get a first term of x^2 in the trinomial is if both binomials have first terms of x. Therefore, write an x in both parentheses.

$$x^2 + 7x + 10 = (x\qquad)(x\qquad)$$

Now, look at the last term, 10. It is the product of the final terms in each binomial. Either $2 \cdot 5 = 10$ or $1 \cdot 10 = 10$.

To determine which pair of factors to use, look at the middle term. It is the sum of the inner product plus the outer product. We are looking for the factors of 10 whose sum is 7 and whose product is 10. They are 5 and 2. Write 5 and 2 in the parentheses.

$$x^2 + 7x + 10 = (x + 5)(x + 2)$$

In the general case,

$$(x + a)(x + b) = x^2 + ax + bx + ab$$
$$= x^2 + (a + b)x + ab \qquad □$$

The last term is the product of the constants.

The middle term is the sum of the constants.

Example 6 □ Factor $x^2 - 7x + 10$.

Notice that this trinomial is almost identical to Example 5. The only difference is the sign of the middle term. Write the trinomial followed by two empty parentheses.

$$x^2 - 7x + 10 = (\qquad)(\qquad)$$

Since there is only one way to get a first term of x^2, use the factors of x^2 to fill in the first term of each binomial.

$$x^2 - 7x + 10 = (x \qquad)(x \qquad)$$

Examine the last term. Look for factors whose product is $+10$. The middle term "$-7x$" says the sum of the factors is -7.

Since their product is positive, both factors must have the same sign.
$(-)(-) = +$ or
$(+)(+) = +$

If both factors have the same sign and their sum is negative, they must both be negative.

The factors we are seeking are -5 and -2.

$$(-5)(-2) = +10 \qquad (-5) + (-2) = -7$$

Write the trial result and multiply to check your answer.

$$x^2 - 7x + 10 = (x - 5)(x - 2) \qquad\qquad □$$

Factoring requires the ability to select combinations of terms to yield a desired sum.

Example 7 □ Find two numbers whose product and sum are as follows.

5, −3
5, −3

a. Product $\quad 15 = \underline{\ 3\ } \cdot \underline{\ 5\ }$
Sum $\qquad 8 = \underline{\ 3\ } + \underline{\ 5\ }$

b. Product $\quad -15 = \underline{\quad} \cdot \underline{\quad}$
Sum $\qquad 2 = \underline{\quad} + \underline{\quad}$

−6, 3   −12, −2
−6, 3   −12, −2

c. Product $\quad -18 = \underline{\quad} \cdot \underline{\quad}$
Sum $\qquad -3 = \underline{\quad} + \underline{\quad}$

d. Product $\quad 24 = \underline{\quad} \cdot \underline{\quad}$
Sum $\qquad -14 = \underline{\quad} + \underline{\quad}$

□

Example 8 □ Factor $x^2 + 2x - 15$.

Rewrite with empty parentheses.

$$x^2 + 2x - 15 = (\qquad)(\qquad)$$

Write in the first factors.

$$x^2 + 2x - 15 = (x \qquad)(x \qquad)$$

Find two factors of -15 whose sum is $+2$.

Both $\qquad (-3)(+5) = -15 \qquad$ and $\qquad (+3)(-5) = -15$

but only

$$(-3) + (+5) = +2$$

Write the result and check.

$$x^2 + 2x - 15 = (x + 5)(x - 3)$$
$$= x^2 + 2x - 15 \qquad \qquad \square$$

Example 9 $\square$ Factor $6x^2 - 5xy + y^2$.
Rewrite with empty parenthesis.

$$6x^2 - 5xy + y^2 = (\qquad)(\qquad)$$

This time there are several ways to get the product $6x^2$.

To obtain a product of y^2, check the signs. Since the last term is positive, we know the signs in the factors are alike. Since the middle term is negative, we know they are both negative. First, fill in the second factor in each parenthesis.

$$6x^2 - 5xy + y^2 = (\quad - y)(\quad - y)$$

Find two factors of $6x^2$ that will combine with the factors of y^2 to yield a middle term of $-5xy$.

Since $(-2x)(-3x) = 6x^2$ and $(-2x) + (-3x) = -5x$

$$6x^2 - 5xy + y^2 = (3x - y)(2x - y) \qquad \qquad \square$$

Find two numbers whose product and sum are as follows.

1. Product   6   **2.** Product   6   **3.** Product -6 **4.** Product -6
     Sum   5   Sum -5 Sum   1   Sum -1

5. Product   15   **6.** Product   10   **7.** Product -12 **8.** Product -15 **9.** Product -18
     Sum -8 Sum   7   Sum   1   Sum -2 Sum -3

Factor the following trinomials.

10. $x^2 + 11x + 28$ **11.** $x^2 - 11x + 24$ **12.** $x^2 + 2x - 24$ **13.** $b^2 - 14b + 24$

14. $y^2 - 13y + 30$ **15.** $x^2 - 12x + 32$ **16.** $a^2 - 7a + 12$ **17.** $x^2 - 13x - 30$

18. $x^2 + 8x - 48$ **19.** $x^2 + 12x + 20$ **20.** $m^2 + 12m + 27$ **21.** $x^2 - 2x - 24$

22. $t^2 - t - 42$ **23.** $x^2 + x - 72$ **24.** $x^2 + x - 56$ **25.** $x^2 - 11x - 42$

26. $x^2 + 10x - 56$ **27.** $b^2 - 8b - 65$ **28.** $x^2 - 6x - 72$ **29.** $z^2 + 4z - 77$

30. $x^2 - 18x + 65$ **31.** $y^2 - 17y + 42$ **32.** $x^2 - 17x + 66$ **33.** $x^2 - 6x - 55$

34. $x^2 + 15x + 44$ **35.** $a^2 - 15a + 26$ **36.** $x^2 - 18x + 45$ **37.** $x^2 - 11x + 28$

38. $y^2 - 10y + 21$ **39.** $x^2 + 5x - 36$ **40.** $x^2 - x - 72$ **41.** $x^2 - 6x - 27$

42. $x^2 - x - 20$ **43.** $x^2 + 15x + 26$ **44.** $x^2 + 16x + 28$ **45.** $x^2 - 13x + 36$

46. $x^2 - 16x + 39$ **47.** $a^2 - 17a - 38$ **48.** $a^2 - 10a - 39$ **49.** $x^2 + 13x - 30$

50. $x^2 + 4x - 32$ **51.** $x^2 + 15x + 36$ **52.** $x^2 + 21x + 38$ **53.** $x^2 + 3x - 40$

54. $x^2 + x - 42$ **55.** $x^2 - 15x + 44$ **56.** $x^2 - 14x + 45$ **57.** $x^2 + 15x + 50$

58. $x^2 + 15x + 54$ **59.** $x^2 - 4x - 60$ **60.** $x^2 - 14x - 51$ **61.** $x^2 + 2x - 48$

62. $x^2 + 2x - 63$ **63.** $x^2 + 20x + 64$ **64.** $x^2 + 19x + 70$ **65.** $x^2 - 14x + 48$

■ 6.2   Factoring More Difficult Trinomials

In the previous section, most of the trinomials you were asked to factor had an x^2 term whose coefficient was 1. This is not always the case.

     In general, trinomials in one variable can be written in the form $ax^2 + bx + c$. It is possible to factor trinomials like this if you can find factors of the product $a \cdot c$ whose sum is b.

Example 10 □ Factor $3x^2 + 7x - 20$.

Examine the first and the third terms of this trinomial. If either of these terms has only one set of factors, place these factors in parentheses.

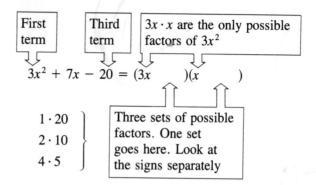

| First term | | Third term | | $3x \cdot x$ are the only possible factors of $3x^2$ |

$$3x^2 + 7x - 20 = (3x \qquad)(x \qquad)$$

$1 \cdot 20$
$2 \cdot 10$
$4 \cdot 5$

Three sets of possible factors. One set goes here. Look at the signs separately

Examine the sign of the third term in $3x^2 + 7x - 20$.

The third term is -20. Therefore the signs of the factored form must be different.

At this point we know the factors of $3x^2 + 7x - 20$ are

$$(3x + \ )(x - \ ) \qquad \text{or} \qquad (3x - \ )(x + \ )$$

Experiment with possible combinations of factors that yield -20 to find the pair that produces the correct middle term.

$$3x^2 + 7x - 20 \overset{?}{=} (3x + 1)(x - 20)$$

$+1x$
$-60x$
$\overline{-59x}$

Look at this result so you can make your next attempt better.

The larger product is negative. The middle term, $+7x$, says that the larger product should be positive. Therefore, try reversing the signs in the factored form.

$$3x^2 + 7x - 20 \overset{?}{=} (3x - 1)(x + 20)$$

$-x$
$+60x$
$\overline{+59x}$

The middle term has the correct sign but is too large. Try smaller factors of 20 such as 4 and 5.

$$3x^2 + 7x - 20 \overset{?}{=} (3x - 4)(x + 5)$$

$-4x$
$+15x$
$\overline{+11x}$

The middle term is still too large. Reverse the factors.

$$3x^2 + 7x - 20 \stackrel{?}{=} (3x + 5)(x - 4)$$

$$5x$$
$$-12x$$
$$-7x$$

The middle term is the right magnitude now but has the wrong sign. Reverse the signs.

$$3x^2 + 7x - 20 \stackrel{?}{=} (3x - 5)(x + 4)$$

$$-5x$$
$$12x$$
$$+7x$$

Yea!

Multiply to check the result.

$$(3x - 5)(x + 4) = 3x^2 + 7x - 20 \qquad \square$$

Example 11 □ Factor $7x^2 - 71x + 10$.

Notice that the only way to get a product of $7x^2$ with positive factors is $7x \cdot x$. However, there are several different combinations whose product is $+10$. Fill in the x terms first.

$$7x^2 - 71x + 10 = (7x \qquad)(x \qquad)$$

Examine the signs.

+ sign in 3rd term	→ Signs are alike

− sign in 2nd term	→ Both signs are −

$$7x^2 - 71x + 10 = (7x - \qquad)(x - \qquad)$$

Experiment with possible combinations of factors that yield $+10$. Since the middle term has a large magnitude, try combinations with one large factor first.

$$10 \cdot 1 = 10$$

$$1 \cdot 10 = 10$$

$$7x^2 - 71x + 10 \overset{?}{=} (7x - 10)(x - 1)$$

Not large enough.

Reverse the factors.

$$7x^2 - 71x + 10 \overset{?}{=} (7x - 1)(x - 10)$$

OK!

Check using the FOIL method.

$$(7x - 1)(x - 10) = 7x^2 - 71x + 10 \qquad \square$$

Example 12 □ Factor $16 - 26x + 3x^2$.

It is easier to recognize factors if we always write trinomials in descending order before we attempt to factor them.

$$16 - 26x + 3x^2 = 3x^2 - 26x + 16$$

There is only one way to get a product of $3x^2$.

$$3x^2 - 26x + 16 = (3x \qquad)(x \qquad)$$

To get the correct middle term use factors of -2 and -8 for 16.

$$3x^2 - 26x + 16 = (3x - 2)(x - 8)$$

Check by using the FOIL method.

$$(3x - 2)(x - 8) = 3x^2 - 26x + 16$$

□

Factor the following.

1. $2x^2 + x - 1$

2. $2x^2 - x - 1$

3. $3x^2 - 5x - 2$

4. $3x^2 + 5x - 2$

5. $7y^2 - 19y + 10$

6. $7y^2 + 3y - 10$

7. $5x^2 + 41x + 8$

8. $5x^2 - 39x - 8$

9. $2 - 7x + 3x^2$

10. $2 + 7x + 3x^2$

11. $3a^2 + 23a + 14$

12. $3a^2 - 23a + 14$

13. $3b^2 - 19b - 14$

14. $3b^2 + 19b - 14$

15. $2x^2 + 13x + 20$

16. $2x^2 + 3x - 2$

17. $1 + 4x + 3x^2$

18. $10 - 17x + 3x^2$

19. $5x^2 + 47x - 30$

20. $5x^2 - 31x + 30$

21. $7a^2 + 12a - 4$

22. $2x^2 - 5x - 12$

23. $2x^2 - 11x + 15$

24. $5x^2 + 17x - 12$

25. $5x^2 - 13x - 6$

26. $7x^2 - 17x - 12$

27. $7x^2 + 25x - 12$

28. $3x^2 + 13x + 12$

29. $3x^2 + 5x - 12$

30. $2x^2 - x - 21$

31. $2x^2 + x - 21$

32. $5x^2 + 29x + 20$

33. $5x^2 - 14x - 24$

34. $7x^2 + 31x + 12$

35. $7x^2 + 16x - 15$

36. $2x^2 + 19x + 35$

37. $2x^2 - 7x - 30$

38. $5x^2 + 8x - 21$

39. $5x^2 - 27x - 18$

40. $7x^2 + 26x + 15$

41. $18 + 45x + 7x^2$

42. $35 - 22x + 3x^2$

It isn't always possible to establish the factors of the x^2 term immediately. When that is the case, check to see if the constant term has only one pair of factors.

Example 13 □ Factor $15x^2 + 32x - 7$.

Disregarding the sign for the moment, the only way to get a product of 7, using integers, is $7 \cdot 1$. Hence we can write the factors of 7 immediately.

$$15x^2 + 32x - 7 = (\qquad 7)(\qquad 1)$$

Examine signs.

| − sign in 3rd term | ⟶ Signs are different |

| + sign in 2nd term | ⟶ Larger magnitude product has + sign |

We still don't know whether the negative sign goes with the 7 or the 1. Let's try

$$15x^2 + 32x - 7 \overset{?}{=} (\quad - 7)(\quad + 1)$$

Experiment with possible combinations of factors that yield $15x^2$.

1st Factor	·	2nd Factor
x	·	$15x$
$15x$	·	x
$3x$	·	$5x$
$5x$	·	$3x$

Just jump in and try the first pair.

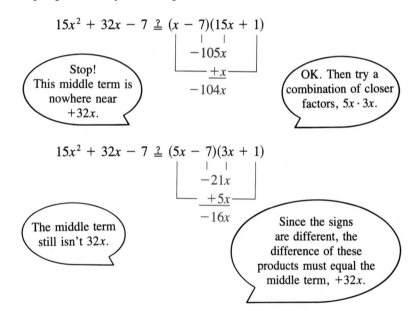

$$15x^2 + 32x - 7 \overset{?}{=} (x - 7)(15x + 1)$$

$$-105x$$
$$+x$$
$$-104x$$

Stop! This middle term is nowhere near $+32x$.

OK. Then try a combination of closer factors, $5x \cdot 3x$.

$$15x^2 + 32x - 7 \overset{?}{=} (5x - 7)(3x + 1)$$

$$-21x$$
$$+5x$$
$$-16x$$

The middle term still isn't $32x$.

Since the signs are different, the difference of these products must equal the middle term, $+32x$.

Reversing first factors

$$15x^2 + 32x - 7 \overset{?}{=} (3x - 7)(5x + 1)$$

$$-35x$$
$$+3x$$
$$-32x$$

That difference is $-32x$. I want $+32x$.

Then just reverse the signs.

Reverse the signs and check your answer.

$$(3x + 7)(5x - 1) = 15x^2 + 32x - 7 \qquad \square$$

Factor the following.

1. $6x^2 + x - 2$ **2.** $6x^2 - x - 2$ **3.** $6x^2 - 7x + 2$ **4.** $6x^2 + 7x + 2$

5. $6x^2 - 7x - 3$ **6.** $6x^2 - 17x - 3$ **7.** $10x^2 - 27x + 5$ **8.** $10x^2 + 49x - 5$

9. $3 - 11y + 8y^2$ **10.** $3 + 10y + 8y^2$ **11.** $9x^2 - 12x - 5$ **12.** $9x^2 + 26x - 3$

13. $7x^2 - 12x + 5$ **14.** $11x^2 - 8x - 3$ **15.** $7x^2 + 12x + 5$ **16.** $11x^2 - 14x + 3$

17. $12a^2 + 17a - 5$ **18.** $12a^2 - 4a - 5$ **19.** $5 + 17a + 12a^2$ **20.** $5 - 19a + 12a^2$

21. $12a^2 - 32a + 5$ **22.** $9x^2 + 9x + 2$ **23.** $10x^2 + 9x + 2$ **24.** $6x^2 + 11x + 3$

25. $20x^2 + 19x + 3$ **26.** $12x^2 + 31x + 7$ **27.** $12x^2 + 25x + 7$ **28.** $6x^2 + 35x + 11$

29. $8x^2 + 26x + 11$ **30.** $15x^2 - 7x - 2$ **31.** $15x^2 + x - 2$ **32.** $20x^2 + 7x - 3$

33. $35x^2 - 8x - 3$ **34.** $12x^2 - 25x - 7$ **35.** $20x^2 + 31x - 7$ **36.** $42x^2 - 29x - 5$

37. $20x^2 + 51x - 11$ **38.** $15x^2 - 28x - 11$ **39.** $12x^2 - 31x + 7$ **40.** $15x^2 - 26x + 7$

41. $5 - 26x + 24x^2$ **42.** $5 - 24x + 16x^2$

Occasionally there are several possible combinations of factors for both the first and the third terms of the trinomial. In this case you simply have to try a few more guesses. However, you are not completely in the dark and, as you gain experience, you'll find it takes fewer tries until you have the trinomial factored or until you know that it's impossible to factor. Consider the following example.

Example 14 □ Factor $20x^2 + 37x - 18$.

Since there are several combinations of factors that will produce 18 and several combinations whose product is $20x^2$, start with a guess at the first terms.

$$20x^2 + 37x - 18 = (5x \qquad)(4x \qquad)$$

This guess may be wrong, but at least I'll know what doesn't work.

You still know something about the signs.

Examine the signs.

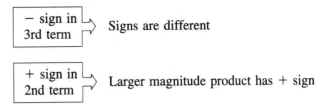

Let's try

$$20x^2 + 37x - 18 \overset{?}{=} (5x + \quad)(4x - \quad)$$

Experiment with combinations of factors that produce 18.

$$\left.\begin{array}{c} 1 \cdot 18 \\ 2 \cdot 9 \\ 3 \cdot 6 \end{array}\right\} \text{All yield 18}$$

Try $3 \cdot 6$

$$20x^2 + 37x - 18 \overset{?}{=} (5x + 3)(4x - 6)$$

$$12x$$
$$-30x$$

Difference $\rightarrow$ $-18x$ is way too small

Therefore, try factors that produce larger products.

$$20x^2 + 37x - 18 \overset{?}{=} (5x + 2)(4x - 9)$$

$$8x$$
$$-45x$$

Difference $\rightarrow$ $-37x$

I got $37x$, but with the wrong sign.

Good; now reverse the signs.

$$20x^2 + 37x - 18 \overset{?}{=} (5x - 2)(4x + 9)$$

Check:

$$(5x - 2)(4x + 9) = 20x^2 + 37x - 18 \qquad \square$$

Example 15 □ Factor $42x^2 - 13x - 40$.

Since the sign of the constant term is negative, we know that the signs in the binomials will be different.

There are many combinations of factors for both 42 and 40, but, since the middle term is moderate, start with factors of about the same size.

$$42x^2 - 13x - 40 = (6x +)(7x -)$$
$$= (6x + 4)(7x - 10)$$

Remember this is only a first guess.

Product with smaller absolute value $\qquad$ $28x$

Product with larger absolute value $\qquad$ $-60x$

Since the middle term is negative arrange the signs so this sum will be negative.

$\overline{-32x}$

Not right. Try again.

$$42x^2 - 13x - 40 = (6x + 5)(7x - 8)$$

Why did you pick 5 and 8?

Since $|-32|$ is large compared to $|-13|$ I went looking for factors of 40 that were close to the same size.

□

Problem Set 6.2C

Factor the following.

1. $6x^2 + 19x + 10$

2. $6x^2 - 19x + 10$

3. $6x^2 + 11x - 10$

4. $6x^2 - 11x - 10$

5. $6x^2 + 13x - 28$

6. $12a^2 - 8a - 15$

7. $4x^2 + 17x - 15$

8. $20y^2 + 13y - 15$

9. $4t^2 + 4t - 15$

10. $20x^2 - 9x - 18$

11. $24x^2 + 47x - 21$

12. $24x^2 + 10x - 21$

13. $4a^2 - 21a - 18$

14. $4a^2 - 9a - 9$

15. $8x^2 + 17x - 21$

16. $12x^2 - 7x - 49$

17. $24x^2 + 14x - 3$

18. $16x^2 + 46x + 15$

19. $6 - 121x + 20x^2$

20. $20 + 29b + 9b^2$

21. $8a^2 - 2a - 15$

22. $4x^2 + 13x - 12$

23. $4x^2 - 4x - 35$

24. $6x^2 + 13x + 6$

25. $6x^2 + 23x + 20$

26. $8x^2 - 18x - 5$

27. $15 - 22x + 8x^2$

28. $10 - 21x + 9x^2$ **29.** $9x^2 - 6x - 8$ **30.** $12x^2 + 23x + 10$

31. $12x^2 - 19x - 21$ **32.** $9x^2 - 31x + 12$ **33.** $15 - 34x + 16x^2$

34. $3 + 16x + 16x^2$ **35.** $16x^2 - 53x + 15$ **36.** $20x^2 + 44x + 21$

37. $20x^2 - 97x - 15$ **38.** $20x^2 - 23x - 21$ **39.** $24x^2 + 31x + 10$

40. $24x^2 - 50x + 21$ **41.** $24x^2 + 38x + 15$ **42.** $24x^2 + 41x + 12$

Example 16 □ Factor $12x^2 - 16x - 60$.

> There are all kinds of factors for 12 and 60!

> True, but remember that step one in factoring is always to remove the common factor. For this case each term contains a 4.

The first step to factor this problem is to remove the common factor using the distributive property.

$$12x^2 - 16x - 60 = 4(3x^2 - 4x - 15)$$

Now it is easier to factor the trinomial

$$= 4(3x + 5)(x - 3) \qquad \square$$

Example 17 □ Factor $10 - 6x^2 - 11x$.
First rearrange in descending order

$$10 - 6x^2 - 11x = -6x^2 - 11x + 10$$

Notice that the second-degree term has a negative coefficient. The easiest way to factor this example is to first factor -1 from each term.

$$-6x^2 - 11x + 10 = -1[6x^2 + 11x - 10]$$
$$= -1[(3x - 2)(2x + 5)]$$
$$= -1(3x - 2)(2x + 5) \qquad \square$$

> Why not write this as $(2 - 3x)(2x + 5)$?

> You could. But, when you work with algebraic fractions it is usually easier to write the factors in descending order with factors of -1 explicitly identified.

Factor the following.

1. $2x^2 + 5x + 3$

2. $2x^2 + 5x - 3$

3. $-2x^2 + 5x + 3$

4. $4x^2 + 6x + 2$

5. $12x^2 - 18x + 6$

6. $10x^2 + 27x + 5$

7. $-10x^2 + 49x + 5$

8. $6x^2 + 29x + 28$

9. $20a^2x^2 + 56a^2x + 15a^2$

10. $-24b^2 + 56b + 10$

11. $32ax^2 + 44ax + 12a$

12. $8x^2 - 10x + 3$

13. $-24ax^2 - 6ax + 45a$

14. $-9ay^2 + 8ay + 20a$

15. $9y^2 + 12y - 5$

16. $24a^2 - 26a - 15$

17. $-12a^2x^2 - 20a^2x + 8a^2$

18. $18x^2y^3 - 39x^2y^2 - 15x^2y$

19. $27x^2 + 18x - 24$

20. $3y + 2xy - 8x^2y$

21. $90 - 15y - 50y^2$

22. $30ax^2 + 22ax + 4a$

23. $24x^4 - 2x^3 - 40x^2$

24. $-12x^5 + 8x^4 + 15x^3$

25. $36ay^3 + 54ay^2 - 10ay$

26. $48x^3y - 112x^2y + 60xy$

27. $21a^2y - 5a^2y^2 - 6a^2y^3$

28. $-5x^2y + 32x^3y - 12x^4y$

29. $36a^4y^2 - 33a^3y^2 + 6a^2y^2$

30. $30a^4x + 65a^3x - 25a^2x$

31. $-24a^4y^3 - 28a^3y^3 + 80a^2y^3$

32. $-9xy^4 - 21xy^3 + 60xy^2$

33. $35a^2 + 13a^2x - 12a^2x^2$

34. $21x^2 - 10x^3 - 24x^4$

35. $36x^2y^3 - 12x^2y^2 - 80x^2y$

36. $72a^2x^3 - 6a^2x^2 - 6a^2x$

Factoring by Grouping

There is a technique that cuts down on the guessing required to factor trinomials. It depends on a process called **factoring by grouping.** Here are a few examples.

Example 18 □ Factor $x^2 - 3x + 4x - 12$ by grouping.
First think of the expression to be factored as a sum of two binomials

$$x^2 - 3x + 4x - 12 = x^2 - 3x \quad + 4x - 12$$

Now remove the common factor from each binomial

$$= x(x - 3) + 4(x - 3)$$

Factor the common binomial which is $(x - 3)$

$$= (x - 3)(x + 4) \qquad \square$$

Example 19 □ Factor $4x^2 + 7x - 20x - 35$.
First divide the expression into two groups

$$4x^2 + 7x - 20x - 35 = 4x^2 + 7x \quad - 20x - 35$$

Factor each group

$$= x(4x + 7) - 5(4x + 7)$$

Factor the binomial

$$= (4x + 7)(x - 5) \qquad \square$$

Example 20 □ Factor $30x^2 - 18x - 25x + 15$.
Make two groups

$$30x^2 - 18x - 25x - 15 = 30x^2 - 18x \quad - 25x + 15$$

Remove common factors

$$= 6x(5x - 3) - 5(5x - 3)$$
$$= (5x - 3)(6x - 5) \qquad \square$$

Problem Set 6.2E

Factor the following by grouping.

1. $x^2 - 2x + 3x - 6$

2. $x^2 - 4x + x - 4$

3. $x^2 - 2x - 3x + 6$

4. $x^2 + 2x - 5x - 10$

5. $2x^2 - 2x + x - 1$

6. $3x^2 - 2x + 6x - 4$

7. $4x^2 - 5x - 8x + 10$

8. $5x^2 + 2x - 15x - 6$

9. $6x^2 - 9x + 8x - 12$

10. $12x^2 - 8x - 15x + 10$

11. $3x^2 - x + 6x - 2$

12. $8x^2 + 6x - 20x - 15$

13. $10x^2 + 15x + 2x + 3$

14. $5x^2 - 4x - 15x + 12$

15. $12x^2 - 10x + 6x - 5$

16. $20x^2 - 28x - 15x + 21$

17. $12x^2 + 2x - 6x - 1$

18. $18x^2 + 15x - 12x - 10$

19. $12x^2 - 20x + 15x - 25$

20. $8x^2 + 28x + 6x + 21$

21. $12x^2 - 10x - 18x + 15$

Here is how to use factoring by grouping to cut down on guessing.

In general, trinomials in one variable can be written in the form $Ax^2 + Bx + C$, where A, B, and C are constants. It is possible to factor trinomials like this if you can find factors of the product $A \cdot C$ whose sum is B.

Let's look at Example 14 again.

Example 21 □ Factor $20x^2 + 37x - 18$.

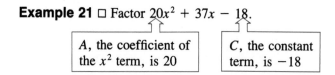

| A, the coefficient of the x^2 term, is 20 | C, the constant term, is -18 |

First find the product of $A \cdot C$. In this example $20(-18) = -360$.

Now we want to find two factors of $A \cdot C$ whose sum is B, the coefficient of the middle term. In this example we are looking for factors of -360 whose sum is $+37$.

List the factors of -360.

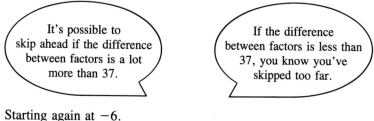

$$-1 \cdot 360$$
$$-2 \cdot 180$$
$$-3 \cdot 120$$
$$-4 \cdot 90$$

Wait a minute. This could take forever!

To speed up the process, notice that we want a positive and a negative factor whose sum is $+37$. Therefore the difference in the absolute values of the factors must be 37.

It's possible to skip ahead if the difference between factors is a lot more than 37.

If the difference between factors is less than 37, you know you've skipped too far.

Starting again at -6.

$$-6 \cdot 60$$
$$-8 \cdot 45 \quad \Leftarrow \boxed{\text{Bingo! } -8 + 45 = +37}$$

Next rewrite the middle term as a sum of two terms using -8 and 45 for coefficients.

$$20x^2 + 37x - 18 = 20x^2 - 8x + 45x - 18$$

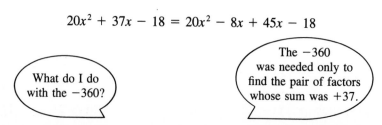

What do I do with the -360?

The -360 was needed only to find the pair of factors whose sum was $+37$.

Next remove the common factor from the first two terms and the last two terms.

$$= 4x(5x - 2) + 9(5x - 2)$$

$$= (4x + 9)(5x - 2) \qquad \text{Factoring the common binomial } (5x - 2)$$

□

Example 22 □ Factor $3x^2 - x - 24$.

The product is $(3)(-24) = -72$.

The coefficient of the middle term (-1) is small. Hence, we will start with two factors that are nearly the same.

$$(8)(-9) \quad \xleftarrow{} \boxed{\text{Sum is } -1}$$

Rewrite the middle term using the coefficients we have found.

$$3x^2 - x - 24 = 3x^2 + 8x - 9x - 24$$

$$= x(3x + 8) - 3(3x + 8)$$

$x - 3, 3x + 8$

$$= (\underline{})(\underline{})$$

□

I could have factored this by inspection.

Fine, do it. Inspection is usually faster, but this method is surer.

Example 23 □ Factor $3x^2 - 34x - 24$.

The product is $(3)(-24) = -72$.

The coefficient of the middle term (-34) is quite large. We will start with one large factor and one small factor.

Rewrite the middle term using the coefficients found.

(1) (-72)

(2) (-36) \qquad Sum is -34

$$3x^2 - 34x - 24 = 3x^2 + 2x - 36x - 24$$

$3x + 2$

$$= x(3x + 2) - 12(\underline{})$$

$(x - 12)(3x + 2)$

$$= \underline{}$$

□

Example 24 □ Factor $3x^2 - 73x + 24$.

The product is $(3)(24) = +72$.

−73

We are now looking for factors of 72 whose sum is ⎯⎯⎯⎯⎯

$$(1)(72) = 72$$
$$(-1)(-72) = 72$$
$$(-1) + (-72) = -73$$

Now we can rewrite the middle term using the coefficients we have found.

$$3x^2 - 73x + 24 = 3x^2 - 1x - 72x + 24$$

3x − 1

$$= x(\underline{\hspace{1cm}}) - 24(3x - 1)$$

(x − 24)(3x − 1)

$$= \underline{\hspace{3cm}}$$

Check by the FOIL method.

$$(x - 24)(3x - 1) = 3x^2 - 73x + 24$$ □

Problem Set 6.2F

Factor the following. Check your answers.

1. $3x^2 + 7x + 2$
2. $3x^2 - 7x + 2$
3. $3x^2 + 5x + 2$
4. $3x^2 - 5x + 2$

5. $3x^2 + x - 2$
6. $3x^2 - x - 2$
7. $3x^2 - 5x - 2$
8. $3x^2 + 5x - 2$

9. $8x^2 + 14x + 3$
10. $8x^2 + 10x - 3$
11. $8x^2 - 10x - 3$
12. $8x^2 - 14x + 3$

13. $8x^2 + 10x + 3$
14. $8x^2 - 10x + 3$
15. $8x^2 + 2x - 3$
16. $8x^2 - 2x - 3$

17. $7y^2 - 17y + 10$
18. $7y^2 + 17y + 10$
19. $7y^2 + 3y - 10$
20. $7y^2 - 3y - 10$

21. $7y^2 - 19y + 10$
22. $5x^2 + 41x + 8$
23. $8 - 25x + 3x^2$
24. $14 + 23a + 3a^2$

25. $3b^2 - 19b - 14$
26. $2x^2 + 13x + 20$
27. $3x^2 - 17x - 6$
28. $5x^2 + 47x - 30$

29. $3b^2 + 19b - 14$
30. $3x^2 - 14x + 15$
31. $6x^2 + 7x - 20$
32. $6x^2 - 5x - 6$

33. $6y^2 + 23y - 4$
34. $4a^2 + 4a - 3$
35. $7x^2 + 19x + 10$
36. $7y^2 - 27y + 18$

37. $8a^2 - 23a - 3$
38. $6a^2 + 19a - 20$
39. $6x^2 + 11x - 10$
40. $6x^2 + 25x + 25$

41. $6x^2 + 5x - 14$
42. $6x^2 + 23x + 15$
43. $5x^2 - 23x + 12$
44. $3x^2 + 19x + 28$

45. $5x^2 - 22x + 8$
46. $4x^2 + 8x - 5$
47. $9x^2 + 6x - 8$
48. $2x^2 + 15x + 18$

49. $3x^2 + 17x + 10$
50. $4x^2 - 23x + 15$
51. $6 - 25x + 4x^2$
52. $10 + 19x + 6x^2$

53. $6x^2 + 17x + 12$ **54.** $8x^2 - 2x - 1$ **55.** $8x^2 - 6x - 9$ **56.** $8x^2 + 9x - 14$

57. $8x^2 - 19x - 15$ **58.** $10 + 13x + 4x^2$ **59.** $12 + 19x + 5x^2$ **60.** $7x^2 + 19x - 6$

61. $7x^2 - 11x - 6$ **62.** $6a^2 + 13a - 28$ **63.** $8a^2 + 2a - 21$ **64.** $9x^2 + 26x - 3$

■ 6.3　　　　　Additional Factoring Techniques

Factoring Trinomial Squares

6.3A　Square of a Binomial

The square of a binomial is the square of the first term plus twice the product of the two terms plus the square of the second term.

$$(a + b)^2 = a^2 + 2ab + b^2$$
$$(a - b)^2 = a^2 - 2ab + b^2$$

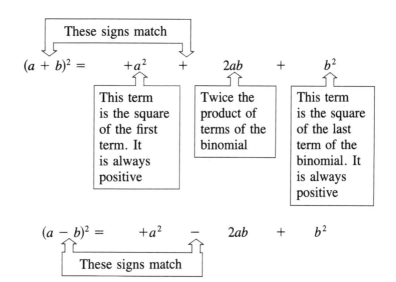

6.3B　Factoring a Perfect Square Trinomial

To factor perfect square trinomials use the general formulas:

$$a^2 + 2ab + b^2 = (a + b)^2$$
$$a^2 - 2ab + b^2 = (a - b)^2$$

where a and b are the numbers that were squared to give a^2 and b^2.

Example 25 □ Study how the following perfect square trinomials are factored.

$$x^2 - 10x + 25 = (x - 5)^2$$
$$x^2 + 10x + 25 = (x + 5)^2$$
$$4x^2 + 12x + 9 = (2x + 3)^2$$

How did you handle the $4x^2$?

$4x^2$ is the square of $2x$, and 9 is the square of 3. The middle term is $2(2x \cdot 3)$ or $12x$. Therefore, we have a perfect square trinomial.

$$9x^2 + 6xy + y^2 = (3x + y)^2 \qquad \square$$

Problem Set 6.3A

Factor the following.

1. $x^2 + 4x + 4$

2. $x^2 - 4x + 4$

3. $x^2 - 8x + 16$

4. $x^2 + 8x + 16$

5. $x^2 + 14x + 49$

6. $x^2 - 14x + 49$

7. $9x^2 - 6x + 1$

8. $9x^2 + 6x + 1$

9. $4x^2 + 20x + 25$

10. $4x^2 - 20x + 25$

11. $x^2 + 2xy + y^2$

12. $x^2 + 4xy + 4y^2$

13. $9x^2 + 12xy + 4y^2$

14. $9x^2 - 12xy + 4y^2$

15. $x^2y^2 - 4xyz + 4z^2$

16. $y^2 + 6y + 9$

17. $a^2 - 6a + 9$

18. $4x^2 - 4x + 1$

19. $4x^2 + 4x + 1$

20. $4x^2 - 28x + 49$

21. $4x^2 + 28x + 49$

22. $a^2 - 4ab + 4b^2$

23. $x^2 + 4xy + 4y^2$

24. $16x^2 - 8xy + y^2$

25. $16a^2 + 8ab + b^2$

26. $25a^2 + 20ab + 4b^2$

27. $25x^2 - 20xy + 4y^2$

28. $x^2y^2 - 6axy + 9a^2$

29. $x^4y^2 - 6a^2x^2y + 9a^4$

30. $4x^2y^4 + 12w^2xy^2 + 9w^4$

Factoring a Difference of Two Squares

Two binomials that are alike except for signs are called **a sum and difference of two terms.** For example, $(x + 4)$ and $(x - 4)$ are a sum and difference of two terms. Multiplying the sum and difference above using the FOIL method, we get

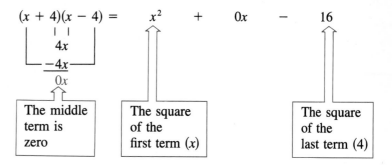

Example 26 □ Multiply $(a + b)(a - b)$.

$$(a + b)(a - b) = \quad a^2 \qquad - \qquad b^2$$

| Square of a | Minus sign | Square of b |

□

6.3C Product of the Sum and Difference of Two Terms

In the product of a sum and difference, the middle term is zero. The result is the difference of the squares of the terms.

$$(a + b)(a - b) = a^2 - b^2$$

Therefore,

6.3D Factoring the Difference of Two Squares

The factors of the difference of two squares are the sum and difference of the numbers being squared.

$$a^2 - b^2 = (a + b)(a - b)$$

Example 27 □ Factor $x^2 - 9$.

$$x^2 - 9 = (x)^2 - (3)^2$$
$$= (x + 3)(x - 3)$$

| Sum | Difference |

□

Example 28 □ Factor $y^2 - 1$.

y, 1

$$y^2 - 1 = (\underline{})^2 - (\underline{})^2$$
$$= (y - 1)(y + 1)$$

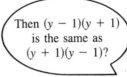

Why didn't we write $(y + 1)(y - 1)$?

You could.
The commutative property of multiplication says it does not matter in what order we multiply factors.

Then $(y - 1)(y + 1)$ is the same as $(y + 1)(y - 1)$?

Yes.

□

Example 29 □ Factor $4x^2 - 25y^2$.

$2x$, $5y$

$$4x^2 - 25y^2 = (\underline{})^2 - (\underline{})^2$$
$$= (2x + 5y)(2x - 5y)$$

□

Next we'll demonstrate that the pattern, not the actual exponents, is the important thing.

Example 30 □ Factor $a^{2n} - b^{4n}$.
Note a^{2n} is $(a^n)^2$.

$$a^{2n} - b^{4n} = (a^n - b^{2n})(a^n + b^{2n})$$

□

The pattern, not the actual exponents, is the important thing.

Example 31 □ Factor $x^4 - 16y^4$.
This can be treated as a difference of two squares if we view things properly.

$$x^4 - 16y^4 = (x^2)^2 - (4y^2)^2$$
$$= (x^2 + 4y^2)(x^2 - 4y^2)$$

That's a difference of two squares.

Since $x^2 - 4y^2$ is still a difference of two squares, we can factor further.

$$x^4 - 16y^4 = (x^2 + 4y^2)(x - 2y)(x + 2y)$$

Can we factor $x^2 + 4y^2$?

No! It's the sum of two squares. See Example 32.

□

Chapter 6 ■ Factoring

Example 32 □ Factor $x^2 + y^2$.

To get x^2, the first term of each binomial must be an x, and to get y^2, the second term of each binomial must be a y.

$$(x \quad y)(x \quad y)$$

If we use opposite signs in the factors, we have the difference of two squares.

$$(x + y)(x - y) = x^2 - y^2$$

If both signs are positive or both are negative, we will get a middle term.

$$(x - y)(x - y) = x^2 - 2xy + y^2$$
$$(x + y)(x + y) = x^2 + 2xy + y^2$$

Since $x^2 + y^2$ does not have a middle term, it cannot be factored. □

Problem Set 6.3B

Factor the following expressions.

1. $x^2 - 4$

2. $x^2 - 25$

3. $x^2 - 36$

4. $4y^2 - 1$

5. $9x^2 - 4$

6. $9x^2 - 16$

7. $y^2 - 49$

8. $a^2 - 64$

9. $16x^2 - 1$

10. $9y^2 - 1$

11. $4x^2 - 25$

12. $9y^2 - 16$

13. $x^2 - y^2$

14. $9a^2 - b^2$

15. $x^2 - 16y^2$

16. $4x^2 - 9y^2$

17. $25x^2 - 4y^2$

18. $49a^2 - 16x^2$

19. $16a^2 - b^2$

20. $25x^2 - y^2$

21. $9x^2 - 25y^2$

22. $36x^2 - 49y^2$

23. $16a^2 - 25b^2$

24. $25s^2 - 36t^2$

25. $(a^2b)^2 - 4$

26. $(ab)^2 - (xy)^2$

27. $x^2y^2 - r^2s^2$

28. $4a^2b^2 - 9x^4y^4$

29. $(x^3y)^2 - 25$

30. $(x^2y^2)^2 - 36$

31. $9a^2b^2 - 4x^4y^4$

32. $25x^4y^4 - 36a^2b^2$

33. $16x^4 - y^4$

34. $a^8 - b^4$

35. $a^8 - b^8$

36. $144a^2 - 49$

37. $a^4 - b^8$

38. $x^8 - y^4$

39. $x^4 - 16y^4$

40. $36x^4y^4 - a^4b^4$

41. $a^{4n} - b^{6n}$

42. $16x^4y^2 - w^4$

43. $16x^4y^4 - w^8$

44. $x^{4n} - 9y^{8n}$

45. $x^{6n} - y^{8n}$

46. $a^{4n} - b^{10n}$

47. $a^8 - 9b^4c^4$

48. $x^{8n} - 16y^{4n}$

Sum and Difference of Two Cubes

Consider the following two products

Product One
$a^2 + \quad ab + b^2$
$\underline{\qquad a - b\qquad}$
$a^3 + a^2b + ab^2$
$\underline{\quad - a^2b - ab^2 - b^3}$
$a^3 \qquad\qquad\quad -b^3$

Product Two
$a^2 - \quad ab + b^2$
$\underline{\qquad a + b\qquad}$
$a^3 - a^2b + ab^2$
$\underline{\quad + a^2b - ab^2 + b^3}$
$a^3 \qquad\qquad\quad +b^3$

These products give us two more easily recognized expressions that can be factored.

6.3E   Factoring the Sum or Difference of Two Cubes

If a and b represent any algebraic expression then

$$a^3 + b^3 = (a + b)(a^2 - ab + b^2)$$
$$a^3 - b^3 = (a - b)(a^2 + ab + b^2)$$

Example 33 □ Factor $x^3 - y^3$.

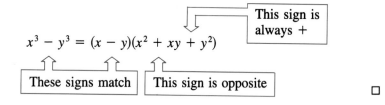

$$x^3 - y^3 = (x - y)(x^2 + xy + y^2)$$

This sign is always +

These signs match | This sign is opposite

□

Example 34 □ Factor $8x^3 + y^3$.

$$8x^3 + y^3 = (2x)^3 + y^3$$
$$= (2x + y)(4x^2 - 2xy + y^2)$$

□

Example 35 □ Factor $p^6 + 64$.

$$p^6 + 64 = (p^2)^3 + (4)^3$$
$$= (p^2 + 4)[(p^2)^2 - 4p^2 + 4^2]$$
$$= (p^2 + 4)(p^4 - 4p^2 + 16)$$

Can we factor this further?

No. Neither $p^2 + 4$ nor $p^4 - 4p^2 + 16$ is factorable. Look at the next example for a similar expression with different results.

□

Example 36 □ Factor $p^6 - 64$.

$$p^6 - 64 = (p^2)^3 - 4^3$$
$$= (p^2 - 4)(p^4 + 4p^2 + 16)$$
$$= (p - 2)(p + 2)(p^4 + 4p^2 + 16)$$

□

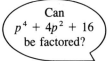

Can $p^4 + 4p^2 + 16$ be factored?

Yes, but it is hard to determine the factors. Check the next example.

Example 37 □ Factor $p^6 - 64$.

Again? We just did that one.

Yes, but this time think of p^6 as $(p^3)^2$.

$$p^6 - 64 = (p^3)^2 - 8^2$$
$$= (p^3 - 8)(p^3 + 8)$$
$$= (p - 2)(p^2 + 2p + 4)(p + 2)(p^2 - 2p + 4)$$
$$= (p - 2)(p + 2)(p^2 + 2p + 4)(p^2 - 2p + 4)$$

Compare the results of this example with the previous one and notice that

$$p^4 + 4p^2 + 16 = (p^2 + 2p + 4)(p^2 - 2p + 4)$$

This is not at all obvious. You should be aware that to factor as the difference of two squares is usually more productive than to factor as the difference of two cubes.

□

Factor the following expressions.

1. $y^3 + 27$ **2.** $x^3 + 8$ **3.** $27x^3 - 8$ **4.** $8a^3 - 27$ **5.** $27a^3 + 8$

6. $8x^3 + 27$ **7.** $27x^3 + 64$ **8.** $64a^3 - 27$ **9.** $a^3 + (2b)^3$ **10.** $a^3b^3 + 125$

11. $125a^3 - c^3d^3$ **12.** $8x^6 - y^3$ **13.** $x^3 + (3y)^3$ **14.** $(2x)^3 + (3y)^3$ **15.** $x^3y^3 - 64a^3$

16. $64a^3b^3 - x^6y^6$ **17.** $64y^6 + 125x^6$ **18.** $27a^6 + b^6c^6$ **19.** $64x^6 - y^6$

20. $a^6b^6 - 64c^6$ **21.** $64x^6 + y^6$ **22.** $a^6 + 64b^6c^6$ **23.** $(ab)^3 - c^6$

24. $(xy)^3 - (wz)^3$ **25.** $64a^6 + b^6$ **26.** $125x^6 + 8y^6$ **27.** $27a^6 + 64b^6$

28. $8a^6b^6 - 27x^6$ **29.** $64a^6 - 125x^6y^6$ **30.** $(xy)^6 - a^3$ **31.** $(xy)^3 - 8a^6$

32. $8a^6b^6 + 27x^6$ **33.** $27x^6y^3 + 8a^3b^6$ **34.** $(xy)^3 + (3a)^6$ **35.** $(2a)^3 + (wy)^6$

36. $(4ab)^3 - (3xy)^6$ **37.** $(ab)^6 + (2c)^6$ **38.** $x^3 - 0.001$ **39.** $x^3 + 0.125$

40. $x^6 - 0.008$ **41.** $x^6 - 27y^3$ **42.** $a^3b^3 - 8c^3$ **43.** $a^3b^6 + 27c^6$

44. $a^6 - 0.125$ **45.** $x^6 - 0.001$ **46.** $a^3 + 0.008b^3$ **47.** $x^6 + 0.125y^3$

48. $(2a)^3 - 0.027$ **49.** $0.064 - (3x)^3$

■ 6.4   Other Useful Factoring Techniques

We are interested in factoring because if we can write an equation as a product set equal to zero we can solve it. Factoring also allows us to simplify many expressions. Unfortunately most polynomials are not factorable. Nevertheless, we would like to help you spot those that are.

Factoring Trinomial Type Expressions

Sometimes a complex expression can be written in the form of a trinomial we can factor. In order to factor a trinomial into two binomials we must write it so that the exponent of the first term is twice the exponent of the second term. Using a box to represent the variable, the pattern is as follows.

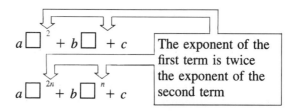

It may be possible to factor expressions that fit the pattern above.

Example 38 □ Factor $y^4 + 6y^2 + 5$.

$$y^4 + 6y^2 + 5 = (\ y)^2 + 6(\ y)^1 + 5$$

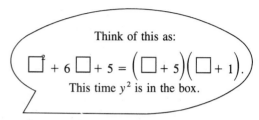

Think of this as:

$$\square^2 + 6\,\square + 5 = \left(\square + 5\right)\left(\square + 1\right).$$

This time y^2 is in the box.

Factoring the trinomial with y^2 as a variable yields

$$y^4 + 6y^2 + 5 = (y^2 + 5)(y^2 + 1)$$ □

Example 39 □ Factor the following.

p^5, p^5

a. $7p^{10} - 11p^5 - 6 = 7\square^2 - 11\square - 6$

p^5, p^5

$$= \left(\square - 2\right)\left(7\square + 3\right)$$

$$= (p^5 - 2)(7p^5 + 3)$$

b. $15x^2y^2 - 14xy - 8 = (5xy + 2)(3xy - 4)$

c. $(x - 2)^2 + 3(x - 2) - 4 = [(x - 2) - 1][(x - 2) + 4]$

$$= (x - 3)(x + 2)$$

Sometimes it is easier to visualize a pattern by substituting a letter instead of a box for a complicated expression. □

Factoring by
Substitution

Example 40 □ Factor $x^6 - 9x^3 + 8$.

$$\text{If we let} \quad u = x^3$$
$$\text{then} \quad u^2 = x^6$$

Substitute these values of x^3 and x^6 in the original algebraic expression.

$$x^6 - 9x^3 + 8 = u^2 - 9u + 8$$

Factoring

$$= (u - 8)(u - 1)$$

Substituting $u = x^3$ back again

$$= (x^3 - 8)(x^3 - 1)$$
$$= (x - 2)(x^2 + 2x + 4)(x - 1)(x^2 + x + 1)$$ □

Example 41 □ Factor $x^{-4} - 5x^{-2} + 4$.

Since the negative exponents appear to complicate the problem, make the following substitution.

$$\text{Let} \quad u = x^{-2}$$
$$u^2 = x^{-4}$$

Substituting

$$x^{-4} - 5x^{-2} + 4 = u^2 - 5u + 4$$

Factoring

$$= (u - 4)(u - 1)$$

Substituting $u = x^{-2}$ back again

$$= (x^{-2} - 4)(x^{-2} - 1)$$

Substitution can be used one more time

$$\text{Let} \quad w = x^{-1}$$
$$w^2 = x^{-2}$$

Substituting

$$(x^{-2} - 4)(x^{-2} - 1) = (w^2 - 4)(w^2 - 1)$$
$$= (w - 2)(w + 2)(w - 1)(w + 1)$$

Substituting $w = x^{-1}$ back again

$$= (x^{-1} - 2)(x^{-1} + 2)(x^{-1} - 1)(x^{-1} + 1) \qquad \square$$

When do I use u or w to substitute?

It doesn't matter which letter you use for the substitution. Pick any letter that you will not confuse with the original variable; u, v, and w are common variables for substitution.

Problem Set 6.4A

Factor the following expressions completely.

1. $y^4 + 3y^2 + 2$

2. $x^4 + 5x^2 + 4$

3. $x^4 - 5x^2 + 4$

4. $x^4 - 5x^2 - 6$

5. $6x^4 + x^2 - 2$

6. $8y^4 + 2y^2 - 3$

7. $x^4 - 9x^2 + 8$

8. $x^6 + 5x^3 + 4$

9. $x^6 - 9x^3 + 8$

10. $x^6 + 9x^3 + 8$

11. $x^4 - 8x^2 + 16$

12. $x^4 + 8x^2 + 16$

13. $16x^4 - 81$

14. $16x^4 - 72x^2 + 81$

15. $8x^6 - 10x^3 - 3$

16. $8x^6 + 31x^3 - 4$

17. $x^4 + x^2 - 20$

18. $x^4 - 8x^2 - 9$

19. $2x^6 - 5x^3 - 12$

20. $3x^6 + 11x^3 - 20$

21. $2x^4 - 17x^2 - 9$

22. $3x^4 - 7x^2 - 20$

23. $3x^6 + x^3 - 2$

24. $2x^6 - 13x^3 - 24$

25. $4x^4 - 25x^2 + 36$

26. $9a^4 - 13a^2 + 4$

27. $24x^6 + 67x^3 + 8$

28. $27x^6 - 53x^3 - 2$

29. $81x^4 - 72x^2 + 16$

30. $16x^4 - 8x^2 + 1$

31. $8x^6 - 65x^3 + 8$

32. $8x^6 + 65x^3 + 8$

33. $2(x + 1)^2 + 3(x + 1) + 1$

34. $6(2x + 1)^2 + 7(2x + 1) - 3$

35. $3(3x - 2)^2 - 11(3x - 2) - 4$

36. $6(3x - 4)^2 + 5(3x - 4) - 6$

37. $(x^2 + 5x)^2 - 2(x^2 + 5x) - 24$

38. $(x^2 - 3x)^2 - 2(x^2 - 3x) - 8$

39. $(2x^2 + 3)^2 - (2x^2 + 3) - 2$

40. $(3x^2 - 4x)^2 - 3(3x^2 - 4x) - 4$

41. $(x^2 - 5x)^2 - 36$

42. $(x^2 + 2x)^2 - 11(x^2 + 2x) + 24$

43. $(3x^2 + 2x)^2 - 9(3x^2 + 2x) + 8$

44. $(2x^2 + 7x)^2 + 2(2x^2 + 7x) - 24$

45. $3x^{-2} - x^{-1} - 2$

46. $6x^{-2} + 13x^{-1} - 5$

47. $10x^{-2} + 11x^{-1} - 6$

48. $5x^{-2} + 9x^{-1} - 2$

Advanced Factoring by Grouping

Before we extend the idea of factoring by grouping, let us use the ideas from Section 6.2 to work three examples.

Example 42 ☐ Factor $ax + ay - xb - yb$.

$$ax + ay - xb - yb = (ax + ay) + (-bx - by) \quad \text{Grouping}$$
$$= a(x + y) - b(x + y) \qquad \text{Factoring each group}$$
$$= (a - b)(x + y) \qquad \text{After factoring } (x + y)$$
☐

Example 43 ☐ Factor $3x + 12 + x^2y + 4xy$.

$$3x + 12 + x^2y + 4xy = (3x + 12) + (x^2y + 4xy) \qquad \text{Grouping}$$
$$= 3(x + 4) + xy(x + 4) \qquad \text{Factor each group}$$
$$= (3 + xy)(x + 4) \qquad \text{Factoring } (x + 4)$$
☐

Example 44 ☐ Factor $4x^2 + 7xy - 20x - 35y$.

$$4x^2 + 7xy - 20x - 35y = (4x^2 + 7xy) + (-20x - 35y)$$
$$= x(4x + 7y) - 5(4x + 7y)$$
$$= (x - 5)(4x + 7y) \qquad \text{☐}$$

So far we have been making groups of two terms each. Sometimes the appropriate group is a trinomial.

Example 45 ☐ Factor $x^2 - 10x + 25 - 16b^2$.

$$x^2 - 10x + 25 - 16b^2 = (x^2 - 10x + 25) - 16b^2 \qquad \text{Factoring the trinomial}$$

$$= (x - 5)^2 - 16b^2 \qquad \text{A difference of squares}$$

$$= (x - 5 + 4b)(x - 5 - 4b) \qquad \text{☐}$$

Example 46 ☐ Factor $x^2 - y^2 - 4x + 4$.

Notice that removing the common factor of 4 from the last two terms leads nowhere. Let's rearrange terms so there is a trinomial in x.

$$x^2 - y^2 - 4x + 4 = x^2 - 4x + 4 - y^2 \qquad \text{Rearranging}$$
$$= (x - 2)^2 - y^2 \qquad \text{Factoring the trinomial}$$
$$= [(x - 2) + y][(x - 2) - y] \qquad \text{Difference of squares}$$
$$= (x + y - 2)(x - y - 2) \qquad \text{Rearranging} \quad \text{☐}$$

Example 47 □ Factor $3x^3 + 9x^2 + 8x + 24$.

Taking the first three terms as a trinomial and removing the common factor of x

$$3x^3 + 9x^2 + 8x + 24 = x(3x^2 + 9x + 8) + 24$$

This leads nowhere. Try grouping two binomials.

$$3x^2 + 9x^2 + 8x + 24 = (3x^3 + 9x^2) + (8x + 24)$$

$x + 3, x + 3$

$$= 3x^2(\underline{\hspace{1cm}}) + 8(\underline{\hspace{1cm}})$$
$$= (3x^2 + 8)(x + 3) \qquad □$$

Problem Set 6.4B

Factor the following expressions.

1. $xy - 4y - 2x + 8$

2. $xy - 3x + y - 3$

3. $ax + 4x - ay - 4y$

4. $2xy - y - 2x^2 + x$

5. $30r^2 + 5r^3 - 12rs - 2r^2s$

6. $2a^2b - 6a^2 - b^2 + 3b$

7. $ab + 4a - 3b - 12$

8. $st + 6t - 5s - 30$

9. $ax + 4a - bx - 4b$

10. $rt - 5r + st - 5s$

11. $12x^3 - 9x^2 - 4x^2y + 3xy$

12. $2ab^3 + 4b^4 - 5ab^2 - 10b^3$

13. $c^2x - d^2x + c^2y - d^2y$

14. $ax^2 - ay^2 - bx^2 + by^2$

15. $2r^2x - 2s^2x + r^2y - s^2y$

16. $3ax^2 - 3ay^2 + bx^2 - by^2$

17. $ax^2 - 4ay^2 + bx^2 - 4by^2$

18. $4a^2x - b^2x - 4a^2y + b^2y$

19. $x^3a + y^3a - 2x^3 - 2y^3$

20. $a^3x - b^3x + 3a^3 - 3b^3$

21. $x^3a^2 - x^3b^2 + y^3a^2 - y^3b^2$

22. $a^3x^2 - b^3x^2 - a^3y^2 + b^3y^2$

23. $a^3x - b^3x - 5a^3 + 5b^3$

24. $rx^3 - ry^3 + 2x^3 - 2y^3$

25. $8ax^3 - ay^3 - 32x^3 + 4y^3$

26. $bx^3 + x^3 - 27by^3 - 27y^3$

27. $32a^3x^2 - 8a^3y^2 - 4b^3x^2 + b^3y^2$

28. $a^5 - 8a^2b^3 - 4a^3b^2 + 32b^5$

29. $3ax^2 + 3ay^2 + 4x^2 + 4y^2$

30. $2a^2x + 2b^2x - 3a^2 - 3b^2$

31. $a^2x^2 - 4a^2 + b^2x^2 - 4b^2$

32. $a^2x^2 - b^2x^2 - 4a^2 + 4b^2$

33. $6ax^2 + by^2 + 2ay^2 + 3bx^2$

34. $12a^2x - 3b^2y + 9b^2x - 4a^2y$

35. $a^2x^2 - 9y^2 + a^2y^2 - 9x^2$

36. $2a^2x^2 - 16b^2 + b^2x^2 - 32a^2$

37. $a^2x^2 + 16b^2 - 4b^2x^2 - 4a^2$

38. $4x^2 + 9a^2y^2 - 36y^2 - a^2x^2$

39. $x^2 + 4x + 4 - 9b^2$

40. $x^2 - 6x + 9 - 4a^2$

41. $x^2 - 9a^2 - 2xy + y^2$

42. $x^2 - 25b^2 + 2xy + y^2$

43. $16x^2 - a^2 + 2ab - b^2$

44. $25a^2 - x^2 - 2xy - y^2$

45. $4x^2 - 4a^2 - 4x + 1$

46. $x^2 - 6xy + 9y^2 - 9a^2$

47. $16x^2 - a^2 - 4b^2 - 4ab$

48. $9a^2 - y^2 + 4xy - 4x^2$

Factoring Completely

This chapter began by saying that the first step in factoring is to use the distributive property to remove common factors. To factor an expression completely, you also use the distributive property to remove any common factors first.

Example 48 □ Factor $25a^3 - 40a^2 - 20a$.

Each term has a common factor of _____

5a

$5a^2 - 8a - 4$

$$25a^3 - 40a^2 - 20a = 5a(\underline{\hspace{2cm}})$$ Remove common factor of $5a$

$$= 5a(5a + 2)(a - 2)$$ Factor the trinomial   □

6.4   Definition of Factored Completely

An expression is **factored completely** if it is written as a product of polynomials, such that:

1. The coefficients in each polynomial factor are integers.
2. No polynomial factor can be factored further by using integer coefficients.

Example 49 □ Factor completely $4x^2y^2 - 4x^4y^2$.

You may be tempted to treat this as a difference of squares. But it's always best to remove common monomial factors first. Each term has a common monomial factor of _____

$4x^2y^2$

$$4x^2y^2 - 4x^4y^2 = 4x^2y^2(1 - x^2)$$ Remove common factor $4x^2y^2$

$$= 4x^2y^2(1 + x)(1 - x)$$ Factor the difference of two squares   □

Example 50 □ Factor completely $5x^6 + 10x^5 - 40x^3 - 80x^2$.

$$5x^6 + 10x^5 - 40x^3 - 80x^2$$

$$= 5x^2(x^4 + 2x^3 - 8x - 16)$$ Remove common factor $5x^2$

$$= 5x^2[x^3(x + 2) - 8(x + 2)]$$ Factor by grouping

$$= 5x^2[(x + 2)(x^3 - 8)]$$ Remove common factor $(x + 2)$

$$= 5x^2(x + 2)(x - 2)(x^2 + 2x + 4)$$ Factor difference of cubes   □

Problem Set 6.4C

Factor completely.

1. $9ax^2 - 15ax - 6a$

2. $10x^4 - 15x^3 - 45x^2$

3. $4x^2 - 16$

4. $9y^2 - 81$

5. $128x^4 - 72x^2$

6. $72a^2b^2x^2 - 32a^2b^2$

7. $81x^4 - 16y^8$

8. $4x^4 - 64y^4$

9. $6x^4 - 15x^3 - 36x^2$

10. $-20a^3b - 85a^2b + 75ab$

11. $4x^2 - 64$

12. $6x^4 - 216$

13. $54x^4 - 24x^2$

14. $64a^4b^2 - 4a^2b^4$

15. $16x^8 - 81y^4$

16. $162x^5 - 2xy^8$

17. $30x^4 + 5x^3 - 60x^2$

18. $56ax^5 + 56ax^4 - 105ax^3$

19. $36x^4y^2 + 42x^3y^3 - 120x^2y^4$

20. $36x^4y^2 - 3x^3y^3 - 60x^2y^4$

21. $12x^6 + 24x^5 + 16x^4 + 32x^3$

22. $5x^8 - 5x^6 - 5x^5 + 5x^3$

23. $24x^4y^3 - 81xy^8$

24. $32x^7 - 128x^5 + 4x^4 - 16x^2$

25. $135x^9y + 40x^3y^4$

26. $x^5 - 4x^3y^4 + 8x^2y^3 - 32y^7$

27. $20x^5y^2 - 60x^4y^2 + 5x^2y^2 - 15xy^2$

28. $12x^4y - 6x^3y^2 + 18x^2y^2 - 9xy^3$

29. $24x^4y - 12x^3y^2 + 16x^3y^3 - 8x^2y^4$

30. $2ab^2x^4 - 2ab^2x^3 + 16ab^2x - 16ab^2$

31. $3b^3x^4 - 6b^3x^3 - 24b^3x + 48b^3$

32. $8x^5 - 72x^3y^2 - x^2y^3 + 9y^5$

33. $63ax^4 + 63ax^3 - 54ax^2$

34. $11bxy^6 - 22by^4 + 33by^3$

35. $42tx^5 - 49tx^4 - 21tx^3$

36. $48ab^4 + 84ab^3 + 18ab^2$

37. $4x^7 - 8x^5 + 32x^4 - 64x^2$

38. $ax^8 - 9ax^6 - 27ax^5 + 243ax^3$

39. $3x^5 - 81x^2y^4$

40. $32x^6 - 4x^3y^5$

41. $30x^4y + 35x^3y^3 - 15x^2y^5$

42. $36x^3y^3 + 30x^2y^5 - 36xy^7$

43. $108x^5 + 4x^2y^3 - 27x^3y^4 - y^7$

44. $8x^5 - 72x^3y^4 - x^2y^3 + 9y^7$

45. $3abx^4 - 9abx^3 - 81abx + 243ab$

46. $a^5x^2 + 3a^4x^2 + 27a^2x^2 + 81ax^2$

47. $12x^5y^2 - 6x^4y^3 + 48x^2y^5 - 24xy^6$

48. $48x^6y^2 - 16x^5y^3 + 12x^3y^5 - 4x^2y^6$

Chapter 6 □ Key Ideas

6.1 1. **Factors** are algebraic expressions that are multiplied together.
  2. The **distributive property** changes a product into a sum.
  3. **Factoring** changes a sum into a product.
  4. To **factor a monomial from a polynomial** remove any monomial factors that are common to every term.
  5. To **factor a trinomial** that has one as the coefficient of the squared term, find two numbers whose product is the constant term and whose sum is the coefficient of the first-degree term.

6.2 1. The **general form** of a trinomial is $ax^2 + bx + c$.
  2. When factoring a trinomial of the form $ax^2 + bx + c$:
    a. the product of the first term is ax^2
    b. the product of the last term is c
    c. the sum of the inner product and the outer product is bx
  3. Polynomials with four terms are generally factored, if possible, by grouping.
  4. A trinomial of the form $Ax^2 + Bx + C$, where A, B, and C are constants, can frequently be factored if you can find factors of the product $A \cdot C$ whose sum is B.

6.3 **1.** To **factor a trinomial square:**
$$a^2 + 2ab + b^2 = (a + b)^2 \quad \text{OR} \quad a^2 - 2ab + b^2 = (a - b)^2$$
2. To **factor the difference of two squares:**
$$a^2 - b^2 = (a + b)(a - b)$$
3. A **sum of squares** like $a^2 + b^2$ is not factorable.
4. To **factor the sum of two cubes:**
$$a^3 + b^3 = (a + b)(a^2 - ab + b^2)$$
5. To **factor the difference of two cubes:**
$$a^3 - b^3 = (a - b)(a^2 + ab + b^2)$$

6.4 **1.** To factor expressions that are quadratic in form, write the expression in the form $a(\quad)^2 + b(\quad) + c$ where the same expression appears in each of the parentheses.
2. Expressions that contain four or more terms are generally factored by grouping. In general, each group may contain two or three terms.
3. To factor an expression completely
   a. use the distributive property to remove any common factors;
   b. apply one of the following methods to the remaining factors: 1) the difference of two squares, 2) the sum or difference of two cubes, 3) trinomials, or 4) grouping.

Chapter 6   Review Test

Factor the following completely.

1. $6a^2x^5 - 8ax^4 + 12a^3x^3$

2. $-6x^4 - 12x^3 + 18x$

3. $x^2 + x - 12$

4. $ax^4 - 10ax^3 + 24ax^2$

5. $2x^2 + 13x + 15$

6. $6a^4b - a^3b - 2a^2b$

7. $6x^2 - x - 12$

8. $10ax^4 + 3ax^3 - 18ax^2$

9. $16x^2 - 25$

10. $27x^3 - 48x$

11. $4x^2 + 20x + 25$

12. $16x^4 - 40x^2 + 25$

13. $9x^4 - 24x^3y + 16x^2y^2$

14. $4x^4y^2 - 20x^2y^3 + 25y^4$

15. $8x^3 - 27$

16. $125x^3 - 64y^6$

17. $x^4 - 10x^2 + 9$

18. $x^4 - 16y^4$

19. $x^6 + 7x^3 - 8$

20. $x^6 - 26x^3y^3 - 27y^6$

21. $(x + 2)^2 + (x + 2) - 6$

22. $2(2x - 1)^2 + 7(2x - 1) - 4$

23. $(x^2 + 5x)^2 - 2(x^2 + 5x) - 24$

24. $(x^2 - 3x)^2 - 2(x^2 - 3x) - 8$

25. $ax + 2bx - 2ay - 4by$

26. $ax + 7a - 3bx - 21b$

27. $x^3 + x^2 - 9x - 9$

28. $2x^3 - 8x + 3x^2 - 12$

7 Rational Expressions

Contents

Preview

This chapter will build on many of the skills you learned in the last two chapters to perform operations on rational expressions. Rational expressions are algebraic fractions that have numerators and denominators that are polynomials.

The first three sections of this chapter will develop skills manipulating rational expressions. The fourth section will show how to solve equations that involve rational expressions. The final section of this chapter will give you a chance to apply the skills you have developed.

■ 7.1   Multiplication and Division of Rational Expressions

A rational expression is an algebraic fraction made up of polynomials. Since every polynomial, no matter how complex, evaluates to a single rational number, all the rules in Chapter 2 for operations with rational numbers apply.

Multiplication of Rational Expressions

Algebraic fractions are multiplied the same way as arithmetic fractions are multiplied.

7.1A   Rule: Multiplication of Rational Expressions

To multiply two rational expressions, multiply their numerators to get the numerator of the product; then multiply the denominators to get the denominator of the product.

$$\frac{a}{b} \cdot \frac{c}{d} = \frac{a \cdot c}{b \cdot d}$$

where a, b, c, d represent polynomials with b and d not equal to zero.

Example 1 □ Multiply $\dfrac{2}{3} \cdot \dfrac{5}{7}$.

$$\frac{2}{3} \cdot \frac{5}{7} = \frac{10}{21}$$

□

Example 2 □ Multiply $\dfrac{2}{3} \cdot \dfrac{a}{b}$.

$$\frac{2}{3} \cdot \frac{a}{b} \quad = \quad \frac{2 \cdot a}{3 \cdot b} = \frac{2a}{3b}$$

| These are two fractions with one multipli- cation sign | This is one fraction with two multipli- cation signs |

□

Example 3 □ Multiply $\dfrac{1}{1} \cdot \dfrac{x}{y}$.

$$\frac{1}{1} \cdot \frac{x}{y} = \frac{1 \cdot x}{1 \cdot y} = \frac{x}{y}$$

This doesn't look like much, but it's one of the best procedures we have: $\dfrac{1}{1} = 1$.

Recall that the identity element for multiplication is 1, and multiplying a number by 1 does not change the number. □

Example 4 □ Multiply $\dfrac{4}{4} \cdot \dfrac{x}{y}$.

$$\frac{4}{4} \cdot \frac{x}{y} = \frac{4 \cdot x}{4 \cdot y} = \frac{4x}{4y}$$ □

Hey! Isn't that the same routine as Example 3?

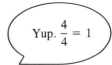

Yup. $\dfrac{4}{4} = 1$

7.1B Definition: Equality of Rational Expressions

If a, b, c, d are polynomials with b, $d \neq 0$, then

$$\frac{a}{b} = \frac{c}{d}$$

if $\dfrac{c}{d}$ equals $1 \cdot \dfrac{a}{b}$.

Example 4 illustrates this definition of equality.
Since $\dfrac{4}{4}$ is 1, we can say that $\dfrac{4x}{4y} = \dfrac{x}{y}$.

Reducing Fractions to Lowest Terms

7.1C Definition: Lowest Terms

A rational expression is reduced to **lowest terms** when the numerator and the denominator contain no common factors.

Rational expressions are reduced in the same manner as arithmetic fractions. We use the old "multiply by 1" technique.

We know that $\dfrac{y}{y} = 1$. Let's use this to reduce the fraction $\dfrac{a \cdot y}{b \cdot y}$.

$$\frac{a \cdot y}{b \cdot y} = \frac{a}{b} \cdot \frac{y}{y}$$

$$= \frac{a}{b} \cdot 1$$

$$= \frac{a}{b}$$

This is just using the multiply by 1 technique backwards.

Rather than write all this, we usually just divide out like factors in the numerator and denominator.

$$\frac{a \cdot y}{b \cdot y} = \frac{a \cdot \overset{1}{\cancel{y}}}{b \cdot \underset{1}{\cancel{y}}} = \frac{a}{b}$$

Sometimes the word **cancel** is used to describe dividing out, but remember, we aren't eliminating anything. We are really just identifying multiplication by 1.

Example 5 □ Reduce $\dfrac{x^4 y^2 z}{x^2 y z^3}$.

First we identify the multiplications by 1.

$$\frac{x^4 y^2 z}{x^2 y z^3} = \boxed{\frac{x^2}{x^2}} \cdot \frac{x^2}{1} \cdot \boxed{\frac{y}{y}} \cdot \frac{y}{1} \cdot \boxed{\frac{z}{z}} \cdot \frac{1}{z^2}$$

$$= 1 \cdot \frac{x^2}{1} \cdot 1 \cdot \frac{y}{1} \cdot 1 \cdot \frac{1}{z^2}$$

$$= \frac{x^2 y}{z^2}$$

Short-cut:

$$\frac{x^4 y^2 z}{x^2 y z^3} = \frac{\overset{x^2\,y\ 1}{\cancel{x^4}\,\cancel{y^2}\,\cancel{z}}}{\underset{1\ \ 1\ z^2}{\cancel{x^2}\,\cancel{y}\,\cancel{z^3}}} = \frac{x^2 y}{z^2}$$

The $x^2 y$ is the product of the remaining factors in the numerator, and the z^2 is the product of the remaining factors in the denominator.

□

Example 6 □ Reduce $\dfrac{3x + 6}{12}$.

The first step is to express this fraction so that both the numerator and denominator are products.

$$\boxed{\text{Sum of terms}} \Rightarrow \dfrac{3x + 6}{12} = \dfrac{3(x + 2)}{12} \Leftarrow \boxed{\text{Product of factors}}$$

$$= \dfrac{3(x + 2)}{3 \cdot 4}$$

Why $3 \cdot 4$ instead of $2 \cdot 6$ in the denominator?

$$= \dfrac{3}{3} \cdot \dfrac{x + 2}{4}$$

$$= 1 \cdot \dfrac{x + 2}{4}$$

Because $\dfrac{3}{3} = 1$, and I saw it coming.

$$= \dfrac{x + 2}{4}$$

The short way:

$$\boxed{\text{Factors}}$$

$$\dfrac{3x + 6}{12} = \dfrac{3 \cdot \overbrace{(x + 2)}}{12} \qquad \text{Factor}$$

$$= \dfrac{\overset{1}{\cancel{3}}(x + 2)}{\underset{4}{\cancel{12}}} \qquad \text{Divide out common factors}$$

$$= \dfrac{x + 2}{4} \qquad\qquad\qquad\qquad\qquad □$$

If dividing out is so neat, why not divide the 2 into the 4 in $\dfrac{x + 2}{4}$?

Because the 2 in the numerator is a term (part of an addition) not a factor (part of a multiplication).

Let's demonstrate with numbers why you can't cancel 2.

Let $x = 5$ in $\dfrac{x + 2}{4}$. Then

$$\dfrac{5 + 2}{4} = \dfrac{7}{4}.$$

If the 2 is divided out, then

$$\frac{5 + \overset{1}{\cancel{2}}}{\underset{2}{\cancel{4}}} \overset{?}{=} \frac{6}{2}$$

Be careful that you divide out only **factors. Terms** can't be divided out.

$\frac{6}{2}$ is not equal to $\frac{7}{4}$.

Be careful that you divide out only factors and not terms.

Example 7 □ Reduce $\frac{x + 6}{2x + 12}$.

Both the numerator and the denominator first must be pure products to reduce this fraction. Factor a 2 from the denominator.

$$\frac{x + 6}{2x + 12} = \frac{(x + 6)}{2(x + 6)}$$

The bar in a fraction is a grouping symbol just like parentheses. So $\underline{x + 6}$ means $(x + 6)$

$$= \frac{1}{2} \cdot \frac{x + 6}{x + 6}$$

$$= \frac{1}{2} \cdot 1$$

$$= \frac{1}{2}$$

The short way:

$$\frac{x + 6}{2x + 12} = \frac{x + 6}{2(x + 6)} \qquad \text{Factor}$$

$$= \frac{\overset{1}{\cancel{(x + 6)}}}{2\underset{1}{\cancel{(x + 6)}}} \qquad \text{Divide out common factor}$$

$$= \frac{1}{2} \qquad\qquad\qquad\qquad □$$

Aren't you supposed to say $2x + 12 \neq 0$?

Yes, the denominator of any fraction can never be zero because division by zero is undefined. However, in order to avoid cluttering the book we make the following statement.

For the remainder of this book, assume that all variables in denominators are restricted so that the denominator cannot equal zero.

Example 8 □ Reduce $\dfrac{x^2 + 2x - 3}{x^2 + 4x - 5}$.

Both the numerator and denominator must be factored before you can divide out.

$$\frac{x^2 + 2x - 3}{x^2 + 4x - 5} = \frac{(x - 1)(x + 3)}{(x - 1)(x + 5)} \qquad \text{Factor}$$

$$= \frac{\overset{1}{\cancel{(x - 1)}}(x + 3)}{\underset{1}{\cancel{(x - 1)}}(x + 5)} \qquad \text{Divide out}$$

$\dfrac{x + 3}{x + 5}$

$$= \underline{\hspace{2cm}} \qquad \qquad □$$

Example 9 □ Reduce $\dfrac{x - 5}{x + 10}$.

$$\frac{x - 5}{x + 10}$$

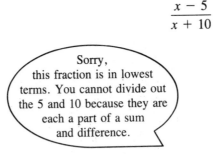

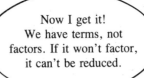

□

Multiplication by Negative One

Multiplying a polynomial by -1 changes all the signs in the polynomial. Notice the effect of multiplying the following difference by -1. Applying the distributive property we obtain:

$$\boxed{\text{Sign changed}}$$

$$(-1)(+a - b) = -a + b = b - a$$

$$\boxed{\text{Sign changed}}$$

If we apply the distributive property in reverse, that is factor out -1 from the expression $b - a$, we have

$$b - a = (-1)(-b + a)$$

$$= (-1)(a - b)$$

Example 10 □ Factor -1 from each of the following.

a. $(a - b) = -1(b - a)$

$y - x$ **b.** $(x - y) = -1(\underline{\hspace{2cm}})$

$b - 2a$ **c.** $2a - b = -1(\underline{\hspace{2cm}})$

$4q - 3p$ **d.** $3p - 4q = -1(\underline{\hspace{2cm}})$

$x - 4$ **e.** $4 - x = -1(\underline{\hspace{2cm}})$ □

Example 11 □ Reduce $\dfrac{3 - x}{x^2 - 7x + 12}$.

$$\frac{3 - x}{x^2 - 7x + 12} = \frac{3 - x}{(x - 4)(x - 3)} \qquad \text{Factor}$$

$$= \frac{(3 - x)}{(x - 4)(x - 3)} \qquad \begin{array}{l}\text{There are no common factors}\\ \text{to divide out}\end{array}$$

Isn't $3 - x$ the same as $x - 3$?

No!! The signs are different. But we can write $3 - x$ as $-1(x - 3)$.

Now we have a common factor to divide out.

$$\frac{3 - x}{(x - 4)(x - 3)} = \frac{-1(x - 3)}{(x - 4)(x - 3)}$$

$$= \frac{-1(\overset{1}{\cancel{x - 3}})}{(x - 4)(\underset{1}{\cancel{x - 3}})}$$

$$= \frac{-1}{x - 4}$$

$$= \frac{-1}{x - 4} \cdot \frac{-1}{-1}$$

$$= \frac{1}{4 - x}$$

Why multiply by 1, written as $\dfrac{-1}{-1}$?

To reduce the number of negative signs in the answer. However, $\dfrac{-1}{x - 4}$ and $\dfrac{1}{4 - x}$ are all correct answers.   □

Reduce to lowest terms.

1. $\dfrac{2x + 8}{8}$

2. $\dfrac{15}{3x - 3}$

3. $\dfrac{3x - 3y}{6y - 6x}$

4. $\dfrac{xy}{x - xy}$

5. $\dfrac{ax - ay}{ax + ay}$

6. $\dfrac{x^3y - x^2y^2}{xy^2 - yx^2}$

7. $\dfrac{x^3 - 2x^2}{x^4 - 2x^3}$

8. $\dfrac{x^5y + x^4y^3}{5x^3y^2 + 5x^2y^4}$

9. $\dfrac{x^5y^3 - x^4y^4}{x^3y^4 - x^2y^5}$

10. $\dfrac{a^3b^4 + 2a^2b^5}{a^4b^3 + 2a^3b^4}$

11. $\dfrac{a^5b - 2a^4b^2}{2a^2b^3 - a^3b^2}$

12. $\dfrac{ab^3 - a^2b^2}{a^3b - a^2b^2}$

13. $\dfrac{4x^3y + 4x^2y^2}{xy^4 + x^2y^3}$

14. $\dfrac{a^3b^3 + a^2b^4}{a^3b^2 + a^4b}$

15. $\dfrac{2x^2y^4 - 2x^3y^3}{x^4y - x^3y^2}$

16. $\dfrac{x^5 - 3x^4}{3xy - x^2y}$

17. $\dfrac{x^2 - 8x + 12}{x^2 + 3x - 10}$

18. $\dfrac{2a^2 - 9a + 4}{2a^2 + 5a - 3}$

19. $\dfrac{3x^2 + 4xy + y^2}{2x^2 - xy - 3y^2}$

20. $\dfrac{x^3 - xy^2}{x^2y - xy^2}$

21. $\dfrac{2x^2 - 5x - 12}{2x^2 + 11x + 12}$

22. $\dfrac{6a^2 + 5a - 4}{3a^2 - 2a - 8}$

23. $\dfrac{2x^2 - xy - y^2}{x^2 + xy - 2y^2}$

24. $\dfrac{2a^2 - 5ab + 2b^2}{2a^2 - 3ab - 2b^2}$

25. $\dfrac{x^4 + xy^3}{x^4 - x^2y^2}$

26. $\dfrac{8a^4b^2 - 27ab^5}{4a^4b - 12a^3b^2 + 9a^2b^3}$

27. $\dfrac{ax^2y + 2x^2y + axy^2 + 2xy^2}{a^2xy - axy - 6xy}$

28. $\dfrac{ax^2 - ay^2 + bx^2 - by^2}{ax - ay + bx - by}$

29. $\dfrac{4x^3y^2 - 25xy^4}{6x^4y - 13x^3y^2 - 5x^2y^3}$

30. $\dfrac{x^3 + 2x^2 - 4x - 8}{x^2 + 4x + 4}$

31. $\dfrac{x^4 - xy^3}{x^2y - y^3}$

32. $\dfrac{27x^3 - 1}{36x^5 + 12x^4 + 4x^3}$

33. $\dfrac{a^2xy^2 - a^2y^3 + axy^3 - ay^4}{a^2y - y^3}$

34. $\dfrac{b^4x + b^3xy + 2b^3x^2 + 2b^2x^2y}{b^4x - 4b^2x^3}$

35. $\dfrac{a^2x - a^2y - b^2x + b^2y}{ax - ay - bx + by}$

36. $\dfrac{a^4 - a^3b - ab^3 + b^4}{b^3 - ab^2 - a^2b + a^3}$

Multiplication of Algebraic Fractions

To save work when multiplying algebraic fractions, divide out all possible factors first.

Example 12 ☐ Multiply $\dfrac{a^2 - 1}{a^2 + a} \cdot \dfrac{a^3}{a^2 + 2a - 3}$.

$$\dfrac{a^2 - 1}{a^2 + a} \cdot \dfrac{a^3}{a^2 + 2a - 3}$$

$$= \dfrac{(a + 1)(a - 1)}{a(a + 1)} \cdot \dfrac{a^3}{(a - 1)(a + 3)} \qquad \text{Factor completely}$$

$$= \dfrac{\overset{1}{\cancel{a}}(\overset{}{\cancel{a + 1}})(\overset{1}{\cancel{a - 1}})}{\underset{1}{\cancel{a}}(\cancel{a + 1})} \cdot \dfrac{\overset{a^2}{\cancel{a^3}}}{(\underset{1}{\cancel{a - 1}})(a + 3)} \qquad \text{Divide out common factors}$$

$\dfrac{a^2}{a + 3}$

$$= \underline{\qquad\qquad} \qquad \text{Multiply} \qquad\qquad ☐$$

Example 13 ☐ Multiply $\dfrac{a^3 + 27}{a^2 - 16} \cdot \dfrac{a^2 - a - 20}{a^2 - 2a - 15}$.

$a - 5, a + 4$

$\dfrac{a^2 - 3a + 9}{a - 4}$

$$\dfrac{a^3 + 27}{a^2 - 16} \cdot \dfrac{a^2 - a - 20}{a^2 - 2a - 15} = \dfrac{(a + 3)(a^2 - 3a + 9)}{(a - 4)(a + 4)} \cdot \dfrac{(\underline{\quad})(\underline{\quad})}{(a - 5)(a + 3)}$$

$$= \underline{\qquad\qquad} \qquad\qquad\qquad\qquad ☐$$

Problem Set 7.1B

Multiply and reduce to lowest terms.

1. $\dfrac{x^2}{y^3} \cdot \left(-\dfrac{y^4}{x^3 y^2}\right)$

2. $\dfrac{x^4}{4y^3} \cdot \dfrac{24y^3}{3x^3}$

3. $-\dfrac{8m^2}{9n} \cdot \dfrac{36n^4}{4m^4}$

4. $\dfrac{1}{y - x} \cdot (x - y)$

5. $\dfrac{2 - a}{2} \cdot \dfrac{a^2}{a^2 - 2a}$

6. $\dfrac{3x - 12}{7x + 21} \cdot \dfrac{7x - 28}{4x - 16}$

7. $\dfrac{x^2}{x^3 - x^2} \cdot (1 - x)$

8. $\dfrac{ay^2 - y^3}{y} \cdot \dfrac{a}{a - y}$

9. $(a - b)^2 \cdot \dfrac{1}{b - a}$

10. $(x^2 - 9) \cdot \dfrac{1}{3 - x}$

11. $\dfrac{4 - a}{a} \cdot \dfrac{a^2}{a^2 - 16}$

12. $\dfrac{a^3}{a^3 - a^4} \cdot (a^2 - 1)$

13. $\dfrac{2x - 6}{5x + 10} \cdot \dfrac{15x + 30}{3x - 9}$

14. $\dfrac{4x - 12}{2x + 6} \cdot \dfrac{3x^2 - 27}{6x - 18}$

15. $\dfrac{a^3 x - a^4}{x^3 + ax^2} \cdot \dfrac{x^2 + ax}{ax - a^2}$

16. $\dfrac{x}{x-y} \cdot \dfrac{x^2 - y^2}{xy}$

17. $\dfrac{a}{a-b} \cdot \dfrac{a^2 - b^2}{ab}$

18. $\dfrac{4x - 16}{12x^3} \cdot \dfrac{8x}{x^2 - 5x + 4}$

19. $\dfrac{x^3}{4x^3 + 16x^2} \cdot \dfrac{x^2 + x - 12}{x^3 - 12x^2}$

20. $\dfrac{x^2 - 25x}{x^2 - 25} \cdot \dfrac{3x^2 - 13x - 10}{3x^2 + 2x}$

21. $\dfrac{x^3 - 8}{2x^2 + 5x - 3} \cdot \dfrac{2x^2 + 3x - 2}{x^2 - 4}$

22. $\dfrac{2ax + 2bx + a + b}{6x^2 - x - 2} \cdot \dfrac{9x - 6}{a^2 + 2ab + b^2}$

23. $\dfrac{y^2 - x^2}{y^2 + x^2} \cdot \dfrac{x^2 + 2xy}{x^2 + xy - 2y^2}$

24. $\dfrac{8x^3 + 27y^3}{2x^2 - xy - 3y^2} \cdot \dfrac{2x^2 - 5xy + 3y^2}{2x^2 + xy - 3y^2}$

25. $\dfrac{x^2 - 16}{x^2 + 8x + 16} \cdot \dfrac{x^2 + 7x + 12}{5x + 15}$

26. $\dfrac{2ax + 2ay - bx - by}{y^2 - x^2} \cdot \dfrac{x^3 + y^3}{b^2 - ab - 2a^2}$

27. $\dfrac{2x - x^2}{3x + 9} \cdot \dfrac{x^2 + x - 6}{x^2 - 4x + 4}$

28. $\dfrac{3x^2 + 11x + 6}{3x^2 - 4x - 4} \cdot \dfrac{x^2 - 6x + 8}{2x^2 - 9x + 4}$

29. $\dfrac{9x^2 + 6xy + 4y^2}{10x^2 - 3xy - y^2} \cdot \dfrac{15x^2 - 7xy - 2y^2}{8y^3 - 27x^3}$

30. $\dfrac{4 + 12y}{4x} \cdot \dfrac{x^2 - xy}{3y^2 - 3x^2}$

31. $\dfrac{x^3 + x^2 - 4x - 4}{3x^2 - 7x + 2} \cdot \dfrac{3x^2 + 5x - 2}{x^2 + 3x + 2}$

32. $\dfrac{1 - x}{3x - 12} \cdot \dfrac{3x^2 - 12}{3x^2 + 3x - 6}$

33. $\dfrac{54a^3 - 128b^3}{16a^2 - b^2} \cdot \dfrac{16a^2 + 8ab + b^2}{12a^2 - 13ab - 4b^2}$

34. $\dfrac{2x^2 - 6x - 8}{x^2 - x - 2} \cdot \dfrac{x^2 - 5x + 6}{3x^2 - 6x - 9}$

35. $\dfrac{x^2 - y^2}{2x^2 + 3xy + y^2} \cdot \dfrac{x^3 + x^2y + xy^2}{x^3 - y^3}$

36. $\dfrac{x^3 + y^3}{x^2 - y^2} \cdot \dfrac{3x^2 - 2xy - y^2}{x^2y^2 - xy^3 + y^4}$

37. $\dfrac{x^2 - 9}{2xy^2} \cdot \dfrac{4x^2y + 8xy^2}{x^2 - 3x + 2xy - 6y}$

38. $\dfrac{4x^2y}{x^2 - 4} \cdot \dfrac{2x^2 - 3x - 2}{4x^2y^2 + 2xy^2}$

39. $\dfrac{ax - bx + ay - by}{a^2 - b^2} \cdot \dfrac{ax + 2ay + bx + 2by}{ax - bx + ay - by}$

40. $\dfrac{x^2 - y^2}{ax + bx - ay - by} \cdot \dfrac{ax - 2ay + bx - 2by}{x^2 + 3xy + 2y^2}$

41. $\dfrac{y^3 + y^2 - 9y - 9}{3y^2 - 8y - 3} \cdot \dfrac{3y^2 - 5y - 2}{y^2 + 4y + 3}$

42. $\dfrac{2a^2 + a - 3}{a^2 + a - 2} \cdot \dfrac{a^3 + 2a^2 - 4a - 8}{2a^2 + 7a + 6}$

43. $\dfrac{2 - x}{4x - 12} \cdot \left(\dfrac{-2x^2 + 9x - 9}{2x^2 - 7x + 6} \right)$

44. $\dfrac{18y - 12}{2y^2 - 6y - 8} \cdot \dfrac{16 - y^2}{9y^2 + 30y - 24}$

45. $\dfrac{2x^3 + 9x^2 + 9x}{2x^2 - 7x - 15} \cdot \dfrac{4x^2 + 13x - 12}{9x^3 - 9x^4 - 4x^5}$

46. $\dfrac{a^3 + 2a^2 - 3a}{a^2 + 5a + 6} \cdot \dfrac{2a^2 + 9a + 10}{5a^2 - 3a^3 - 2a^4}$

47. $\dfrac{x^3 + y^3}{2x^2 + 5xy + 3y^2} \cdot \dfrac{12x^3 + 18x^2y}{2x^3y - 2x^2y^2 + 2xy^3}$

48. $\dfrac{4a^2 + 4ab + b^2}{2a^2 - ab - b^2} \cdot \dfrac{3a^2 - 4ab + b^2}{2a^2 + 3ab + b^2}$

49. $\dfrac{4x^2 - 4xy + y^2}{2x^2 - 3xy - 2y^2} \cdot \dfrac{2x^2 + 3xy + y^2}{2x^2 + xy - y^2}$

50. $\dfrac{16x^3 - 54y^3}{4x^2 + 12xy + 9y^2} \cdot \dfrac{2x^2 + 5xy + 3y^2}{2x^2 - xy - 3y^2}$

51. $\dfrac{a^3 + a^2 - a - 1}{2a^2 - 3a + 1} \cdot \dfrac{2a^2 + 3a - 2}{a^2 + 3a + 2}$

52. $\dfrac{4a - 20}{2a^2 - 5a - 12} \cdot \dfrac{3a^2 - 7a - 20}{18a^2 - 60a - 150}$

Inverses and Identity Elements

Before we define division of rational expressions we must recall some terms that are used in the definition of division.

Remember that:

1. Zero is the identity element for addition.

$$a + 0 = a$$

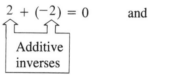

Identity element

2. Two expressions are additive inverses of each other if their sum is zero.

$$2 + (-2) = 0 \qquad \text{and} \qquad -3x + 3x = 0$$

Additive inverses   Additive inverses

3. Subtraction of a number is defined as addition of the additive inverse of the number.

$$a - b = a + (-b)$$

We use similar properties for multiplication:

4. One is the identity element for multiplication.

$$a \cdot 1 = a$$

Identity element

5. Two expressions are multiplicative inverses of each other if their product is 1.

$$2 \cdot \frac{1}{2} = 1 \qquad \text{and} \qquad \frac{x}{3} \cdot \frac{3}{x} = 1$$

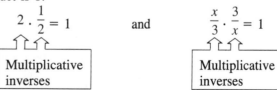

Multiplicative inverses   Multiplicative inverses

6. Division by a number is equivalent to multiplication by the multiplicative inverse of the number.

$$a \div b = a \cdot \frac{1}{b} \qquad \text{where } b \neq 0.$$

A comparison of addition and multiplication follows.

	Addition	Multiplication
Identity Element	0	1
Identity Law	$0 + a = a$	$1 \cdot a = a$
Inverses	$a + (-a) = 0$	$a \cdot \dfrac{1}{a} = 1$, where $a \neq 0$
How the Inverse Operation Is Defined	$a - b = a + (-b)$	$a \cdot b = a \cdot \dfrac{1}{b}$, where $b \neq 0$

Example 14 □ Give the multiplicative inverse of the following.

$\dfrac{1}{3}, \dfrac{4}{-1}, \dfrac{5s}{-r}$

$$3 \longrightarrow \underline{} \qquad \frac{-1}{4} \longrightarrow \underline{} \qquad \frac{-r}{5s} \longrightarrow \underline{}$$

$b - 3, \dfrac{1}{x + y}, \dfrac{3x - y}{x + 2y}$

$$\frac{1}{b - 3} \longrightarrow \underline{} \qquad x + y \longrightarrow \underline{} \qquad \frac{x + 2y}{3x - y} \longrightarrow \underline{}$$

In this example we used arrows instead of equal signs because a number does not equal its multiplicative inverse.   □

Division of Rational Expressions

Using the multiplicative inverse concept, we can give a rule for division of rational expressions.

7.1D   Rule: Division of Algebraic Fractions

To divide by a rational expression, multiply by its multiplicative inverse.

Example 15 □ Divide $\dfrac{3a}{4} \div \dfrac{15}{a^3}$.

$\dfrac{a^3}{15}$

| Change of operation |

$\dfrac{3a}{4} \div \dfrac{15}{a^3} = \dfrac{3a}{4} \cdot \underline{\qquad}$ Multiply by the multiplicative inverse

| Multiplicative inverses |

$$= \dfrac{\overset{1}{\cancel{3}a}}{4} \cdot \dfrac{a^3}{\underset{5}{\cancel{15}}}$$ Divide out common factors

$\dfrac{a^4}{20}$

$$= \underline{\qquad}$$ Multiply   □

Division Involving Factoring

Example 16 □ Divide $\dfrac{2x^2 + 7x + 6}{2x^2 - x - 15} \div \dfrac{6x + 9}{x^2 - 9}$.

$\dfrac{x^2 - 9}{6x + 9}$

$$\dfrac{2x^2 + 7x + 6}{2x^2 - x - 15} \div \dfrac{6x + 9}{x^2 - 9} = \dfrac{2x^2 + 7x + 6}{2x^2 - x - 15} \cdot \underline{\qquad\qquad}$$

Multiply by the
multiplicative inverse

$2x + 3, x + 2$

$$= \dfrac{(\underline{\quad})(\underline{\quad})}{(2x + 5)(x - 3)} \cdot \dfrac{(x + 3)(x - 3)}{3(2x + 3)}$$

Factor

$$= \dfrac{\overset{1}{\cancel{(2x + 3)}}(x + 2)}{(2x + 5)\underset{1}{\cancel{(x - 3)}}} \cdot \dfrac{(x + 3)\overset{1}{\cancel{(x - 3)}}}{3\underset{1}{\cancel{(2x + 3)}}}$$

Divide out
common factors

$\dfrac{(x + 2)(x + 3)}{3(2x + 5)}$

$$= \underline{\qquad\qquad}$$ Multiply   □

We frequently leave answers like the one above in factored form. Not only is it less work, but it is often more useful.

Perform the indicated operations. Reduce to lowest terms.

1. $\dfrac{x}{y} \div \dfrac{x}{y - x}$

2. $(y - x) \div (x - y)$

3. $\dfrac{1 - x}{x^2} \div (x - 1)$

4. $\dfrac{x - 3}{x^3} \div \dfrac{x + 3}{x^2 + 3x}$

5. $\dfrac{5x - 5}{5x + 5} \div \dfrac{6x - 6}{6x + 1}$

6. $\dfrac{3a - 6}{3a^2 - 6a} \div \dfrac{3}{a^3 + 3a^2}$

7. $\dfrac{x - y}{y^2} \div (y - x)$

8. $(a - b) \div \dfrac{b - a}{a^2}$

9. $\dfrac{4a - 28}{2a^2 - 10a} \div \dfrac{14x - 2ax}{3a - 15}$

10. $\dfrac{2y - 12}{3y^2 - 6y} \div \dfrac{4y + 12}{6y - 12}$

11. $\dfrac{3x + 1}{3x + 3} \div \dfrac{6x + 6}{4x + 4}$

12. $\dfrac{2 - x}{x^3} \div \dfrac{x - 2}{2x}$

13. $\dfrac{xy - x^2}{x^2} \div \dfrac{xy - y^2}{y^2}$

14. $\dfrac{x^3 - 8}{x^2 - 4x + 4} \div \dfrac{x^2 + 2x + 4}{2x^2 - x - 6}$

15. $\dfrac{x^2 - 25}{x^2 - 5} \div (x - 5)$

16. $\dfrac{a^2 + b^2}{b^2 - a^2} \div \dfrac{a + b}{a - b}$

17. $\dfrac{x^2 + 6x + 9}{6x + 6} \div \dfrac{x + 3}{3x + 3}$

18. $\dfrac{x^2 - 2x - 3}{x^2 + 3x + 2} \div \dfrac{x + 3}{4x + 8}$

19. $\dfrac{2x^2 - 2y^2}{3x^2 - xy - 4y^2} \div \dfrac{2x^2 - xy - y^2}{6x^2 - 5xy - 4y^2}$

20. $\dfrac{x^2 + 5x + 6}{x^3 + 3x} \div \dfrac{2x^2 + 5x + 2}{2x^3 + x^2}$

21. $\dfrac{4x^3 - 4x^2 + x}{8x^2 - 4x} \div \dfrac{4x^3 - 2x^2}{4x^3}$

22. $\dfrac{x^3 + 3x^2 - x - 3}{5x^2 + 4x - 1} \div \dfrac{x^3 + 2x^2 - 3x}{x^3 + 2x^2 + x}$

23. $\dfrac{b^2 - a^2}{a + b} \div \dfrac{4a^2 - 6ab + 2b^2}{a^2 - 2ab}$

24. $(8x^2 + 16x) \div \dfrac{4x^2 - 4x - 24}{4x^2 - 11x - 3}$

25. $\dfrac{y^2 - xy}{2x^3} \div \dfrac{x^2 - xy}{6y^2}$

26. $\dfrac{2a^2 - 2ab}{6ab} \div \dfrac{b^3 - ab^2}{3a^2b}$

27. $\dfrac{x^2 - 1}{x^2 + 4x + 4} \div \dfrac{x^2 + 2x - 3}{2x^2 + 5x + 2}$

28. $\dfrac{x^2 - 3}{x^2 - 9} \div \dfrac{x}{x - 3}$

29. $\dfrac{x^2 + 8x + 16}{x^2 + 6x + 8} \div \dfrac{x^2 + 2x - 8}{x^2 - x - 6}$

30. $\dfrac{4x^2 + 4x + 1}{3x^2 + 10x + 3} \div \dfrac{2x^2 - 5x - 3}{x^2 - 9}$

31. $\dfrac{y^2 - x^2}{3x^2 - 7xy + 2y^2} \div \dfrac{2x^2 - xy - y^2}{6x^2 + xy - y^2}$

32. $\dfrac{x^3 + y^3}{2x^2 - 3x + 2xy - 3y} \div \dfrac{2x^3y - 2x^2y^2 + 2xy^3}{12x^3y - 18x^2y}$

33. $\dfrac{8a^3 - b^3}{4a^2 - 4ab + b^2} \div \dfrac{4a^2 + 4ab + b^2}{4a^2 - b^2}$

34. $\dfrac{3x - y}{3x + y} \div \dfrac{9x^2 - 6xy + y^2}{27x^3 + y^3}$

35. $\dfrac{x^3 + 3x^2 - 4x - 12}{x^3 - 3x^2 - 10x} \div \dfrac{x^2 + 5x + 6}{x^4 - 4x^3 - 5x^2}$

36. $\dfrac{y^2 - x^2}{x^3 - y^3} \div \dfrac{3x^2 + 2xy - y^2}{3x^2 + 5xy - 2y^2}$

37. $\dfrac{y^4 + y^3}{y} \div \dfrac{2y^2 + 3y + 1}{2y^2 - 3y - 2}$

38. $\dfrac{2x^2 - xy - y^2}{x^2 - y^2} \div \dfrac{8x^3 + y^3}{4x^2 - 4xy + y^2}$

39. $\dfrac{a^3 - 8b^3}{a^2 + 4ab + 4b^2} \div \dfrac{2a^3b + 4a^2b^2 + 8ab^3}{a^2 + ab - 2b^2}$

40. $\dfrac{2ax + 2bx + ay + by}{3a^2 - ab - 4b^2} \div \dfrac{6x^2 + xy - y^2}{9ax - 12bx - 3ay + 4by}$

41. $\dfrac{4x - x^2}{4x^2 - 20x + 16} \div \dfrac{2x^2 - 4x}{6x^2 - 6}$

42. $\dfrac{x^3 + x}{x^2} \div \dfrac{x^4}{x^3 + x^2} \cdot \dfrac{x^3}{x^2 + x}$

43. $\dfrac{x^3 - x^2y}{x^2 - 6xy + 9y^2} \div \dfrac{x^2 - 2xy + y^2}{xy^2 - 3y^3} \cdot \dfrac{x^3 - 27y^3}{x^3 - 3x^2y + 9xy^2}$

44. $\dfrac{x^3}{ax + 3a} \cdot \dfrac{2x^2 + 8x}{x^3 - 3x^2} \div \dfrac{x^2 + 4x}{x^3 - 9x}$

45. $\dfrac{x^3 - 2x^2 - 3x}{x^2 - 2x - 8} \div \dfrac{9 - 6x + x^2}{x^3 - 7x^2 + 12x} \cdot \dfrac{x^2 + 5x + 4}{x^3 + 4x^2}$

46. $\dfrac{x^2 + 2x - 3}{x^3 - x^2} \cdot \dfrac{2x^2 + 4x + 2}{4x^2 - 12x - 16} \div \dfrac{3x^2 + 12x + 9}{x^3 + x^2 - 20x}$

47. $\dfrac{125a^3 - 27b^3}{25a^2 - 10ab - 3b^2} \div \dfrac{10a^2 - 11ab + 3b^2}{5a^2 - 8ab + 3b^2} \cdot \dfrac{10a^2 - 3ab - b^2}{3a^2 - 2ab - b^2}$

48. $\dfrac{y^4}{y^3 + y} \div \dfrac{3y^2 - 2y - 1}{y^2 - 1} \cdot \dfrac{6y^2 - y - 1}{2y^3 - 2y^2}$

49. $\dfrac{a^2 - 2a - 8}{a^3 - 4a^2} \cdot \dfrac{a^2 - 4a - 5}{a^2 + 5a + 6} \div \dfrac{a^2 - 6a + 5}{a^4 + 3a^3}$

50. $\dfrac{64x^3 - 27y^3}{16x^2 - 8xy - 3y^2} \div \dfrac{16x^2 + 24xy + 9y^2}{4x^2 + 3xy + 8x + 6y} \cdot \dfrac{16x^2 + 16xy + 3y^2}{x^2 - 3x - 10}$

Recall from Chapter 1 the rule for addition of fractions.

7.2A   Rule: Addition of Fractions

To add two algebraic fractions with a common denominator, add the numerators and place the result over the common denominator.

$$\frac{a}{c} + \frac{b}{c} = \frac{a + b}{c}$$

Example 17 □ Add $\dfrac{a + 4}{a} + \dfrac{a - 1}{a}$.

$$\frac{a + 4}{a} + \frac{a - 1}{a} = \frac{a + 4 + a - 1}{a} \qquad \text{Addition of fractions}$$

$$= \frac{2a + 3}{a} \qquad \text{Combine like terms} \qquad □$$

The rule for subtraction of fractions with common denominators was also given in Chapter 1.

7.2B   Rule: Subtraction of Fractions

To subtract a fraction, add its additive inverse.

$$\frac{a}{c} - \frac{b}{c} = \frac{a}{c} + \frac{-b}{c}$$

Example 18 □ Subtract $\dfrac{x}{x + 1} - \dfrac{1}{x + 1}$.

$$\frac{x}{x + 1} - \frac{1}{x + 1} = \frac{x}{x + 1} + \frac{-1}{x + 1}$$

$$= \frac{x + (-1)}{x + 1}$$

$$= \frac{x - 1}{x + 1}$$

When a fraction is negative, associate the negative sign with the numerator.

It looks like it would be easier to just subtract numerators.

That's true for simple cases, but with complex numerators it's better to use the definition of subtraction.

□

Example 19 □ Subtract $\dfrac{6a + 1}{a - b} - \dfrac{9a - 3}{a - b}$.

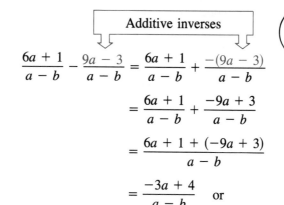

$$\frac{6a + 1}{a - b} - \frac{9a - 3}{a - b} = \frac{6a + 1}{a - b} + \frac{-(9a - 3)}{a - b}$$

Two expressions are additive inverses if their sum is zero.

$$= \frac{6a + 1}{a - b} + \frac{-9a + 3}{a - b} \qquad \text{Remove parentheses}$$

$$= \frac{6a + 1 + (-9a + 3)}{a - b} \qquad \text{Add numerators}$$

$$= \frac{-3a + 4}{a - b} \quad \text{or}$$

$$= \frac{4 - 3a}{a - b}$$

The preferred style is to avoid negative signs in the first term. This style uses one less symbol.

□

Example 20 □ Subtract $\dfrac{3x - y}{x - y} - \dfrac{x + y}{x - y}$.

$$\frac{3x - y}{x - y} - \frac{x + y}{x - y} = \frac{3x - y}{x - y} + \frac{-(x + y)}{x - y}$$

$-x - y$

$$= \frac{3x - y}{x - y} + \frac{\overline{}}{x - y} \qquad \text{Remove parentheses}$$

$$= \frac{3x - y + (-x - y)}{x - y}$$

$2x - 2y$

$$= \frac{\overline{}}{x - y} \qquad \text{Combine like terms}$$

$$= \frac{2(x - y)}{x - y}$$

$$= 2$$

If we factor the numerator, we can reduce the answer.

Right.

□

Add or subtract as indicated and reduce to lowest terms.

1. $\dfrac{3}{x+3} + \dfrac{6}{x+3}$

2. $\dfrac{y}{y-4} - \dfrac{y+3}{y-4}$

3. $\dfrac{x+4}{x-2} + \dfrac{x-8}{x-2}$

4. $\dfrac{2x+3}{3x+4} - \dfrac{x+1}{3x+4}$

5. $\dfrac{2y}{y-3} - \dfrac{y+3}{y-3}$

6. $\dfrac{3x+4}{2x+3} - \dfrac{x+1}{2x+3}$

7. $\dfrac{4x-1}{5x-2} + \dfrac{x-1}{5x-2}$

8. $\dfrac{7x-3}{2x-1} - \dfrac{3x-1}{2x-1}$

9. $\dfrac{a-8}{a^2-1} + \dfrac{a+10}{a^2-1}$

10. $\dfrac{3y+2}{y^2-9} - \dfrac{y+8}{y^2-9}$

11. $\dfrac{5x-14}{x^2-x-12} + \dfrac{2-2x}{x^2-x-12}$

12. $\dfrac{y+y^3}{y^2-9} - \dfrac{y^3+3}{y^2-9}$

13. $\dfrac{x^2+4x+3}{x^2-x-6} + \dfrac{x^2-3}{x^2-x-6}$

14. $\dfrac{5b-7}{b^2-4} + \dfrac{9-4b}{b^2-4}$

15. $\dfrac{3-x}{x^2-16} + \dfrac{2x-7}{x^2-16}$

16. $\dfrac{4x+3}{x^2+2x-15} - \dfrac{3x-2}{x^2+2x-15}$

17. $\dfrac{5x-8}{2x^2-7x+3} + \dfrac{7-3x}{2x^2-7x+3}$

18. $\dfrac{7x-5}{10x^2+3x-4} - \dfrac{2x-9}{10x^2+3x-4}$

19. $\dfrac{3x-2}{21x^2+23x+6} + \dfrac{4x+5}{21x^2+23x+6}$

20. $\dfrac{5x+4}{18x^2-3x-10} + \dfrac{x-9}{18x^2-3x-10}$

Addition and Subtraction of Algebraic Fractions Without Common Denominators

We can add or subtract fractions only if their denominators are identical. In cases where two fractions do not have identical denominators, we make the denominators identical by using the multiply by 1 routine.

Example 21 □ Add $\dfrac{3}{x} + \dfrac{4}{y}$.

For algebraic fractions we build a common denominator that contains every factor used in any denominator of the fractions to be added. In this case $x \cdot y$ has both x and y as factors so we'll use it for our common denominator.

$$\dfrac{3}{x} + \dfrac{4}{y} = \dfrac{3}{x}\dfrac{y}{y} + \dfrac{4}{y}\dfrac{x}{x}$$

The expression by which you multiply each fraction to get common denominators is always equal to one

$$= \dfrac{3y}{xy} + \dfrac{4x}{xy}$$

$$= \dfrac{3y+4x}{xy}$$ □

1. Find a common denominator in this way:
   a. Factor each denominator.
   b. Write each different factor that appears in any denominator.
   c. Raise each factor to the highest power it has in any single denominator. The result of this process is the least common denominator (LCD).
2. Convert all fractions to be added to equivalent fractions with a common denominator by supplying the missing factors as 1's.
3. Add the numerators of the equivalent fractions and place the result over the common denominator.
4. Reduce to lowest terms.

Example 22 □ Subtract $\dfrac{5}{6pq^2} - \dfrac{7}{9p^3q}$.

To subtract, first convert to an addition problem and then add:

$$\frac{5}{6pq^2} - \frac{7}{9p^3q} = \frac{5}{6pq^2} + \frac{-7}{9p^3q}$$

Find the common denominator.

Factor each denominator: $6pq^2 = 2 \cdot 3 \cdot p \cdot q^2$
$9p^3q = 3^2 \cdot p^3 \cdot q$

Write the different factors: $2 \cdot 3 \cdot p \cdot q$

Raise each factor to its highest power in any one denominator.

$2 \cdot 3^2 \cdot p^3 \cdot q^2$

$$LCD = \underline{\hspace{1cm}} \cdot \underline{\hspace{1cm}} \cdot \underline{\hspace{1cm}} \cdot \underline{\hspace{1cm}}$$

Convert to fractions with common denominators:

$$\frac{5}{6pq^2} + \frac{-7}{9p^3q} = \frac{5}{2 \cdot 3 \cdot pq^2} + \frac{-7}{3 \cdot 3 \cdot p^3q}$$

$$= \frac{5}{2 \cdot 3 \cdot pq^2} \cdot \boxed{\frac{3p^2}{3p^2}} + \frac{-7}{3 \cdot 3 \cdot p^3q} \cdot \boxed{\frac{2q}{2q}}$$

Missing factors

$$= \frac{15p^2}{18p^3q^2} + \frac{-14q}{18p^3q^2}$$

If you leave the denominators in factored form, it's easier to see what is needed.

$\dfrac{15p^2 - 14q}{18p^3q^2}$

Add: $= \underline{\hspace{2cm}}$ □

Example 23 □ Add $\dfrac{x}{x^2 - 49} + \dfrac{7}{7 - x}$.

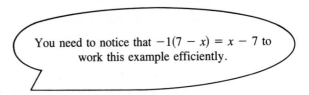

You need to notice that $-1(7 - x) = x - 7$ to work this example efficiently.

Multiplying the second fraction by $\dfrac{-1}{-1}$ will reduce the number of factors in the common denominator.

$$\frac{x}{x^2 - 49} + \frac{7}{(7 - x)} \cdot \frac{-1}{-1} = \frac{x}{x^2 - 49} + \frac{-7}{x - 7}$$

Find the common denominator.
Factor:

$$x^2 - 49 = (x - 7)(x + 7)$$

$x - 7$ is already factored.

List factors: $(x - 7)(x + 7)$

Raise to powers: LCD $= (x - 7)(x + 7)$

Each factor is used only once in this problem.

Convert to common denominators:

$\dfrac{x + 7}{x + 7}$

$$\frac{x}{x^2 - 49} + \frac{-7}{x - 7} = \frac{x}{(x - 7)(x + 7)} + \frac{-7}{x - 7} \cdot \frac{(\underline{\qquad})}{(\underline{\qquad})}$$

$$= \frac{x}{(x - 7)(x + 7)} + \frac{-7x - 49}{(x - 7)(x + 7)}$$

Add: $$= \frac{x - 7x - 49}{(x - 7)(x + 7)}$$

$$= \frac{-6x - 49}{(x - 7)(x + 7)}$$ □

Chapter 7 ■ Rational Expressions

▶ Pointers to Speed Addition of Rational Expressions

It is possible to streamline building the common denominator.

Example 24 □ Subtract $\dfrac{2x}{3x^2 - 27} - \dfrac{x + 1}{x^2 - x - 6}$.

Write the problem as an addition with all denominators factored. Leave space to fill in factors missing from the common denominator.

$$\frac{2x}{3x^2 - 27} - \frac{x + 1}{x^2 - x - 6}$$

$$= \frac{2x}{3(x - 3)(x + 3)}\left(\right) + \frac{-(x + 1)}{(x - 3)(x + 2)}\left(\right)$$

Compare the factors in each denominator. Use the space to supply the missing factors so that all denominators have the same factors.

$$= \frac{2x}{3(x - 3)(x + 3)}\left(\frac{x + 2}{x + 2} \right) + \frac{-(x + 1)}{(x - 3)(x + 2)}\left(\frac{(3)(x + 3)}{(3)(x + 3)} \right)$$

Now, multiply out each numerator and indicate the sum of the numerators over the common denominator.

> Keep this negative sign outside the bracket until you've done all multiplication inside

$$= \frac{2x^2 + 4x - 3[(x + 1)(x + 3)]}{3(x - 3)(x + 3)(x + 2)}$$

$$= \frac{2x^2 + 4x - 3[x^2 + 4x + 3]}{3()()()}$$

Leaving the denominator blank or writing empty parentheses indicates there was no change from the line above

$$= \frac{2x^2 + 4x - 3x^2 - 12x - 9}{}$$

$$= \frac{-x^2 - 8x - 9}{3(x - 3)(x + 3)(x + 2)}$$

The numerator won't factor, so no reductions are possible

Be sure to write the denominator on the final line. Normally, we leave it in factored form. □

Example 25 □ Subtract $\dfrac{5x + 2}{4x - 8} - \dfrac{2x + 5}{x^2 - x - 2}$.

Convert to an addition problem.

> If you leave the negative sign outside the parentheses, it will be easier to convert to equivalent fractions.

$$\frac{5x + 2}{4x - 8} - \frac{2x + 5}{x^2 - x - 2} = \frac{5x + 2}{4x - 8} + \frac{-(2x + 5)}{x^2 - x - 5}$$

Factor denominators:

$$= \frac{5x + 2}{4(x - 2)} + \frac{-(2x + 5)}{(x - 2)(x + 1)}$$

Convert to equivalent fractions with common denominators:

$\dfrac{x + 1}{x + 1}, \dfrac{4}{4}$

$$= \frac{5x + 2}{4(x - 2)} \cdot \frac{(\underline{\quad})}{(\underline{\quad})} + \frac{-(2x + 5)}{(x - 2)(x + 1)} \cdot \frac{(\underline{\quad})}{(\underline{\quad})}$$

$$= \frac{5x^2 + 7x + 2}{4(x - 2)(x + 1)} + \frac{-(8x + 20)}{4(x - 2)(x + 1)}$$

Add:
$$= \frac{5x^2 + 7x + 2 - 8x - 20}{4(x - 2)(x + 1)}$$

$5x^2 - x - 18$

$$= \frac{(\underline{\quad\quad})}{4(x - 2)(x + 1)}$$

When the numerator is factorable, factor it. Reduce the fraction if possible:

$$= \frac{(5x + 9)\overset{1}{(x - 2)}}{4\underset{1}{(x - 2)}(x + 1)}$$

$$= \frac{5x + 9}{4(x + 1)} \qquad\qquad □$$

Problem Set 7.2B

Add or subtract the following. Reduce your answers to lowest terms.

1. $\dfrac{7}{x^2} - \dfrac{6}{xy}$

2. $\dfrac{1}{x + 3} + \dfrac{1}{3}$

3. $\dfrac{1}{x - 2} - \dfrac{2}{3x}$

4. $\dfrac{4}{x + 4} - \dfrac{2}{x - 4}$

5. $\dfrac{8}{x-4} + \dfrac{4}{4-x}$

6. $5 - \dfrac{5x}{x+5}$

7. $\dfrac{3x}{3x-1} + 4$

8. $\dfrac{3}{x^2+3x} + \dfrac{1}{x+3}$

9. $\dfrac{1}{x+2} + \dfrac{1}{2}$

10. $\dfrac{2}{x-3} + \dfrac{3}{x}$

11. $\dfrac{1}{5-x} - \dfrac{2}{x+5}$

12. $\dfrac{3}{6-x} - \dfrac{2}{x+6}$

13. $2 - \dfrac{4x}{3x+4}$

14. $\dfrac{5x}{2x-3} - 4$

15. $\dfrac{3}{4(y-4)} - \dfrac{3x}{y(y-4)}$

16. $\dfrac{1}{9-3x} + \dfrac{1}{x^2-3x}$

17. $\dfrac{3}{x-2} + \dfrac{3x}{(x-2)(x-4)}$

18. $\dfrac{6}{3x-12} - \dfrac{12}{x^2-6x+8}$

19. $\dfrac{x}{x^3-8} - \dfrac{1}{x^2+2x-8}$

20. $\dfrac{3x-4}{2x-4} + \dfrac{4-5x}{x^2+2x-8}$

21. $\dfrac{12}{x^2-2xy-3y^2} - \dfrac{3}{x^2+xy}$

22. $\dfrac{1}{x^2-3x-18} - \dfrac{x}{x^3+27}$

23. $\dfrac{2}{m^2+4m+4} + \dfrac{1}{m^2-4}$

24. $\dfrac{4}{x^2+6x-7} - \dfrac{3}{x^2+8x+7}$

25. $\dfrac{2}{5(x-5)} - \dfrac{2}{x(x-5)}$

26. $\dfrac{2}{12-3x} + \dfrac{1}{x^2-4x}$

27. $\dfrac{2}{x-3} + \dfrac{x-1}{(x-3)(x-4)}$

28. $\dfrac{x+5}{4(x-3)} - \dfrac{2}{(x-3)(x+5)}$

29. $\dfrac{1}{2x-10} - \dfrac{x-1}{x^2-2x-15}$

30. $\dfrac{2}{4x-8} - \dfrac{x-2}{x^2-5x+6}$

31. $\dfrac{2x}{x^3-1} - \dfrac{1}{x^2+3x-4}$

32. $\dfrac{2}{x^2-3x-10} - \dfrac{x}{x^3+8}$

33. $\dfrac{8}{x^2-4xy+3y^2} - \dfrac{5}{x^2-3xy}$

34. $\dfrac{6}{x^2+xy-6y^2} - \dfrac{3}{x^2-2xy}$

35. $\dfrac{4x+9}{3x+6} + \dfrac{2x+5}{x^2+x-2}$

36. $\dfrac{x}{x^2-16} - \dfrac{1}{2x-8}$

7.2 ■ Addition and Subtraction of Rational Expressions

37. $\dfrac{1}{x^2 + 5x - 50} - \dfrac{x}{125 - x^3}$

38. $\dfrac{x}{x^2 - 15x + 36} - \dfrac{1}{9 - 3x}$

39. $1 - \dfrac{x^2 + 4x}{x^2 + 5x + 4}$

40. $\dfrac{9x - 81}{x^2 - 3x} - 3$

41. $2 - \dfrac{24 - 8x}{x^2 - 2x}$

42. $\dfrac{2}{x^2 - x - 2} - \dfrac{8}{x^3 + x^2 - 4x - 4}$

43. $\dfrac{3x + 4}{3x + 9} + \dfrac{2x - 5}{x^2 + 2x - 3}$

44. $\dfrac{5x - 4}{4x - 8} + \dfrac{3x - 2}{x^2 + 2x - 8}$

45. $\dfrac{2x}{x^2 - 9} - \dfrac{3}{2x - 6}$

46. $\dfrac{4x}{x^2 - 16} - \dfrac{2}{3x - 12}$

47. $\dfrac{2}{x^2 + 2x - 15} - \dfrac{x}{27 - x^3}$

48. $\dfrac{3}{x^2 - x - 6} - \dfrac{6}{x^3 - x^2 - 9x + 9}$

49. $2 - \dfrac{x^2 + 4x}{x^2 - 5x + 4}$

50. $3 - \dfrac{2x^2 + 8x}{x^2 + 2x - 8}$

51. $\dfrac{9x - 12}{x^2 - 4x} - 3$

52. $\dfrac{4x - 10}{x^2 - 6x} - 2$

■ **7.3**   **Complex Fractions**

> **7.3   Definition: Complex Fractions**
>
> A **complex fraction** is a fraction that contains fractions in its numerator or denominator or both.

For example,

$$\dfrac{\frac{1}{2}}{\frac{1}{3}}, \qquad \dfrac{\frac{1}{x}}{2}, \qquad \dfrac{1 + \frac{1}{x}}{\frac{1}{2}}, \qquad \dfrac{\frac{1}{2} + x}{\frac{1}{3} + \frac{x}{3}}$$

are complex fractions.

    Now, the problem is to simplify these complex fractions completely. That is, we want to write them as a single fraction. There are two ways to do this.

Example 26 □ Simplify $\dfrac{\frac{1}{x}}{\frac{1}{y}}$.   ⟵ Numerator   ⟵ Denominator

Method I—Division

Follow the directions in the problem.

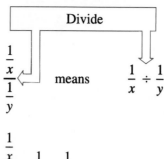

Divide

$$\dfrac{\dfrac{1}{x}}{\dfrac{1}{y}} \quad \text{means} \quad \dfrac{1}{x} \div \dfrac{1}{y}$$

Remember that the bar line of the fraction means "divide."

$$\dfrac{\dfrac{1}{x}}{\dfrac{1}{y}} = \dfrac{1}{x} \div \dfrac{1}{y}$$

To divide, invert the denominator and multiply.

$$= \dfrac{1}{x} \cdot \dfrac{y}{1}$$

$$= \dfrac{y}{x}$$

Method II—LCD Method

Look at the two fractions that make up the numerator and the denominator of our complex fraction, that is, $\dfrac{1}{x}$ and $\dfrac{1}{y}$.

The LCD of $\dfrac{1}{x}$ and $\dfrac{1}{y}$ is xy.

Multiply the numerator and the denominator by the LCD.

$$\dfrac{\dfrac{1}{x}}{\dfrac{1}{y}} = \dfrac{\dfrac{1}{x} \cdot \dfrac{xy}{xy}}{\dfrac{1}{y}}$$

When we multiply by $\dfrac{xy}{xy}$, which equals 1, we don't change the value of the original problem.

$$= \dfrac{\dfrac{1}{x} \cdot \dfrac{xy}{1}}{\dfrac{1}{y} \cdot \dfrac{xy}{1}}$$

$$= \dfrac{y}{x}$$

Why multiply by the LCD?

So that the denominators of the two fractions will divide out and our answer will become a single fraction.

□

Example 27 □ Simplify $\dfrac{\dfrac{1}{2x} + x}{\dfrac{1}{3x} - \dfrac{x}{3}}$.

Method I

The problem itself tells us what to do. First, add the terms in the numerator. Second, subtract the terms in the denominator. Third, divide the numerator by the denominator.

Add in the numerator:

$$\frac{1}{2x} + x = \frac{1}{2x} + \frac{x}{1} \cdot \boxed{\frac{2x}{2x}}$$

$\dfrac{1 + 2x^2}{2x}$

$$= \underline{\qquad}$$ ⟵ Now the numerator is a single fraction

Subtract in the denominator:

$\dfrac{1 - x^2}{3x}$

$$\frac{1}{3x} - \frac{x}{3} = \underline{\qquad}$$ ⟵ And the denominator is a single fraction

Divide:

$$\frac{\dfrac{1}{2x} + x}{\dfrac{1}{3x} - \dfrac{x}{3}} = \frac{\dfrac{1 + 2x^2}{2x}}{\dfrac{1 - x^2}{3x}}$$

$$= \frac{1 + 2x^2}{2x} \div \frac{1 - x^2}{3x}$$

$\dfrac{3 + 6x^2}{2 - 2x^2}$

$$= \frac{1 + 2x^2}{2x} \cdot \frac{3x}{1 - x^2}$$

$$= \underline{\qquad}$$

Method II

Now we will work the same problem using the LCD method.

Simplify: $\dfrac{\dfrac{1}{2x} + x}{\dfrac{1}{3x} - \dfrac{x}{3}} =$

$$\frac{\dfrac{1}{2x} + x}{\dfrac{1}{3x} - \dfrac{x}{3}} = \frac{\dfrac{1}{2x} + \dfrac{x}{1}}{\dfrac{1}{3x} - \dfrac{x}{3}}$$ The denominator of x is 1

The common denominator of all the denominators is $6x$.

$$= \cfrac{\dfrac{1}{2x} + \dfrac{x}{1} \cdot \dfrac{6x}{1}}{\dfrac{1}{3x} - \dfrac{x}{3} \cdot \dfrac{6x}{1}}$$

This is multiplication by one

$$= \cfrac{\dfrac{1}{2x} \cdot \dfrac{6x}{1} + \dfrac{x}{1} \cdot \dfrac{6x}{1}}{\dfrac{1}{3x} \cdot \dfrac{6x}{1} - \dfrac{x}{3} \cdot \dfrac{6x}{1}}$$

Distributive property

$$= \frac{3 + 6x^2}{2 - 2x^2}$$

Simplifying fractions □

Example 28 □ Simplify $\cfrac{\dfrac{y}{x+y} - \dfrac{x}{x-y}}{\dfrac{x}{x-y} + \dfrac{y}{x+y}}$.

The easiest way to work this example is to multiply both numerator and denominator by $(x - y)(x + y)$.

$$\cfrac{\dfrac{y}{x+y} - \dfrac{x}{x-y}}{\dfrac{x}{x-y} + \dfrac{y}{x+y}} = \cfrac{\dfrac{y}{x+y} \dfrac{(x-y)(x+y)}{1} - \dfrac{x}{x-y} \dfrac{(x-y)(x+y)}{1}}{\dfrac{x}{x-y} \dfrac{(x-y)(x+y)}{1} + \dfrac{y}{x+y} \dfrac{(x-y)(x+y)}{1}}$$

$$= \frac{y(x - y) - x(x + y)}{x(x + y) + y(x - y)}$$

$$= \frac{xy - y^2 - x^2 - xy}{x^2 + xy + xy - y^2}$$

$$= \frac{-1(y^2 + x^2)}{x^2 + 2xy - y^2}$$ □

Simplify the following complex fractions. Reduce your answers to lowest terms.

1. $\dfrac{\frac{3}{4}}{\frac{5}{8}}$

2. $\dfrac{1 + \frac{1}{3}}{\frac{2}{3}}$

3. $\dfrac{3 - \frac{3}{8}}{3 + \frac{1}{2}}$

4. $\dfrac{\frac{a^2}{b}}{\frac{b^2}{a}}$

5. $\dfrac{1 + \frac{3}{4}}{2 - \frac{1}{2}}$

6. $\dfrac{5 - \frac{5}{8}}{2 + \frac{3}{4}}$

7. $\dfrac{6 + \frac{1}{10}}{5 - \frac{2}{5}}$

8. $\dfrac{4 + \frac{3}{4}}{2 - \frac{5}{6}}$

9. $\dfrac{\frac{1}{x} + 4}{\frac{5}{2x} + \frac{5}{x}}$

10. $\dfrac{1 - \frac{1}{3y}}{2 + \frac{1}{2y}}$

11. $\dfrac{1 + \frac{1}{a}}{a - \frac{1}{a}}$

12. $\dfrac{a + \frac{1}{a^2}}{1 + \frac{1}{a}}$

13. $\dfrac{x^3 - \frac{1}{x}}{x + \frac{1}{x}}$

14. $\dfrac{\frac{a}{b^2}}{3 - \frac{a}{b^2}}$

15. $\dfrac{\frac{x}{y} + 1}{\frac{x^2}{y^2} - 1}$

16. $\dfrac{\frac{1}{x} - \frac{1}{y}}{\frac{y}{x} - \frac{x}{y}}$

17. $\dfrac{2 + \frac{1}{2y}}{1 - \frac{1}{3y}}$

18. $\dfrac{4 + \frac{1}{5x}}{6 + \frac{1}{2x}}$

19. $\dfrac{a - \frac{4}{a}}{1 - \frac{2}{a}}$

20. $\dfrac{1 - \frac{1}{a}}{1 - \frac{1}{a^2}}$

21. $\dfrac{\frac{2}{x} + 3}{\frac{3}{2x} + \frac{1}{x}}$

22. $\dfrac{3 - \frac{1}{x}}{\frac{1}{3x} - \frac{2}{x}}$

23. $\dfrac{1 - \frac{a}{b}}{1 - \frac{a^2}{b^2}}$

24. $\dfrac{\frac{1}{x} + \frac{1}{y}}{\frac{x}{y^2} - \frac{1}{x}}$

25. $\dfrac{\frac{a^2 + b^2}{a^2 - b^2}}{\frac{a - b}{a + b} + \frac{a + b}{a - b}}$

26. $\dfrac{\frac{a^2}{a - b} - a}{\frac{b^2}{b - a} - b}$

27. $\dfrac{\frac{1}{x^2} - \frac{1}{y^2}}{\frac{1}{x} - \frac{1}{y}}$

28. $\dfrac{\frac{1}{x + y} - \frac{1}{x - y}}{\frac{1}{x - y} + \frac{1}{x + y}}$

29. $\dfrac{\frac{2}{x + 2} + \frac{6}{x - 2}}{\frac{4}{x + 2} - \frac{2}{x - 2}}$

30. $\dfrac{x - 1 + \frac{2}{x + 2}}{x - 3 + \frac{4}{x + 2}}$

31. $\dfrac{x + 2 - \frac{4}{x - 1}}{x - 4 + \frac{2}{x - 1}}$

32. $\dfrac{1 + \frac{y^2}{x^2 - y^2}}{x - \frac{x^2}{x - y}}$

33. $\dfrac{1 - \dfrac{y^3}{x^3 + y^3}}{y^2 - \dfrac{y^3}{x + y}}$

34. $\dfrac{\dfrac{x-1}{x+1} - \dfrac{x+1}{x-1}}{\dfrac{x-1}{x+1} + \dfrac{x+1}{x-1}}$

35. $\dfrac{2 - \dfrac{3}{x} - \dfrac{2}{x^2}}{2 - \dfrac{5}{x} - \dfrac{3}{x^2}}$

36. $\dfrac{\dfrac{6}{x^2} - \dfrac{7}{x} - 3}{\dfrac{4}{x^2} - \dfrac{8}{x} + 3}$

37. $\dfrac{\dfrac{ab}{a^2 - b^2}}{\dfrac{a-b}{a+b} - \dfrac{a+b}{a-b}}$

38. $\dfrac{\dfrac{a}{a-b} - 1}{\dfrac{b}{b-a} - 1}$

39. $\dfrac{\dfrac{2}{x+3} + \dfrac{4}{x-3}}{\dfrac{6}{x+3} - \dfrac{4}{x-3}}$

40. $\dfrac{\dfrac{3}{x+4} - \dfrac{6}{x-4}}{\dfrac{9}{x+4} + \dfrac{6}{x-4}}$

41. $\dfrac{a - \dfrac{a^2}{a-b}}{2 + \dfrac{2b}{a^2-b^2}}$

42. $\dfrac{a - \dfrac{a^2}{a-2b}}{1 - \dfrac{a^2}{a^2-4b^2}}$

43. $\dfrac{\dfrac{x}{y} - 4 + \dfrac{4y}{x}}{\dfrac{x}{y} - \dfrac{4y}{x}}$

44. $\dfrac{\dfrac{x}{y} + 1 + \dfrac{y}{x}}{\dfrac{x^2}{y} - \dfrac{y^2}{x}}$

45. $\dfrac{x - 2 - \dfrac{6}{x+3}}{x - 1 + \dfrac{-12}{x+3}}$

46. $\dfrac{x + 4 + \dfrac{6}{x-3}}{x + 1 - \dfrac{12}{x-3}}$

47. $\dfrac{2 - \dfrac{9}{x} + \dfrac{4}{x^2}}{2 + \dfrac{5}{x} - \dfrac{3}{x^2}}$

48. $\dfrac{\dfrac{4}{x^2} + \dfrac{8}{x} + 3}{\dfrac{6}{x^2} + \dfrac{7}{x} - 3}$

■ 7.4   Solving Equations with Fractions

To solve an equation with fractions, the easiest thing to do is to make it into an equation without fractions. That's done by eliminating the denominators. All we need to do is find something to divide out the denominators. We need an expression that has every denominator as one of its factors. The simplest expression with this property is the least common denominator (LCD) of the fractions in the equation. So, to eliminate the fractions, we multiply both sides of the equation by the least common denominator.

Example 29 □ Solve $\dfrac{2}{3x} + 3 = \dfrac{5}{6x}$ for x.

6x

The LCD is _____ .
Multiply both sides by $6x$.

$$6x\left(\dfrac{2}{3x} + 3\right) = \left(\dfrac{5}{6x}\right) \cdot 6x$$

$$6x \cdot \dfrac{2}{3x} + 6x \cdot 3 = \dfrac{5}{6x} \cdot 6x \qquad \text{Distributive Property}$$

$$\overset{2}{\cancel{6x}} \cdot \dfrac{2}{\cancel{3x}} + 6x \cdot 3 = \dfrac{5}{\cancel{6x}} \cdot \overset{1}{\cancel{6x}} \qquad \text{Simplify and solve for } x$$

$$4 + 18x = 5$$

$$18x = 1$$

$\dfrac{1}{18}$

$$x = \text{_____}$$

Check.

When should I check my answers?

All solutions **should be** checked, but checking **must** be part of the problem when there is a variable in the denominator.

$$\frac{2}{3x} + 3 = \frac{5}{6x}$$

$$\frac{2}{3\left(\dfrac{1}{18}\right)} + 3 \stackrel{?}{=} \frac{5}{6\left(\dfrac{1}{18}\right)} \qquad \text{Replace } x \text{ with } \frac{1}{18}$$

$$\frac{2}{\dfrac{1}{6}} + 3 \stackrel{?}{=} \frac{5}{\dfrac{1}{3}} \qquad \text{Simplify}$$

$$2 \cdot \frac{6}{1} + 3 \stackrel{?}{=} 5 \cdot \frac{3}{1}$$

$$12 + 3 \stackrel{?}{=} 15$$

$$15 = 15 \qquad\qquad\qquad \Box$$

Example 30 $\Box$ Solve $\dfrac{4}{y + 2} = \dfrac{1}{y - 1}$ for y.

The LCD is $(y + 2)(y - 1)$.

Multiply both sides by the LCD.

$$(y + 2)(y - 1) \cdot \left[\frac{4}{y + 2}\right] = \left[\frac{1}{y - 1}\right] \cdot (y + 2)(y - 1)$$

$$(y + 2)(y - 1) \cdot \frac{4}{y + 2} = \frac{1}{y - 1} \cdot (y + 2)(y - 1)$$

$$4y - 4 = y + 2$$

$$3y = 6$$

$$y = 2$$

Check.

$$\frac{4}{y + 2} = \frac{1}{y - 1}$$

$$\frac{4}{(\underline{}) + 2} \stackrel{?}{=} \frac{1}{(\underline{}) - 1} \qquad \text{Replace } y \text{ with } (2)$$

2, 2

$$\frac{4}{4} = \frac{1}{1}$$

$$1 = 1 \qquad\qquad\qquad \Box$$

Example 31 □ Solve $\dfrac{a + 2}{a - 2} + 6 = \dfrac{7a}{a + 2}$ for a.

Multiply both sides by the LCD, which is $(a - 2)(a + 2)$.

$$(a - 2)(a + 2)\left[\dfrac{a + 2}{a - 2} + 6\right] = \left[\dfrac{7a}{a + 2}\right](a - 2)(a + 2)$$

$$(a - 2)(a + 2)\dfrac{(a + 2)}{(a - 2)} + (a - 2)(a + 2)6 = \dfrac{7a(a - 2)(a + 2)}{(a + 2)}$$

$$a^2 + 4a + 4 + 6a^2 - 24 = 7a^2 - 14a$$

$$4a - 20 = -14a \quad \text{Collect like terms}$$

$$-20 = -18a$$

$$\dfrac{-20}{-18} = a$$

$$\dfrac{10}{9} = a$$

Check.

$$\dfrac{a + 2}{a - 2} + 6 = \dfrac{7a}{a + 2}$$

$$\dfrac{\left(\dfrac{10}{9}\right) + 2}{\left(\dfrac{10}{9}\right) - 2} + 6 \stackrel{?}{=} \dfrac{7 \cdot \left(\dfrac{10}{9}\right)}{\left(\dfrac{10}{9}\right) + 2} \qquad \text{Replace } a \text{ with } \dfrac{10}{9}$$

$$\dfrac{\dfrac{28}{9}}{-\dfrac{8}{9}} + 6 \stackrel{?}{=} \dfrac{\dfrac{70}{9}}{\dfrac{28}{9}}$$

$$\dfrac{28}{9} \cdot \dfrac{9}{-8} + 6 \stackrel{?}{=} \dfrac{70}{9} \cdot \dfrac{9}{28}$$

$$-\dfrac{7}{2} + \dfrac{12}{2} \stackrel{?}{=} \dfrac{10}{4}$$

$$\dfrac{5}{2} \stackrel{?}{=} \dfrac{5}{2}$$

□

Example 32 □ Solve $\dfrac{x + 3}{x - 3} + 3 = \dfrac{6}{x - 3}$ for x.

The LCD is $x - 3$.
Multiply both sides by $x - 3$.

$$(x - 3)\left[\dfrac{x + 3}{x - 3} + 3\right] = \left[\dfrac{6}{x - 3}\right] \cdot (x - 3)$$

$$\overset{1}{(x - 3)} \cdot \dfrac{x + 3}{\underset{1}{x - 3}} + (x - 3) \cdot 3 = \dfrac{6}{\underset{1}{x - 3}} \cdot \overset{1}{(x - 3)} \qquad \text{Distribute and simplify}$$

$$x + 3 + 3x - 9 = 6$$

$$4x - 6 = 6$$

$$4x = 12$$

$$x = 3$$

Check.

$$\dfrac{x + 3}{x - 3} + 3 = \dfrac{6}{x - 3}$$

$$\dfrac{(3) + 3}{(3) - 3} + 3 \overset{?}{=} \dfrac{6}{(3) - 3} \qquad \text{Replace } x \text{ with } (3)$$

$$\dfrac{6}{-} + 3 \overset{?}{=} \dfrac{6}{-}$$

Be careful; you cannot divide by zero.

0, 0

Since we can't divide by zero, 3 cannot be considered a solution to the above equation.

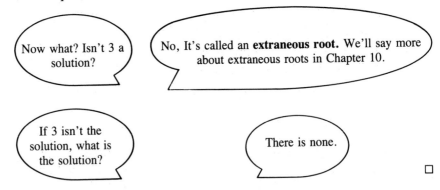

Now what? Isn't 3 a solution?

No, It's called an **extraneous root.** We'll say more about extraneous roots in Chapter 10.

If 3 isn't the solution, what is the solution?

There is none.

□

When the equations have variables in the denominator, it is particularly important to check your solution in the original equation. In Example 32 the solution $x = 3$ makes the original equation meaningless because we can't divide by zero. Since $x = 3$ is not allowed, we say that the equation has no solution.

Chapter 7 ■ Rational Expressions

Problem Set 7.4

Solve for the variable. Check your answers.

1. $\dfrac{x}{2} = \dfrac{5}{2} - \dfrac{3x}{4}$

2. $\dfrac{2x}{5} = \dfrac{3x}{5} + 3$

3. $\dfrac{6}{5x} = \dfrac{8}{5} + \dfrac{6}{x}$

4. $\dfrac{2}{3x} = \dfrac{2}{3} + \dfrac{1}{x}$

5. $\dfrac{1}{2x} = \dfrac{1}{6} - \dfrac{1}{3x}$

6. $\dfrac{4}{x} = \dfrac{5}{6} + \dfrac{3}{2x}$

7. $\dfrac{4}{x} + \dfrac{2}{x} = \dfrac{2}{3}$

8. $\dfrac{2}{x} - \dfrac{1}{6} = \dfrac{1}{3} - \dfrac{1}{x}$

9. $\dfrac{x}{2} - 1 = \dfrac{x}{6}$

10. $\dfrac{x}{2} - \dfrac{5}{4} = \dfrac{3x}{2}$

11. $\dfrac{5}{4} + \dfrac{9}{4x} = \dfrac{1}{2}$

12. $\dfrac{1}{2} + \dfrac{1}{x} = \dfrac{5}{4x}$

13. $\dfrac{2x}{2} - \dfrac{1}{4} = \dfrac{5x}{2}$

14. $\dfrac{2x}{4} + 1 = \dfrac{3x}{2}$

15. $\dfrac{1}{2} + \dfrac{1}{3x} = \dfrac{1}{3} - \dfrac{2}{3x}$

16. $\dfrac{1}{2} + \dfrac{3}{2x} = 1 - \dfrac{1}{2x}$

17. $\dfrac{x + 1}{2} = 4 + \dfrac{x - 1}{3}$

18. $\dfrac{x - 1}{5} = 1 + \dfrac{2x - 3}{3}$

19. $\dfrac{3x}{2x - 2} = 2$

20. $\dfrac{7x}{3x - 4} = 3$

21. $\dfrac{5}{x - 5} = \dfrac{x}{x - 5}$

22. $\dfrac{9}{x - 3} - \dfrac{5}{x - 3} = 2$

23. $\dfrac{4x}{3 - x} = 3 - \dfrac{5}{x - 3}$

24. $\dfrac{13}{x - 1} = \dfrac{x}{x - 1} + 3$

25. $\dfrac{4}{x + 2} = \dfrac{6}{x}$

26. $\dfrac{3}{y - 2} = \dfrac{4}{y}$

27. $\dfrac{x + 2}{3} = 1 + \dfrac{x - 2}{2}$

28. $\dfrac{x + 5}{2} = 3 - \dfrac{2x - 4}{5}$

29. $\dfrac{3x}{2x - 5} = 3$

30. $\dfrac{9x}{5x - 4} = 2$

31. $\dfrac{7}{x - 6} = \dfrac{x}{x - 6} + \dfrac{1}{x - 6}$

32. $\dfrac{7}{x + 4} - \dfrac{5}{x + 4} = 2$

33. $\dfrac{5}{y - 2} = \dfrac{4}{y}$

34. $\dfrac{11}{a + 5} = \dfrac{3}{a}$

35. $\dfrac{7}{y - 4} - \dfrac{3}{y - 4} = 2$

36. $\dfrac{3}{x - 4} + \dfrac{x}{4 - x} = \dfrac{1}{4 - x}$

37. $\dfrac{x}{x - 4} = \dfrac{4}{x - 4}$

38. $\dfrac{6}{y + 1} = \dfrac{3}{y + 1} + \dfrac{1}{2}$

39. $\dfrac{3x}{5x - 3} + \dfrac{1}{3} = \dfrac{27}{5x - 3}$

40. $\dfrac{5x}{3x - 4} + 1 = \dfrac{10x}{3x - 4}$

41. $\dfrac{3x}{2x - 1} - \dfrac{2}{7} = \dfrac{-7}{2x - 1}$

42. $\dfrac{x}{x - 5} + \dfrac{1}{5 - x} = 1$

43. $\dfrac{x+2}{2x-3} = 2 + \dfrac{2}{2x-3}$

44. $\dfrac{1}{3y-5} = \dfrac{2}{4y+5}$

45. $\dfrac{2x}{4x-7} - 2 = \dfrac{4}{4x-7}$

46. $\dfrac{5}{y-2} = \dfrac{3}{y-2} + \dfrac{1}{2}$

47. $\dfrac{2}{x-3} + \dfrac{x}{3-x} = \dfrac{1}{3-x}$

48. $\dfrac{x+2}{2x+3} = \dfrac{3}{4}$

49. $\dfrac{2x}{4x-3} + 2 = \dfrac{4}{4x-3}$

50. $\dfrac{3}{5x-7} + 3 = \dfrac{6x}{5x-7}$

51. $\dfrac{7}{y+2} = \dfrac{4}{y+2} + \dfrac{3}{5}$

52. $\dfrac{8}{y-6} - \dfrac{1}{3} = \dfrac{7}{y-6}$

53. $-\dfrac{4}{6a-7} + 2 = \dfrac{3a}{6a-7}$

54. $\dfrac{5}{7a+5} + 3 = \dfrac{6}{7a+5}$

55. $\dfrac{x+2}{3x+4} = \dfrac{3}{5}$

56. $\dfrac{2x-1}{6x-1} = \dfrac{3}{8}$

57. $\dfrac{6}{5x-8} = \dfrac{9}{3x+6}$

58. $\dfrac{8}{3x-3} = \dfrac{6}{4x+3}$

59. $\dfrac{x}{x-7} + \dfrac{3}{7-x} = 1$

60. $\dfrac{3x}{5x-1} + \dfrac{1}{3} = \dfrac{9}{5x-1}$

■ 7.5   Applications

Number Problems

To solve number problems we often have to write equations with algebraic fractions.

Example 33 □ If the same number is added to the numerator and denominator of the fraction $\dfrac{3}{7}$, the result is equivalent to $\dfrac{3}{4}$. Find the number.

Make a model.

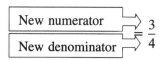

Express each component in algebraic terms.
Let $x =$ the number to be added.

$3 + x$ The original numerator was 3. The new numerator is _____

7 The original denominator was _____

$7 + x$ The new denominator is _____

            Chapter 7 ■ Rational Expressions

Translate the model into an equation.

$$\frac{3 + x}{7 + x} = \frac{3}{4}$$

4(7 + x)

The LCD is _____

$$4(7 + x)\frac{3 + x}{7 + x} = \frac{3}{4} \cdot 4(7 + x)$$

$$4(\overset{1}{\cancel{7 + x}})\frac{3 + x}{\underset{1}{\cancel{7 + x}}} = \frac{3}{\underset{1}{\cancel{4}}} \cdot \overset{1}{\cancel{4}}(7 + x)$$

$$4(3 + x) = 3(7 + x)$$

$$12 + 4x = 21 + 3x$$

$$x = 9$$

The number to be added is 9. □

Example 34 □ The denominator of a fraction exceeds the numerator by 3. If the numerator is decreased by 2 and the denominator is increased by 4, the value of the fraction is $\frac{1}{4}$. Find the value of the original fraction.

A model for the problem is

$$\frac{\text{New numerator}}{\text{New denominator}} = \frac{1}{4}$$

Let x = original numerator

y = original denominator

x − 2

_____ = new numerator

y + 4

_____ = new denominator

Use these variables to set up an equation from the model.

$$\frac{x - 2}{y + 4} = \frac{1}{4}$$

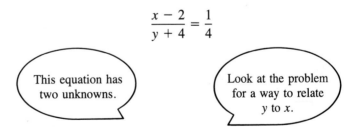

This equation has two unknowns.

Look at the problem for a way to relate y to x.

The problem states, "The denominator of a fraction exceeds the numerator by 3." Therefore, we can write a second equation

$$y = x + 3$$

Substituting $(x + 3)$ for y in the first equation yields

$$\frac{1}{4} = \frac{x - 2}{(x + 3) + 4}$$

$$\frac{1}{4} = \frac{x - 2}{x + 7}$$

$$\overset{1}{\cancel{4}}(x + 7)\frac{1}{\cancel{4}} = \frac{(x - 2)}{\cancel{(x + 7)}}(4)(\cancel{x + 7})^{1}$$

$$x + 7 = 4x - 8$$

$$15 = 3x$$

$$5 = x$$

Now that we know x, the original numerator, we can find y, the original denominator.

$$y = x + 3$$

$$y = 5 + 3$$

$$y = 8$$

The original fraction was $\dfrac{5}{8}$.

Testing the result against the problem:

$$\frac{5 - 2}{8 + 4} = \frac{3}{12}$$

$$= \frac{1}{4} \qquad \text{which is correct.} \qquad \square$$

Example 35 $\square$ Find two numbers so that one number is five times the other and the sum of their reciprocals is $\dfrac{2}{5}$.

Let x = smaller number $\qquad\qquad \dfrac{1}{x}$ = reciprocal of smaller number
$\quad\;\; y$ = larger number $\qquad\qquad\quad \dfrac{1}{y}$ = reciprocal of larger number

Writing a system of equations:

$$\frac{1}{x} + \frac{1}{y} = \frac{2}{5} \quad \Longleftarrow \quad \boxed{\text{The sum of their reciprocals is } \dfrac{2}{5}}$$

$$y = 5x \quad \Longleftarrow \quad \boxed{\text{One is five times the other}}$$

Solving the first equation by substituting $5x$ for y:

$$\frac{1}{x} + \frac{1}{(5x)} = \frac{2}{5}$$

$$5x\left(\frac{1}{x} + \frac{1}{5x}\right) = \left(\frac{2}{5}\right)(5x) \qquad \text{Multiply by } 5x \text{ to clear the denominators}$$

$$5 + 1 = 2x$$

$$6 = 2x$$

$$3 = x$$

Since $y = 5x$, the numbers are 3 and 15.
Checking

$$\frac{1}{3} + \frac{1}{15} = \frac{2}{5}$$

True.

☐

Combined Rate Problems

Many applications of algebra involve two people or machines working on the same task at different rates.

Example 36 ☐ Suppose that a garden hose can fill a swimming pool in 30 hours.

$\dfrac{15}{30}$

     What part of the pool would be filled in 15 hours? _____

$\dfrac{1}{30}$

     What part of the pool could the hose fill in one hour? _____

A pool equipped with a 2-inch filler pipe can be filled in 12 hours by the pipe alone.

$\dfrac{1}{12}$

     What part of the pool would be filled in one hour by the pipe? _____

Now suppose the pool is filled using both the garden hose and the filler pipe at the same time. How long will it take to fill the pool?

Less than the 12 hours the pipe requires.

True. To find out how long, let's consider what has happened by the end of one hour.

$\dfrac{1}{30}$

     What part of the pool did the hose fill? _____

$\dfrac{1}{12}$

     What part of the pool did the pipe fill? _____

Let t = time required to fill the pool by the hose and the pipe working together. The part of the pool filled in one hour is then $\dfrac{1}{t}$.

Now we can write an equation to represent the situation at the end of one hour.

Part of pool filled by hose		Part of pool filled by pipe		Part of pool filled by both
$\dfrac{1}{30}$	$+$	$\dfrac{1}{12}$	$=$	$\dfrac{1}{t}$

To solve for t, multiply by the LCD.

$$(60t)\left(\frac{1}{30} + \frac{1}{12}\right) = \frac{1}{t}(60t)$$

$$2t + 5t = 60$$

$$7t = 60$$

$$t = 8\frac{4}{7} \qquad \text{hours required to fill the pool} \qquad \square$$

Example 37 □ The printing plant where Mark works has a weekly job of printing programs for the summer theater. The new press can do the job in 6 hours. The older press required 9 hours to print the programs alone. One Friday, the theater delivered the material late and Mark put the job on both presses. How long did it take both presses working together to complete the job? Imagine what had happened by the end of one hour. Let x = time required for both presses to do the job.

Part done by fast press in one hour		Part done by slow press in one hour		Part done by both presses in one hour
$\dfrac{1}{6}$	$+$	$\dfrac{1}{9}$	$=$	$\dfrac{1}{x}$

$$18x\left(\frac{1}{6} + \frac{1}{9}\right) = \left(\frac{1}{x}\right)18x$$

$$\frac{18x}{6} + \frac{18x}{9} = \frac{18x}{x}$$

$$3x + 2x = 18$$

$$5x = 18$$

$$x = 3.6 \text{ hours} \qquad \square$$

Chapter 7 ■ Rational Expressions

Example 38 □ A satellite can run either its television relay for 6 hours or its heater for 2 hours before it drains its batteries. The satellite is also equipped with a solar panel that can completely recharge its batteries in 3 hours. If the batteries are fully charged, how long can all three systems (television relay, heater, and solar panel) be operated before the batteries are drained?

Let t = time all three systems can operate together.

Energy put in by solar panel	$-$	Energy used by television	$-$	Energy used by heater	$=$	Net change in energy of the satellite

At the end of one hour the situation is

$$\frac{1}{3} - \frac{1}{6} - \frac{1}{2} = \frac{1}{t}$$

$$6t\left(\frac{1}{3} - \frac{1}{6} - \frac{1}{2}\right) = \left(\frac{1}{t}\right)6t$$

$$2t - t - 3t = 6$$

$$-2t = 6$$

$$t = -3$$

How can it operate for −3 hours?

Answers do not always come out the way we expect. In this case there is a reasonable physical interpretation. Read on.

We set up our model to represent the net change in energy of the system. What we found was that after one hour the net gain was $\frac{1}{-3}$ or $\frac{-1}{3}$ of the system's capacity. In other words the system lost $\frac{1}{3}$ of its energy. Therefore, all three systems can operate for three hours before the batteries are drained.

□

Set up an equation for each problem and solve.

1. If the same number is subtracted from both the numerator and denominator of the fraction $\frac{19}{22}$, the result is equivalent to $\frac{4}{5}$. Find the number.

2. The denominator of a fraction is 4 times the numerator. If the numerator is multiplied by 3 and the denominator is increased by 5, the resulting fraction is $\frac{3}{5}$. What is the original fraction?

3. The denominator of a fraction is 6 more than the numerator. Reduced to lowest terms the fraction is $\frac{4}{7}$. Find the original fraction.

4. The numerator of a fraction is 3 times the denominator. If the numerator is increased by 6 and the denominator is decreased by 1, the fraction is $\frac{9}{2}$. What is the original fraction?

5. One number is 3 times the other, and the sum of their reciprocals is $\frac{1}{6}$. Find the numbers.

6. The numerator exceeds the denominator by 4. If the numerator is decreased by 5, and the denominator is increased by 6, the value of the resulting fraction is $\frac{1}{2}$. What is the value of the original fraction?

7. The denominator of a fraction exceeds the numerator by 3. Reduced to lowest terms, the fraction is $\frac{3}{4}$. Find the original fraction.

8. One number is one third the other, and the sum of their reciprocals is $\frac{1}{3}$. Find the numbers.

9. The denominator of a fraction exceeds the numerator by 4. If the numerator is multiplied by 2 and the denominator is increased by 10, the resulting fraction is $\frac{2}{3}$. Find the original fraction.

10. The numerator of a fraction is 4 times the denominator. If the numerator is increased by 6 and the denominator is multiplied by 3, the resulting fraction is $\frac{5}{3}$. Find the original fraction.

11. The denominator of a fraction exceeds the numerator by 4. If the numerator is multiplied by 3 and the denominator is decreased by 2, the value of the fraction is $\frac{7}{3}$. What is the value of the original fraction?

12. Jerry can do the gardening at a small park in 8 hours. The same work takes his younger brother Jack 12 hours. How long would it take the brothers to do the job if they worked together?

13. In a printing shop, John, who has little experience, can typeset a job in 9 hours. Jim, who is experienced, can finish the same job in 6 hours. Working together, how long would it take them to do the job?

14. On a camping trip Sue found that the battery in her trailer would last 24 hours if she used her lights only. However, if she used her small T.V. also, the battery would last only 18 hours. How long would the battery last if she used her T.V. only?

15. At a car wash, Jim and his brother can wash a car in 15 minutes. Mary and her sister can wash a similar car in 12 minutes. If the four people worked together, how long would it take them to wash a car?

16. A typist can complete a typing job in 8 hours. A more experienced friend can type the same amount of material in 6 hours. How long would it take them to complete the job, if they worked together?

17. A file clerk can do the month's filing in 8 hours. Two file clerks can do the job in 5 hours. How long would it take the second file clerk to do the job working alone?

18. Jeff can fill his swimming pool in 12 hours using the regular filler pipe. He can also fill the pool with his garden hose in 30 hours. The pool can drain in 15 hours. Jeff was in a hurry to fill the pool so he used the regular filler pipe and garden hose; however, he forgot to close the drain pipe. How long did it take him to fill the pool?

19. Jae, an experienced teaching assistant can grade a set of papers in the Mathematics Learning Center in 45 minutes, whereas, Kim, a less experienced person, can grade a similar set of papers in 75 minutes. How long would it take Jae and Kim to grade the papers if they worked together?

20. In a tire shop it takes John 45 minutes to install a new set of tires. Jim, an experienced fast worker can install a similar set of tires in 30 minutes. Working together, how long would it take John and Jim to install a set of tires?

Chapter 7 □ Key Ideas

7.1 **1.** To **multiply two rational expressions**

$$\frac{a}{b} \cdot \frac{c}{d} = \frac{a \cdot c}{b \cdot d}$$ where a, b, c, and d represent polynomials with b and d not equal to zero.

2. To **reduce a rational expression to lowest terms,** remove all common factors from the numerator and denominator. Divide only common factors and not terms.

3. To **multiply algebraic fractions**
  a. factor all numerators and denominators completely;
  b. divide out all possible factors in any numerator and denominator.

4. To **divide by an algebraic expression,** multiply by its multiplicative inverse

$$\frac{a}{b} \div \frac{c}{d} = \frac{a}{b} \cdot \frac{d}{c}$$

7.2 **1. Addition of fractions** with a common denominator

$$\frac{a}{c} + \frac{b}{c} = \frac{a+b}{c}$$

2. Subtraction of fractions with a common denominator

$$\frac{a}{c} - \frac{b}{c} = \frac{a}{c} + \frac{-b}{c}$$

$$= \frac{a-b}{c}$$

3. To add or subtract fractions without a common denominator:
   a. Find a common denominator, preferably the lowest common denominator.
   b. Convert all fractions to be added or subtracted to equivalent fractions with a common denominator.
   c. Add or subtract the numerators of the equivalent fractions and place the result over the common denominator.
   d. Reduce to lowest terms.

7.3 **1.** To **simplify complex fractions**
   a. Division method
      1. Simplify and write both numerator and denominator as single fractions.
      2. To divide, multiply the numerator by the reciprocal of the denominator.
   b. LCD method
      1. Find the LCD of all fractions in the numerator and denominator.
      2. Multiply the numerator and denominator of the given fraction by the LCD.
      3. Reduce to lowest terms.

7.4 **1.** To **solve equations with fractions:**
   a. Multiply both sides of the equation by the LCD of all fractions in the equation.
   b. Simplify and solve the resulting equation.
   c. Check all solutions in the original equation. (Division by zero is not possible.)

7.5 Many applications may result in problems with algebraic fractions. The applications in this chapter include number problems and combined rate problems.

Reduce. **(7.1)**

1. $\dfrac{x^2 - 8x}{x^2}$

2. $\dfrac{8x^2 - 50y^2}{8x^2 - 18xy - 5y^2}$

3. $\dfrac{27x^3 + 8y^3}{9x^2 + 12xy + 4y^2}$

4. $\dfrac{x^3 + 3x^2 - 4x - 12}{x^3 + x^2 - 6x}$

Perform the indicated operations. Reduce all answers to lowest terms. **(7.1)**

5. $\dfrac{5x - 10}{6x + 18} \cdot \dfrac{3x^2 + 18x + 27}{5x^2 + 5x - 30}$

6. $\dfrac{x^3 - 27y^3}{x^2 - 9y^2} \cdot \dfrac{x^2 + xy - 6y^2}{x^2 + 3xy + 9y^2}$

7. $\dfrac{8y^2 + 2xy - 3x^2}{27x^3 - 64y^3} \div \dfrac{3x^2 - 2xy - 8y^2}{9x^2 + 12xy + 16y^2}$

8. $\dfrac{x^3 - x + 3x^2 - 3}{2x^2 + 5x - 3} \div \dfrac{3x^2 - x - 4}{3x^2 - 7x + 4}$

9. $\left(\dfrac{ax - 2ay - bx + 2by}{6a^3 + 8a^2b} \div \dfrac{x^3 - 8y^3}{3a^2 + 7ab + 4b^2} \right) \cdot \dfrac{x^2 + 2xy + 4y^2}{a^2 - b^2}$

Add or subtract. Reduce your answers to lowest terms. **(7.2)**

10. $\dfrac{4}{x^2 + 4x} + \dfrac{1}{x + 4}$

11. $\dfrac{y}{y^3 - 27} - \dfrac{1}{y^2 + 3y - 18}$

12. $1 - \dfrac{x^2 + 2x}{x^2 + 3x + 2}$

13. $\dfrac{x}{x^2 - 12x + 20} - \dfrac{1}{8 - 4x}$

14. $\dfrac{5}{x + 3} - \dfrac{4x - 13}{x^2 + x - 6}$

Simplify the following complex fractions. Reduce your answers to lowest terms. **(7.3)**

15. $\dfrac{\dfrac{a}{b} + 1}{\dfrac{a^2}{b^2} - 1}$

16. $\dfrac{x^2 - \dfrac{x^3}{x - y}}{\dfrac{y^3}{y - x} - y^2}$

17. $\dfrac{x - 3 - \dfrac{6}{x + 2}}{x + 4 + \dfrac{1}{x + 2}}$

Solve for the variable. Check your answers.

18. $\dfrac{8x}{5x - 3} = 7$

19. $\dfrac{5}{2y + 3} = \dfrac{7}{4y - 3}$

20. $\dfrac{x - 3}{3x - 4} = 4 - \dfrac{9}{3x - 4}$

Set up an equation for each of the following and solve.

21. The numerator of a fraction is one third of the denominator. If the numerator is decreased by 2 and the denominator is divided by 2, the value of the fraction is $\dfrac{1}{2}$. What is the value of the original fraction?

22. At a pie shop Roberto can assemble 50 pies for baking in 3 hours. It takes Jim 5 hours to assemble 50 pies. If they work together, how long would it take Roberto and Jim working together to assemble 50 pies?

8 Rational Exponents and Radicals

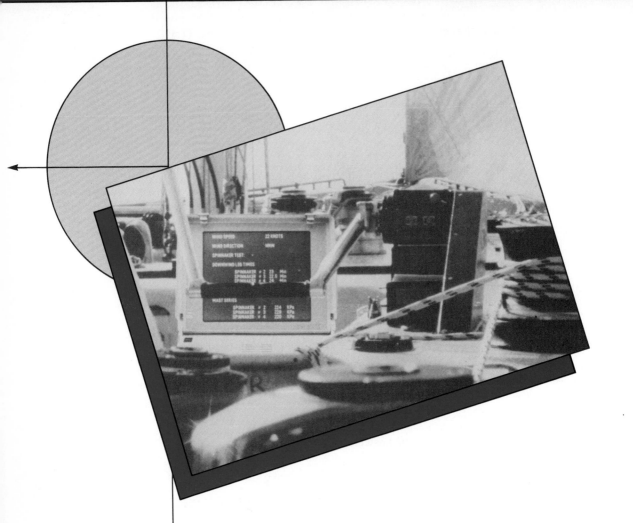

Contents

Preview

Earlier chapters dealt with equations that could be represented as straight lines. Later chapters will deal with equations whose graphs are curved lines. Equations of curved lines all have terms that are raised to some power. The solutions to these equations require the ability to manipulate radicals or expressions with rational exponents. The equations that tell how to predict the maximum speed of a sailboat as a function of its length are one example of how radical expressions are used.

This chapter will build with radicals and rational exponents the skills you will need to do well in later chapters. As you progress through the chapter, bear in mind that radicals and fractional exponents are just two different forms of notation for the same idea. Also remember that even though some of the expressions appear complex, they can be manipulated by applying a few basic rules. The laws of exponents don't change. Remember, exponents act on a single number. But, also recall that any algebraic expression, no matter how complex, represents a single number. Therefore the laws of exponents remain the same no matter how complicated an expression appears.

■ 8.1　Exponents and Radicals

Chapter 5 defined the laws of exponents for integer exponents. They are repeated here for your convenience.

Laws of Exponents

Multiplication of Powers	$a^x \cdot a^y = a^{x+y}$
Division of Powers	$\dfrac{a^x}{a^y} = a^{x-y}$
Power to a Power	$(a^x)^y = a^{x \cdot y}$
Product to a Power	$(a^x b^y)^z = a^{xz} \cdot b^{yz}$
Quotient to a Power	$\left(\dfrac{a^x}{a^y}\right)^z = \dfrac{a^{xz}}{a^{yz}}$
Zero Exponent	$a^0 = 1 \quad a \neq 0$
Negative Exponent	$a^{-x} = \dfrac{1}{a^x}$

The laws of exponents were developed with the understanding that all exponents would be integers.

More advanced math courses can prove that the laws of exponents hold for exponents that are any real number. We must, however, insist that, except for a few special cases, the bases are positive numbers.

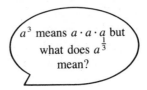

a^3 means $a \cdot a \cdot a$ but what does $a^{\frac{1}{3}}$ mean?

Assuming that the laws of exponents do apply to real exponents, let's see if we can give a reasonable interpretation to exponents that are not integers but are rational numbers.

Suppose that $a^x \cdot a^x = a^1$.

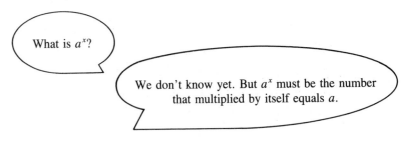

What is a^x?

We don't know yet. But a^x must be the number that multiplied by itself equals a.

From the laws of exponents

    if $a^x \cdot a^x = a^1$

                $a^{x+x} = a^1$

    then $x + x = 1$ Since the bases are equal, the

                  $2x = 1$ exponents must be equal.

$$x = \frac{1}{2}$$

Using the rule for multiplication of powers, we can say $a^{\frac{1}{2}} \cdot a^{\frac{1}{2}} = a$.

8.1A Definition: Square Root

$a^{\frac{1}{2}}$ is one of two equal factors whose product is a. $a^{\frac{1}{2}}$ is referred to as the **second root of a** or the **square root of a.**

$$a^{\frac{1}{2}} \cdot a^{\frac{1}{2}} = a$$

Chapter 8 ■ Rational Exponents and Radicals

Example 1 □ $9^{\frac{1}{2}} = 3$ because $3 \cdot 3 = 9$

                $4^{\frac{1}{2}} = 2$ because $2 \cdot 2 = 4$ □

Computers generally require that if the exponent is not an integer, then the base must be a positive number. As long as we keep this restriction, all the rules of integer exponents hold exactly.

What's $7^{\frac{1}{2}}$?

That's some irrational number greater than 2 and less than 3.

When we can't use decimals to express exactly what a square root is, we use another symbol to express the square root.

$$7^{\frac{1}{2}} \text{ is one way we indicate the square root of 7}$$

and $7^{\frac{1}{2}} \cdot 7^{\frac{1}{2}} = 7$

$$\left(213^{\frac{1}{2}}\right)\left(213^{\frac{1}{2}}\right) = 213$$

Because $213^{\frac{1}{2}}$ is the number that multiplied by itself equals 213.

$$16^{\frac{1}{2}} \cdot 16^{\frac{1}{2}} = 16$$

There are two possible ways to interpret the value of $16^{\frac{1}{2}}$. They are 4 and -4 because

$$(4)(4) = 16$$
$$(-4)(-4) = 16$$

There are indeed two square roots of 16. But, since we would like a symbol to stand for only one number, we say $a^{\frac{1}{2}}$ is the principal or positive square root of a.

$$16^{\frac{1}{2}} = +4$$

To indicate the negative square root of 16, we would write

$$-16^{\frac{1}{2}} = -4$$

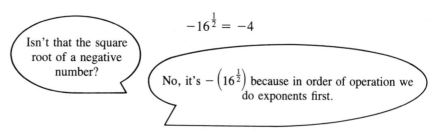

Isn't that the square root of a negative number?

No, it's $-\left(16^{\frac{1}{2}}\right)$ because in order of operation we do exponents first.

We use other exponents of the form $\dfrac{1}{n}$.

$$a^{\frac{1}{3}} \cdot a^{\frac{1}{3}} \cdot a^{\frac{1}{3}} = a^{\frac{3}{3}}$$
$$= a$$

and

$$a^{\frac{1}{4}} \cdot a^{\frac{1}{4}} \cdot a^{\frac{1}{4}} \cdot a^{\frac{1}{4}} = a^{\frac{4}{4}}$$
$$= a$$

8.1B   Definition: nth Root of a

$a^{\frac{1}{n}}$ for any natural number n is called the **nth root of a.** $\left(a^{\frac{1}{n}}\right)$ used as a factor n times gives a product of a.

Example 2 □

a. $16^{\frac{1}{4}} = 2$ because $2 \cdot 2 \cdot 2 \cdot 2 = 16$

b. $27^{\frac{1}{3}} = 3$ because $3 \cdot 3 \cdot 3 = 27$

c. $64^{\frac{1}{3}} = 4$ because $4 \cdot 4 \cdot 4 = 64$

d. $243^{\frac{1}{5}} = 3$ because $3 \cdot 3 \cdot 3 \cdot 3 \cdot 3 = 243$ □

Notice in Definition 8.1B the exponent was of the form $\dfrac{1}{n}$.

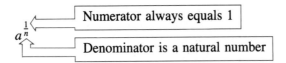

$a^{\frac{1}{n}}$ Numerator always equals 1

Denominator is a natural number

What would powers of the form $a^{\frac{m}{n}}$ mean?
We can use the laws of exponents to write a definition.

8.1C   Property of Exponents: for $a > 0$

$$a^{\frac{m}{n}} = \left(a^{\frac{1}{n}}\right)^{m} \qquad \text{or} \qquad a^{\frac{m}{n}} = (a^{m})^{\frac{1}{n}}$$

Exponents of the form $\dfrac{m}{n}$ require a two-step simplification.

Example 3 □

$$64^{\frac{3}{2}} = \left(64^{\frac{1}{2}}\right)^3$$
$$= 8^3$$
$$= 512$$

Could that have been $(64^3)^{\frac{1}{2}}$?

Yes, but $(64^3)^{\frac{1}{2}} = 262144^{\frac{1}{2}}$
$= 512$
which is more arithmetic than I want to do.

□

8.1D   Rule: Multiplication of Powers

To multiply two powers with the same base, add the exponents.

$$a^x \cdot a^y = a^{x+y}$$

Example 4 □

a. $16^{\frac{1}{2}} \cdot 16^{\frac{1}{2}} = 16^1$

b. $x^{\frac{1}{2}} \cdot x^{\frac{1}{3}} = x^{\frac{1}{2}+\frac{1}{3}} = x^{\frac{5}{6}}$

c. $y^{\frac{1}{3}} \cdot y^{-\frac{1}{3}} = y^0 = 1$

d. $3a^{-2} \cdot 2a^{-\frac{1}{2}} = 6a^{-\frac{5}{2}}$ □

8.1E   Rule: Division of Powers

To divide two powers with the same base, subtract the exponents.

$$\frac{a^x}{a^y} = a^{x-y} \qquad a \neq 0$$

Example 5 □

a. $\dfrac{16^{\frac{1}{2}}}{16^{\frac{1}{4}}} = 16^{\frac{1}{2}-\frac{1}{4}}$

$= 16^{\frac{1}{4}}$
$= 2$

b. $\dfrac{x^{\frac{3}{4}}}{x^2} = x^{\frac{3}{4}-2}$

$= x^{\frac{3}{4}-\frac{8}{4}}$

$= x^{-\frac{5}{4}} = \dfrac{1}{x^{\frac{5}{4}}}$

Why make the last step of part B?

c. $\dfrac{p^{-\frac{2}{7}}}{p^{-\frac{5}{7}}} = p^{-\frac{2}{7} - \left(-\frac{5}{7}\right)} = p^{\frac{3}{7}}$

Sometimes we prefer answers without negative exponents.

□

8.1F Rule: Power Raised to a Power

To raise a power to a power, multiply exponents.

$$(a^x)^y = a^{x \cdot y}$$

Example 6 □

a. $\left(y^{\frac{1}{2}}\right)^2 = y^{\frac{2}{2}}$
$= y$

b. $\left(x^{\frac{2}{3}}\right)^4 = x^{\frac{8}{3}}$

c. $\left(y^4 z^{-\frac{1}{2}}\right)^3 = y^{12} z^{-\frac{3}{2}}$

d. $\left(y^{-\frac{2}{3}}\right)^{-\frac{3}{4}} = y^{\left(-\frac{2}{3}\right) \cdot \left(-\frac{3}{4}\right)}$
$= y^{\frac{1}{2}}$ □

Frequently the power to a power rule can be used to simplify expressions with fractional exponents.

Recall $\qquad x^{\frac{m}{n}} = (x^m)^{\frac{1}{n}} \qquad$ or $\qquad \left(x^{\frac{1}{n}}\right)^m$

Example 7 □

a. $8^{\frac{2}{3}} = \left(8^{\frac{1}{3}}\right)^2$ Take the cube root of 8 first.
$= (2)^2$
$= 4$

c. $64^{\frac{5}{3}} = \left(64^{\frac{1}{3}}\right)^5$
$= (4)^5$
$= 1024$

b. $100^{-\frac{3}{2}} = \left(100^{\frac{1}{2}}\right)^{-3}$ $100^{\frac{1}{2}} = 10$
$= (10)^{-3}$
$= \dfrac{1}{10^3}$
$= \dfrac{1}{1000}$ □

Problem Set 8.1A

Simplify the following expressions. Write answers without negative exponents.

1. $49^{\frac{1}{2}}$ **2.** $125^{\frac{1}{3}}$ **3.** $16^{\frac{1}{4}}$ **4.** $32^{\frac{1}{5}}$

5. $8^{\frac{2}{3}}$ **6.** $16^{-\frac{3}{4}}$ **7.** $64^{-\frac{2}{3}}$ **8.** $4^{\frac{3}{2}}$

9. $27^{-\frac{4}{3}}$ **10.** $4^{\frac{5}{2}}$ **11.** $9^{\frac{5}{2}}$ **12.** $8^{-\frac{5}{3}}$

13. $25^{-\frac{3}{2}}$

14. $16^{-\frac{5}{4}}$

15. $8^{-\frac{2}{3}}$

16. $32^{-\frac{6}{5}}$

17. $9^{-\frac{3}{2}}$

18. $27^{-\frac{5}{3}}$

19. $32^{\frac{3}{5}}$

20. $64^{\frac{5}{6}}$

21. $16^{\frac{3}{2}}$

22. $125^{\frac{2}{3}}$

23. $125^{-\frac{1}{3}}$

24. $27^{-\frac{2}{3}}$

25. $9^{-\frac{7}{2}}$

26. $4^{-\frac{7}{2}}$

27. $32^{-\frac{4}{5}}$

28. $27^{-\frac{4}{3}}$

29. $a^{\frac{1}{2}} \cdot a^{\frac{2}{3}}$

30. $x^{\frac{3}{5}} \cdot x^{\frac{1}{5}}$

31. $y^{\frac{2}{3}} \cdot y^{-\frac{2}{3}}$

32. $x^{\frac{3}{4}} \cdot x^{-\frac{1}{3}}$

33. $4a^{\frac{2}{3}} \cdot 6a^{\frac{5}{6}}$

34. $5y^{-\frac{1}{4}} \cdot 6y^{-\frac{3}{5}}$

35. $-4x^{-\frac{2}{3}} \cdot 8y^{-\frac{3}{4}}$

36. $\dfrac{x^{\frac{2}{3}}}{x^2}$

37. $\dfrac{x^{-2}}{x^{\frac{1}{2}}}$

38. $\dfrac{\left(x^{\frac{1}{2}}\right)\left(y^{-\frac{2}{3}}\right)}{x^{-1}y^3}$

39. $a^{\frac{2}{5}} \cdot a^{-\frac{7}{5}}$

40. $a^{\frac{3}{4}} \cdot a^{\frac{2}{3}}$

41. $x^{\frac{3}{5}} \cdot x^{-\frac{1}{3}}$

42. $y^{\frac{3}{4}} \cdot y^{-\frac{5}{6}}$

43. $-3x^{\frac{5}{3}} \cdot 4x^{-\frac{3}{4}}$

44. $-4y^{-\frac{2}{3}} \cdot 6y^{\frac{5}{6}}$

45. $\dfrac{x^{\frac{1}{2}}}{x^{\frac{2}{3}}}$

46. $\dfrac{x^{\frac{3}{4}}}{x^{-2}}$

47. $\dfrac{x^{\frac{2}{3}} \cdot y^{-\frac{3}{4}}}{x^{\frac{1}{2}} \cdot y^{-\frac{5}{6}}}$

48. $\dfrac{x^{-2} \cdot y^{-\frac{1}{3}}}{x^{\frac{3}{2}} \cdot y^3}$

49. $\left(\dfrac{a^{\frac{3}{4}}}{b^{\frac{1}{2}}}\right)^4$

50. $\left(\dfrac{x^{\frac{4}{3}}}{y^{-\frac{2}{3}}}\right)^3$

51. $\left(\dfrac{a^{-\frac{2}{3}}}{b^2}\right)^{-\frac{3}{2}}$

52. $\left(8a^{-\frac{2}{3}}b^3\right)^{-\frac{2}{3}}$

53. $\left(\dfrac{4x^{-\frac{2}{3}}}{9y^{-\frac{1}{2}}}\right)^{-\frac{1}{2}}$

54. $\left(8x^3y^{-6}\right)^{\frac{2}{3}}$

55. $\left(\dfrac{x^{\frac{1}{2}}y^{-\frac{2}{3}}}{16}\right)^{-\frac{3}{4}}$

56. $\left(a^{\frac{1}{3}}b^{-\frac{3}{4}}\right)^{12}$

57. $\left(\dfrac{a^{\frac{2}{3}}}{b^{\frac{1}{4}}y^{\frac{1}{6}}}\right)^{-12}$

58. $\left(\dfrac{x^{-\frac{5}{6}}y^{\frac{1}{4}}}{a^3}\right)^{-8}$

59. $\dfrac{a^{-\frac{3}{4}}b^2}{a^{\frac{5}{3}}b^{-\frac{2}{3}}}$

60. $\dfrac{x^{\frac{2}{3}}y^4}{x^{\frac{5}{6}}y^{\frac{2}{5}}}$

61. $\left(\dfrac{x^{\frac{3}{4}}}{y^{\frac{2}{3}}}\right)^{12}$

62. $\left(\dfrac{a^{-\frac{1}{2}}}{b^{-\frac{2}{3}}}\right)^6$

63. $\left(\dfrac{a^{-\frac{4}{3}}}{b^4}\right)^{\frac{3}{4}}$

64. $\left(\dfrac{x^{-\frac{2}{3}}}{y^{-4}}\right)^{-\frac{3}{2}}$

65. $\left(32x^{-\frac{5}{3}}y^5\right)^{-\frac{2}{5}}$

66. $\left(\dfrac{16x^{-\frac{4}{3}}}{9y^{-2}}\right)^{-\frac{1}{2}}$

67. $\left(27x^6y^{-3}\right)^{\frac{2}{3}}$

68. $\left(\dfrac{a^{\frac{1}{3}}y^{-\frac{3}{2}}}{8}\right)^{-\frac{2}{3}}$

69. $\left(a^{-\frac{2}{3}}b^{-\frac{5}{6}}\right)^{12}$

70. $\left(\dfrac{x^{\frac{3}{4}}}{y^{-\frac{1}{3}}a^{\frac{5}{6}}}\right)^{-12}$

71. $\left(\dfrac{x^{-\frac{3}{5}}y^{\frac{1}{2}}}{a^3}\right)^{-10}$

72. $\dfrac{a^{-\frac{1}{6}}b^3}{a^{\frac{4}{3}}b^{-\frac{4}{3}}}$

Until now we have insisted that with expressions of the form $a^{\frac{m}{n}}$ the base be a positive number. For a few special cases it is possible to define powers when the base is negative. Let's examine these special cases.

Case I. $a^{\frac{m}{n}}$ where $n = 1$

In this case the exponent is an integer and we are dealing with a^m where m may be any integer and the base a may be any positive or negative number.

Example 8 □

a. $2^3 = 8$ **b.** $(-2)^3 = -8$ **c.** $2^{-3} = \dfrac{1}{8}$

d. $(-2)^{-3} = \dfrac{1}{(-2)^3}$ **e.** $(-24)^0 = 1$

$\phantom{(-2)^{-3}} = \dfrac{1}{-8}$

$\phantom{(-2)^{-3}} = \dfrac{-1}{8}$ □

Case II. $a^{\frac{m}{n}}$ where $m = 1$ and $n = $ an even integer, $n \neq 0$

In this case, we have $a^{\frac{1}{n}}$. We are asking for an even root of a number. If the base is positive, we can evaluate $a^{\frac{1}{n}}$.

In general there are n different values for $a^{\frac{1}{n}}$. However, most of these values are complex numbers. How to find all of them is usually part of a trigonometry course.

Consider $4^{\frac{1}{2}}$.

$$\text{Both} \qquad (+2)(+2) = +4$$
$$(-2)(-2) = +4$$

Therefore, there are two values for $4^{\frac{1}{2}}$. We call the positive value **the principal square root** of 4.

In general, if the base is positive,

$a^{\frac{1}{2}}$ refers to the positive square root of a, and

$-a^{\frac{1}{2}}$ refers to the negative square root of a.

If the base a is negative, we cannot find any even real roots.

In other words $(-4)^{\frac{1}{2}}$ is not a real number.

How about
$(-2)(+2)$
$= -4$?

Remember, an nth root requires n identical factors.
So -2 is not identical to $+2$.

Even roots of negative numbers are not real numbers.

Case III. $a^{\frac{m}{n}}$ **where** $m = 1$ **and** n **is an odd integer**

This time, we are looking at powers like $8^{\frac{1}{3}}$, $a^{\frac{1}{5}}$, $a^{-\frac{1}{7}}$, $b^{-\frac{1}{9}}$.

These are called **odd roots.**

We can find odd roots of both positive and negative numbers.

When n is an odd number, if a is positive, $a^{\frac{1}{n}}$ is positive, and if a is negative, $a^{\frac{1}{n}}$ is negative.

Example 9 □

a. $+8^{\frac{1}{3}} = +2$ **b.** $(-8)^{\frac{1}{3}} = -2$

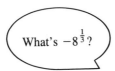

What's $-8^{\frac{1}{3}}$?

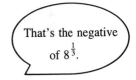

That's the negative
of $8^{\frac{1}{3}}$.

c. $(-27)^{\frac{1}{3}} = -3$ because $(-3)(-3)(-3) = -27$

d. $-27^{\frac{1}{3}} = -(27^{\frac{1}{3}}) = -(3)$ **e.** $27^{\frac{1}{3}} = 3$

f. $32^{\frac{1}{5}} = 2$ **g.** $(-32)^{\frac{1}{5}} = -2$

h. $(32)^{-\frac{1}{5}} = \dfrac{1}{32^{\frac{1}{5}}} = \dfrac{1}{2}$ **i.** $(-32)^{-\frac{1}{5}} = \dfrac{1}{(-32)^{\frac{1}{5}}} = \dfrac{1}{-2} = \dfrac{-1}{2}$

A negative
exponent means the
same power in the
denominator.

It doesn't
have anything to do
with the sign of the
denominator.

□

Here is a summary of the special cases for $a^{\frac{m}{n}}$.
We always require that $a \neq 0$.

If $n = 0$, we have division by zero and nothing is defined.

If $n = 1$ and m is any integer, the rules for integer exponents apply.

If $m = 0$ and $n \neq 0$, we have $a^0 = 1$.

If $m = 1$, we have the following.

	a is positive	a is negative
n is even	$a^{\frac{1}{n}}$ is the positive nth root of a $-a^{\frac{1}{n}}$ is the negative nth root of a	not defined
n is odd	$a^{\frac{1}{n}}$ is positive and the only real nth root of a	$a^{\frac{1}{n}}$ is negative and the only real nth root of a

Let's look at another case that deserves special attention.

$$\text{What is } (x^n)^{\frac{1}{n}}?$$

If we multiply the exponents

$$(x^n)^{\frac{1}{n}} = x^{\frac{n}{n}} = x$$

But consider $x = -4$ and $n = +2$

$$[(-4)^2]^{\frac{1}{2}} = 16^{\frac{1}{2}}$$
$$= 4 \quad \Leftarrow \boxed{\text{This is not } x = -4}$$

How about $x = -4$ and $n = 3$?

$$[(-4)^3]^{\frac{1}{3}} = (-64)^{\frac{1}{3}}$$
$$= -4 \quad \Leftarrow \boxed{\text{This is equal to } x = -4}$$

8.1G   Definition: nth Root of x^n

$$(x^n)^{\frac{1}{n}} = x \qquad \text{if } n \text{ is odd}$$
$$(x^n)^{\frac{1}{n}} = |x| \qquad \text{if } n \text{ is even}$$

Example 10 □

a. $(3^2)^{\frac{1}{2}} = 3$　　　**b.** $[(-3)^2]^{\frac{1}{2}} = 3$

c. $(2^3)^{\frac{1}{3}} = 2$　　　**d.** $\left((-2)^3\right)^{\frac{1}{3}} = -2$

> Compare Example 10a with 10b. Then notice the difference between 10c and 10d.

□

Problem Set 8.1B

Simplify the following.

1. 2^4　　**2.** $(-2)^4$　　**3.** 2^5　　**4.** $(-2)^5$　　**5.** $27^{\frac{1}{3}}$　　**6.** $(-27)^{\frac{1}{3}}$

7. $8^{\frac{1}{3}}$　　**8.** $8^{-\frac{1}{3}}$　　**9.** $16^{\frac{1}{2}}$　　**10.** $(-16)^{\frac{1}{2}}$　　**11.** $-(10)^0$　　**12.** $(-10)^0$

13. $(-8)^{\frac{1}{3}}$　　**14.** $(-8)^{-\frac{1}{3}}$　　**15.** $-(-8)^{\frac{1}{3}}$　　**16.** $-(-8)^{-\frac{1}{3}}$　　**17.** -3^4　　**18.** $(-3)^4$

19. $(-8)^{\frac{1}{3}}$　　**20.** $(-8)^{-\frac{1}{3}}$　　**21.** -5^{-2}　　**22.** $(-5)^{-2}$　　**23.** $(-8)^0$　　**24.** -8^0

25. $-(-25)^{\frac{1}{2}}$　**26.** $-25^{\frac{1}{2}}$　　**27.** $-(-27)^{\frac{1}{3}}$　**28.** $-27^{\frac{1}{3}}$　　**29.** $(-32)^{-\frac{1}{5}}$　**30.** $-(32)^{-\frac{1}{5}}$

31. $-64^{\frac{1}{2}}$　　**32.** $(-64)^{\frac{1}{2}}$　　**33.** $(4^2)^{\frac{1}{2}}$　　**34.** $[(-4)^2]^{\frac{1}{2}}$　　**35.** $-(4^2)^{\frac{1}{2}}$　　**36.** $-[(-4)^2]^{\frac{1}{2}}$

37. $(2^5)^{\frac{1}{5}}$　　**38.** $[(-2)^5]^{\frac{1}{5}}$　　**39.** $-(2^5)^{\frac{1}{5}}$　　**40.** $-[(-2)^5]^{\frac{1}{5}}$　　**41.** $(9^2)^{\frac{1}{2}}$　　**42.** $[(-9)^2]^{\frac{1}{2}}$

43. $-(9^2)^{\frac{1}{2}}$　　**44.** $-[(-4)^2]^{\frac{1}{2}}$　**45.** $(3^3)^{\frac{1}{3}}$　　**46.** $[(-3)^3]^{\frac{1}{3}}$　　**47.** $-(3^3)^{\frac{1}{3}}$　　**48.** $-[(-3)^{-3}]^{\frac{1}{3}}$

■ **8.2**　　　　　**Radical Notation**

The previous section used exponential notation for roots.
An equivalent notation for roots is called **radical notation.**

$\sqrt{a}$ is equivalent to $a^{\frac{1}{2}}$. It represents the principal or non-negative square root of "a" in the same manner as $a^{\frac{1}{2}}$.

$\sqrt[3]{a}$, read "the cube root of a," is equivalent to $a^{\frac{1}{3}}$.

$\sqrt[n]{a}$, read "the nth root of a," is equivalent to $a^{\frac{1}{n}}$.

The symbol $\sqrt{}$ is referred to as a **radical sign.** The number under the radical sign is called the **radicand.** The number shown here $\sqrt[\nwarrow]{}$ in the radical sign is called the **index.**

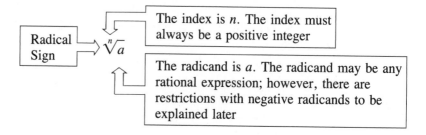

$\sqrt[3]{64}$ is read "the cube root of 64." $\sqrt[3]{64} = 4$ because $4 \cdot 4 \cdot 4 = 64$

Radicals with indexes greater than 3 do not have special names.

$\sqrt[4]{625}$ is read "the fourth root of 625."

8.2A   Definition: Radical Notation

When $a > 0$ and m and n are integers

$$\text{since} \qquad a^{\frac{m}{n}} = (a^m)^{\frac{1}{n}} \qquad \text{and} \qquad a^{\frac{m}{n}} = \left(a^{\frac{1}{n}}\right)^m$$

we define

$$\sqrt[n]{a^m} = a^{\frac{m}{n}} \qquad \text{and} \qquad (\sqrt[n]{a})^m = a^{\frac{m}{n}}$$

> $\sqrt{4}$ could be $+2$ or -2. Which do you mean?

> The notation $\sqrt[n]{a}$ refers to the principal root (the non-negative root) when n is even. Therefore, $\sqrt{4}$ represents $+2$.

Example 11 ☐ Write in exponential form.

$$\sqrt[3]{x^2} = x^{\frac{2}{3}}$$

$$(\sqrt[3]{x})^2 = x^{\frac{2}{3}}$$

$$\sqrt[5]{(a + b)^4} = \underline{\qquad\qquad}$$

$(a + b)^{\frac{4}{5}}$

> Does that mean $\sqrt[n]{x^m} = (\sqrt[n]{x})^m$?

> Yes, but the restrictions in the table on page 322 apply.

☐

Any expression involving radicals can be simplified by converting the expression to exponential notation. However, we frequently work with expressions in radical notation.

Since any expression of the form $\sqrt[n]{a^m}$ is equivalent to $a^{\frac{m}{n}}$, the same restrictions in m and n apply. We will develop the properties of radicals from the equivalent exponential property.

8.2B   Multiplication Law for Radicals

$$\text{Because} \qquad a^{\frac{1}{n}} \cdot b^{\frac{1}{n}} = (ab)^{\frac{1}{n}}$$

$$\sqrt[n]{a}\; \sqrt[n]{b} = \sqrt[n]{ab}$$

The product of two roots is the root of the product.

Example 12 □

a. $\sqrt{4}\; \sqrt{9} = \sqrt{4 \cdot 9}$

b. $\sqrt[3]{x}\; \sqrt[3]{y} = \sqrt[3]{x \cdot y}$

c. $\sqrt{x + 3}\; \sqrt{x - 3} = \sqrt{x^2 - 9}$

> Notice that the index of both radicals must be identical.

□

Example 13 □ Find each of the following.

a. $\sqrt[3]{-27} = -3$

b. $\sqrt[3]{125} = $ _____

c. $\sqrt[3]{-64} = $ _____

d. $\sqrt[6]{64} = $ _____

e. $\sqrt[5]{-32} = $ _____

f. $\sqrt{\dfrac{1}{16}} = $ _____

g. $\sqrt[3]{-\dfrac{1}{27}} = $ _____

h. $\sqrt[4]{\dfrac{16}{81}} = $ _____

□

As you work with roots and factoring you will find it handy to be able to recognize the powers in this table.

n	n^2	n^3	n^4	n^5	n^6
2	4	8	16	32	64
3	9	27	81	243	
4	16	64	256	1024	
5	25	125	625		
6	36	216			
7	49	343			

The multiplication law for radicals can be used in reverse order. For instance, $\sqrt{a^3}$ can be viewed as $\sqrt{a^2 \cdot a} = \sqrt{a^2} \cdot \sqrt{a}$ which is a $\sqrt{a}$. Using the same reasoning $\sqrt{50}$ can be viewed as $\sqrt{25 \cdot 2} = \sqrt{25} \cdot \sqrt{2}$ which is $5\sqrt{2}$.

5

−4, 2

−2, $\dfrac{1}{4}$

$-\dfrac{1}{3}, \dfrac{2}{3}$

Example 14 □ Simplify $\sqrt{20a^3}$.

$$\sqrt{20a^3} = \sqrt{4 \cdot 5 \cdot a \cdot a^2}$$
$$= \sqrt{4}\,\sqrt{5}\,\sqrt{a}\,\sqrt{a^2}$$
$$= 2 \cdot \sqrt{5} \cdot \sqrt{a} \cdot a$$
$$= 2 \cdot a \cdot \sqrt{5}\,\sqrt{a}$$
$$= 2a\sqrt{5a} \qquad \qquad \square$$

Could I write 20 as $2 \cdot 10$?

Yes, but it won't lead anywhere. We want to find factors that are perfect squares.

Example 15 □ Simplify $\sqrt[4]{32x^6y^5}$.

$$\sqrt[4]{32x^6y^5} = \sqrt[4]{2 \cdot 16 \cdot x^4 \cdot x^2 \cdot y^4 \cdot y}$$
$$= \sqrt[4]{16}\,\sqrt[4]{x^4}\,\sqrt[4]{y^4}\,\sqrt[4]{2 \cdot x^2y}$$
$$= 2xy\sqrt[4]{2x^2y} \qquad \qquad \square$$

Example 16 □ Simplify $\sqrt[4]{a^4 + b^4}$.

$\sqrt[4]{a^4 + b^4}$ is in simplified form. $\qquad \square$

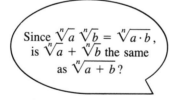

Since $\sqrt[n]{a}\,\sqrt[n]{b} = \sqrt[n]{a \cdot b}$, is $\sqrt[n]{a} + \sqrt[n]{b}$ the same as $\sqrt[n]{a + b}$?

No! Check below.

Example 17 □

10, 10

a. $\sqrt{4} \cdot \sqrt{25} = $ _____   **b.** $\sqrt{4 \cdot 25} = $ _____ $\qquad \square$

Example 18 □

7, 5

a. $\sqrt{16} + \sqrt{9} = $ ____   **b.** $\sqrt{16 + 9} = $ _____

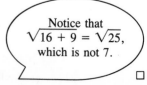

Notice that $\sqrt{16 + 9} = \sqrt{25}$, which is not 7.

$\qquad \square$

8.2C Division Law for Radicals

Because

$$\frac{a^{\frac{1}{n}}}{b^{\frac{1}{n}}} = \left(\frac{a}{b}\right)^{\frac{1}{n}} \qquad b \neq 0$$

$$\frac{\sqrt[n]{a}}{\sqrt[n]{b}} = \sqrt[n]{\frac{a}{b}} \qquad b \neq 0$$

The quotient of two roots is the root of the quotient.

A good first step in simplifying the quotient of two radicals with the same index is to write the quotient under a single radical and simplify before extracting roots.

Example 19 □

a. $\dfrac{\sqrt{27}}{\sqrt{3}} = \sqrt{\dfrac{27}{3}}$

$\qquad = \sqrt{9}$

$\qquad = 3$

b. $\dfrac{\sqrt[4]{32a^6}}{\sqrt[4]{2a^2}} = \sqrt[4]{\dfrac{32a^6}{2a^2}}$

$\qquad = \sqrt[4]{16a^4}$

$\qquad = 2|a|$

Why the absolute value sign on $|a|$?

In case a is negative. Recall that $\sqrt[4]{(-3)^4} = +3$.

In general if n is even, then $\sqrt[n]{a^n} = |a|$. □

Rationalizing the Denominator

Frequently when we are looking for common denominators, it's easier to write all expressions without radicals in the denominator. The process of eliminating radicals from the denominator is called **rationalizing the denominator.**

Example 20 □ Rationalize $\dfrac{1}{\sqrt[3]{2}}$.

We use two facts to eliminate the $\sqrt[3]{2}$ in the denominator.

Fact 1. $\qquad \sqrt[3]{2}\ \sqrt[3]{2}\ \sqrt[3]{2} = 2$

Fact 2. $\qquad \dfrac{\sqrt[3]{2}}{\sqrt[3]{2}} = 1$

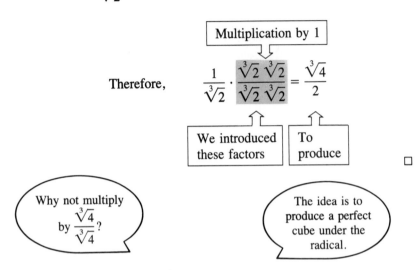

Multiplication by 1

Therefore, $\qquad \dfrac{1}{\sqrt[3]{2}} \cdot \boxed{\dfrac{\sqrt[3]{2}\ \sqrt[3]{2}}{\sqrt[3]{2}\ \sqrt[3]{2}}} = \dfrac{\sqrt[3]{4}}{2}$

We introduced these factors | To produce

□

Why not multiply by $\dfrac{\sqrt[3]{4}}{\sqrt[3]{4}}$?

The idea is to produce a perfect cube under the radical.

Example 21 □ Rationalize $\dfrac{1}{\sqrt[4]{x}}$.

$$\dfrac{1}{\sqrt[4]{x}} = \dfrac{1}{\sqrt[4]{x}} \cdot \boxed{\dfrac{\sqrt[4]{x^3}}{\sqrt[4]{x^3}}}$$

$$= \dfrac{\sqrt[4]{x^3}}{\sqrt[4]{x^4}}$$

$$= \dfrac{\sqrt[4]{x^3}}{|x|}$$

Do I really need the absolute value sign?

In the future you can omit it. We will agree that for the remainder of this book any variable under a radical represents a positive number.

□

8.2D   Rule: A Radical Is Said to Be Simplified If:

1. The radicand does not contain any factor with a power greater than or equal to the index of the radical.
2. No fraction appears under the radical.
3. No radical appears in the denominator of a fraction.

Chapter 8 ■ Rational Exponents and Radicals

Example 22 □ Simplify $\sqrt[3]{x^5}$.

This violates condition one for a simplified radical because 5 is greater than 3.

$$\sqrt[3]{x^5} = \sqrt[3]{x^3}\,\sqrt[3]{x^2}$$
$$= x\sqrt[3]{x^2} \qquad \qquad □$$

Example 23 □ Simplify $\sqrt[4]{\dfrac{a^3}{16}}$.

This violates condition two for a simplified radical because $\dfrac{a^3}{16}$ is a fraction.

$$\sqrt[4]{\frac{a^3}{16}} = \frac{\sqrt[4]{a^3}}{\sqrt[4]{16}}$$
$$= \frac{\sqrt[4]{a^3}}{2} \qquad \qquad □$$

Example 24 □ Simplify $\dfrac{1}{\sqrt{x+y}}$.

This violates condition three for a simplified radical expression because the denominator is a radical.

$$\frac{1}{\sqrt{x+y}} = \frac{1}{\sqrt{x+y}} \cdot \frac{\sqrt{x+y}}{\sqrt{x+y}}$$
$$= \frac{\sqrt{x+y}}{x+y} \qquad \qquad □$$

Example 25 □ Simplify $\sqrt[3]{\dfrac{x^3}{y^5}}$.

$$\sqrt[3]{\frac{x^3}{y^5}} = \frac{\sqrt[3]{x^3}}{\sqrt[3]{y^3}\,\sqrt[3]{y^2}}$$
$$= \frac{x}{y\sqrt[3]{y^2}} \cdot \frac{\sqrt[3]{y}}{\sqrt[3]{y}}$$

Multiply by $\dfrac{\sqrt[3]{y}}{\sqrt[3]{y}}$ to produce a perfect cube under the radical in the denominator

$$= \frac{x\sqrt[3]{y}}{y\sqrt[3]{y^3}}$$
$$= \frac{x\sqrt[3]{y}}{y \cdot y}$$
$$= \frac{x\sqrt[3]{y}}{y^2} \qquad \qquad □$$

Example 26 □ Simplify $\sqrt[4]{\dfrac{3x^5a^3}{243x^{-2}a^7}}$.

First make all simplifications under the radical that are possible.

$$\sqrt[4]{\frac{3x^5a^3}{243x^{-2}a^7}} = \sqrt[4]{\frac{1}{81} \cdot \frac{x^7}{a^4}}$$

$$= \frac{\sqrt[4]{x^4}\,\sqrt[4]{x^3}}{\sqrt[4]{81}\,\sqrt[4]{a^4}}$$

$$= \frac{x\sqrt[4]{x^3}}{3a}$$

Problem Set 8.2

Simplify the following. All variables represent positive numbers.

1. $\sqrt{x^4}$
2. $\sqrt[3]{x^4}$
3. $\sqrt[4]{x^4}$

4. $\sqrt{a^4b^2}$
5. $\sqrt[3]{a^6b^4}$
6. $\sqrt[5]{a^6b^5}$

7. $\sqrt{18x^2y^3}$
8. $\sqrt[3]{32x^4y^7}$
9. $\sqrt{32x^4y^7}$

10. $\sqrt[5]{32x^4y^7}$
11. $-\sqrt{75a^4b}$
12. $-\sqrt[3]{54a^4b^9}$

13. $-\sqrt{50a^5b^6}$
14. $\sqrt{72x^5y^7}$
15. $\sqrt[3]{16x^4y^2}$

16. $-\sqrt[3]{125a^5b^3}$
17. $\sqrt[4]{32x^5y^6}$
18. $-\sqrt[4]{48a^8b^9}$

19. $-\sqrt[5]{64a^6b^7}$
20. $\sqrt[5]{128x^4y^5}$
21. $\sqrt{27a^5b^{10}}$

22. $\sqrt[3]{72x^7y^8}$
23. $\sqrt[4]{64x^7y^9}$
24. $\sqrt{96x^7y^{12}}$

25. $\sqrt[5]{-64x^5y^6}$
26. $\sqrt{3x}\,\sqrt{6x^3}$
27. $\sqrt[3]{x-2}\,\sqrt[3]{x+2}$

28. $\sqrt[3]{-4x^2y}\,\sqrt[3]{16xy^7}$
29. $\sqrt[4]{2x^4}\,\sqrt[4]{4x^2}$
30. $\sqrt[5]{9x^4y}\,\sqrt[5]{27x^2y^7}$

31. $\sqrt[3]{-5x^4y}\,\sqrt[3]{-75xy^3}$
32. $\sqrt[4]{50a^3b}\,\sqrt[4]{200a^4b^5}$
33. $\sqrt{5x^2}\,\sqrt{10x^5}$

34. $\sqrt{6a^3y}\,\sqrt{8a^5y^3}$
35. $\sqrt[3]{4x^2y}\,\sqrt[3]{16x^4y^4}$
36. $\sqrt[3]{9a^4b}\,\sqrt[3]{-6a^2b^5}$

37. $\sqrt[4]{25x^3y^2}\,\sqrt[4]{75x^5y^6}$
38. $\sqrt[4]{18ab^3}\,\sqrt[4]{27a^5b^2}$
39. $\sqrt[5]{-16a^3b^7}\,\sqrt[5]{6a^4b^2}$

40. $\sqrt[5]{-27x^2y^4}\,\sqrt[5]{-9x^6y^3}$
41. $\sqrt{\dfrac{144x^9}{36y^4}}$
42. $\sqrt{\dfrac{81x^7}{25y^3}}$

43. $\sqrt[3]{\dfrac{54x^4y^2}{16xy^5}}$ **44.** $\sqrt[5]{\dfrac{128x^9y}{2x^2y^7}}$ **45.** $\sqrt{\dfrac{98ax^7}{3b^3y^4}}$

46. $\sqrt{\dfrac{12x^5y}{27a^5b^2}}$ **47.** $\sqrt[3]{\dfrac{24x^4y^5}{5a^5b^2}}$ **48.** $\sqrt[5]{-\dfrac{64a^3b^7}{6a^7b}}$

49. $\sqrt[4]{\dfrac{162x^9y^4}{5a^3b^5}}$ **50.** $\sqrt{\dfrac{50a^7x}{72b^3y^2}}$ **51.** $\sqrt[3]{-\dfrac{60x^2b^3}{80a^4y^7}}$

52. $-\sqrt{\dfrac{98a^4y^5}{90b^5x^3}}$ **53.** $\sqrt[4]{\dfrac{81a^5b}{48x^3y^5}}$ **54.** $\sqrt[5]{\dfrac{10ab^6}{45x^2y^7}}$

55. $-\sqrt[3]{-\dfrac{125a^3x^5}{3by^4}}$ **56.** $\sqrt[4]{\dfrac{150x^4y^7}{160a^5b^9}}$ **57.** $\sqrt{\dfrac{50a^3b}{18a^5b^3}}$

58. $\sqrt{\dfrac{75a^5b^3}{12a^6b}}$ **59.** $\sqrt[3]{\dfrac{24a^2b}{15a^4b^5}}$ **60.** $\sqrt[3]{\dfrac{81xy^5}{18x^2y^7}}$

61. $\sqrt[4]{\dfrac{48xy^5}{12x^3y^2}}$ **62.** $\sqrt[4]{\dfrac{162x^3y^2}{10x^4y^5}}$ **63.** $\sqrt[5]{\dfrac{5x^6y^4}{64x^7y^2}}$ **64.** $\sqrt[5]{\dfrac{2a^9b}{128a^4b^8}}$

65. $\sqrt{\dfrac{5a^3b}{98a^6b^7}}$ **66.** $-\sqrt[3]{-\dfrac{10a^3b^2}{40x^4b}}$ **67.** $\sqrt[3]{-\dfrac{54x^2a^4}{5yx^7}}$ **68.** $\sqrt[4]{\dfrac{15x^5a^7}{24y^6b^9}}$

69. $-\sqrt[4]{\dfrac{32x^3a^2}{40x^4b}}$ **70.** $\sqrt[4]{-\dfrac{160a^5b^6}{5yx^7}}$ **71.** $-\sqrt[5]{-\dfrac{10a^7b^3}{24y^6b^9}}$ **72.** $\sqrt[5]{-\dfrac{6a^6b^7}{128x^3b^4}}$

■ 8.3   Addition and Subtraction of Radical Expressions

Radicals are similar if they have the same index and the same radicand.

	Similar			Not Similar	
$\sqrt{a}$	and	$2\sqrt{a}$	$\sqrt{a}$	and	$\sqrt{2a}$
$\sqrt[3]{x}$	and	$2\sqrt[3]{x}$	$\sqrt[3]{x}$	and	$\sqrt{x}$
$-2\sqrt[4]{5x^2}$	and	$3\sqrt[4]{5x^2}$	$\sqrt[4]{5x^2}$	and	$\sqrt{5x}$
$\sqrt{(x+y)}$	and	$5\sqrt{(x+y)}$	$\sqrt{(x+y)}$	and	$\sqrt{5(x+y)}$

It's important that the expressions under the radicals be identical.

Also, the index of both radicals must be identical before they are called similar.

Similar radicals can be combined in the same way like terms are combined.

Example 27 □ Combine $4\sqrt{5} + 3\sqrt{5}$.

$$4\sqrt{5} + 3\sqrt{5} = 7\sqrt{5} \qquad \qquad □$$

Example 28 □ Combine $2\sqrt[3]{4x^2} + \sqrt[3]{4x^2} - 8\sqrt[3]{4x^2}$.

$$2\sqrt[3]{4x^2} + \sqrt[3]{4x^2} - 8\sqrt[3]{4x^2} = -5\sqrt[3]{4x^2} \qquad □$$

Example 29 □ Combine $7\sqrt[3]{x} + \sqrt{5} + 4\sqrt[3]{x} - 3\sqrt{5}$.

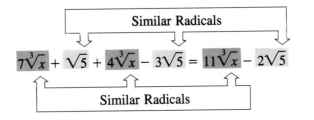

Similar Radicals

$$7\sqrt[3]{x} + \sqrt{5} + 4\sqrt[3]{x} - 3\sqrt{5} = 11\sqrt[3]{x} - 2\sqrt{5}$$

Similar Radicals

□

Example 30 □ Combine $\sqrt{294x} - \sqrt{50x} + \sqrt{150x}$.

These aren't similar radicals.

First simplify each radical.

$$\sqrt{294x} - \sqrt{50x} + \sqrt{150x} = \sqrt{49 \cdot 6x} - \sqrt{25 \cdot 2x} + \sqrt{25 \cdot 6x}$$
$$= 7\sqrt{6x} - 5\sqrt{2x} + 5\sqrt{6x}$$
$$= 12\sqrt{6x} - 5\sqrt{2x} \qquad □$$

Problem Set 8.3

Simplify the following. Assume all variables represent positive numbers.

1. $3\sqrt{5} + 8\sqrt{5}$

2. $\sqrt{8} + \sqrt{18}$

3. $3\sqrt[3]{2x^2} + 4\sqrt{2x^2} - 2\sqrt{2x^2}$

4. $\sqrt{60x^5} + \sqrt{135x^5}$

5. $\sqrt[3]{16x^5} + x\sqrt[3]{54x^2}$

6. $12\sqrt{90x^3} - 3\sqrt{160x^3}$

7. $4x\sqrt[3]{x} - \sqrt[3]{x^4} + \sqrt[3]{x^3}$

8. $\sqrt[4]{5x} - \sqrt[4]{80x} + \sqrt{20x}$

9. $\sqrt{20x} + \sqrt{45x} - \sqrt{80x}$

10. $\sqrt[3]{16x^2} + \sqrt{32x} - \sqrt[3]{54x^2}$

11. $\sqrt{20x^3} - \sqrt[3]{40x} + 5x\sqrt{45x}$

12. $\sqrt[3]{16x^7} - 3x\sqrt[3]{54x^4} + 5x^2\sqrt[3]{128x}$

13. $\sqrt[3]{81x^5} + x\sqrt[3]{24x^2} - \sqrt[3]{192x^5}$

14. $x\sqrt{12x^3} - x^2\sqrt[3]{24x} + \sqrt{48x^5}$

15. $\sqrt[5]{64x^{11}} - x\sqrt[5]{2x^6} + x\sqrt[5]{2x}$

16. $x\sqrt[5]{3x^{12}} - 2x\sqrt[5]{3x^7} + x^2\sqrt[4]{3x}$

17. $x\sqrt[4]{48x^7} - \sqrt[4]{3x^{11}} + 4x\sqrt[5]{3x^8}$

18. $3x\sqrt[4]{64x^5} - 5\sqrt[4]{4x^9} - 4\sqrt[3]{32x^4}$

19. $6\sqrt{20x^7} - 5x\sqrt{45x^5} + 4x^2\sqrt{125x^3}$

20. $4x^4\sqrt{24x} + 3x^2\sqrt{54x^5} - 3x\sqrt{96x^7}$

21. $\dfrac{14\sqrt{2a} - \sqrt{8a} + \sqrt{45a}}{3}$

22. $\dfrac{6y\sqrt{63y^3} + 8\sqrt{28y^5}}{2y}$

23. $2\sqrt[4]{16x^5} + \sqrt[4]{x^5} - x\sqrt[4]{x}$

24. $\sqrt[3]{27x^4} + \sqrt[3]{-8x^4} + x\sqrt[3]{125x}$

25. $7\sqrt{x^5} - x^2\sqrt{x} + 8x\sqrt{x^3}$

26. $\sqrt[5]{64x^7} - x\sqrt[5]{2x} + 3x\sqrt[5]{2x^2}$

27. $5\sqrt{5x^5} - 3x\sqrt{20x^3} - 3x^2\sqrt{45x}$

28. $\sqrt[3]{-27x^4} + 5x\sqrt[3]{x} + \sqrt[3]{8x^2}$

29. $\dfrac{14x\sqrt[5]{x^7} - \sqrt[5]{x^{12}} + 2x^2\sqrt[5]{x^2}}{15}$

30. $\dfrac{2\sqrt[3]{x^4} - x\sqrt[3]{8x}}{4}$

31. $5\sqrt[4]{a^7} - 6a\sqrt[4]{a} - 9a\sqrt[4]{a^3}$

32. $\dfrac{18x\sqrt{3x} - \sqrt{27x^3}}{3}$

33. $\dfrac{9\sqrt[3]{16a^4} - 5a\sqrt[3]{2a} + 6a\sqrt[3]{54a}}{2a}$

34. $\dfrac{\sqrt{28a^3} - 8a\sqrt{7a} + \sqrt{63a}}{3}$

35. $\dfrac{14\sqrt{x^7} - x^3\sqrt{x} + 8x^2\sqrt{x^3}}{7x}$

36. $\dfrac{20x\sqrt{3x} - 5\sqrt{12x^3}}{5x}$

37. $\dfrac{36x^2\sqrt[3]{2x} - 7x\sqrt[3]{16x^4}}{11x}$

38. $\dfrac{8\sqrt[3]{24x^5} + 2x\sqrt[3]{2x^2} - \sqrt[3]{81x^5}}{5x}$

39. $\dfrac{16a\sqrt[5]{a^7} - 3\sqrt[5]{a^{12}} + 6a^2\sqrt[5]{a^2}}{3x}$

40. $\dfrac{3x^2\sqrt[5]{2x^3} - \sqrt[5]{64x^{13}} + 7x\sqrt[5]{2x^8}}{4x^2}$

41. $\dfrac{8x^3\sqrt[4]{3x^3} - \sqrt[4]{48x^{15}} + 2x^2\sqrt[4]{6x^7}}{2x^2}$

42. $\dfrac{11a^2\sqrt[4]{4a} + 2a\sqrt[4]{64a^5} - 3\sqrt[4]{4a^9}}{3x^3}$

43. $\dfrac{18\sqrt{24x^7} + 4x\sqrt{54x^5} - 3x^2\sqrt{96x^3}}{6x^2}$

44. $\dfrac{3x\sqrt{20x^3} - 5\sqrt{45x^5} - 3x^2\sqrt{125x}}{6x}$

45. $21a^3\sqrt[3]{40a^2} - 2\sqrt[3]{135a^{11}} - 3a^2\sqrt[3]{5a^5}$

46. $\dfrac{7\sqrt[3]{16x^7} + 2x\sqrt[3]{2x^4} - 2x^2\sqrt[3]{54x}}{6x^2}$

47. $\dfrac{15x^3\sqrt[5]{3x^4} - x^2\sqrt[5]{96x^9} - 2\sqrt[5]{3x^{19}}}{6x^2}$

48. $\dfrac{9x^2\sqrt[4]{7x^3} - 6\sqrt[4]{112x^{11}} + 12x\sqrt[4]{5x^7}}{3x^2}$

We multiply radical expressions in the same way we multiply polynomials.

Example 31 □ Find the product $\sqrt{3}(\sqrt{x} + \sqrt{6})$.

$$\sqrt{3}(\sqrt{x} + \sqrt{6}) = \sqrt{3}\,\sqrt{x} + \sqrt{3}\,\sqrt{6} \quad \text{Distributive property}$$
$$= \sqrt{3x} + \sqrt{18} \quad\quad\quad \text{Product rule}$$
$$= \sqrt{3x} + 3\sqrt{2} \quad\quad\quad \text{Simplify} \quad\quad □$$

Example 32 □ Multiply $(\sqrt[3]{x} + 2)(\sqrt[3]{x^2} - 3)$.

$$(\sqrt[3]{x} + 2)(\sqrt[3]{x^2} - 3) = \sqrt[3]{x}\,\sqrt[3]{x^2} + \sqrt[3]{x}(-3) + 2\sqrt[3]{x^2} - 6$$
$$= x - 3\sqrt[3]{x} + 2\sqrt[3]{x^2} - 6 \quad\quad □$$

> Notice
> $$\sqrt[3]{x}\,\sqrt[3]{x^2} = \sqrt[3]{x^3}$$
> $$= x.$$

Example 33 □ Find the square of $(2\sqrt{ay} - \sqrt{3})$.
Recall that $(a + b)^2 = a^2 + 2ab + b^2$.

$$\underbrace{(2\sqrt{ay} - \sqrt{3})^2 = (2\sqrt{ay})^2}_{\substack{\text{Square of}\\\text{the first}}} + \underbrace{2(2\sqrt{ay})(-\sqrt{3})}_{\substack{\text{Twice the}\\\text{product}}} + \underbrace{(-\sqrt{3})^2}_{\substack{\text{Square of}\\\text{the second}}} \quad \begin{array}{l}\text{Square of}\\\text{a binomial}\end{array}$$

$$= 4ay - 4\sqrt{3ay} + 3 \quad\quad □$$

Dividing Radicals

There is no technique that will always find the quotient of two radical expressions. Frequently, we eliminate radicals in the denominator by rationalizing the denominator.

8.4 Definition: Conjugate

Two binomials are called **conjugates** if one binomial is the sum and the other is the difference of the same two terms.
    $a + b$ and $a - b$ are conjugates of each other.

Example 34 □ Write the conjugate of each of the following.

$x - 3$

$\sqrt{a} - \sqrt{b}$

$\sqrt{ax - 2}$

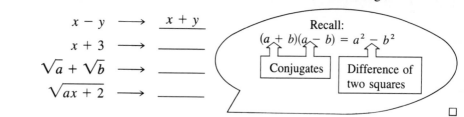

$$x - y \longrightarrow \underline{\quad x + y \quad}$$

$$x + 3 \longrightarrow \underline{\qquad\qquad}$$

$$\sqrt{a} + \sqrt{b} \longrightarrow \underline{\qquad\qquad}$$

$$\sqrt{ax + 2} \longrightarrow \underline{\qquad\qquad}$$

Recall:
$(a + b)(a - b) = a^2 - b^2$

Conjugates Difference of two squares

□

Example 35 □ Rationalize $\dfrac{1}{\sqrt{x} - 2}$.

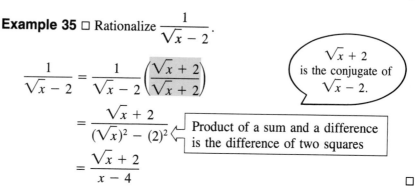

$$\frac{1}{\sqrt{x} - 2} = \frac{1}{\sqrt{x} - 2}\left(\frac{\sqrt{x} + 2}{\sqrt{x} + 2}\right)$$

$\sqrt{x} + 2$ is the conjugate of $\sqrt{x} - 2$.

$$= \frac{\sqrt{x} + 2}{(\sqrt{x})^2 - (2)^2}$$

Product of a sum and a difference is the difference of two squares

$$= \frac{\sqrt{x} + 2}{x - 4}$$

□

Example 36 □ Rationalize the denominator and simplify $\dfrac{\sqrt{x} - 1}{\sqrt{x} - \sqrt{y}}$.

$$\frac{\sqrt{x} - 1}{\sqrt{x} - \sqrt{y}} = \frac{\sqrt{x} - 1}{\sqrt{x} - \sqrt{y}}\left(\frac{\sqrt{x} + \sqrt{y}}{\sqrt{x} + \sqrt{y}}\right)$$

$$= \frac{(\sqrt{x} - 1)(\sqrt{x} + \sqrt{y})}{(\sqrt{x} - \sqrt{y})(\sqrt{x} + \sqrt{y})}$$

$$= \frac{x + \sqrt{xy} - \sqrt{x} - \sqrt{y}}{x - y}$$

That's simplified?

It's the best we can do. Notice there are no radicals in the denominator.

□

Example 37 □ Rationalize the denominator and simplify $\dfrac{x - y}{\sqrt{x} + \sqrt{y}}$.

$$\frac{x - y}{\sqrt{x} + \sqrt{y}} = \frac{x - y}{\sqrt{x} + \sqrt{y}}\left[\frac{\sqrt{x} - \sqrt{y}}{\sqrt{x} - \sqrt{y}}\right]$$

$$= \frac{(x - y)(\sqrt{x} - \sqrt{y})}{x - y}$$

$$= \sqrt{x} - \sqrt{y}$$

□

Multiply the following.

1. $\sqrt{3}(\sqrt{3} - \sqrt{6})$

2. $\sqrt{6x}(\sqrt{18x} - \sqrt{12x})$

3. $\sqrt{5a}(5 - \sqrt{10a})$

4. $\sqrt{3xy}(\sqrt{27x} + \sqrt{18y})$

5. $\sqrt[3]{x}(5 - \sqrt[3]{x^2})$

6. $\sqrt[3]{3}(x - \sqrt[3]{9})$

7. $\sqrt{5x}(\sqrt{10x} - \sqrt{15x})$

8. $\sqrt{3ax}(\sqrt{12a} - \sqrt{15x})$

9. $\sqrt{6ay}(\sqrt{3ay} + \sqrt{6y})$

10. $\sqrt{2xy}(\sqrt{18x} - \sqrt{50y})$

11. $\sqrt[3]{x^2}(\sqrt[3]{3} - \sqrt[3]{x})$

12. $\sqrt[3]{3x^2}(\sqrt[3]{9x} + \sqrt[3]{18x^2})$

13. $(\sqrt{3} - \sqrt{x})(2\sqrt{3} + 3\sqrt{x})$

14. $(\sqrt{2x} - 5)(\sqrt{2x} + 5)$

15. $(\sqrt{3x} - \sqrt{y})^2$

16. $(2 - \sqrt{x})(2 + \sqrt{x})$

17. $(\sqrt[3]{x} - 5)(\sqrt[3]{x^2} + 5)$

18. $(\sqrt{5} - 3\sqrt{y})(\sqrt{5} + 2\sqrt{y})$

19. $(\sqrt[4]{x^3} + 2)(\sqrt[4]{x^3} - 2)$

20. $(5\sqrt{x} - \sqrt{3y})(3\sqrt{x} + \sqrt{3y})$

21. $(\sqrt[3]{2x} - \sqrt[3]{y})^2$

22. $(\sqrt[4]{2x^2} - 3)(\sqrt[4]{8x^2} + 3)$

23. $(\sqrt[4]{3x^2} + 5)^2$

24. $(x - \sqrt{3y})(x + \sqrt{3y})$

25. $(\sqrt{2x} - \sqrt{7})(\sqrt{2x} + \sqrt{7})$

26. $(5 - \sqrt{y})(6 + \sqrt{y})$

27. $(\sqrt[3]{x} - 4)(\sqrt[3]{x^2} + 5)$

28. $(\sqrt[3]{2x} + 5)(\sqrt[3]{4x^2} - 6)$

29. $(\sqrt{3x} + \sqrt{y})^2$

30. $(\sqrt{5x} - \sqrt{6a})^2$

31. $(6\sqrt{x} - 7\sqrt{y})(5\sqrt{x} + 2\sqrt{y})$

32. $(3\sqrt{6} + 4\sqrt{a})(4\sqrt{6} - 3\sqrt{a})$

33. $(5\sqrt{10} + 3\sqrt{2x})(2\sqrt{10} - 4\sqrt{2x})$

34. $(7\sqrt{2x} + 3\sqrt{6a})(8\sqrt{2x} - 3\sqrt{6a})$

35. $(\sqrt[3]{3x^2} - \sqrt[3]{a})^2$

36. $(\sqrt{2a} + \sqrt{3x^2})^2$

Rationalize each denominator and simplify.

37. $\dfrac{1}{7 + \sqrt{5}}$

38. $\dfrac{3}{9 - \sqrt{7}}$

39. $\dfrac{3}{\sqrt{x} + 3}$

40. $\dfrac{5}{\sqrt{xy} - 5}$

41. $\dfrac{6}{\sqrt{5} + \sqrt{2}}$

42. $\dfrac{10}{\sqrt{7} - \sqrt{3}}$

43. $\dfrac{\sqrt{10}}{\sqrt{5} - \sqrt{2}}$

44. $\dfrac{2\sqrt{5}}{\sqrt{5} + \sqrt{3}}$

45. $\dfrac{3\sqrt{15}}{\sqrt{3} + \sqrt{5}}$

46. $\dfrac{3\sqrt{6}}{2\sqrt{3} - \sqrt{2}}$

47. $\dfrac{3}{\sqrt{6} - 5}$

48. $\dfrac{9}{\sqrt{5} + 4}$

49. $\dfrac{2}{\sqrt{y} - 6}$

50. $\dfrac{5}{4 + \sqrt{a}}$

51. $\dfrac{6}{\sqrt{6} - \sqrt{3}}$

52. $\dfrac{7}{\sqrt{5} - \sqrt{6}}$ **53.** $\dfrac{3\sqrt{6}}{\sqrt{6} + \sqrt{2}}$ **54.** $\dfrac{4\sqrt{10}}{\sqrt{5} - \sqrt{10}}$ **55.** $\dfrac{5\sqrt{15}}{\sqrt{5} - \sqrt{3}}$ **56.** $\dfrac{12\sqrt{6}}{\sqrt{3} - \sqrt{6}}$

57. $\dfrac{\sqrt{10} + 1}{2\sqrt{15} - \sqrt{5}}$ **58.** $\dfrac{\sqrt{6} - \sqrt{3}}{3\sqrt{2} + \sqrt{6}}$ **59.** $\dfrac{\sqrt{a}}{\sqrt{a} + \sqrt{b}}$ **60.** $\dfrac{\sqrt{b}}{\sqrt{2b} - \sqrt{a}}$ **61.** $\dfrac{\sqrt{x} + \sqrt{y}}{\sqrt{x} - \sqrt{y}}$

62. $\dfrac{\sqrt{6} - 2}{3\sqrt{18} + \sqrt{3}}$ **63.** $\dfrac{\sqrt{10} + 3}{\sqrt{5} - 3\sqrt{15}}$ **64.** $\dfrac{\sqrt{3} - \sqrt{6}}{\sqrt{6} - 2\sqrt{2}}$ **65.** $\dfrac{\sqrt{a}}{\sqrt{2a} - 2\sqrt{3b}}$

66. $\dfrac{\sqrt{b}}{\sqrt{3b} - \sqrt{2a}}$ **67.** $\dfrac{3\sqrt{x} - \sqrt{y}}{\sqrt{3x} - \sqrt{y}}$ **68.** $\dfrac{\sqrt{3a} - \sqrt{2b}}{\sqrt{3b} - \sqrt{2a}}$ **69.** $\dfrac{\sqrt{5x} - \sqrt{10y}}{\sqrt{10x} - \sqrt{5y}}$

Chapter 8 □ Key Ideas

8.1 **1. Square root of a** is represented by $a^{\frac{1}{2}}$ where $a^{\frac{1}{2}} \cdot a^{\frac{1}{2}} = a$.

2. **The nth root of a** is given by $a^{\frac{1}{n}}$ where n is any natural number.

3. Property of exponents

$$a^{\frac{m}{n}} = \left(a^{\frac{1}{n}}\right)^{m} \qquad \text{or} \qquad a^{\frac{m}{n}} = (a^{m})^{\frac{1}{n}}$$

4. If x and y are any rational number:

   a. $a^{x} \cdot a^{y} = a^{x+y}$ **b.** $\dfrac{a^{x}}{a^{y}} = a^{x-y} \qquad a \neq 0$ **c.** $(a^{x})^{y} = a^{x \cdot y}$

5. **Case I:** $a^{\frac{m}{n}}$ where $n = 1$ becomes a^{m}

   Case II: $a^{\frac{m}{n}}$ where $m = 1$ and n is even, becomes $a^{\frac{1}{n}}$ where $a \geq 0$

   Case III: $a^{\frac{m}{n}}$ where $m = 1$ and n is odd, becomes $a^{\frac{1}{n}}$

   $\qquad\qquad$ for $a > 0$, $a^{\frac{1}{n}} > 0$ $\qquad\qquad$ for $a < 0$, $a^{\frac{1}{n}} < 0$

6. $(x^{n})^{\frac{1}{n}} = x$ if n is odd $\qquad (x^{n})^{\frac{1}{n}} = |x|$ if n is even

8.2 **1.** $\sqrt[n]{a}$ is equivalent to $a^{\frac{1}{n}}$ $\qquad\qquad$ **2.** $\sqrt[n]{a^{m}} = a^{\frac{m}{n}}$ $\quad$ and $\quad (\sqrt[n]{a})^{m} = a^{\frac{m}{n}}$

3. **Multiplication law for radicals** $\qquad$ **4. Division law for radicals**

$$\sqrt[n]{a} \cdot \sqrt[n]{b} = \sqrt[n]{a \cdot b} \qquad\qquad \dfrac{\sqrt[n]{a}}{\sqrt[n]{b}} = \sqrt[n]{\dfrac{a}{b}}, b \neq 0$$

5. In general, if n is even: $\quad \sqrt[n]{a^{n}} = |a|$

   $\qquad\qquad$ if n is odd: $\quad \sqrt[n]{a^{n}} = a$

6. **Simplified radical form**

   a. The radicand does not contain any factor with a power greater than or equal to the index of the radical.

   b. No fraction appears under the radical.

   c. No radical appears in the denominator of a fraction.

8.3 **1. Similar Radicals**

   a. Radicals are similar if they have the same index and the same radicand.

   b. Similar radicals can be added or subtracted.

8.4   **1. Radical expressions are multiplied** in a manner similar to multiplying polynomials.

   2. Conjugates
   The sum and difference of the same two terms are called conjugates of each other:
   $a + b$ and $a - b$ are conjugates of each other.

   3. To rationalize the denominator of a fraction, multiply the numerator and the denominator by the conjugate of the denominator.

Chapter 8   Review Test

Simplify the following. **(8.1)**

1. $(-27)^{\frac{1}{3}}$ **2.** $-(16)^{\frac{1}{4}}$ **3.** $(-64)^{\frac{1}{6}}$ **4.** $(-8)^{\frac{2}{3}}$ **5.** $16^{\frac{5}{4}}$

6. $x^{\frac{1}{3}} \cdot x^{\frac{3}{4}}$ **7.** $\dfrac{x^{\frac{5}{6}} y^{-2}}{x^{\frac{1}{3}} y^{\frac{2}{3}}}$ **8.** $\dfrac{a^3 b^{-\frac{2}{3}}}{a^{-\frac{5}{4}} b^{-\frac{2}{3}}}$ **9.** $-\dfrac{6x^{\frac{7}{3}} y^{-\frac{3}{4}}}{2x^{-2} y^{-3}}$

Simplify the following. **(8.2)**

10. $-\sqrt[5]{-32}$ **11.** $\sqrt[4]{\dfrac{1}{81}}$ **12.** $\sqrt{48a^5}$ **13.** $\sqrt[4]{32a^5 b^6}$ **14.** $\sqrt[3]{-54a^5 b^6}$

15. $\dfrac{\sqrt{64a^3 b^5}}{\sqrt{8ab}}$ **16.** $\sqrt[3]{\dfrac{16x^4 y}{9xy^2}}$ **17.** $\sqrt[5]{\dfrac{2x^6 a^3}{64x^{-4} a^8}}$ **18.** $\sqrt{\dfrac{48x^{-3} y}{16xy^{-4}}}$

Combine the following radicals. **(8.3)**

19. $\sqrt{18} - \sqrt{98} + \sqrt{75}$

20. $\sqrt{12x^3} - x\sqrt{27x} + \sqrt{3x}$

21. $\dfrac{\sqrt{48x^5} - x\sqrt{27x^3} + x^2\sqrt{75x}}{3}$

22. $\dfrac{3\sqrt[3]{16x^4} - x\sqrt[3]{54x} + \sqrt[3]{81x}}{3x}$

Multiply the following. **(8.4)**

23. $\sqrt{5}(\sqrt{5} - \sqrt{10})$ **24.** $\sqrt{10x}(\sqrt{5x} - \sqrt{6x})$ **25.** $\sqrt[3]{4}(y - \sqrt[3]{2})$

26. $\sqrt[3]{2x}(\sqrt[3]{4x^2} - \sqrt[3]{12x})$ **27.** $(\sqrt{3x} - 6)(\sqrt{3x} + 8)$ **28.** $(\sqrt{3x} - \sqrt{5y})^2$

29. $(\sqrt{2} - \sqrt{6y})(2\sqrt{2} + 3\sqrt{6y})$

Rationalize the denominators of the following and simplify. **(8.4)**

30. $\dfrac{2}{\sqrt{6} - \sqrt{5}}$ **31.** $\dfrac{2}{\sqrt{6} - \sqrt{2}}$ **32.** $\dfrac{4\sqrt{6}}{3\sqrt{3} - 2\sqrt{2}}$ **33.** $\dfrac{2\sqrt{x}}{3\sqrt{x} - \sqrt{y}}$

34. $\dfrac{\sqrt{6}}{\sqrt{15} - 2\sqrt{3}}$ **35.** $\dfrac{\sqrt{10} - \sqrt{5}}{\sqrt{2} - \sqrt{5}}$ **36.** $\dfrac{3\sqrt{x} + \sqrt{2y}}{\sqrt{x} - \sqrt{2y}}$

Chapter 8 ■ Rational Exponents and Radicals

Write a statement to illustrate the property named.

1. Commutative property of multiplication. **(1.2)**

$$3 \cdot (a + b) = \underline{\hspace{3cm}}$$

What does the (1.2) mean?

The (1.2) refers to Section 2 in Chapter 1. All numbers in parentheses refer to section numbers. If you have difficulty with a problem, refer to the indicated section for help.

Perform the indicated operations. Write answers in simplest form.

2. $(4x - 3y)^2$ **(5.2)**

3. $(a^2 - 4a + 1)(5a^2 + 2a - 3)$ **(5.2)**

4. $\dfrac{2x^2 - 8}{6x + 18} \cdot \dfrac{3x^2 - 6x - 45}{8 - x^3}$ **(7.1)**

5. $\dfrac{x^2 - 9y^2}{3x^2 - 10xy + 3y^2} \div \dfrac{x^3 + 27y^3}{x^2 - 6xy + 9y^2}$ **(7.1)**

Evaluate the following.

6. $(-15) - |-3| - (-5)$ **(1.1)**

7. $-\dfrac{2}{3}[4 + 3(-2) - 4 - (-3)^3]$ **(1.2)**

8. $\dfrac{6(-5) + 5(-2)}{9 + (-1)}$ **(1.2)**

9. $3a^2 - \dfrac{1}{3}b^3cy$ for $a = -2, b = 3, c = -1, y = \dfrac{1}{3}$ **(1.3)**

10. $6x^2y - 8ax^3 - \dfrac{ab}{x}$ for $x = -2, y = \dfrac{1}{3}, a = \dfrac{1}{2}, b = 8$ **(5.1)**

Solve for the variable.

11. $4(2y - 3) = y - (3y - 17)$ **(2.1)**

12. $\dfrac{x - 3}{8} + \dfrac{5}{6} = \dfrac{2x}{3} - \dfrac{4}{3}$ **(2.1)**

13. $P = 2s - rs$ for r **(2.2)**

14. $\dfrac{7x}{4x - 5} = 3$ **(7.4)**

15. $\dfrac{x - 7}{3x - 4} = 5 - \dfrac{8}{3x - 4}$ **(7.4)**

Solve the following inequalites.

16. $5(6a - 4) > 10$ **(2.3)**

17. $4(2x - 9) - 3(4x + 8) > 12$ **(2.3)**

Solve and graph the following.

18. $|2x - 3| < 5$ **(2.4)**

19. $|3y - 6| \geq 15$ **(2.4)**

20. Find the slope of the line through the points $(-6, 8)$ and $(3, -4)$. **(3.1)**

21. Find the equation of the line with slope $-\dfrac{3}{4}$ and passing through $(0, -4)$. **(3.2)**

22. Find the equation of the line through the points $(-8, 4)$ and $(6, -10)$. **(3.2)**

23. Find the equation of the line parallel to the line $3x - 2y = 4$ and passing through the point $(5, -4)$. **(3.3)**

24. Find the equation of the line perpendicular to the line $5x + 4y = 7$ passing through the point $(-3, 2)$. **(3.3)**

Sketch the graphs of the following equations. **(3.2)**

25. $4x - 3y = -12$ Use the intercepts method.

26. $5x + 2y = 8$ Use the slope and y-intercept method.

Graph the following inequalities.

27. $3(4x - 5) \geq \dfrac{1}{2}(12 - 4x)$ **(2.3)**

28. $|3x - 7| + 5 < 13$ **(2.4)**

Use the addition method to find the simultaneous solution. **(4.1)**

29. $5x + 9y = -19$
$2x - 3y = -12$

Use the substitution method to find the simultaneous solution. **(4.1)**

30. $-3x + y = 17$
$2x - 3y = -23$

Find the simultaneous solution. **(4.3)**

31. $2x + 5y + z = 3$
$x - 4y + 2z = 11$
$4x + 3y - 6z = -3$

Set up two equations and solve.

32. A grocer sells peanuts for \$2.50 a pound and cashews for \$4.25 a pound. How many pounds of each must he mix to obtain a mixture of 20 pounds selling for \$3.20 a pound? **(4.2)**

Set up three equations and solve.

33. A phone booth contains \$13.20 in coins consisting of nickels, dimes, and quarters. The container has 4 times as many dimes as nickels and 8 fewer nickels than quarters. How many of each coin is in the phone booth? **(4.3)**

Chapter 8 ■ Rational Exponents and Radicals

Set up an equation and solve.

34. At the pure water bottling company one machine can produce 180 bottles in 5 hours, and a newer machine can produce 180 bottles in 3 hours. How long would it take the two machines working together to produce 720 bottles of water?　　**(7.5)**

35. The denominator of a fraction exceeds the numerator by 4. If the numerator is multiplied by 4 and the denominator is increased by 3, the resulting fraction is $\dfrac{6}{5}$. What is the original fraction? **(8.5)**

Simplify the following. Write the results with positive exponents only. **(5.1)**

36. $(-3x^3y^{-7})^3(-2x^{-3}y^4)^2$

37. $\left(\dfrac{x^{-3}y^4z^{-1}}{x^{-1}y^{-3}z^3}\right)^{-3}$

38. $-\dfrac{24(3x-1)^0(2x+5)^{-3}}{3(3x-1)^{-5}(2x+5)^{-4}}$

Simplify the following.

39. $-(-32)^{\frac{4}{5}}$　　**(8.1)**

40. $\dfrac{a^4b^{-\frac{2}{3}}}{a^{-\frac{5}{4}}b^{\frac{7}{3}}}$　　**(8.1)**

41. $\dfrac{\sqrt{32x^3y^4}}{\sqrt{8x^4y}}$　　**(8.2)**

42. $\sqrt{\dfrac{24x^{-2}y}{4xy^{-5}}}$　　**(8.2)**

43. $\dfrac{\sqrt{18x^5}-x^2\sqrt{8x}+3x\sqrt{12x^3}}{3x}$　　**(8.3)**

44. Reduce to lowest terms.

$$\dfrac{18x^2-32}{24x^2+16x-64}\quad \textbf{(5.1)}$$

45. Divide. Write your answer with positive exponents only. **(5.2)**

$$\dfrac{-8x^4y+24xy^5-18x^2y^3}{-12xy^4}$$

46. Divide using long division. **(5.3)**

$$(24x^5+2x^4+x^2-2)\div(3x-2)$$

47. Use synthetic division to find the following quotient. **(5.4)**

$$(3x^4+8x^3-5x+2)\div(x+2)$$

Factor the following completely.

48. $8ax^3-18ax^2-35ax$　　**(6.1)**

49. x^4-9y^6　　**(6.3)**

50. x^6-7x^3-8　　**(6.2, 6.3)**

51. $(x^2+2x)^2-11(x^2+2x)+24$　　**(6.2)**

52. x^2-y^2+6x+9　　**(6.4)**

53. $x^3+2x^2-16x-32$　　**(6.4)**

Add or subtract. Reduce all answers to lowest terms. **(7.2)**

54. $\dfrac{y}{y^2 - 10y + 24} + \dfrac{1}{8 - 2y}$

55. $1 - \dfrac{5x - 15}{x^2 + 2x - 15}$

Simplify the following complex fractions. Reduce your answers to lowest terms. **(7.3)**

56. $\dfrac{1 - \dfrac{x}{y}}{\dfrac{x^2}{y^2} - 1}$

57. $\dfrac{x - 4 - \dfrac{8}{x + 3}}{x + 2 + \dfrac{-2}{x + 3}}$

Simplify the following radicals.

58. $\sqrt[3]{32}\,(\sqrt[3]{2} - x)$ **(8.3)**

59. $(\sqrt{2x} - \sqrt{3y})^2$ **(8.4)**

60. $(\sqrt{y} - \sqrt{3x})(\sqrt{y} + \sqrt{3x})$ **(8.4)**

61. $\dfrac{\sqrt{5} - \sqrt{10}}{\sqrt{2} - \sqrt{5}}$ **(8.4)**

9 Complex Numbers, Quadratic Equations

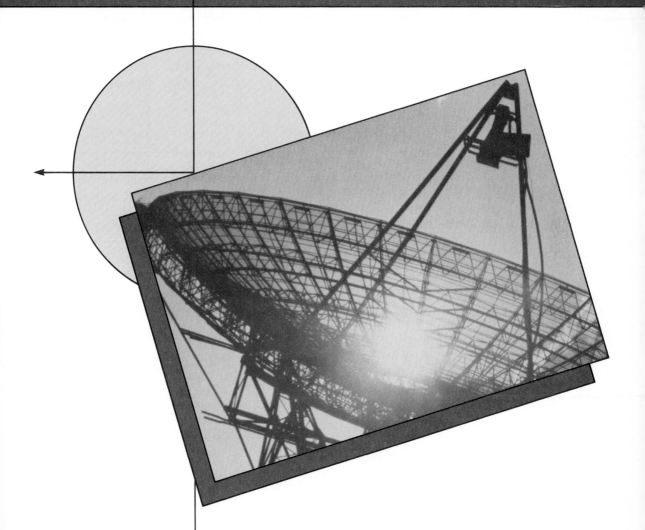

Contents

Many real world applications of mathematics involve quadratic equations. These are the equations that represent the shape of a car's headlight or a radar antenna. They also describe the motion of a ball thrown into the air. Since quadratic equations have maximum or minimum values, they are useful in some business applications to maximize profit or minimize costs. Some solutions of quadratic equations are not real numbers, therefore the first section of this chapter will be devoted to a study of complex numbers.

The second and third sections of this chapter will show you two ways to solve quadratic equations. The fourth section develops the quadratic formula which allows you to solve any quadratic equation. Finally this chapter will show you how to use your calculator and the quadratic formula to solve quadratic equations.

■ 9.1   Complex Numbers

To this point we have said there are no real square roots of negative numbers. That statement is true. However, it is similar to saying there are no integers equal to one divided by four or saying that no man can bear a child. We do divide one by four and children are born. In both cases, the desired result is achieved by extending the set of participants that are involved.

To find the square root of negative numbers we extend our set of numbers. To do this we define a new system of numbers where we define $\sqrt{-1} = i$.

How can you say that out of the blue?

The same way we defined $1 \div 4$ to be $\dfrac{1}{4}$ or $-x$ as the number you add to x to get zero for an answer.

9.1A   Definition: Imaginary Unit

The imaginary unit is a number called i with the property

$$i^2 = -1 \qquad \text{and} \qquad i = \sqrt{-1}$$

9.1B   Definition: Pure Imaginary Numbers

Pure imaginary numbers are numbers of the form

$$bi \qquad \text{where} \qquad b \in \text{reals}$$
$$i = \sqrt{-1}$$

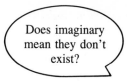

Does imaginary mean they don't exist?

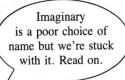

Imaginary is a poor choice of name but we're stuck with it. Read on.

If you think of imaginary as meaning a product of the mind then these numbers exist as much as any other numbers exist. When was the last time you touched $a - 2$ or $a + 2$, for that matter?

9.1C   Powers of i

Some properties follow immediately from the definition of i.

$$i = \sqrt{-1} \qquad i^2 = -1$$

$$i^3 = i^2 \cdot i \qquad i^4 = i^2 \cdot i^2$$

$$= -1 \cdot i \qquad\quad = (-1) \cdot (-1)$$

$$= -i \qquad\qquad = +1$$

Since $i^4 = 1$ then i^4 raised to any power is still equal to 1.

$$i^{12} = i^4 \cdot i^4 \cdot i^4$$

The next example illustrates how to use this idea to simplify powers of i.

Example 1 □ Evaluate i^{35}.
Since $i^4 = 1$ we will identify powers of i^4.

$$i^{35} = i^{32+3}$$

$$= i^{32} + i^3$$

$$= (i^4)^8 \cdot i^3$$

$$= 1^8 \cdot i^3$$

$$= -i$$

Why call i^{35} i^{32+3}?

Because 32 is a multiple of 4 and $i^4 = 1$.

It also follows

Does i always equal $\sqrt{-1}$?

if $\qquad a > 0$

then $\qquad \sqrt{-a} = \sqrt{a(-1)}$

$$= \sqrt{a} \cdot \sqrt{-1}$$

$$= \sqrt{a}\, i$$

Since i^2 is defined as equal to -1, $i = \sqrt{-1}$ is always true.

□

Example 2 □ Find the following square roots.

a. $\sqrt{-25} = \sqrt{25}\,\sqrt{-1}$
$= 5i$

b. $\sqrt{-16} = \sqrt{16}\,\sqrt{-1}$
$= 4i$

c. $\sqrt{-20} = \sqrt{4}\,\sqrt{5}\,\sqrt{-1}$
$= 2\sqrt{5}\,i$

d. $\sqrt{-18} = \sqrt{9}\,\sqrt{2}\,\sqrt{-1}$
$= 3\sqrt{2}\,i$ □

Not only can we raise i to powers, we can also multiply by other constants or variables.

We can also add, subtract, and multiply expressions containing i.

The set of pure imaginary numbers that we have defined is completely different from the set of real numbers.

There is however a set of numbers that includes both. It is the **complex number system.**

9.1D   Definition: The Complex Number System

The complex number system is the set of numbers of the form $a + bi$

where $a, b \in$ real numbers

$i =$ imaginary unit $= \sqrt{-1}$

a is called the real part of $a + bi$
bi is called the pure imaginary part of $a + bi$

The real number system and the imaginary numbers are subsets of the complex number system.

In complex numbers of the form $a + bi$

If $b = 0$, $a \neq 0$ $a + bi$ is a real number

If $a = 0$, $b \neq 0$ $a + bi$ is a pure imaginary number

If $a \neq 0$, $b \neq 0$ $a + bi$ is an imaginary number

9.1E   Definition: Equality of Complex Numbers

$$a + bi = c + di$$

if, and only if, $a = c$ and $b = d$

Example 3 □ Solve $3x + 12i = 21 - 2yi$ for x and y.
By the definition of equality of complex numbers

$$3x = 21 \qquad \text{and} \qquad 12 = -2y$$
$$x = 7 \qquad\qquad\qquad -6 = y \qquad\qquad □$$

Example 4 □ Solve $2x + 3 + 4i = 5 - (y - 1)i$ for x and y.
By the definition of equality of complex numbers

$$2x + 3 = 5 \qquad \text{and} \qquad 4 = -(y - 1)$$
$$2x = 2 \qquad\qquad\qquad 4 = -y + 1$$
$$x = 1 \qquad\qquad\qquad 3 = -y$$
$$\qquad\qquad\qquad\qquad\qquad y = -3 \qquad □$$

For addition and subtraction of complex numbers, we add the real and the imaginary parts separately.

9.1F   Addition of Complex Numbers

$$(a + bi) + (c + di) = (a + c) + (b + d)i$$

Example 5 □ Add $(3 + 2i) + (4 - 5i)$.
Treat these just as you would treat two binomials.

You seem to be treating i as a variable.

Yes, we may treat it as a variable even though it is really a constant.

$$(3 + 2i) + (4 - 5i) = (3 + 4) + (2i - 5i)$$
$$= 7 - 3i$$

Treat subtraction just as you would treat subtraction of binomials.   □

Example 6 □ Subtract $(4 - 3i) - (-10 + 5i)$.

$$(4 - 3i) - (-10 + 5i) = (4 + 10) + (-3 - 5)i$$
$$= 14 - 8i$$

To multiply two complex numbers, use the same procedure as for multiplication of two binomials and then use the fact that $i^2 = -1$.   □

9.1G   Product of Two Complex Numbers

$$(a + bi)(c + di) = ac + bci + adi + bdi^2$$
$$= ac + (bc + ad)i + bd(-1)$$
$$= ac + (bc + ad)i - bd$$
$$= (ac - bd) + (bc + ad)i$$

Example 7 □ Multiply $(3 + 4i)(2 - 5i)$.

$$(3 + 4i)(2 - 5i) = 6 - 15i + 8i - 20i^2$$
$$= 6 - 7i - 20(-1)$$
$$= 6 - 7i + 20$$
$$= 26 - 7i \qquad \square$$

Example 8 □ Multiply $(4 + \sqrt{3}\ i)(2 - \sqrt{3}\ i)$.

$$(4 + \sqrt{3}\ i)(2 - \sqrt{3}\ i) = 8 - 4\sqrt{3}\ i + 2\sqrt{3}\ i - 3i^2$$
$$= 8 - 2\sqrt{3}\ i - 3(-1)$$
$$= 11 - 2\sqrt{3}\ i \qquad \square$$

Example 9 □ Multiply $(2 + \sqrt{-3})(4 - \sqrt{-3})$.

$$(2 + \sqrt{-3})(4 - \sqrt{-3}) = (2 + \sqrt{3}\ i)(4 - \sqrt{3}\ i)$$
$$= 8 + 4\sqrt{3}\ i - 2\sqrt{3}\ i - 3i^2$$
$$= 8 + 2\sqrt{3}\ i + 3$$
$$= 11 + 2\sqrt{3}\ i \qquad \square$$

Why change $\sqrt{-3}$ to $\sqrt{3}\ i$ before multiplying?

Because
$$\sqrt{-3} \cdot \sqrt{-3} \stackrel{?}{=} \sqrt{(-3)(-3)}$$
$$\sqrt{3}\ i \cdot \sqrt{3}\ i \stackrel{?}{=} \sqrt{9}$$
$$\sqrt{9}\ i^2 \stackrel{?}{=} \sqrt{9}$$
$$-3 \neq 3$$

Always change from a negative number under the radical sign to i before multiplying or dividing complex numbers.

To divide a complex number by a real number, use the fact that division by reals is defined as multiplication by the multiplicative inverse.

Example 10 □ Simplify $\dfrac{(6 + 4i)}{2}$.

$$\frac{6 + 4i}{2} = (6 + 4i)\frac{1}{2}$$

$$= \frac{6}{2} + \frac{4}{2}i \qquad \text{Distributive property}$$

$3 + 2i$

$$= \underline{\hspace{2cm}} \qquad\qquad □$$

Division of Complex Numbers

To divide a complex number by a complex number we use a process similar to rationalizing denominators.

Example 11 □ Simplify $\dfrac{5 + 10i}{5i}$.

$$\frac{5 + 10i}{5i} = \frac{5 + 10i}{5i} \cdot \frac{i}{i} \qquad \text{Multiplication by one}$$

$$= \frac{5i + 10i^2}{5i^2}$$

$$= \frac{5i - 10}{-5}$$

$$= -i + 2$$

$$= 2 - i \qquad\qquad □$$

Example 12 □ Simplify $\dfrac{5 + 4i}{1 + 2i}$.

Multiply numerator and denominator by the conjugate of the denominator.

$$\frac{5 + 4i}{1 + 2i} = \frac{5 + 4i}{1 + 2i} \cdot \frac{(1 - 2i)}{(1 - 2i)}$$

$$= \frac{5 - 6i - 8i^2}{1 - 4i^2}$$

$$= \frac{5 - 6i + 8}{1 + 4}$$

Why do we need the last step?

$$= \frac{13 - 6i}{5}$$

$$= \frac{13}{5} - \frac{6}{5}i$$

A complex number has a real part and an imaginary part. We prefer to write each part explicitly.

□

Find the following square roots.

1. $\sqrt{-9}$ **2.** $\sqrt{-25}$ **3.** $-\sqrt{-36}$ **4.** $\sqrt{-125}$ **5.** $-\sqrt{-18}$

6. $-\sqrt{-8}$ **7.** $\sqrt{-75}$ **8.** $-\sqrt{-98}$ **9.** $\sqrt{-72}$ **10.** $\sqrt{-150}$

11. $\sqrt{-32}$ **12.** $\sqrt{-50}$ **13.** $-\sqrt{-27}$ **14.** $-\sqrt{-48}$ **15.** $-\sqrt{-80}$

16. $-\sqrt{-96}$

Find values of x and y for each of the following.

17. $5x - 6yi = 20 + 18i$

18. $4x + 7yi = -12 + 28i$

19. $3x - 9yi = 15 + 3i$

20. $8y + 10i = 24 + 5xi$

21. $10y - 12i = 2 - 3xi$

22. $(2x - 1) + 5i = 9 + (y - 1)i$

23. $(3x + 2) + 18i = -4 - (2y + 4)i$

24. $(4x - 1) + (3y - 2)i = 5 + 8i$

25. $9y + (5x - 1)i = 18 - 10i$

26. $(6x - 4) - 21i = 8 + 7yi$

27. $7x + (2y + 3)i = 28 + (4y - 5)i$

28. $(5x - 1) - 11i = 9 + (7y + 3)i$

29. $(3x + 4) + 7yi = (7x - 8) + 14i$

30. $8x + (3y + 11)i = (8y - 9)i - 24$

31. $(2y + 1) + (3x - 4)i = -7 + (6 - 2x)i$

32. $11 - (4 - 7y)i = (5x - 4) + 11i$

Evaluate each of the following.

33. i^5 **34.** $-i^{19}$ **35.** i^{77} **36.** $-i^{103}$ **37.** i^{87} **38.** i^{73} **39.** i^{89}

40. i^{67} **41.** $-i^{18}$ **42.** $-i^{29}$ **43.** i^{97} **44.** i^{57}

Simplify the following.

45. $(3 + 2i) + (-5 + 6i)$

46. $(-4 + 8i) - (7 - 12i)$

47. $(18 + \sqrt{-8}) + (-8 - 3\sqrt{-2})$

48. $(-6 - \sqrt{-36}) - (7 + 6i)$

49. $[(4 - 3i) - (6 + 7i)] - (3 - \sqrt{-25})$

50. $3 + 2i - [(3 - 6i) - (8 + \sqrt{-36})]$

51. $[(2 + 3i) - (-4 + 5i)] - (8 - 6i)$

52. $(7 - 4i) - [(2 - 6i) - (-7 + 3i)]$

53. $(22 - \sqrt{-4}) - (-4 + \sqrt{-49})$

54. $(26 + \sqrt{-64}) - (-3 - \sqrt{-49})$

55. $[(3 - 2i) + (18 - \sqrt{-36})] - (17 - \sqrt{-81})$

56. $(14 - \sqrt{-25}) - [(7 - \sqrt{-49}) - (23 - 19i)]$

57. $3i(7 - 8i)$

58. $-8i(12 + 3i)$

59. $\sqrt{-25}\,(7 - \sqrt{-49})$

60. $(3 - 4i)^2$

61. $(\sqrt{5} + 2i)(\sqrt{5} - 3i)$

62. $(\sqrt{27} - \sqrt{-2})(\sqrt{3} + \sqrt{-8})$

63. $-6i(8 - 4i)$

64. $-12i(4 - 8i)$

65. $-\sqrt{-36}\,(4 + \sqrt{-49})$

66. $-\sqrt{-81}\,(8 - \sqrt{-16})$

67. $(\sqrt{63} + 3i)(\sqrt{7} + 2i)$

68. $(\sqrt{44} - 8i)(\sqrt{11} + 9i)$

69. $\dfrac{9 - 24i}{3}$

70. $\dfrac{18 - \sqrt{-12}}{2}$

71. $\dfrac{8}{2i}$

72. $\dfrac{5 + 4i}{3i}$

73. $-\dfrac{\sqrt{3} - 2i}{6i}$

74. $\dfrac{39}{2 - i}$

75. $\dfrac{12i}{3 + 2i}$

76. $\dfrac{3i}{5 - \sqrt{-4}}$

77. $\dfrac{5 + 3i}{2 - 2i}$

78. $\dfrac{3 - 4i}{5 + 2i}$

79. $\dfrac{3 - \sqrt{-6}}{3 + \sqrt{-6}}$

80. $\dfrac{6 - \sqrt{-2}}{6 + \sqrt{-2}}$

81. $\dfrac{8i}{2 - \sqrt{2}\,i}$

82. $-\dfrac{3}{3 - \sqrt{-4}}$

83. $\dfrac{24 - \sqrt{-12}}{3}$

84. $\dfrac{27 - \sqrt{-27}}{3}$

85. $\dfrac{42 - \sqrt{-72}}{6}$

86. $\dfrac{48 - \sqrt{-128}}{8}$

87. $\dfrac{6 + 8i}{4i}$

88. $\dfrac{9 - 12i}{3i}$

89. $\dfrac{35}{2 + i}$

90. $\dfrac{40}{3 - i}$

91. $\dfrac{15i}{4 + 5i}$

92. $\dfrac{25i}{3 - 4i}$

93. $\dfrac{36i}{5 - \sqrt{-9}}$

94. $\dfrac{42i}{6 + \sqrt{-6}}$

95. $\dfrac{3 - 7i}{8 + 5i}$

96. $\dfrac{9 + 2i}{5 - 6i}$

97. $\dfrac{5 - \sqrt{-4}}{5 + \sqrt{-4}}$

98. $\dfrac{6 + \sqrt{-9}}{6 - \sqrt{-9}}$

99. $\dfrac{4 - \sqrt{-10}}{4 + \sqrt{-10}}$

100. $\dfrac{5 + \sqrt{-12}}{5 - \sqrt{-12}}$

Complex numbers provide solutions to some equations that otherwise might not have a solution. This section will investigate some methods for solving quadratic equations. These equations have various kinds of solutions such as whole numbers, rational numbers, irrational numbers, and complex numbers.

9.2A   Definition: Quadratic Equation

Equations of the form $ax^2 + bx + c = 0$ are quadratic equations in standard form; a, b, c represent constants. If $a = 0$ there is no second-degree term and hence no quadratic equation.

We can solve many quadratic equations by factoring them into a product of two binomials and applying the following theorem.

9.2B   Product Equal to Zero

If the product of two factors is zero, then one or both of the factors is zero. If $a \cdot b = 0$ then $a = 0$ or $b = 0$ or both a and $b = 0$.

We use this theorem to solve equations.

Example 13 □ Solve $(x - 7)(x + 2) = 0$.
    Since $(x - 7)(x + 2) = 0$

$$\text{either} \qquad x - 7 = 0 \qquad \text{or} \qquad x + 2 = 0$$
$$x = 7 \qquad\qquad\qquad x = -2$$

Testing our solutions in the original equation

$$\text{For} \quad x = 7 \qquad\qquad\qquad \text{For} \quad x = -2$$
$$(x - 7)(x + 2) = 0 \qquad\qquad (x - 7)(x + 2) = 0$$
$$(7 - 7)(7 + 2) \overset{?}{=} 0 \qquad\qquad (-2 - 7)(-2 + 2) \overset{?}{=} 0$$
$$0(9) = 0 \quad \text{True} \qquad\qquad\qquad -9(0) = 0 \quad \text{True} \quad □$$

Example 14 □ Solve $12x^2 + 5x - 2 = 0$.
    Notice that this equation is a sum, not a product. However, after we factor we'll have a product.

$$12x^2 + 5x - 2 = 0$$
$$(3x + 2)(4x - 1) = 0$$

Then $(3x + 2) = 0$ or $4x - 1 = 0$

$3x = -2$ $\qquad\qquad$ $4x = 1$

$x = -\dfrac{2}{3}$ $\qquad\qquad$ $x = \dfrac{1}{4}$

Checking the original equation

For $\quad x = -\dfrac{2}{3}$ $\qquad\qquad\qquad$ For $\quad x = \dfrac{1}{4}$

$12x^2 + 5x - 2 = 0$ $\qquad\qquad$ $12x^2 + 5x - 2 = 0$

$12\left(-\dfrac{2}{3}\right)^2 + 5\left(-\dfrac{2}{3}\right) - 2 \overset{?}{=} 0$ $\qquad$ $12\left(\dfrac{1}{4}\right)^2 + 5\left(\dfrac{1}{4}\right) - 2 \overset{?}{=} 0$

$12\left(\dfrac{4}{9}\right) - \dfrac{10}{3} - 2 \overset{?}{=} 0$ $\qquad\qquad$ $12\left(\dfrac{1}{16}\right) + \dfrac{5}{4} - 2 \overset{?}{=} 0$

$\dfrac{16}{3} - \dfrac{10}{3} - \dfrac{6}{3} \overset{?}{=} 0$ $\qquad\qquad$ $\dfrac{3}{4} + \dfrac{5}{4} - \dfrac{8}{4} \overset{?}{=} 0$

$0 = 0$ $\qquad\qquad\qquad\qquad$ $0 = 0$ $\quad\square$

Example 15 $\square$ Solve $3x^2 + 12x = 0$.
This time use the distributive law to factor.

$$3x^2 + 12x = 0$$

$$3x(x + 4) = 0$$

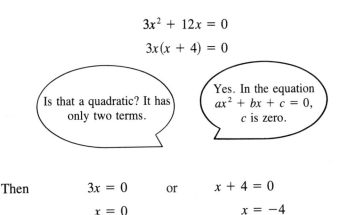

Is that a quadratic? It has only two terms.

Yes. In the equation $ax^2 + bx + c = 0$, c is zero.

Then $\qquad$ $3x = 0$ $\qquad$ or $\qquad$ $x + 4 = 0$

$x = 0$ $\qquad\qquad\qquad$ $x = -4$

The solutions are $x = 0$ or $x = -4$.
Checking in the original equation

$3(0)^2 + 12(0) \overset{?}{=} 0$ $\qquad$ $3(-4)^2 + 12(-4) \overset{?}{=} 0$

$0 + 0 \overset{?}{=} 0$ $\qquad\qquad$ $+48 + (-48) \overset{?}{=} 0$

$0 = 0$ $\qquad\qquad\qquad$ $0 = 0$ $\quad\square$

Example 16 □ Solve $x^2 = 9$.

First rewrite in standard form.

$$x^2 = 9$$

$$x^2 - 9 = 0$$

$$(x - 3)(x + 3) = 0 \qquad \text{Factoring}$$

Therefore $\qquad x - 3 = 0 \qquad$ or $\qquad x + 3 = 0$

$$x = 3 \qquad\qquad\qquad x = -3 \qquad □$$

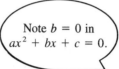

Note $b = 0$ in $ax^2 + bx + c = 0$.

That's ok. But if $a = 0$, it is not a quadratic.

Example 17 □ Solve $\dfrac{3x}{4} - \dfrac{2}{x} = \dfrac{5}{2}$.

First clear the fractions by multiplying by the least common denominator, $4x$.

$$\frac{3x}{4} - \frac{2}{x} = \frac{5}{2}$$

$$4x\left[\frac{3x}{4} - \frac{2}{x}\right] = \left[\frac{5}{2}\right]4x$$

$$3x^2 - 8 = 10x$$

$$3x^2 - 10x - 8 = 0 \qquad \text{Writing in standard form}$$

$$(3x + 2)(x - 4) = 0 \qquad \text{Factoring}$$

Then $\qquad 3x + 2 = 0 \qquad$ or $\qquad x - 4 = 0$

$$3x = -2 \qquad\qquad x = 4$$

$$x = -\frac{2}{3} \qquad\qquad\qquad □$$

You can test the solutions by substituting them in the original equation.

Example 18 □ Solve $2x^2 + (x - 5)^2 + (x - 6)^2 = x + 46$.

Simplify by performing the indicated operations. In this case, square the binomials.

$$2x^2 + (x - 5)^2 + (x - 6)^2 = x + 46$$

$$2x^2 + x^2 - 10x + 25 + x^2 - 12x + 36 = x + 46$$

$$4x^2 - 23x + 15 = 0$$

$$(4x - 3)(x - 5) = 0$$

Then $\qquad$ $4x - 3 = 0$ or $x - 5 = 0$

$$4x = 3 \qquad\qquad x = 5$$

$$x = \frac{3}{4}$$ $\qquad\qquad\qquad\qquad$ □

The ability to solve quadratic equations allows us to solve applications that use a variable squared.

Example 19 □ A patio is 6 yards longer than it is wide. If the total area of the patio is 40 square yards, determine the dimensions of the patio.

The formula for area is Area = length · width.

Let   width = w

length = $w + 6$

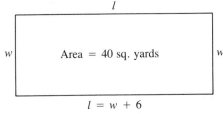

$$A = l \cdot w$$

$$40 = (w + 6)w$$

$$40 = w^2 + 6w$$

$$0 = w^2 + 6w - 40$$

$$0 = (w - 4)(w + 10)$$

Therefore, $w = 4$ or $w = -10$.

Since you can't have a width of -10 in a concrete world, we reject that solution.

If

$$w = 4$$

then the length is

$$l = w + 6$$

$$= 4 + 6$$

$$= 10$$

The patio is 4 yards by 10 yards. $\qquad\qquad$ □

Example 20 □ The ratio of the base of a triangle to its altitude is 5 to 3. If the area of the triangle is 120 sq. cm., find the length of the base.

When we know the ratio of two numbers, problems are easier to work if instead of x we use $3x$ as a variable.

Let $3x$ = altitude of the triangle

$5x$ = base of the triangle

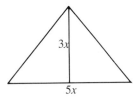

The ratio of $3x$ to $5x$ is 3 to 5.

The area of a triangle is

$$A = \frac{1}{2}b \cdot h$$

$$120 = \frac{1}{2}(5x)(3x)$$

$$120 = \frac{15x^2}{2}$$

$$240 = 15x^2$$

$$16 = x^2$$

$$4 = x$$

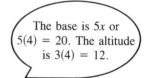

Be careful. Four isn't the length of the base.

The base is $5x$ or $5(4) = 20$. The altitude is $3(4) = 12$.

□

Example 21 □ A company in the inner city has a parking lot that is 40 yards by 60 yards in size. Due to a trend toward smaller cars, it would now like to reduce the total area of the parking lot and replace that area with a grass strip of uniform width around the outside edges of the lot except along the street. How wide should the strip be made in order to have 1200 square yards of parking lot remaining?

We can set up an equation for the new area.

■ Figure 9.1

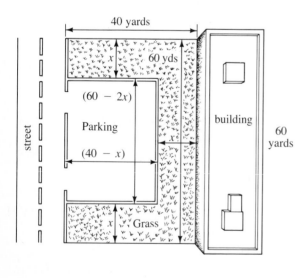

New length $= 60 - 2x$

New width $= 40 - x$

Area $=$ length $\cdot$ width

$$1200 = (60 - 2x)(40 - x)$$

$$1200 = 2400 - 140x + 2x^2$$

$$0 = 2x^2 - 140x + 1200$$

$$0 = x^2 - 70x + 600$$

$$0 = (x - 60)(x - 10)$$

$$x = 60 \qquad \text{or} \qquad x = 10$$

Chapter 9 ■ Complex Numbers, Quadratic Equations

The company can make a 10-yard wide grass strip around the outside of the parking lot.

What happened to the $x = 60$ answer?

The equation has two solutions $x = 10$ and $x = 60$. We reject $x = 60$ since that is wider than the original lot.

$\square$

Problem Set 9.2

Solve the following quadratic equations by factoring.

1. $x^2 - 6x + 5 = 0$

2. $x^2 - 6x + 8 = 0$

3. $2x^2 + 7x - 4 = 0$

4. $3x^2 + 11x + 6 = 0$

5. $8x^2 + 3 = 14x$

6. $4x^2 = 4x + 15$

7. $x^2 = 16$

8. $25x^2 - 60x + 36 = 0$

9. $3x^2 - 27 = 0$

10. $8x^2 - 40x = 0$

11. $x^2 - 4x = 12$

12. $x^2 - 14 = -5x$

13. $3x^2 = 22x - 24$

14. $4x^2 + 17x = -15$

15. $10x^2 + 23x = 5$

16. $6x^2 = 11x + 7$

17. $16x^2 + 25 = 40x$

18. $9x^2 + 30x + 25 = 0$

19. $12x^2 = 18x$

20. $12x^2 = 15x$

21. $5 + \dfrac{14}{x} - \dfrac{3}{x^2} = 0$

22. $10x + 1 - \dfrac{2}{x} = 0$

23. $x = \dfrac{16}{15} + \dfrac{1}{x}$

24. $x + \dfrac{7}{12} = \dfrac{1}{x}$

25. $\dfrac{17}{x} + \dfrac{15}{x^2} = 4$

26. $\dfrac{21}{x^2} = 2 + \dfrac{11}{x}$

27. $x = \dfrac{9}{20} + \dfrac{1}{x}$

28. $x + \dfrac{1}{2} = \dfrac{1}{9x}$

29. $2x^2 + (x - 2)^2 + (x - 3)^2 = 3(x + 1)$

30. $3x^2 + (x + 5)^2 + (x + 1)^2 = x + 38$

31. $(2x - 3)^2 + x = 12$

32. $(3x - 5)^2 + 22x = 45$

33. $3x^2 + (x + 3)^2 + (x - 4)^2 = 11 - 19x$

34. $4x^2 + (x - 2)^2 + (x - 5)^2 = 27 - x$

35. $(3x + 4)^2 = 11 + 6x$

36. $(2x - 1)^2 = 5(x + 2)$

Indicate what the variable represents. Set up an equation and solve for the variable.

37. A swimming pool is 8 meters longer than it is wide. If the total area of the swimming pool is 65 square meters, find the dimensions of the swimming pool.

38. The outside measurement of a picture frame is 20 inches by 8 inches. If the area of the picture is 64 square inches, find the width of the frame. Assume the frame has the same width on all sides.

39. The length of a patio is twice its width. If the total area is 288 square feet, find the dimensions of the patio.

40. A family has an area 60 feet by 48 feet available to build a rectangular swimming pool. The area of the swimming pool is to be 1408 square feet. Find the width of the decking surrounding the pool if it is to be of equal width on all sides.

41. An open box is to be constructed out of a piece of cardboard that measures 44 inches by 38 inches. It is to be made by cutting out squares of equal size from each corner then turning up the remaining pieces to form the sides of the box. If the base of the box contains 432 square inches, how large are the squares cut from the corners of the cardboard?

42. A plot of ground measures 40 feet by 60 feet. Design a rectangular rose garden surrounded by a sidewalk of uniform width to fit in this plot so that the area of the garden is $\frac{5}{8}$ of the area of the entire plot of ground. How wide is the sidewalk?

43. The ratio of the length of a rectangle to its width is 8 to 5. If the area of the rectangle is 1960 square inches. Find the dimensions of the rectangle.

44. The area of a triangle is 14 square centimeters. Find the length of the base and altitude if the base is 3 centimeters longer than the altitude $\left(A = \frac{1}{2}bh\right)$.

45. The square of a positive integer diminished by three times the integer is 18. Find the integer.

46. Three times the square of a positive integer diminished by 16 times the integer is 35. Find the integer.

47. Two times the square of an integer increased by 5 times the integer is 18. Find the integer.

48. The height an object will rise when shot or thrown vertically into the air (disregarding air resistance) is given by the formula $h = vt - 16t^2$, where h is the height, v is the initial velocity, and t is the time in seconds. If an object is shot into the air with an initial velocity of 192 feet per second, find the two times (on the way up and on the way down) that the object is 560 feet off the ground.

49. In problem 48, find the entire time the object was in the air.

50. If a stone is thrown vertically into the air with an initial velocity of 80 feet per second, how long will it take the stone to strike the ground?

51. The square of a positive integer increased by two times the integer is 24. Find the integer.

52. Four times the square of a negative integer increased by 19 times the integer is 5. Find the integer.

53. Problem 48 gives the formula for the height an object will rise when shot or thrown vertically into the air. If an object is shot into the air vertically with an initial velocity of 224 feet per second, find the time that the object is 640 feet above the ground on its way down.

54. Assuming that it takes an object thrown into the air the same amount of time to rise to its maximum height as it does for the object to fall to the ground, how long will it take the object in problem 53 to rise to its maximum height?

55. In problems 53 and 54, find the maximum height the object will rise.

■ **9.3**   **Solving Quadratic Equations by Completing the Square**

The primary rule to remember in equation solving is that you must do the same thing to both sides of the equation. Until now we could add, subtract, multiply, and divide. Now another possibility is to take the same root of both sides of an equation.

> ### 9.3A   Property of Square Roots of Equations
>
> If
> $$a^2 = b$$
> Then either
> $$a = \sqrt{b} \qquad \text{or} \qquad a = -\sqrt{b}$$

This property says that when the square roots of both sides of an equation are taken, two new equations result. That is because

$$(\sqrt{b})^2 = b \qquad \text{and also} \qquad (-\sqrt{b})^2 = b$$

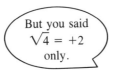

But you said $\sqrt{4} = +2$ only.

Not quite. We said $+2$ was the principal square root. The symbol $\sqrt{4}$ means $+2$. However the symbol $-\sqrt{4} = -2$ indicates the negative square root of four.

Example 22 □ Solve $x^2 = 25$.
By Property 9.3A

$$x = \sqrt{25} \qquad \text{or} \qquad x = -\sqrt{25}$$
$$x = 5 \qquad\qquad\qquad x = -5 \qquad\qquad □$$

Extraction of Roots

The method above is sometimes called **extraction of roots.** It produces the same results as solution by factoring.

Example 23 □ Solve $x^2 = 5$.

$$x = \sqrt{5} \qquad \text{or} \qquad x = -\sqrt{5}$$

A shorthand way to say the same thing is

$$x = \pm\sqrt{5}$$

± is read "plus or minus."

□

Example 24 □ Solve $(x - 2)^2 = 25$.
First take the square root of both sides.

$$(x - 2)^2 = 25$$
$$\sqrt{(x - 2)^2} = \sqrt{25} \qquad \text{or} \qquad \sqrt{(x - 2)^2} = -\sqrt{25}$$
$$x - 2 = 5 \qquad\qquad\qquad x - 2 = -5$$
$$x = 7 \qquad\qquad\qquad\quad x = -3 \qquad □$$

Example 25 □ Solve $(x + 3)^2 - 18 = 0$.

$$(x + 3)^2 - 18 = 0$$
$$(x + 3)^2 = 18 \qquad\qquad\qquad \text{Adding 18 to both sides}$$
$$x + 3 = \sqrt{18} \qquad \text{or} \qquad x + 3 = -\sqrt{18}$$
$$x + 3 = 3\sqrt{2} \qquad\qquad\qquad x + 3 = -3\sqrt{2}$$
$$x = -3 + 3\sqrt{2} \qquad\qquad x = -3 - 3\sqrt{2}$$

These results can be combined in a single statement.

$$x = -3 \pm 3\sqrt{2} \qquad\qquad □$$

Chapter 9 ■ Complex Numbers, Quadratic Equations

Solve each of the following equations by extraction of roots.

1. $x^2 = 49$

2. $x^2 = 11$

3. $x^2 = 12$

4. $x^2 = 48$

5. $(x + 2)^2 = 36$

6. $(x - 3)^2 = 64$

7. $(x - 5)^2 = 25$

8. $(x + 6)^2 = 49$

9. $(x - 7)^2 = 4$

10. $(x + 4)^2 - 16 = 0$

11. $(x + 3)^2 - 25 = 0$

12. $(x + 8)^2 - 9 = 0$

13. $(x - 2)^2 = 25$

14. $(x - 4)^2 = 36$

15. $(x + 5)^2 - 36 = 0$

16. $(x - 6)^2 = 36$

17. $(x + 7)^2 - 64 = 0$

18. $(x - 8)^2 - 49 = 0$

19. $(x + 2)^2 - 4 = 0$

20. $(x - 7)^2 - 49 = 0$

21. $(x - 9)^2 - 81 = 0$

22. $(x + 9)^2 - 64 = 0$

23. $(x - 11)^2 - 4 = 0$

24. $(x + 12)^2 = 36$

25. $(x + 8)^2 = 8$

26. $(x + 7)^2 = 20$

27. $(x - 2)^2 = 18$

28. $(x - 9)^2 - 81 = 0$

29. $(x - 5)^2 - 100 = 0$

30. $(x - 6)^2 - 98 = 0$

31. $(x + 4)^2 - 75 = 0$

32. $(x - 10)^2 = 288$

33. $(x + 11)^2 - 44 = 0$

34. $(x - 1)^2 - 3 = 0$

35. $(x - 7)^2 - 7 = 0$

36. $(x - 3)^2 - 15 = 0$

37. $(x - 7)^2 - 18 = 0$

38. $(x + 1)^2 - 12 = 0$

39. $(x + 3)^2 - 24 = 0$

40. $(x - 4)^2 - 32 = 0$

41. $(x + 5)^2 - 50 = 0$

42. $(x - 8)^2 - 72 = 0$

43. $(x + 9)^2 - 45 = 0$

44. $(x + 10)^2 - 125 = 0$

45. $(x - 10)^2 - 75 = 0$

46. $(x + 11)^2 - 96 = 0$

47. $(x - 12)^2 - 124 = 0$

48. $(x + 12)^2 - 117 = 0$

Completing the Square

We still have a problem. What do we do with a quadratic equation that is not factorable? An extension of the extraction of roots method, called **completing the square,** allows us to solve any quadratic equation.

First we need to learn how to complete a square. Recall the rule for squaring a binomial.

9.3B Squaring a Binomial

The square of a binomial is the square of the first term plus twice the product of the terms plus the square of the second term.

$$(a + b)^2 = a^2 + 2ab + b^2$$

Example 26 □ Square the following binomials.

$$(x + 1)^2 = x^2 + 2x + 1$$

4

$$(x + 2)^2 = x^2 + 4x + \underline{\hspace{1cm}}$$

Now let's study what's happening. □

Example 27 □ Square the following binomials.

6, 9

$$(x + 3)^2 = x^2 + \underline{\hspace{1cm}} x + \underline{\hspace{1cm}}$$

3

The constant term of the binomial that was squared is _____

6

The coefficient of the middle term of the trinomial is _____

9

The third term of the trinomial is _____ □

Look at another.

Example 28 □ Square the following binomial.

−8, 16

$$(x - 4)^2 = x^2 + \underline{\hspace{1cm}} x + \underline{\hspace{1cm}}$$

−4

The constant term of the binomial is _____

−8

The coefficient of the middle term is _____

The binomial constant is half the middle term of the trinomial.

What is the relationship between the constant of the binomial and the middle term of the trinomial? _____ □

Example 29 □ Square the following binomial.

−12, 36

$$(x - 6)^2 = x^2 + \underline{\hspace{1cm}} x + \underline{\hspace{1cm}}$$

−12

The coefficient of the middle term of the trinomial is _____

−6

Half of −12 is _____

−6

The constant term of the binomial is _____

36

The square of −6 is _____

36

The third term of the trinomial is _____ □

Perfect Square Trinomials

Trinomials that are perfect squares of a binomial have a constant term that is equal to the square of one-half the coefficient of the middle term. If you are given the first two terms of a trinomial in standard form, you can provide the third term required to make it a perfect square of a binomial.

Example 30 □ Provide the missing term in each trinomial below so that the result is a perfect square of a binomial.

$$x^2 + 10x + \underline{25} \qquad = (x + 5)^2 \qquad \boxed{25 \text{ is } \left(\frac{1}{2} \cdot 10\right)^2}$$

9, $x + 3$ $x^2 + 6x + \underline{\hspace{1cm}} = (\underline{\hspace{2cm}})^2$

64, $x - 8$ $x^2 - 16x + \underline{\hspace{1cm}} = (\underline{\hspace{2cm}})^2$

36, $x + 6$ $x^2 + 12x + \underline{\hspace{1cm}} = (\underline{\hspace{2cm}})^2$

4, $x - 2$ $x^2 - 4x + \underline{\hspace{1cm}} = (\underline{\hspace{2cm}})^2$ □

> This works only if the coefficient of the x^2 term is 1.

> The constant term of the binomial is always half the coefficient of the middle term of the trinomial.

> The constant term of the binomial always has the same sign as the middle term of the trinomial.

Completing the Square

> This process of supplying the missing third term so that a trinomial is a perfect square of a binomial is called **completing the square**.

Problem Set 9.3B

Provide the missing third term so that the result is a perfect square of a binomial.

1. $x^2 + 4x +$ _____

2. $x^2 + 10x +$ _____

3. $x^2 - 8x +$ _____

4. $x^2 + 36x +$ _____

5. $x^2 - 18x +$ _____

6. $x^2 - 12x +$ _____

7. $x^2 + x +$ _____

8. $x^2 + 3x +$ _____

9. $x^2 - 5x +$ _____

10. $x^2 - 7x +$ _____

11. $x^2 - 11x +$ _____

12. $x^2 + 13x +$ _____

13. $x^2 - 2x +$ _____

14. $x^2 - 6x +$ _____

15. $x^2 + 14x +$ _____

16. $x^2 + 16x +$ _____

17. $x^2 + 7x +$ _____

18. $x^2 + 9x +$ _____

19. $x^2 + 11x +$ _____

20. $x^2 + 15x +$ _____

21. $x^2 - 17x +$ _____

22. $x^2 + 5x +$ _____

23. $x^2 - 13x +$ _____

24. $x^2 + 7x +$ _____

Solving Quadratics by Completing the Square

The method of completing the square can be used to solve quadratic equations.

Example 31 □ Solve $x^2 - 8x + 12 = 0$ using the method of completing the square. Write the equation with the constant term on the right side of the equation.

$$x^2 - 8x + 12 = 0$$
$$x^2 - 8x = -12$$

Determine what must be added to the left-hand side to make it a perfect square.

−4

What is half the coefficient of $-8x$?_____

16

What is $(-4)^2$?_____

16

What must be added to the left-hand side to make it a perfect square?_____

Since we must do the same thing to both sides of the equation, add 16 to both sides.

$$x^2 - 8x + 16 = -12 + 16$$

Next write the left member as a binomial squared.

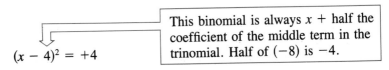

$(x - 4)^2 = +4$ | This binomial is always x + half the coefficient of the middle term in the trinomial. Half of (-8) is -4.

Now take the square root of both sides.

$$x - 4 = \sqrt{4} \qquad \text{or} \qquad x - 4 = -\sqrt{4}$$
$$x - 4 = 2 \qquad\qquad\qquad x - 4 = -2$$
$$x = +6 \qquad\qquad\qquad x = 2 \qquad\qquad □$$

Could the equation $x^2 - 8x + 12 = 0$ have been factored to find the solution?

Yes, it could. We just wanted to show the method. Some equations don't factor easily. After we have learned several methods, we usually pick the easiest method to solve a particular equation.

Chapter 9 ■ Complex Numbers, Quadratic Equations

Example 32 □ Solve $x^2 + 9x + 10 = 0$.

This is unfactorable. To use the method of completing the square, first write the equation with the constant on the right side.

$$x^2 + 9x + 10 = 0$$

$$x^2 + 9x \qquad = -10$$

Leave space here

We want to make the left side a perfect square.

The coefficient of the x term is _____ .

9

Half that coefficient is _____ .

$\dfrac{9}{2}$

Squaring $\dfrac{9}{2}$ yields _____ .

$\dfrac{81}{4}$

Now add $\dfrac{81}{4}$ to both sides of the equation to complete the square.

$$x^2 + 9x + \boxed{\frac{81}{4}} = -10 + \boxed{\frac{81}{4}}$$

$$\left(x + \frac{9}{2}\right)^2 = -\frac{40}{4} + \frac{81}{4}$$

$$\left(x + \frac{9}{2}\right)^2 = \frac{41}{4}$$

$$x + \frac{9}{2} = \pm\sqrt{\frac{41}{4}}$$

$$x + \frac{9}{2} = \pm\frac{\sqrt{41}}{2}$$

$$x = -\frac{9}{2} \pm \frac{\sqrt{41}}{2}$$

$$x = \frac{-9 \pm \sqrt{41}}{2} \qquad\qquad □$$

Even though the answer above is an exact answer, it is messy to use. For most purposes we can use a calculator to find an adequate decimal approximation.

$$\frac{-9 + \sqrt{41}}{2} \approx -1.298 \qquad\qquad \frac{-9 - \sqrt{41}}{2} \approx -7.702$$

Example 33 □ Solve $2x^2 + 2x - \dfrac{15}{2} = 0$.

First get the constant term on the right side.

$$2x^2 + 2x - \frac{15}{2} = 0$$

$$2x^2 + 2x \qquad = \frac{15}{2}$$

For the method of completing the square to work, the coefficient of the x^2 term must be 1. Therefore, multiply both sides of the equation by $\dfrac{1}{2}$.

$$\frac{2x^2}{2} + \frac{2x}{2} \qquad = \frac{15}{2} \cdot \frac{1}{2}$$

$$x^2 + \quad x \qquad = \frac{15}{4}$$

1

$\dfrac{1}{2}$

$\dfrac{1}{4}$

Now, the coefficient of the x term is _____ .

One-half the coefficient of the x term is _____ .

The square of half the coefficient of x is _____ .

$\dfrac{1}{4}$

The quantity to add to both sides of the equation in order to complete the square is _____

Half the coefficient of the middle term of the perfect square is $\dfrac{1}{2}$:

$\left(x + \dfrac{1}{2}\right)^2 = x^2 + x + \dfrac{1}{4}$.

$$x^2 + x + \frac{1}{4} = \frac{15}{4} + \frac{1}{4}$$

$$\left(x + \frac{1}{2}\right)^2 = \frac{16}{4}$$

$$\left(x + \frac{1}{2}\right)^2 = 4$$

$$x + \frac{1}{2} = 2 \qquad \text{or} \qquad x + \frac{1}{2} = -2$$

$$x = \frac{3}{2} \qquad\qquad\qquad x = -\frac{5}{2}$$

□

Example 34 □ Solve $2x^2 + 14x + 9 = 0$ by completing the square.
Write the equation with the constant term on the right side.

$$2x^2 + 14x + 9 = 0$$

$$2x^2 + 14x \qquad = -9$$

Make the coefficient of the x^2 term a "$+1$" by dividing both sides of the equation by the coefficient of x^2, which is 2.

$$\frac{2x^2}{2} + \frac{14x}{2} \qquad = -\frac{9}{2}$$

$$x^2 + \quad 7x \qquad = -\frac{9}{2}$$

> Can't I multiply both sides by $\frac{1}{2}$?

> Of course, division is defined as multiplication by the multiplicative inverse.

Complete the square by adding the square of half the coefficient of the x term to both sides.

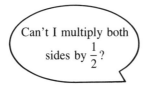

$$x^2 + 7x + \left(\frac{7}{2}\right)^2 = -\frac{9}{2} + \left(\frac{7}{2}\right)^2$$

$$\left(x + \frac{7}{2}\right)^2 = -\frac{9}{2} + \frac{49}{4}$$

$$\left(x + \frac{7}{2}\right)^2 = -\frac{9\,(2)}{2\,(2)} + \frac{49}{4} \qquad \text{Find a common denominator}$$

$$\left(x + \frac{7}{2}\right)^2 = -\frac{18}{4} + \frac{49}{4}$$

$$\left(x + \frac{7}{2}\right)^2 = \frac{31}{4}$$

Take the square root of both sides

$$x + \frac{7}{2} = \frac{\sqrt{31}}{2} \qquad \text{or} \qquad x + \frac{7}{2} = -\frac{\sqrt{31}}{2}$$

$$x = -\frac{7}{2} + \frac{\sqrt{31}}{2} \qquad\qquad x = -\frac{7}{2} - \frac{\sqrt{31}}{2}$$

$$x \approx -0.716 \qquad\qquad\qquad x \approx -6.284 \qquad\quad □$$

Example 35 □ Solve $2x^2 + 10x + 14 = 0$.

First write the constant on the right.

$$2x^2 + 10x + 14 = 0$$
$$2x^2 + 10x \qquad = -14$$

Divide both sides by 2 to get an x^2 coefficient of 1.

$$x^2 + 5x \qquad = -7$$

Add $\left(\dfrac{5}{2}\right)^2$ to both sides.

$$x^2 + 5x + \frac{25}{4} = -7 + \frac{25}{4}$$

$$\left(x + \frac{5}{2}\right)^2 = -\frac{28}{4} + \frac{25}{4}$$

$$\left(x + \frac{5}{2}\right)^2 = -\frac{3}{4}$$

Taking the square root of both sides

$$\sqrt{\left(x + \frac{5}{2}\right)^2} = \pm\frac{\sqrt{-3}}{\sqrt{4}}$$

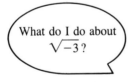

What do I do about $\sqrt{-3}$?

It's possible for quadratic equations to have complex roots.

$$x + \frac{5}{2} = +\frac{\sqrt{3}\,i}{2} \qquad \text{or} \qquad x + \frac{5}{2} = -\frac{\sqrt{3}\,i}{2}$$

$$x = -\frac{5}{2} + \frac{\sqrt{3}\,i}{2} \qquad\qquad\qquad x = -\frac{5}{2} - \frac{\sqrt{3}\,i}{2}$$

Since this is our first answer with a complex solution, we will demonstrate that it is indeed a solution of the original equation. It's a little easier if we write

$$x = -\frac{5}{2} \pm \frac{\sqrt{3}\,i}{2} \qquad \text{as} \qquad x = \frac{-5 \pm \sqrt{3}\,i}{2}$$

Substituting $x = \dfrac{-5 + \sqrt{3}\,i}{2}$ in the equation.

$$2x^2 + 10x + 14 = 0$$

$$2\left(\frac{-5 + \sqrt{3}\,i}{2}\right)^2 + 10\left(\frac{-5 + \sqrt{3}\,i}{2}\right) + 14 \stackrel{?}{=} 0$$

$$2\frac{(25 - 10\sqrt{3}\,i + 3i^2)}{4} - 25 + 5\sqrt{3}\,i + 14 \stackrel{?}{=} 0$$

$$\frac{25}{2} - \frac{10\sqrt{3}\,i}{2} + \frac{3(-1)}{2} + 5\sqrt{3}\,i - 11 \stackrel{?}{=} 0$$

$$\frac{25}{2} - \frac{10\sqrt{3}\,i}{2} - \frac{3}{2} + \frac{10\sqrt{3}\,i}{2} - 11 \stackrel{?}{=} 0$$

$$\frac{22}{2} - 11 - \frac{10\sqrt{3}\,i}{2} + \frac{10\sqrt{3}\,i}{2} \stackrel{?}{=} 0$$

$$0 = 0 \qquad \square$$

Some scientific calculators allow you to use complex numbers directly. Such a calculator would simplify the demonstration above.

Problem Set 9.3C

Solve the following quadratic equations by completing the square.

1. $x^2 - 6x + 8 = 0$

2. $x^2 - 2x + 24 = 0$

3. $x^2 - 4x - 12 = 0$

4. $x^2 + 6x - 27 = 0$

5. $x^2 + 5x - 14 = 0$

6. $x^2 - 5x - 36 = 0$

7. $x^2 + 3x = 28$

8. $x^2 - 44 = 7x$

9. $x^2 + 2x - 15 = 0$

10. $x^2 - 7x + 12 = 0$

11. $x^2 + 9x + 20 = 0$

12. $x^2 + 4x = 12$

13. $x^2 + 21 = 10x$

14. $x^2 - 5x = 24$

15. $x^2 - 7x = 18$

16. $x^2 + 2x = 48$

17. $3x^2 + 12x = 0$

18. $2x^2 - 10x = 0$

19. $3x^2 - 12 = 0$

20. $2x^2 - 32 = 0$

21. $3x^2 - 24 = 0$

22. $5x^2 - 60 = 0$

23. $3x^2 + 12 = 0$

24. $4x^2 + 36 = 0$

25. $4x^2 - 24 = 0$

26. $5x^2 - 40 = 0$

27. $6x^2 - 42 = 0$

28. $7x^2 - 35 = 0$

29. $4x^2 + 20x = 0$

30. $5x^2 + 30 = 0$

31. $7x^2 + 42 = 0$

32. $6x^2 + 48 = 0$

33. $3x^2 - 5x - 2 = 0$

34. $2x^2 + 11x + 12 = 0$

35. $4x^2 - 11x = 6$

36. $5x^2 - 8 = 6x$

37. $6x^2 - 5x - 6 = 0$

38. $8x^2 + 14x - 15 = 0$

39. $2x^2 - 6x - 3 = 0$

40. $3x^2 + 9x + 8 = 0$

41. $5x^2 + x - 3 = 0$

42. $2x^2 - 3x + 4 = 0$

43. $3x^2 - 4x - 3 = 0$

44. $3x^2 + 6x + 5 = 0$

45. $4x^2 - 3x - 2 = 0$

46. $4x^2 - 6x + 3 = 0$

47. $2x^2 + 4x - 7 = 0$

48. $3x^2 - 6x - 8 = 0$

49. $3x^2 + 6x + 7 = 0$

50. $2x^2 - 6x + 8 = 0$

51. $4x^2 - 12x - 3 = 0$

52. $4x^2 + 4x - 3 = 0$

53. $2x^2 - 7x + 4 = 0$

54. $2x^2 + 5x + 1 = 0$

55. $3x^2 + 9x + 5 = 0$

56. $4x^2 - 3x + 6 = 0$

57. $4x^2 - 5x + 5 = 0$

58. $5x^2 + 15x + 9 = 0$

59. $3x^2 - 4x + 3$

60. $2x^2 - 6x + 7 = 0$

■ 9.4　　　The Quadratic Formula

Recall from Section 9.2 that the standard form of a quadratic equation is $ax^2 + bx + c = 0$ where a, b, and c are constants and $a \neq 0$.

In the equation $2x^2 + 6x - 7 = 0$, $a = 2$, $b = 6$, and $c = -7$.

Any quadratic equation can be written in standard form. Next we will derive a formula that will give the solutions of any quadratic equation in terms of its coefficients a, b, and c.

To derive the quadratic formula, we must solve for x in the general form of a quadratic equation by completing the square.

Subtract c from both sides of the equation.

$$ax^2 + bx + c = 0$$
$$ax^2 + bx = -c$$

To make the coefficient of x equal to 1 divide both sides by a.

$$\frac{ax^2}{a} + \frac{bx}{a} = -\frac{c}{a}$$

$$x^2 + \frac{b}{a}x = -\frac{c}{a}$$

The coefficient of the x term is $\dfrac{b}{a}$.

One-half the coefficient of the x term is $\dfrac{1}{2} \cdot \dfrac{b}{a} = \dfrac{b}{2a}$.

The square of $\dfrac{b}{2a}$ is $\dfrac{b^2}{4a^2}$.

Adding $\dfrac{b^2}{4a^2}$ to both sides completes the square of the left-hand side.

$$x^2 + \frac{b}{a}x + \frac{b^2}{4a^2} = \frac{b^2}{4a^2} + \frac{-c}{a}$$

Write the left side as a binomial squared

$$\left(x + \frac{b}{2a}\right)^2 = \frac{b^2}{4a^2} + \frac{-c}{a}$$

Make the denominators alike on the right side

$$\left(x + \frac{b}{2a}\right)^2 = \frac{b^2}{4a^2} + \frac{-c}{a}\left(\frac{4a}{4a}\right)$$

Add the fractions on the right

$$\left(x + \frac{b}{2a}\right)^2 = \frac{b^2 - 4ac}{4a^2}$$

Take the square root of both sides

$$\sqrt{\left(x + \frac{b}{2a}\right)^2} = \pm\sqrt{\frac{b^2 - 4ac}{4a^2}}$$

$$x + \frac{b}{2a} = \pm\frac{\sqrt{b^2 - 4ac}}{2a}$$

Subtract $\dfrac{b}{2a}$ from both sides

$$x = -\frac{b}{2a} \pm \frac{\sqrt{b^2 - 4ac}}{2a}$$

Write the result over the common denominator

$$x = \frac{-b \pm \sqrt{b^2 - 4ac}}{2a}$$

The solutions to the quadratic equation are

$$x = \frac{-b + \sqrt{b^2 - 4ac}}{2a} \qquad \text{and} \qquad x = \frac{-b - \sqrt{b^2 - 4ac}}{2a}$$

9.4A   Quadratic Formula

The roots of any quadratic equation in the form $ax^2 + bx + c = 0$ are

$$x = \frac{-b \pm \sqrt{b^2 - 4ac}}{2a}$$

where $a \neq 0$.

Example 36 □ Solve $2x^2 - 13x - 15 = 0$ by using the quadratic formula.
In $2x - 13x - 15 = 0, a = 2, b = -13, c = -15$.
Write the quadratic formula.

$$x = \frac{-b \pm \sqrt{b^2 - 4ac}}{2a}$$

Substitute the values of a, b, c.

$$x = \frac{-(-13) \pm \sqrt{(-13)^2 - 4(2)(-15)}}{2(2)}$$

$$= \frac{13 \pm \sqrt{169 + 120}}{4}$$

$$= \frac{13 \pm \sqrt{289}}{4}$$

I could have solved that by factoring.

$$= \frac{13 \pm 17}{4}$$

$$x = \frac{30}{4} \quad \text{or} \quad x = -\frac{4}{4}$$

$$x = \frac{15}{2} \qquad\qquad x = -1$$

The next is one you can't factor. However, the formula always works.

□

Example 37 □ Solve $x^2 - 6x = 13$.
First write the equation in standard form.

$$x^2 - 6x - 13 = 0 \qquad a = 1, b = -6, c = -13$$

$$x = \frac{-b \pm \sqrt{b^2 - 4ac}}{2a}$$

Substituting for a, b, c

$$x = \frac{-(-6) \pm \sqrt{(-6)^2 - 4(1)(-13)}}{2(1)}$$

$$= \frac{6 \pm \sqrt{36 + 52}}{2}$$

$$= \frac{6 \pm \sqrt{88}}{2}$$

$$= \frac{6 \pm 2\sqrt{22}}{2}$$

$$= \frac{2(3 \pm \sqrt{22})}{2}$$

$$= 3 \pm \sqrt{22}$$

□

Chapter 9 ■ Complex Numbers, Quadratic Equations

Example 38 ☐ Solve $x^2 + x + 2 = 0$ using the quadratic formula.
Since $a = 1$, $b = 1$, $c = 2$

$$x = \frac{-b \pm \sqrt{b^2 - 4ac}}{2a}$$

$$= \frac{-(1) \pm \sqrt{(1)^2 - 4(1)(2)}}{2(1)}$$

$$= \frac{-1 \pm \sqrt{1 - 8}}{2}$$

$$= \frac{-1 \pm \sqrt{-7}}{2}$$

$$= -\frac{1}{2} \pm \frac{\sqrt{7}}{2}i$$

☐

> This equation has complex solutions.

You can verify that these are the solutions to the equation by substituting them in the original equation.

▶ Pointers About the Quadratic Formula

$$x = \frac{-b \pm \sqrt{b^2 - 4ac}}{2a}$$

Notice it appears that the quadratic formula is undefined if $a = 0$. That is true. However, this is not a problem since we do not have a quadratic equation if $a = 0$. Also, notice that the quadratic formula is really two equations because of the $\pm$ sign. This means we will usually have two roots to a quadratic equation. The number of roots and whether they are real or complex, is determined by what is under the radical sign.

$b^2 - 4ac$ is called the **discriminant** of a quadratic equation because it allows us to discriminate between the possible solutions.
If $b^2 - 4ac > 0$ then $\sqrt{b^2 - 4ac}$ yields two real solutions.

If $b^2 - 4ac = 0$ then we are adding and subtracting 0 to $-\frac{b}{2a}$. Hence there is only one real solution to the equation: $-\frac{b}{2a}$.

If $b^2 - 4ac < 0$ then we are taking the square root of a negative number and there will be two complex solutions to the equation.

CAUTION: The division bar extends across the entire formula. Everything above the bar is divided by 2a.

Example 39 □ Use the discriminant to determine the nature of the roots in Examples 37 and 38.

(Ex 37)
$$x^2 - 6x - 13 = 0$$
$$b^2 - 4ac = (-6)^2 - 4(1)(-13)$$
$$= 36 + 52$$
$$= 88$$

Since the discriminant equals 88, which is a positive number, there are two real roots.

(Ex 38)
$$x^2 + x + 2 = 0$$
$$b^2 - 4ac = 1^2 - 4(1)(2)$$
$$= 1 - 8$$
$$= -7$$

Since the discriminant equals -7, which is a negative number, there are two complex roots. □

Example 40 □ Determine the nature of the roots of the equation $x^2 - 6x + 9 = 0$.

$$b^2 - 4ac = (-6)^2 - 4(1)(9)$$
$$= 36 - 36$$
$$= 0$$

Since the discriminant is zero, there is one real root. □

Problem Set 9.4

Solve the following equations by using the quadratic formula.

1. $x^2 + x - 6 = 0$

2. $x^2 + 6x + 8 = 0$

3. $2x^2 - 5x - 3 = 0$

4. $4x^2 - 5x - 6 = 0$

5. $6x^2 + 5x - 4 = 0$

6. $8x^2 + 6x - 9 = 0$

7. $3x^2 + 6x = 0$

8. $2x^2 - 10x = 0$

9. $9x^2 + 4x = 0$

10. $5x^2 - 8x = 0$ **11.** $x^2 - 2x - 8 = 0$ **12.** $x^2 - 2x - 15 = 0$

13. $x^2 + 11x - 12 = 0$ **14.** $5x^2 - 16x + 12 = 0$ **15.** $4x^2 - 4x - 15 = 0$

16. $9x^2 + 9x - 4 = 0$ **17.** $6x^2 - x - 15 = 0$ **18.** $12x^2 - x - 6 = 0$

19. $3x^2 - 9x = 0$ **20.** $4x^2 + 6x = 0$ **21.** $3x^2 - 4 = 0$

22. $5x^2 - 9 = 0$ **23.** $2x^2 + 9 = 0$ **24.** $3x^2 + 8 = 0$

25. $3x^2 - 8 = 0$ **26.** $2x^2 - 9 = 0$ **27.** $4x^2 + 9 = 0$

28. $3x^2 + 4 = 0$ **29.** $2x^2 + 4x - 3 = 0$ **30.** $3x^2 - 5x + 4 = 0$

31. $4x^2 - 8x - 3 = 0$ **32.** $5x^2 + x + 1 = 0$ **33.** $3x^2 + 7 = 2x$

34. $2x^2 + 3x = 3$ **35.** $6x^2 = 2x - 3$ **36.** $5x^2 - 4x = 1$

37. $4x^2 - 3 = 2x$ **38.** $5x^2 + 4 = 5x$ **39.** $3x^2 - 6x - 2 = 0$

40. $2x^2 - 4x - 7 = 0$ **41.** $4x^2 + 3x + 1 = 0$ **42.** $5x^2 + 2x + 3 = 0$

43. $2x^2 + 5x = 6$ **44.** $3x^2 + 2x = 5$ **45.** $5x^2 - 2x = -2$

46. $6x^2 + 1 = x$ **47.** $3x^2 + 1 = 4x$ **48.** $4x^2 + 3 = 6x$

Use the discriminant to determine how many roots exist for each of the following equations and if the roots are real or complex.

49. $x^2 - 4x + 4 = 0$ **50.** $x^2 + 5x - 3 = 0$ **51.** $4x^2 + 12x + 9 = 0$

52. $3x^2 + 2 = x$ **53.** $x^2 - 5x = 6$ **54.** $3x^2 + 4x = 0$

55. $4x^2 - 11 = 0$ **56.** $5x^2 + 6 = 0$ **57.** $5x^2 - 4x + 1 = 0$

58. $9x^2 - 30x + 25 = 0$ **59.** $2x^2 = 3x + 1$ **60.** $3x^2 = 4x - 2$

61. $3x^2 - 4x + 7 = 0$ **62.** $2x^2 + 5x - 3 = 0$ **63.** $x^2 - 10x + 25 = 0$

64. $2x^2 + 3x + 4 = 0$ **65.** $4x^2 - x = 3$ **66.** $9x^2 + 12x + 4 = 0$

67. $6x^2 - 4 = 3x$ **68.** $5x^2 - 23 = 4x$ **69.** $7x^2 + 8x + 4 = 0$

70. $5x^2 + 4 = 20x$ **71.** $16x^2 + 9 = 24x$ **72.** $3x^2 + 4 = 2x$

In algebra textbooks, most numbers work out to be integers or familiar fractions. Part of this is tradition from the days before calculators were available. But, there are two other good reasons why mathematics books use "nice" numbers.

1. The authors want to help you develop your powers of estimation and general familiarity with mental computation. You could say we are practicing advanced arithmetic so that you will be able to recognize efficient ways of simplifying numerical relationships on sight.
2. Frequently messy arithmetic just obscures the algebraic idea that we are trying to teach. Nevertheless, in the real world many problems do not have "nice" answers. Therefore a calculator approach is very useful. Solutions to quadratic equations is a good example.

Before we evaluate the quadratic formula on a calculator we will look at a few features of most scientific calculators that make them more convenient to use than basic four-function calculators.

Your calculator is programmed to follow the rules for order of operation as outlined in Chapter 1.

Briefly your calculator should follow this order:

1. Operations within parentheses.
2. Operations that raise quantities to powers like $\boxed{x^2}$, $\boxed{y^x}$, $\boxed{\sqrt{x}}$.
3. Multiplication and division.
4. Addition and subtraction.

To confirm that your calculator follows the rules for order of operation, evaluate $4^2 - 2 \cdot (-3)$ by doing the following steps:

Press: $4\boxed{x^2}$
Press: $\boxed{-}$ 2
Press: $\boxed{\times}$ 3 $\boxed{+/-}\boxed{=}$

The display should indicate 22.
On your calculator you must be careful with division.

Example 41 □ Evaluate $\dfrac{12}{2 \cdot 3}$ on your calculator.

I can do that in my head.

I know you can. We're illustrating the order in which your calculator performs operations.

Press: 12 $\boxed{\div}$ 2 $\boxed{\div}$ 3 $\boxed{=}$
You should see 2 on the display.

I read that problem as 12 divided by two times three. Why doesn't 12 ⬜÷ 2 ⬜× 3 work?

That's $\dfrac{12}{2} \cdot 3 = 18.$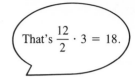

The problem is really 12 divided by the quantity $2 \cdot 3$. If your calculator has parentheses, you can use them to evaluate this expression.

Press: 12 ⬜÷ ⬜(2 ⬜× 3 ⬜) ⬜=

The display should show 2. We'll talk more about parentheses later. ☐

It is possible to evaluate the expression

$$\frac{-(-0.5) \pm \sqrt{(-0.5)^2 - 4(7.3)(-8.1)}}{2(7.3)}$$

by writing intermediate results on a piece of scratch paper. But it's more efficient if you take advantage of the power of your calculator.

 Using Your Calculator

A useful key on your calculator is the memory key. Most scientific calculators have at least one memory. If your calculator has more than one memory, we strongly suggest you spend the time with your manual to learn how to use the memories.

Your calculator probably has two keys. One of these keys stores the number on the display in memory; it is probably labeled STO or $x \rightarrow m$. To retrieve a value from memory there is a key labeled RCL or RM for recall memory.

Try: $2 + 3$ ⬜STO or ⬜$x \rightarrow m$ A small m should appear on the display.

Now press: ⬜+ The value of 5 should appear.

Now press: ⬜RCL or ⬜RM 3 should appear from memory.

The first thing we did was $2 + 3$. Why didn't 5 come out of memory?

The ⬜STO or ⬜$x \rightarrow m$ key moves the contents of the display to the memory immediately. It does not have any suspended operations.

If you want to store an intermediate value in memory, press the ⬜= key to get the intermediate value on the display, then press ⬜STO

The memory is like a tape recording. Using it does not destroy the stored value. Each time you press ⬜RM or ⬜RCL, the current value in memory will return to the display.

Then how do I get rid of a value in memory?

It depends on the calculator. Pressing the CLR or AC key may do it. You can also STO a zero in memory.

It usually isn't necessary to clear a value from memory because the next value you enter into memory replaces the previous value.

Your calculator probably also has a $m+$ or SUM key. This key adds the current display value to the value in memory. If it looks as if nothing happened, that is because the value in memory, not the value on the display, was changed.

Try: 5   or   6   or
      STO   SUM
      $x \rightarrow m$ $m+$

This sequence should produce an 11 in memory. To see it on the display, press RCL or RM.

Now let's look at a quadratic equation where your calculator can be a real help.

Example 42 □ Solve $7.3x^2 - 0.5x - 8.1 = 0$.
Applying the quadratic formula:

$$a = 7.3, b = -0.5, c = -8.1$$

$$x = \frac{-(-0.5) \pm \sqrt{(-0.5)^2 - 4(7.3)(-8.1)}}{2(7.3)}$$

There are many ways to evaluate this expression by using a calculator. We will evaluate the radical, store that result in memory, and use the stored value to avoid rekeying.

Press: .5 x^2 ← No need to enter the negative sign since the result is positive

You could let the calculator handle the signs in $-4(7.3)(-8.1)$, but it's faster if you determine the sign of the product and enter the appropriate operation; in this case it's $+$.

Press: $+$ 4 $\times$ 7.3 $\times$ 8.1 $=$
The display should indicate 236.77.

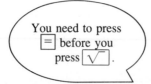
You need to press $=$ before you press $\sqrt{}$.

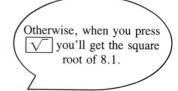
Otherwise, when you press $\sqrt{}$ you'll get the square root of 8.1.

Press $\boxed{\sqrt{}}$ to get a display 15.387332.

Now store the value of the radical for later use.

Press: $\boxed{\text{STO}}$

To obtain the positive sum press: $\boxed{+}.5\ \boxed{=}\ \boxed{\div}\ 2\ \boxed{\div}\ 7.3\ \boxed{=}$.

Since $-(-0.5) =$ .5, it's shorter to press $+.5$.

The $\boxed{=}$ after .5 is necessary. Otherwise, the calculator will divide $\dfrac{.5}{2 \cdot 7.3}$ instead of $\dfrac{.5 + 15.378}{2 \cdot 7.3}$.

The value of the first root should be displayed as 1.0887735.

Now find the value of the second root.

Press: $.5\ \boxed{-}\ \boxed{\text{RCL}}\ \boxed{=}$

The display should be -14.887332.

Press: $\boxed{\div}\ 2\ \boxed{\div}\ 7.3\ \boxed{=}$

The second root displayed should be: -1.0196803. □

Example 43 □ Solve $3.3x^2 + 7x + 8 = 0$.

Applying the quadratic formula:

$$a = 3.3,\ b = 7,\ c = 8$$

$$x = \frac{-7 \pm \sqrt{7^2 - 4(3.3)(8)}}{2(3.3)}$$

First evaluate the discriminant.

Press: $7\ \boxed{x^2}\ -\ 4\ \boxed{\times}\ 3.3\ \boxed{\times}\ 8\ \boxed{=}$ to get -56.6.

Notice that the discriminant is negative. Your calculator probably doesn't take roots of negative numbers.

You will have to take the square root of the absolute value of the quantity under the radical and indicate i in your answer.

Before you take the square root of the discriminant, press $\boxed{+/-}$ to get a positive value on the display.

Press: $\boxed{+/-}\ \boxed{\sqrt{}}$. The display should be 7.5232971.

Now it's best if you divide the formula into real and imaginary parts.

$$\frac{-7 \pm \sqrt{-56.6}}{2(3.3)} = \frac{-7}{2(3.3)} \pm \frac{\sqrt{56.6}\, i}{2(3.3)}$$

First, evaluate the imaginary part.

From previous work, the display should be $\sqrt{56.6} \approx 7.5232971$.

Press: $\boxed{\div}$ 2 $\boxed{\div}$ 3.3 $\boxed{=}$

The magnitude of the imaginary part is now displayed as 1.1398935.

Next we evaluate the real part of the solution.

Press: 7 $\boxed{+/-}$ $\boxed{\div}$ 2 $\boxed{\div}$ 3.3 $\boxed{=}$ to display -1.0606061.

The solutions are:

$$x \approx -1.0606061 \pm 1.1398935i \qquad \square$$

> Do I need all those decimal places?

> If the decimal values given in the original equation are exact values, even seven decimal places are only an approximation of the answer.

▶ Pointers About Accuracy

Some numbers are considered absolutely accurate; others are by their nature approximations.
   Numbers that are absolutely accurate are:

counting numbers. The number of desks in the room can be known exactly.

numbers that result from theory. The area of a triangle is exactly $\frac{1}{2}$ (base)(height).

   On the other hand, all measurements are necessarily approximations. It is either usually very difficult or very expensive to measure to more than four-place accuracy.
   In the previous example you were asked to solve: $3.3x^2 + 7x - 8 = 0$.
   Since this is a mathematics book you might infer that we mean exactly $3.3x^2$. However, the usual practice is to assume that in numbers written with decimals, the accuracy of the last place written is questionable.
   From now on in this text we will assume that any answer is no more accurate than the least accurate piece of data used in its calculation.
   In doing operations with calculators, we will not round off intermediate answers, instead we will use the full accuracy of the calculator up to the final answer. Then we will round the final answer to the accuracy of the measured data.
   When we show intermediate results in the text we will generally show four decimal places so that you can compare your calculator display with the textbook.
   For more information about accuracy and rounding, see the appendix.

Use the quadratic formula to solve the following equations. Give answers correct to three decimal places.

1. $0.3x^2 + 2x - 0.7 = 0$

2. $0.9x^2 - 0.3x - 4 = 0$

3. $0.5x^2 - 0.4x - 0.8 = 0$

4. $0.6x^2 - 0.9x - 0.7 = 0$

5. $4x^2 - 1.2x + 0.09 = 0$

6. $3x^2 - 3.4x + 0.06 = 0$

7. $0.08x^2 - 5.6x + 13 = 0$

8. $0.09x^2 - 2.4x + 16 = 0$

9. $-0.9x^2 + 23.4x + 1.2 = 0$

10. $-1.2x^2 + 36.9x + 5.8 = 0$

11. $3.1x^2 - 0.8x + 4.2 = 0$

12. $5.4x^2 - 0.23x + 8.4 = 0$

13. $0.64x^2 - 0.48x + 0.09 = 0$

14. $0.45x^2 - 0.37x + 0.08 = 0$

15. $0.15x^2 - 0.13x + 0.08 = 0$

16. $0.16x^2 - 0.24x + 0.09 = 0$

17. $3.01x^2 + 5.86x + 7.45 = 0$

18. $7.09x^2 + 4.64x - 3.47 = 0$

19. $7.80x^2 - 5.92x - 3.85 = 0$

20. $4.03x^2 + 6.74x + 8.92 = 0$

21. $-0.6x^2 + 234.5x + 21.7 = 0$

22. $-0.7x^2 - 346.9x + 31.8 = 0$

23. $3.8400x^2 + 4.5120x - 6.500 = 0$

24. $25.64x^2 + 4.743x + 5.800 = 0$

25. $341.8x^2 - 1056x + 51.43 = 0$

26. $981.4x^2 - 321.6x - 3.812 = 0$

27. $8.920x^2 - 0.0542x - 34.50 = 0$

28. $72.90x^2 + 8.425x + 2.847 = 0$

29. $1.2200x^2 + 2.7000x - 4.2000 = 0$

30. $2.1341x^2 + 7.4195x - 5.2000 = 0$

Chapter 9 □ Key Ideas

9.1   1. The **imaginary unit** is called i. It has the property: $i^2 = -1$ and $i = \sqrt{-1}$.
      2. **Pure imaginary numbers** are numbers of the form bi where $b \in$ Reals and $i = \sqrt{-1}$.
      3. **Powers of i** $i = i$ $i^2 = -1$ $i^3 = -i$ $i^4 = 1$

4. A **complex number** is of the form:

$a + bi$, where $a, b \in$ real numbers

5. Equality of complex numbers

$a + bi = c + di$, if and only if $a = c$ and $b = d$

6. Addition of complex numbers

$(a + bi) + (c + di) = (a + c) + (b + d)i$

7. Product of two complex numbers

$(a + bi)(c + di) = (ac - bd) + (bc + ad)i$

8. To divide by a complex number

Multiply the numerator and denominator of the fraction by the conjugate of the denominator.

9.2 **1. Quadratic equation**—The standard form of a quadratic equation is:

$ax^2 + bx + c = 0$, where $a \neq 0$

2. Product equal to zero

If $a \cdot b = 0$, then $a = 0$ or $b = 0$ or both equal zero.

3. To solve quadratic equations
  a. set the equation equal to zero;
  b. factor the equation completely;
  c. set each factor equal to zero;
  d. solve each factor set equal to zero for the variable.

4. Many applications can be solved by using equations of degree higher than one.

9.3 **1. Property of square roots of equations**

If $a^2 = b$, then either $a = \sqrt{b}$ or $a = -\sqrt{b}$.

2. Solving equations by **extraction of roots** method:

extract the indicated root of both sides of an equation.

3. Completing the square depends upon the square of a binomial.

$(a + b)^2 = a^2 + 2ab + b^2$

4. To complete the square:

$(x + a)^2 = x^2 + 2ax + \underline{\hspace{2cm}}$

The number to be squared is one half the coefficient of x, namely $\dfrac{1}{2}(2a) = a$

$\left[\dfrac{1}{2}(2a)\right]^2 = a^2$ hence, $x^2 + 2ax + \underline{a^2} = (x + a)^2$

5. To solve a quadratic equation by the method of **completing the square:**
   a. make the coefficient of the squared term equal to one by dividing all terms by the coefficient of the squared term;
   b. write the equation with the constant term on the right side;
   c. complete the square on the left side;
   d. add the same quantity to the right side as was added to the left side when completing the square;
   e. use the method of **extraction of roots** to solve for the variable.

9.4 The quadratic formula

1. Write the equation in standard form.

$$ax^2 + bx + c = 0$$

2. The roots of a quadratic equation in standard form are:

$$x = \frac{-b \pm \sqrt{b^2 - 4ac}}{2a} \qquad \text{where } a \neq 0$$

3. $b^2 - 4ac$ is the **discriminant** of a quadratic equation.
   a. If $b^2 - 4ac > 0$, the equation has two real solutions.
   b. If $b^2 - 4ac = 0$, the equation has one real solution.
   c. If $b^2 - 4ac < 0$, the equation has two complex solutions.

9.5

1. Many applications of the quadratic formula are facilitated by using a calculator to perform the arithmetic.

2. The memory key of the calculator can be used to reduce the number of key strokes to compute a solution to a quadratic equation.

Chapter 9 Review Test

Evaluate the following. **(9.1)**

1. $\sqrt{-25}$ 2. $\sqrt{-81}$ 3. $\sqrt{-18}$ 4. $\sqrt{-75}$ 5. i^3 6. i^2 7. i^5

8. i^{12}

Solve for x and y. **(9.1)**

9. $3x + 4yi = 9 - 12i$ 10. $(x - 3) - 5yi = 12 + (3y + 24)i$

Simplify the following. **(9.1)**

11. $(7 - 5i) + (-8 + 3i)$ 12. $(-3i + 4) - (6 - 9i)$ 13. $(2 + 8i)(7 - 4i)$

14. $(-6 + \sqrt{-3})(4 + 3\sqrt{-3})$ 15. $\dfrac{8 - 12i}{4}$ 16. $\dfrac{10 + 5i}{5i}$

17. $\dfrac{5 + 4i}{2 - 3i}$ 18. $\dfrac{8 - \sqrt{-8}}{3 + \sqrt{-18}}$

Solve for x by factoring. **(9.2)**

19. $3x^2 = 27$ **20.** $4x^2 - 64 = 0$ **21.** $2x^2 - 9x - 18 = 0$

22. $x + \dfrac{7}{6} = \dfrac{10}{3x}$ **23.** $2x^2 + (x - 3)^2 + (x + 2)^2 = 2x + 28$

Solve the following problems. State in words exactly what your variable represents. Set up an equation and solve. **(9.2)**

24. The area of a picture is 96 square inches. It is set in a frame of uniform width of two inches. If the total area of the picture and frame is twice the area of the picture alone, what are the dimensions of the picture frame?

25. The sum of the squares of two positive consecutive odd integers is 202. What are the integers?

Solve for x. **(9.3)**

26. $2x^2 = 32$ **27.** $3x^2 + 54 = 0$ **28.** $(x - 4)^2 = 49$ **29.** $(x + 5)^2 = 18$

Solve for x by completing the square. **(9.3)**

30. $x^2 - 7x - 18 = 0$ **31.** $3x^2 + 5x = 2$ **32.** $2x^2 = 4x - 5$

Solve the following quadratic equations by using the quadratic formula. **(9.4)**

33. $x^2 + 3x - 10 = 0$ **34.** $2x^2 - 15 = x$ **35.** $3x^2 = 4x + 3$

36. $4x^2 - 5x + 2 = 0$

Use the discriminant to determine the nature of the roots of the following quadratic equations. **(9.4)**

37. $4x^2 - 20x + 25 = 0$ **38.** $3x^2 - 8x = 16$ **39.** $2x^2 + 5x + 7 = 0$

40. $5x^2 + 1 = 3x$

Use your calculator and the quadratic formula to solve the following equations. Assume all numbers are the result of measurement. Express your final answer to the accuracy justified by the values given in the problems. **(9.5)**

41. $0.8x^2 + 4x + 0.7 = 0$ **42.** $0.64x^2 - 1.44x + 0.81 = 0$

43. $0.72x^2 + 0.56x + 1.43 = 0$ **44.** $3.5800x^2 + 4.8000x - 5.8450 = 0$

10 Quadratic Equations, Variation

Contents

The last chapter showed how to solve any quadratic equation. This chapter will extend your equation solving skills to equations containing radicals and some higher order equations that can be treated as quadratic equations.

The final two sections of this chapter will help you apply the skills you have developed with radicals and equation solving to some real world applications. The techniques used in variation are the first steps taken by scientists as they write laws that describe physical, chemical, biological, and economic relationships.

The mathematics used in the motion problems of this chapter is the fundamental mathematics involved in describing the flight of a rocket to the moon or the path of a car on a highway.

■ 10.1   Equations Containing Radicals

To solve equations containing radicals we usually square both sides of the equation to eliminate the radical.

Example 1 □ Solve $\sqrt{x} = 3$.

$$\sqrt{x} = 3$$
$$(\sqrt{x})^2 = (3)^2 \qquad \text{Squaring both sides}$$
$$x = 9 \qquad\qquad\qquad □$$

It isn't always this easy. Frequently we have to rearrange the equation to isolate the radical on one side.

Example 2 □ Solve $\sqrt{x + 11} - 2x + 6 = 0$.

$$\sqrt{x + 11} = 2x - 6 \qquad \text{Isolate the radical}$$
$$(\sqrt{x + 11})^2 = (2x - 6)^2 \qquad \text{Square both sides}$$
$$x + 11 = 4x^2 - 24x + 36$$
$$0 = 4x^2 - 25x + 25$$
$$0 = (4x - 5)(x - 5)$$

Then $\qquad 4x - 5 = 0 \qquad$ or $x - 5 = 0$

$$4x = 5 \qquad\qquad\qquad x = 5$$

$$x = \frac{5}{4}$$

Testing both of these roots in the original equation first test $x = \dfrac{5}{4}$.

$$\sqrt{x + 11} - 2x + 6 = 0$$

$$\sqrt{\frac{5}{4} + 11\left(\frac{4}{4}\right)} - 2\left(\frac{5}{4}\right) + 6 \stackrel{?}{=} 0$$

$$\sqrt{\frac{49}{4}} - \frac{10}{4} + 6 \stackrel{?}{=} 0$$

$$\frac{7}{2} - \frac{5}{2} + 6 \stackrel{?}{=} 0$$

$$\frac{1}{1} + 6 \neq 0$$

What happened?

The answer $x = \dfrac{5}{4}$ is called an **extraneous root**. Read below.

When you multiply both sides of an equation by a variable or square both sides of an equation, you may introduce false roots.

Now test $x = 5$.

$$\sqrt{x + 11} - 2x + 6 = 0$$

$$\sqrt{5 + 11} - 2(5) + 6 \stackrel{?}{=} 0$$

$$\sqrt{16} - 10 + 6 \stackrel{?}{=} 0$$

$$4 - 4 = 0 \qquad x = 5 \text{ is a solution.} \quad \square$$

If it is possible to gain invalid solutions to an equation, you might ask, "Can I lose valid solutions to an equation?"

The following shows what can happen if you divide both sides of the equation by an expression containing the variable.

Consider $x\sqrt{x} - 2\sqrt{x} = 0$.

If we factor $\sqrt{x}$

$$\sqrt{x}(x - 2) = 0$$

and

$$\sqrt{x} = 0 \qquad \text{or} \qquad x = 2$$

$$x = 0$$

But if we multiply both sides by $\dfrac{1}{\sqrt{x}}$

$$x\sqrt{x} - 2\sqrt{x} = 0$$

becomes

$$\frac{1}{\sqrt{x}} x \sqrt{x} - \frac{1}{\sqrt{x}} 2\sqrt{x} = \frac{1}{\sqrt{x}} \cdot 0$$

$$x - 2 = 0$$

$$x = 2$$

> We lose the root $x = 0$ because we divided both sides by zero.

Caution: If you square both sides of an equation or multiply both sides of an equation by a variable, test all solutions for extraneous roots. If you divide both sides of an equation by an expression containing a variable, you may lose roots.

Example 3 □ Solve $\sqrt{x - 9} = \dfrac{6\sqrt{x}}{x}$.

Here is one radical on each side of the equation. It's possible to start by squaring both sides.

$$\sqrt{x - 9} = \frac{6\sqrt{x}}{x}$$

$$x - 9 = \frac{36x}{x^2} \qquad \text{Square both sides}$$

$$x - 9 = \frac{36}{x} \qquad \text{Reduce}$$

$$x^2 - 9x = 36 \qquad \text{Clear fractions}$$

$$x^2 - 9x - 36 = 0$$

$$(x - 12)(x + 3) = 0 \qquad \text{Factor}$$

$$x = 12 \quad \text{or} \quad x = -3$$

Testing both possible solutions:

For $x = 12$

$$\sqrt{x - 9} = \frac{6\sqrt{x}}{x}$$

$$\sqrt{12 - 9} \stackrel{?}{=} \frac{6\sqrt{12}}{12}$$

$$\sqrt{3} \stackrel{?}{=} \frac{\sqrt{12}}{2}$$

$$\sqrt{3} = \frac{2\sqrt{3}}{2}$$

$$\sqrt{3} = \sqrt{3}$$

For $x = -3$

$$\sqrt{x - 9} = \frac{6\sqrt{x}}{x}$$

$$\sqrt{-3 - 9} = \frac{6\sqrt{-3}}{-3}$$

$$\sqrt{-12} = -2\sqrt{-3}$$

$$2\sqrt{3}\,i \neq -2\sqrt{3}\,i$$

Extraneous root

It checks. □

Chapter 10 ■ Quadratic Equations, Variation

A Pointer About Extraneous Roots

Here are two reasons why extraneous roots are generated.

    I. Multiplication of both sides of an equation by zero

    II. A logic error

Let's look at each case.

Case I:

If you multiply both sides of an equation by zero, you introduce extraneous roots. Here are two ways to multiply by zero without realizing that you are.

1. Multiply both sides of an equation by a variable.

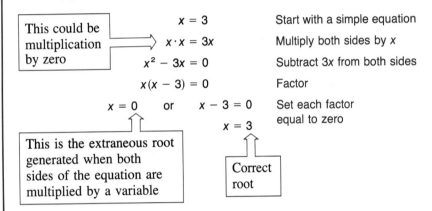

This could be multiplication by zero	

$x = 3$ Start with a simple equation

$x \cdot x = 3x$ Multiply both sides by x

$x^2 - 3x = 0$ Subtract 3x from both sides

$x(x - 3) = 0$ Factor

$x = 0$ or $x - 3 = 0$ Set each factor equal to zero

$x = 3$

This is the extraneous root generated when both sides of the equation are multiplied by a variable

Correct root

2. Multiply both sides of an equation by an expression containing a variable.

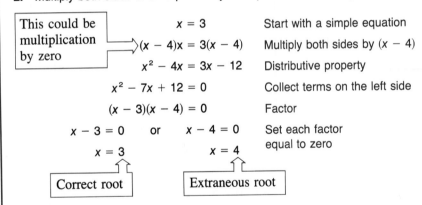

This could be multiplication by zero	

$x = 3$ Start with a simple equation

$(x - 4)x = 3(x - 4)$ Multiply both sides by $(x - 4)$

$x^2 - 4x = 3x - 12$ Distributive property

$x^2 - 7x + 12 = 0$ Collect terms on the left side

$(x - 3)(x - 4) = 0$ Factor

$x - 3 = 0$ or $x - 4 = 0$ Set each factor equal to zero

$x = 3$ $x = 4$

Correct root   Extraneous root

Case II:

The logic error involves assuming that, if a statement is true, its converse is true. This may or may not be the case with **if** statements.

What's a converse?

A statement made by interchanging the **if** and **then** clauses of an if-then statement.

Sample **if** statement: If it rains, the sidewalks are wet. Converse of the statement above: If the sidewalks are wet, it has rained.

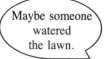
Maybe someone watered the lawn.

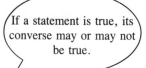
If a statement is true, its converse may or may not be true.

Now let's look at how we can fall into this logic trap and generate an extraneous root by squaring both sides of an equation.

$$x = 3$$ Start with a simple equation

$$x^2 = 9$$ Square both sides

$$x^2 - 9 = 0$$ Subtract 9 from both sides

$$(x - 3)(x + 3) = 0$$ Factor

$$x - 3 = 0 \quad \text{or} \quad x + 3 = 0$$ Set each factor equal to 0

$$x = 3 \qquad\qquad x = -3$$

Correct root

This solution does not check in the original equation. It is an extraneous root

Where is the logic error?

The statement "If $x = -3$ then $x^2 = 9$" is an **if** statement. The converse of this statement "If $x^2 = 9$ then $x = -3$" is not necessarily true; x could also equal $+3$.

The best way to avoid extraneous roots is to test all solutions of an equation.

Example 4 □ Solve $\sqrt{2x - 5} - \sqrt{x - 2} = 2$.

If you square both sides of this, you'll have a mess. A better strategy is to isolate one radical.

$$\sqrt{2x - 5} - \sqrt{x - 2} = 2$$

$$\sqrt{2x - 5} = \sqrt{x - 2} + 2$$

$$(\sqrt{2x - 5})^2 = (\sqrt{x - 2} + 2)^2 \qquad \text{Square both sides}$$

$$2x - 5 = (\sqrt{x - 2})^2 + 4\sqrt{x - 2} + 4$$

$$2x - 5 = x - 2 + 4\sqrt{x - 2} + 4$$

Chapter 10 ■ Quadratic Equations, Variation

Next simplify by collecting like terms.

$$x - 7 = 4\sqrt{x - 2}$$ Isolate the radical

$$(x - 7)^2 = (4\sqrt{x - 2})^2$$ Square again

$$x^2 - 14x + 49 = 16(x - 2)$$

$$x^2 - 14x + 49 = 16x - 32$$

$$x^2 - 30x + 81 = 0$$

$$(x - 27)(x - 3) = 0$$

$$x = 27 \text{ or } x = 3 \text{ are possible roots.}$$

Testing in the original equation

For $x = 27$

$$\sqrt{2x - 5} - \sqrt{x - 2} = 2$$
$$\sqrt{2(27) - 5} - \sqrt{27 - 2} \overset{?}{=} 2$$
$$\sqrt{54 - 5} - \sqrt{25} \overset{?}{=} 2$$
$$7 - 5 = 2$$
OK

For $x = 3$

$$\sqrt{2x - 5} - \sqrt{x - 2} = 2$$
$$\sqrt{2(3) - 5} - \sqrt{3 - 2} \overset{?}{=} 2$$
$$\sqrt{6 - 5} - \sqrt{1} \overset{?}{=} 2$$
$$0 \neq 2$$
Extraneous root   □

Problem Set 10.1

Solve for the variable. Check all solutions.

1. $\sqrt{x + 7} = 3$

2. $\sqrt{x + 4} + 4 = 7$

3. $\sqrt{x + 5} - x + 1 = 0$

4. $\sqrt{7 - x} + 7 = 3x$

5. $\sqrt{2x + 1} = 3$

6. $\sqrt{3x + 4} + 1 = 5$

7. $\sqrt{3x + 1} = x - 1$

8. $\sqrt{4x - 3} = 2x - 3$

9. $\sqrt{x + 6} = 2$

10. $\sqrt{x + 7} + 3 = 5$

11. $\sqrt{x + 8} - x = 6$

12. $\sqrt{x + 5} - 2x = 4$

13. $\sqrt{3x + 10} + 4 = 6$

14. $\sqrt{2x + 15} - 1 = 2$

15. $\sqrt{3x + 6} = x + 2$

16. $\sqrt{4x + 1} = 4x - 5$

17. $\sqrt{3x + 1} = \sqrt{5x - 9}$

18. $\sqrt{4x + 5} = \sqrt{6x - 5}$

19. $\sqrt{3x + 13} = \sqrt{4x + 4}$

20. $\sqrt{2x + 13} = \sqrt{5x + 19}$

21. $\sqrt{x - 5} = \dfrac{6\sqrt{x}}{x}$

22. $\sqrt{x + 12} = \dfrac{8\sqrt{x}}{x}$

23. $\sqrt{2x + 7} = \dfrac{2\sqrt{x}}{x}$

24. $\sqrt{3x + 22} = \dfrac{4\sqrt{x}}{x}$

25. $\sqrt{x - 6} = \dfrac{4\sqrt{x}}{x}$

26. $\sqrt{x - 30} = \dfrac{8\sqrt{x}}{x}$

27. $\sqrt{3x - 22} = \dfrac{4\sqrt{x}}{x}$

28. $\sqrt{2x - 7} = \dfrac{3\sqrt{x}}{x}$

29. $\sqrt{3x + 1} - \sqrt{x + 1} = 2$

30. $\sqrt{3x + 4} - \sqrt{x - 3} = 3$

31. $\sqrt{2x + 4} = 2 + \sqrt{x - 2}$

32. $\sqrt{2x + 1} = \sqrt{x - 3} + 2$

33. $\sqrt{2x + 6} + \sqrt{x + 1} = 2$

34. $\sqrt{5x - 1} - \sqrt{2x - 1} = 3$

35. $\sqrt{5x + 3} = \dfrac{6\sqrt{x}}{x}$

36. $\sqrt{5x - 4} = \dfrac{3\sqrt{x}}{x}$

37. $\sqrt{3x + 10} - \sqrt{x + 3} = 3$

38. $\sqrt{5x + 9} - 2 = \sqrt{3x + 1}$

39. $\sqrt{2x + 2} - \sqrt{x - 3} = 2$

40. $\sqrt{3x + 7} - \sqrt{x + 3} = 2$

41. $\sqrt{2x + 13} = 5 - \sqrt{x + 6}$

42. $\sqrt{3x + 7} = 4 - \sqrt{x + 5}$

43. $\sqrt{5x + 1} - \sqrt{2x + 2} = 2$

44. $\sqrt{7x + 7} - \sqrt{3x - 2} = 3$

45. $\sqrt{2x + 15} + \sqrt{x + 7} = 5$

46. $\sqrt{3x + 22} - \sqrt{x + 3} = 3$

47. $\sqrt{8x + 9} = \sqrt{2x + 7} + 4$

48. $\sqrt{7x - 3} = 2 + \sqrt{4x + 1}$

■ 10.2 **Quadratic Type Equations**

Many equations that are not quadratic equations can be solved using the techniques we use for quadratic equations.

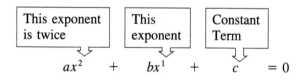

$$ax^2 \;+\; bx^1 \;+\; c \;= 0 \qquad \text{is a quadratic equation.}$$

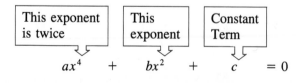

$$ax^4 \;+\; bx^2 \;+\; c \;= 0 \qquad \text{is a quadratic type equation.}$$

Example 5 □ Solve $x^4 - 7x^2 + 12 = 0$.

This is a quadratic type equation since it can be expressed as $(x^2)^2 - 7(x^2) + 12 = 0$. Therefore, we can factor it.

$$x^4 - 7x^2 + 12 = 0$$

$$(x^2 - 4)(x^2 - 3) = 0$$

Therefore $\quad x^2 - 4 = 0 \quad$ or $\qquad\qquad\qquad x^2 - 3 = 0$

$$(x - 2)(x + 2) = 0 \qquad\qquad (x - \sqrt{3})(x + \sqrt{3}) = 0$$

$$x = 2 \quad \text{or} \quad x = -2 \qquad\qquad x = \sqrt{3} \quad \text{or} \quad x = -\sqrt{3}$$

All four solutions will satisfy the original equation. □

Example 6 □ Solve $x^{\frac{2}{3}} + 6x^{\frac{1}{3}} - 16 = 0$.

Notice $\dfrac{2}{3}$ is twice $\dfrac{1}{3}$ so we may treat this as a quadratic type.

$$x^{\frac{2}{3}} + 6x^{\frac{1}{3}} - 16 = 0$$

$$(x^{\frac{1}{3}} + 8)(x^{\frac{1}{3}} - 2) = 0$$

Therefore $\quad x^{\frac{1}{3}} + 8 = 0 \qquad\qquad$ or $\qquad\qquad x^{\frac{1}{3}} - 2 = 0$

$$x^{\frac{1}{3}} = -8 \qquad\qquad\qquad\qquad x^{\frac{1}{3}} = 2$$

$$x = -512 \;\; \text{Cube both} \qquad\qquad x = 8 \;\; \text{Cube both}$$
$$\text{sides} \qquad\qquad\qquad\qquad \text{sides}$$

Test these solutions by substituting them in the original equation. You can also use substitution to *solve* Example 6.

$$\text{Let} \qquad u = x^{\frac{1}{3}} \qquad \text{then} \qquad u^2 = x^{\frac{2}{3}}$$

Making these substitutions

$$x^{\frac{2}{3}} + 6x^{\frac{1}{3}} - 16 = 0$$

becomes

$$u^2 + 6u - 16 = 0$$

$$(u + 8)(u - 2) = 0$$

$$u = -8 \qquad \text{or} \qquad u = 2$$

Resubstituting $u = x^{\frac{1}{3}}$

$$x^{\frac{1}{3}} = -8 \qquad \text{or} \qquad x^{\frac{1}{3}} = 2$$

$$x = -512 \qquad\qquad\qquad x = 8 \qquad \square$$

Example 7 □ Solve $x^{-6} - 64 = 0$.

Since $x^{-6} = x^{-3} \cdot x^{-3}$ we can treat this as a difference of two squares.

$$x^{-6} - 64 = 0$$

$$(x^{-3} - 8)(x^{-3} + 8) = 0$$

Then $\qquad x^{-3} - 8 = 0 \quad$ or $\quad x^{-3} + 8 = 0$

$$x^{-3} = 8 \qquad\qquad x^{-3} = -8$$

$$[x^{-3}]^{-1} = 8^{-1} \qquad [x^{-3}]^{-1} = (-8)^{-1}$$

> Raise both sides to the -1 power.

$$x^3 = \frac{1}{8} \qquad\qquad x^3 = -\frac{1}{8}$$

$$x = \frac{1}{2} \qquad\qquad x = -\frac{1}{2}$$

□

Example 8 □ Solve $x^{-2} - x^{-1} - 2 = 0$.

$$x^{-2} - x^{-1} - 2 = 0$$

$$(x^{-1} - 2)(x^{-1} + 1) = 0$$

$$x^{-1} - 2 = 0 \quad \text{or} \quad x^{-1} + 1 = 0$$

$$x^{-1} = 2 \qquad\qquad x^{-1} = -1$$

$$(x^{-1})^{-1} = 2^{-1} \qquad (x^{-1})^{-1} = (-1)^{-1}$$

> Raise both sides to the -1 power.

$$x = \frac{1}{2} \qquad\qquad x = -1$$

□

Example 8A □ The example above can also be solved by substituting a "dummy variable" for x^{-1}.

Let $\qquad\qquad\qquad u = x^{-1}$

Then $\qquad\qquad x^{-2} - x^{-1} - 2 = 0$

becomes $\qquad\qquad u^2 - u - 2 = 0$

$$(u - 2)(u + 1) = 0 \qquad\qquad \text{Factoring}$$

Therefore: $\qquad\qquad u = 2 \qquad$ or $\qquad u = -1$

Resubstituting for u: $\qquad x^{-1} = 2 \qquad\qquad x^{-1} = -1$

$$\frac{1}{x} = 2 \qquad\qquad \frac{1}{x} = -1$$

$$x = \frac{1}{2} \qquad\qquad x = -1$$

□

Solve for x.

1. $x^4 - 9x^2 + 20 = 0$

2. $x^4 - 9x^2 + 8 = 0$

3. $x^4 - 13x^2 + 36 = 0$

4. $x^4 - 5x^2 + 4 = 0$

5. $x^4 + 4x^2 - 45 = 0$

6. $x^4 - 5x^2 - 36 = 0$

7. $x^4 - x^2 - 12 = 0$

8. $4x^4 + 23x^2 - 6 = 0$

9. $x^{\frac{2}{3}} - 6x^{\frac{1}{3}} + 8 = 0$

10. $x^{\frac{2}{3}} - 2x^{\frac{1}{3}} - 15 = 0$

11. $2x^{\frac{2}{3}} + 7x^{\frac{1}{3}} = 15$

12. $3x^{\frac{2}{3}} - 14x^{\frac{1}{3}} = 24$

13. $x^4 - 12x^2 + 27 = 0$

14. $x^4 - 17x^2 + 16 = 0$

15. $x^4 - 13x^2 + 12 = 0$

16. $x^4 - 20x^2 + 36 = 0$

17. $x^4 - 17x^2 - 60 = 0$

18. $x^4 - 23x^2 - 108 = 0$

19. $9x^4 - 44x^2 - 5 = 0$

20. $4x^4 - 15x^2 - 4 = 0$

21. $x^{-3} - 27 = 0$

22. $x^{-3} + 8 = 0$

23. $x^{-3} + 64 = 0$

24. $x^{-3} - 125 = 0$

25. $(x - 2)^4 - 3(x - 2)^2 = 4$

26. $(x + 3)^4 + 8(x + 3)^2 = 9$

27. $(x + 4)^4 - 15(x + 4)^2 - 16 = 0$

28. $(x - 5)^4 - 5(x - 5)^2 - 36 = 0$

29. $x^{-2} - 3x^{-1} + 2 = 0$

30. $x^{-2} + x^{-1} - 20 = 0$

31. $4x^{-4} - 17x^{-2} + 4 = 0$

32. $9x^{-4} - 13x^{-2} + 4 = 0$

33. $x^{-2} - 4x^{-1} - 12 = 0$

34. $x^{-2} - 2x^{-1} - 15 = 0$

35. $4x^{-4} - 37x^{-2} + 9 = 0$

36. $9x^{-4} - 37x^{-2} + 4 = 0$

37. $(2x^2 - 3x)^2 - 11(2x^2 - 3x) + 18 = 0$

38. $(3x^2 + 4x)^2 - 3(3x^2 + 4x) - 4 = 0$

39. $(3x^2 - x)^2 - 14(3x^2 - x) + 40 = 0$

40. $(2x^2 - 5x)^2 - 10(2x^2 - 5x) - 24 = 0$

41. $x^{\frac{1}{2}} - 6x^{\frac{1}{4}} + 8 = 0$

42. $6x^{\frac{1}{2}} - 11x^{\frac{1}{4}} + 3 = 0$

43. $x^{\frac{1}{2}} - 7x^{\frac{1}{4}} + 12 = 0$

44. $x^{\frac{1}{2}} - 3x^{\frac{1}{4}} + 2 = 0$

45. $10x^{\frac{1}{2}} - 17x^{\frac{1}{4}} + 3 = 0$

46. $12x^{\frac{1}{2}} - 17x^{\frac{1}{4}} + 6 = 0$

47. $6x^{\frac{2}{3}} + x^{\frac{1}{3}} - 12 = 0$

48. $15x^{\frac{2}{3}} + 7x^{\frac{1}{3}} - 4 = 0$

10.2 ■ Quadratic Type Equations

Science often involves observing a situation, noticing a pattern, and writing an equation that describes what is happening.

Direct Variation

Let's think about buying gasoline. There is a direct connection between the number of gallons you buy and the cost of filling your tank. Double the amount of fuel and you double the cost. Triple the amount of fuel and you triple the cost. This is an example of direct variation.

10.3A   Definition: Direct Variation

If x is related to y so that increasing the absolute value of x produces an increase in the absolute value of y according to the formula

$$y = kx$$

y is said to vary directly as x.
k is called the constant of proportionality or the constant of variation.

Some examples of direct variation are:

The circumference of a circle varies directly with the length of its radius.

The cost of gasoline varies directly with the number of gallons purchased.

The amount of sales tax paid is directly proportional to the value of the goods purchased.

Example 9 □ The amount paid for hamburger varies directly as the number of pounds purchased. If you can purchase two and a half pounds of hamburger for $4.50, how much will it cost to buy six pounds of hamburger?
   An equation that expresses the relationship is

$$C = kN \qquad \text{where} \qquad C = \text{Cost}$$
$$N = \text{Number of pounds}$$
$$k = \text{Constant of proportionality}$$

In this example the constant of proportionality will be the price per pound of hamburger.

The set of values we know initially is called the **initial conditions.** In this case the initial conditions are two and a half pounds of hamburger cost $4.50. Substituting the initial conditions

$$C = kN$$

becomes

$$4.50 \text{ dollars} = k\, 2.5 \text{ pounds}$$

Solving for k

$$\frac{4.50 \text{ dollars}}{2.5 \text{ pounds}} = k$$

$$1.80 \frac{\text{dollars}}{\text{pound}} = k$$

Now that we know the constant of proportionality, we can solve for the cost of any amount of hamburger.

The equation that specifies the price of hamburger is

$$C = 1.80 \frac{\text{dollars}}{\text{pound}} \cdot N$$

In this case we want the cost of 6 pounds of hamburger.

$$C = 1.80 \frac{\text{dollars}}{\text{pound}} \cdot 6 \text{ pounds}$$

$$C = 10.80 \text{ dollars} \qquad \qquad \square$$

Example 10 □ Hooke's Law says that the force required to stretch a spring beyond its rest length varies directly with the distance the spring is stretched. Write an equation to express this relationship.

■ Figure 10.1

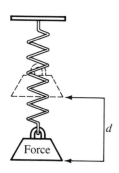

$F = kD$ where F = Force

D = Distance stretched

k = Constant of proportionality or spring constant

□

Example 11 □ Determine the spring constant of a garage door spring that is stretched two feet by a force of 300 pounds. Then find how much force would be required to stretch the same spring four feet.

$$F = kD$$

$$300 \text{ lbs} = k(2 \text{ ft})$$ Substituting values

$$\frac{300 \text{ lbs}}{2 \text{ ft}} = k$$

$$k = 150 \text{ lbs/ft is the spring constant.}$$

Now that we know the spring constant, we can fill in the equation and find the force required to stretch the spring four feet.

$$F = kD$$

$$F = \left(150 \frac{\text{lbs}}{\text{ft}}\right)D$$

$$F = \left(150 \frac{\text{lbs}}{\cancel{\text{ft}}}\right)(4 \cancel{\text{ft}})$$ Substituting known values

$$F = 600 \text{ lbs}$$ □

Why did you divide out the feet?

Things like $\frac{\text{lbs}}{\text{ft}}$ are called dimensions. They tell us what kind of quantity we are dealing with. Read the pointer which follows.

▶ **Pointer for Better Understanding**

In mathematics we deal with pure numbers. People that use mathematics usually are concerned with numbers that have units or a dimension attached.

$$6 \text{ ft}$$
$$28 \frac{\text{miles}}{\text{gal}}$$ } are numbers with dimensions.
$$400 \text{ kilowatt hours}$$

It's easy to see where the term **dimensions** comes from if you consider measurement.

Length is measured in feet.   = ft

Area is measured in square feet.   = ft^2

Volume is measured in cubic feet.   = ft^3

Chapter 10 ■ Quadratic Equations, Variation

It makes no sense to say

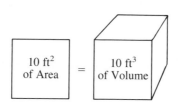 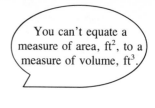

The dimensions on both sides of an equation must agree if the equal sign is true.

Notice how this works in some formulas you are familiar with.

Volume

The volume of something 6 ft long, 3 ft wide, and 2 ft high is written

$$V = L \cdot W \cdot H$$
$$36 \text{ ft}^3 = (6 \text{ ft})(3 \text{ ft})(2 \text{ ft})$$

ft · ft · ft = ft³ just as when you multiply variables.

Interest

What is the amount of interest you pay on a loan of $100 borrowed at 12% for six months?

By 12% interest we mean you pay $12 per $100 borrowed per year.

$$\text{A rate of 12\% interest means } \left(\frac{\frac{12 \text{ dollars}}{100 \text{ dollars}}}{1 \text{ year}} \right) = \frac{12 \text{ dollars}}{100 \text{ dollars}} \cdot \frac{1}{\text{year}}$$

$$= \frac{12}{100} \cdot \frac{1}{\text{year}}$$

The formula with dimensions is

$$I = PRT$$
$$= 100 \text{ dollars} \cdot \left(\frac{12 \text{ dollars}}{100 \text{ dollars}} \cdot \frac{1}{\text{year}} \right) \cdot \frac{1}{2} \text{year}$$
$$= 6 \text{ dollars}$$

Notice that you pay your interest in dollars, not years, ounces, or miles.

In variation problems, the constant of proportionality must have the proper dimensions to make both sides of the equation equal.

Consider Hooke's Law.

$$F = kD$$

In order to make F represent a force measured in pounds, k must be in $\frac{\text{lbs}}{\text{ft}}$

and D must be in ft. The formula won't work if D is in cm and k is in $\frac{\text{tons}}{\text{mile}}$.

Inverse Variation

On a trip of 100 miles you could decrease the time to your destination by increasing your speed. This is an example of inverse variation.

10.3B   Definition: Inverse Variation

If x and y are related in such a way that increasing absolute values of x produce decreasing absolute values for y according to the equation

$$y = \frac{k}{x}$$

y is said to vary inversely as x.

Example 12 □ Notice how inverse and direct variation are in contrast.

Inverse Variation:
   The time to drive a given distance is given by the formula

$$t = \frac{d}{r}$$ ← If distance is held constant, d serves as the constant of variation

Notice that as r increases, t decreases.
For a 100-mile trip

$$r = 20 \text{ mph} \qquad t = \frac{100}{20} = 5 \text{ hrs}$$

$$r = 30 \text{ mph} \qquad t = \frac{100}{30} = 3.3 \text{ hrs}$$

But, notice that increasing r by 50% did not cause a 50% reduction in t.

Direct Variation:
   If you hold speed constant, the time to your destination varies directly as the length of the trip. Increasing the distance by 50% increases the time by 50%.

$$T = kD$$

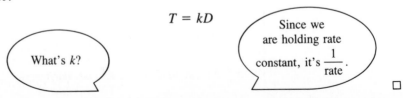

What's k?

Since we are holding rate constant, it's $\frac{1}{\text{rate}}$.

□

Joint Variation

The distance formula is an example of joint variation. The distance you travel varies directly as the speed and it also varies directly as the time spent driving.

10.3C Definition: Joint Variation

If the absolute value of y increases as the absolute value of the product of two or more variables increases according to the formula

$$y = kabc$$

y is said to vary jointly with a, b, c.

Direct, inverse, and joint variation are not limited to first-degree equations. A quantity may vary as the square, cube, or square root of another quantity in any combination.

Equation	English Statement
$y = kx$	y varies directly as x
$y = \dfrac{k}{x}$	y varies inversely as x
$E = IR$	E varies directly as R
$A = k\acute{R}^2$	A varies directly as R^2
$y = \dfrac{k}{\sqrt[3]{x}}$	y varies inversely as the cube root of x
$y = \dfrac{kab^2}{\sqrt{c}}$	y varies jointly as a and the square of b, and y varies inversely with the square root of c

Determining the Constant of Proportionality

Example 13 □ The distance an object falls (if you discount wind resistance) is directly proportional to the square of the time it falls.
   Written as a direct variation we have

$$s = kt^2 \qquad \text{where} \qquad s = \text{distance}$$

$$k = \text{constant of proportionality}$$

$$t = \text{time in seconds}$$

By measuring the distance an object falls in a given time, scientists calculate the constant of variation.

Find the constant of variation if it takes 2 seconds for an object to fall 64 feet.

$$s = kt^2$$

$$64 \text{ ft} = k(2 \text{ sec})^2$$

$$64 \text{ ft} = k \cdot 4 \text{ sec}^2$$

$$16\frac{\text{ft}}{\text{sec}^2} = k$$

Using this constant we can calculate how far an object will fall in three seconds.

$$s = 16\frac{\text{ft}}{\text{sec}^2} \cdot t^2$$

$$s = 16\frac{\text{ft}}{\text{sec}^2} \cdot (3 \text{ sec})^2 \qquad \text{Substituting 3 sec for } t$$

$$s = 16\frac{\text{ft}}{\text{sec}^2} \cdot 9 \text{ sec}^2$$

$$s = 144 \text{ ft} \qquad\qquad\qquad \square$$

Example 14 $\square$ A bowling ball rolls off a building at 1024 feet above the ground. How long will it take to hit the ground? (Ignore wind resistance.)
Using the formula from Example 13

$$s = 16\frac{\text{ft}}{\text{sec}^2} \cdot t^2$$

Substituting the known value for distance, we can solve for time.

$$1024 \text{ ft} = 16\frac{\text{ft}}{\text{sec}^2} \cdot t^2$$

$$\frac{1}{\overset{}{\underset{1}{16}} \frac{\text{sec}^2}{\text{ft}}} \overset{64}{1024} \text{ ft} = 16\frac{\text{ft}}{\text{sec}^2} \cdot t^2 \cdot \frac{1}{16} \frac{\text{sec}^2}{\text{ft}} \qquad \text{Multiply both sides by } \frac{1 \text{ sec}^2}{16 \text{ ft}}$$

$$64 \text{ sec}^2 = t^2$$

$$8 \text{ sec} = t \qquad\qquad\qquad \square$$

Example 15 □ Because the density of air is less at higher altitudes, the airspeed indicators of an airplane indicate a speed below their true airspeed.

$$T = kI \qquad \text{where} \qquad T = \text{true airspeed}$$
$$I = \text{indicated airspeed}$$

If an airplane indicates a speed of 90 knots when it is really flying at 110 knots, what is its true airspeed when it indicates a speed of 135 knots? First find k.

$$T = kI$$
$$110 \text{ knots} = k(90 \text{ knots})$$
$$\frac{11}{9} = k$$

This time k has no dimensions.

Now use k in the equation to find the desired speed.

$$T = \frac{11}{9}I$$

$$T = \frac{11}{\cancel{9}}(\cancel{135}^{\,15} \text{ knots})$$

$$T = 11(15 \text{ knots})$$

$$T = 165 \text{ knots} \qquad\qquad □$$

Problem Set 10.3

Translate the given statement into a mathematical equation, using k as the constant of variation.

1. s varies directly as t.

2. t varies inversely as w.

3. G varies directly as the square of h.

4. V is directly proportional to u.

5. y varies directly as x and inversely as the square of w.

6. H varies jointly as a and b.

7. L varies jointly as m and n, and inversely as q.

8. The volume occupied by a gas varies directly as the temperature (T) and inversely as the pressure (P).

9. The distance a falling object falls varies directly as the square of the time (t) it has fallen.

10. The resistance (R) of an electrical wire varies directly as the length (l) and inversely as the square of the diameter (d).

11. The area of a circle varies directly as the square of the radius.

12. The area of a triangle varies jointly as the length of the base and the length of its altitude.

13. The volume of a right circular cylinder varies jointly as the square of its radius and its height.

14. The circumference of a circle varies as its radius.

15. The volume of a cone varies jointly as the radius of its base squared and its height.

16. A circle of radius 6 inches has an area of 36π square inches. Find the constant of variation; then find the area of a circle with radius 9 feet. (Refer to problem 11.)

17. The area of a triangle is 30 square inches. The base is 12 inches in length and its altitude is 5 inches. Find the constant of variation, then find the area of a triangle with base 7 inches and altitude 4 inches. (Refer to problem 12.)

18. The volume of a right cylinder with radius 4 inches and height 5 inches is 80π cubic inches. Find the constant of variation; then find the volume of a cylinder with radius 6 inches and height 8 inches. (Refer to problem 13.)

19. The circumference of a circle with radius 9 centimeters is 18π centimeters. Find the constant of variation; then find the circumference of a circle with radius 15 inches. (Refer to problem 14.)

20. A cone with radius 5 centimeters and height 6 centimeters has a volume of 50π cubic centimeters. Find the constant of variation; then find the volume of a cone with radius 4 inches and height 9 inches. (Refer to problem 15.)

Solve the following problems by setting up a mathematical equation, using k as the constant of variation.

21. Find k if it takes 4 seconds for an object to fall 78.4 meters. (See problem 9.)

22. The intensity of light varies inversely as the square of the distance from the source. If the intensity of a certain light is 180 candlepower at a distance of 12 feet from the source, find the candlepower of the same type of light at a distance 18 feet from its source.

23. If the intensity of a certain light is 120 candlepower at a distance of 9 feet, find the distance from the source of a light with intensity of 270 candlepower. (See problem 22.)

24. Problem 10 gives information concerning the resistance of an electrical wire. If a certain type of wire 50 feet long with a diameter of $\dfrac{1}{32}$ inch has a resistance of 16 ohms, how much resistance would a 160-foot piece of the same type of wire have?

25. Using the same type of wire as in problem 24, find the diameter of a wire that is 75 feet long and has a resistance of 15 ohms.

26. Refer to problem 10. A 10-foot piece of wire with a diameter of 0.01 inch has a resistance of 20 ohms. What length of the same type of wire with a diameter of 0.02 inches will have a resistance of 10 ohms?

27. The surface area of a sphere varies as the square of its radius. A sphere of radius 5 inches has a surface area of 100π square inches. Find the constant of variation. Use this constant of variation to find the surface area of a sphere with a radius of 6 inches.

28. The volume of a sphere varies as the cube of its radius. A sphere of radius 3 inches has a volume of 36π cubic inches. Find the constant of variation. Use this constant to find the volume of a sphere with a radius of 6 inches.

29. The pressure (P) on a submerged object in any fluid is directly proportional to the depth (h) of the object. A flat plate is submerged in a fluid. The pressure is 280 lb/ft^2 when the plate is 4 feet below the surface. Find the pressure per square foot when the plate is submerged at a depth of 6 feet.

30. The pressure on an object submerged in water is directly proportional to the depth at which it is submerged. 312 lb/ft^2 is exerted on an object submerged to a depth of 5 feet. Find the pressure per square foot on the object if it is submerged to a depth of 20 feet.

31. The length a spring stretches is directly proportional to the force applied. A force of 12 pounds is required to stretch a spring 20 inches. How much force is required to stretch the spring 15 inches?

32. The length of skid marks a car makes when braking is directly proportional to the square of the speed of the car. k represents the coefficient of friction. If a car traveling 25 mph leaves skid marks 30 feet long, how long would the skid marks be if the car is traveling 40 mph?

33. Using the value of k found in problem 32, find the velocity of a car that left skid marks 120 feet long.

34. The velocity of an automobile leaving a highway and vaulting down a canyon is directly proportional to horizontal distances traveled from the edge of the cliff and inversely proportional to the square root of the height from "take off" point to the landing point measured vertically. A car traveling 34 miles per hour left a highway and struck the ground below at a point measured vertically 36 feet and horizontally 75 feet from the edge of the road. A second car left a roadway and landed in a canyon below. The officer determined the horizontal distance to be 189 feet and a vertical drop of 81 feet. Find the velocity of the car.

35. The weight of a man on the earth is directly proportional to his weight on the moon. If a 192-lb man weighs 32 lbs on the moon, how much would a man weighing 168 lbs on the earth weigh on the moon?

Our knowledge of algebra allows us to handle many real-life problems that deal with travel and motion. The skills used in this section are used to navigate aircraft around the world. The basic formula for all motion problems is given below.

10.4   Distance Formula

$$\text{Distance} = \text{Rate} \cdot \text{Time}$$
$$d \quad = \quad r \; \cdot \; t$$

Example 16 □ Find the distance a train covers if it travels at 55 miles per hour for 2 hours.

$$d = r \cdot t$$

$$d = \left(55\frac{\text{miles}}{\text{hr}}\right)(2 \text{ hr})$$

$$= 110 \text{ miles} \qquad\qquad\qquad □$$

▶ **Pointers for Help with Problem Solving**

Using Variables

We can use variables in expressions to explicitly state relationships that are expressed in English phrases.

Many of these ideas are extensions of the basic idea in Section 2.5. If you know the value of one unit, you can determine the value of any number of units.

Here is some practice in using variables to represent the quantities used in motion problems.

Example 17 □ If d represents the distance driven by a woman, twice her distance is $2d$.

$\dfrac{d}{2}$ Half her distance is _____ .

$d - 3$ Three miles less than her distance is _____ .

$6 + 2d$ Six miles more than twice her distance is _____ .

The remaining distance she must drive to complete a 200-mile

$200 - d$ trip is _____ .   □

Example 18 □ If a man drives at a speed of r miles per hour,

$2r$ Twice his speed is _____ .

$r - 20$ A speed 20 miles per hour slower is _____ .

$r + 30$ A speed 30 miles per hour faster is _____ .   □

55 miles

110 miles

55*t* miles

55(*t* + 2) miles

(55 − *x*) mph

2(55 − *x*) miles

Example 19 ☐ A car is traveling on an interstate highway at 55 mph.

In one hour it travels ——————— .

In two hours it travels ——————— .

In *t* hours it travels ——————— .

In *t* + 2 hours it travels ——————— .

A car going *x* mph slower is traveling at a rate of ——————— .

In 2 hours the slower car travels ——————— .   ☐

Example 20 ☐ Travel in an airplane flown at a constant speed took 5 hours at 200 mph.

After flying for two hours how many hours of flying remained? ———————

The distance remaining to be covered after the first two

hours is ——————— .

Another airplane flying *x* mph faster flies at a rate of ——————— .

If the faster aircraft flies for *t* hours it will cover a distance of ——————— .   ☐

3 hrs

$3 \text{ hr}\left(200\,\dfrac{\text{miles}}{\text{hr}}\right)$
$= 600 \text{ miles}$

$(200 + x)\dfrac{\text{miles}}{\text{hr}}$

$(200 + x)t$ miles

Pointers for Problem Solving

Visualization

Frequently a sketch helps to visualize the relationships in a problem.

An anxious father left St. Louis to visit his ill son in Kansas City, 257 miles away. The father was driving west on I-40 at 55 mph. Meanwhile, unknown to him, at the same time an ambulance left Kansas City with his son and headed east toward St. Louis at 70 mph. After three hours, how far apart were the son and father?

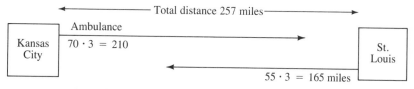

The sketch helps us realize that the car and the ambulance have passed each other. Now to add to this example,

A car with the patient's mother left Kansas City a half hour after the ambulance in the example above. If the mother was driving at 50 mph, how far behind the ambulance was she 2 hours after she left Kansas City?

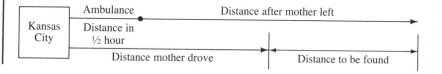

■ Figure 10.3

■ Figure 10.4

Example 21 ☐ An airplane flying 300 miles per hour left Chicago for Los Angeles. Two hours later a jet flying at 550 miles per hour left Chicago traveling the same route. How long will it take the jet to catch up with the slower airplane?

If possible draw a sketch of what is happening.

Shortly after the jet takes off, the picture looks like this.

■ Figure 10.5

When the jet overtakes the small plane, the picture will look like this.

■ Figure 10.6

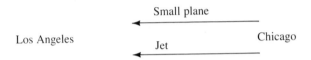

There is one basic formula, $r \cdot t = d$, and two aircraft are involved. We can build a chart to keep track of the elements involved in the problem.

	r	$\cdot$	t	$=$	d
Plane					
Jet					

From sentence one, we know the rate, or speed, of the plane. (Enter this value in the chart.) From sentence two we know the rate, or speed, of the jet. (Enter this value on the chart.)

	r	$\cdot$	t	$=$	d
Plane	300				
Jet	550				

We know that the jet left two hours later; therefore, the jet will be in the air two hours less than the plane, when they meet. Therefore, if the plane is in the air t hours, the jet is in the air $(t - 2)$ hours. (Enter these values in the chart under t.)

	r	$\cdot$	t	$=$	d
Plane	300		t		
Jet	550		$t - 2$		

We have a formula that relates the variables for distance, rate, and time. $(d = r \cdot t)$ Therefore, we can calculate the third variable based on the information in the chart.

Chapter 10 ■ Quadratic Equations, Variation

Why not get the value of the third variable from the problem?

Frequently it isn't there, but we have enough information on the chart to determine it.

Since $d = r \cdot t$, the distance for the plane is $300t$. The distance for the jet is $550(t - 2)$. Now complete the chart.

	r	$\cdot$	t	$=$	d
Plane	300		t		$300t$
Jet	550		$t - 2$		$550(t - 2)$

Next look for some statement of equality in the problem. Frequently this statement can be built around the last chart column that you filled in. When the jet overtakes the smaller airplane, they both will be the same distance from Chicago.

Distance by jet = Distance by small plane

$$550(t - 2) = 300t$$
$$550t - 1100 = 300t$$
$$250t = 1100$$
$$t = \frac{1100}{250}$$
$$t = 4\frac{2}{5} \text{ hours}$$

Wait a minute. Can't a jet fly clear across the country in $4\frac{2}{5}$ hours?

Right! Always test your answer against your experience. Read on to see what's wrong here.

As a final step in any problem, always be sure you answered the question asked by the problem. The problem asked, how long did the jet fly? The value we chose for t was the time the small plane flew. The jet flew $t - 2$ hours or $4\frac{2}{5} - 2 = 2\frac{2}{5}$ hours.

□

Example 22 □ On a river with a current of 6 mph, a boat travels 8 miles upstream in the same time it can travel 18 miles downstream. Find the speed of the boat in still water.

Start with a chart.

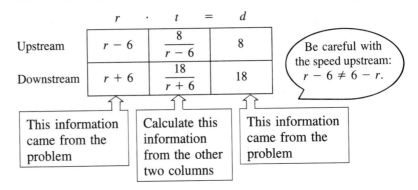

	r	$\cdot$	t	$=$	d
Upstream	$r - 6$		$\dfrac{8}{r - 6}$		8
Downstream	$r + 6$		$\dfrac{18}{r + 6}$		18

Be careful with the speed upstream: $r - 6 \neq 6 - r$.

This information came from the problem

Calculate this information from the other two columns

This information came from the problem

How do I know whether to use $r - 6$ or $6 - r$?

I frequently test the relationship with numbers I make up out of the air. Say the boat goes 30 mph with a current of 8 mph. Its speed downstream is $30 + 8 = 38$ mph. Its speed upstream is $30 - 8 = 22$ mph. So using r instead of 30 mph, we have $r + 8 =$ speed downstream and $r - 8$ speed upstream.

Time Upstream = Time Downstream   Problem tells us this

$$\frac{8}{r - 6} = \frac{18}{r + 6}$$ Read these values for time directly from the chart

$$8(r + 6) = 18(r - 6)$$

$$8r + 48 = 18r - 108$$

$$156 = 10r$$

$$r = 15.6 \text{ mph}$$ The rate of the boat in still water   □

Example 23 □ A car traveled 450 miles at a constant speed. If the driver had driven 5 mph slower, the trip would have taken one hour longer. How fast was the car traveling?

First make a chart.

	r	$\cdot$	t	$=$	d
Given speed	r		t		450
5 mph slower	$r - 5$		$t + 1$		450

In both cases the distance is equal.

$$r \cdot t = (r - 5)(t + 1)$$

Here we have one equation but two unknowns. However, the top line of the chart tells us

$$r \cdot t = 450$$

Therefore

$$t = \frac{450}{r}$$

Substituting in our first equation for t

$$\frac{450}{r} \cdot r = (r - 5)\left(\frac{450}{r} + 1\right)$$

$$450 = (r - 5)\left(\frac{450}{r} + 1\right)$$

$$450r = (r - 5)\left(\frac{450}{r} + 1\right)r \qquad \text{Multiply both sides by } r$$

$$450r = (r - 5)(450 + r)$$

$$450r = r^2 + 445r - 2250$$

$$0 = r^2 - 5r - 2250 \qquad \text{If you have trouble factoring,}$$
$$0 = (r - 50)(r + 45) \qquad \text{use the quadratic formula and}$$
$$\text{your calculator.}$$

$$r - 50 = 0 \text{ or } r + 45 = 0$$

$$r = 50 \qquad r = -45 \Longleftarrow \boxed{\text{Reject this impossible rate}} \qquad \square$$

Problem Set 10.4

1. How far will a family travel in an 8-hour day if they average 52 miles per hour?

2. How long will it take a salesperson to travel 275 miles, traveling at an average speed of 50 miles per hour?

3. An airplane traveled 1140 miles in four and three-fourths hours. What was the speed of the airplane?

4. A salesperson left Dallas, Texas, for Chicago flying his plane at 160 miles per hour. One hour later his competitor flew in the company jet at 400 miles per hour. How long will it take the jet to overtake the slower airplane?

5. The Brown family left Chicago for Miami, traveling by car at 52 miles per hour. Five hours later the Smith family left in their airplane traveling over the same route at 130 miles per hour. How long after the Smith family left did they overtake the Brown family?

6. Jose and Maria each rented an airplane. They left the same airport at the same time going in the same direction. Jose traveled at 140 miles per hour, whereas Maria traveled 178 miles per hour. After three and one half hours, how far apart were the two airplanes?

7. Bob and Louise left Chicago in opposite directions. Louise drove 8 miles per hour faster than Bob. After 3 hours they were 300 miles apart. How fast did each drive?

8. Merv took his family to the beach for an outing. He drove 54 miles per hour to the beach. After spending 3.5 hours at the beach, the family returned home by the same route but at 36 miles per hour (due to heavy traffic). How far did the family live from the beach if they were gone from home a total of 4.5 hours?

9. A canoeist can travel 6 miles upstream on a river in the same time he can travel 15 miles downstream. If the speed of the current is 3 miles per hour, how fast can the canoeist travel in still water?

10. A family rented a truck to move all of their belongings from Los Angeles to the San Francisco area, a distance of 400 miles. If they had contracted with a moving company to move their furniture, they could have driven the family car at 10 miles per hour faster and the trip would have been made in 2 hours less time. How fast did they travel with the truck?

11. A vacationing family traveled by car. On a particular day they were able to travel 240 miles the first part of the day, but for the balance of the day they traveled 10 miles per hour slower through the mountains, traveling another 150 miles. If a total of 7 hours were spent traveling that day, how fast did they travel on each part of the trip?

12. A salesperson flies at the rate of 120 miles per hour on a business trip. He must rent a car to go from the airport to his client. He drives the car at 50 miles per hour. The total trip of 445 miles takes 4 hours. How much time did he spend in the airplane? How far did he travel in the car?

13. Dean and his wife traveled from Los Angeles to Vancouver, British Columbia, by jet in two hours. One hour was spent transferring from the airport to the port where they boarded a ship for Alaska. The jet traveled 20 times as fast as the ship. Seven hours after leaving Los Angeles, they had traveled a total of 1320 miles. How fast did they travel by jet?

14. John left Denver, Colorado, by airplane traveling eastward. One hour later Mel left Denver traveling west by airplane at a rate 20 miles per hour faster than John. Three hours after Mel left Denver, they were 1180 miles apart. How fast did each one fly?

15. On a river a boat can travel 45 miles downstream in the same amount of time it can travel 27 miles upstream. If the boat can travel 12 miles per hour in still water, what is the speed of the stream?

16. Two bicyclists, Maria and Jose, left Denver, Colorado, at the same time traveling in opposite directions. Jose rode his bicycle 4 miles per hour faster than Maria, and after $2\frac{1}{2}$ hours they were 80 miles apart. How fast was each traveling?

17. Two airplanes leave an airport at the same time, flying in the same direction, one going one and one half times as fast as the other. At the end of 2 hours they are 120 miles apart. How fast was each airplane flying?

18. Tyrone drove his car from his home to an adjoining town at the rate of 50 miles per hour. Returning to his home by the same route, he could go only 45 miles per hour. If his total traveling time was one hour and sixteen minutes, how far was his home from the town?

19. The captain of a steamship on the Mississippi River noted that it took one and six tenths times longer to travel 90 miles against the stream as downstream. If the steamer travels 20 miles per hour in still water, what is the rate of the current of the river?

20. A salesman driving from Los Angeles to San Diego took 45 minutes to get out of the Los Angeles area. After leaving the Los Angeles area, he was able to maintain a steady speed 24 miles per hour faster than through the city. If the entire trip of 125 miles took two and three-fourths hours, how fast did he drive through Los Angeles?

Chapter 10 □ Key Ideas

10.1 **1.** To **solve equations containing radicals** (square roots) square both sides of the equation to eliminate the radicals.

 2. **All roots must be tested,** since squaring both sides of an equation may introduce extraneous roots.

10.2 **1.** General form of a quadratic equation is:
$$ax^2 + bx + c = 0$$

 2. A **quadratic type** equation is an equation that can be written in the form:
$$a(\quad)^2 + b(\quad) + c = 0$$
Solve the equation for the quantity in the blank space, then for the variable itself.

10.3 **1.** In the equation $y = kx$, y is said to vary **directly** as x. k is called the constant of proportionality or the constant of variation.

 2. In the equation $y = \dfrac{k}{x}$, y is said to vary **inversely** as x. k is the constant of variation.

 3. In the equation $y = kabc$, y is said to vary **jointly** with a, b, and c.

10.4 The **distance formula** is given by:
$$d = rt \text{ which stands for distance} = \text{rate} \cdot \text{time}$$

Solve for the variable. (**10.1**)

1. $\sqrt{3x + 4} = 4$ **2.** $\sqrt{x + 6} + 2x = 9$ **3.** $\sqrt{2x + 2} = x - 3$

4. $\sqrt{x - 6} = \dfrac{4 \cdot \sqrt{x}}{x}$ **5.** $\sqrt{4x + 5} - \sqrt{2x + 3} = 2$ **6.** $\sqrt{3x + 4} - \sqrt{2x + 6} = 1$

Solve for the variable. (**10.2**)

7. $x^4 - 29x^2 + 100 = 0$ **8.** $2x^4 - 7x^2 - 9 = 0$

9. $2x^{\frac{2}{3}} + x^{\frac{1}{3}} - 6 = 0$ **10.** $(x + 3)^2 - 4(x + 3) = 12$

11. $3x^{-2} - 10x^{-1} - 8 = 0$ **12.** $12x^{-4} - 11x^{-2} + 2 = 0$

13. $x^{\frac{1}{2}} - 7x^{\frac{1}{4}} + 12 = 0$ **14.** $(3x^2 + 5x)^2 - 14(3x^2 + 5x) + 24 = 0$

Determine the constant of variation k. Set up a mathematical relation. Solve for the variable. (**10.3**)

15. The volume of a sphere varies directly as the cube of the radius. The volume of a sphere of radius 3 cm is $36\pi \text{cm}^3$. Find the volume of a sphere with radius 6 cm.

16. The length of the skid mark of an automobile sliding to a stop is directly proportional to the square of the velocity. It was found that a car traveling 60 miles per hour left a skid mark 172 feet in length when attempting to stop. Find the length of the skid marks of a car traveling 50 miles per hour.

17. The strength of a wooden beam with a rectangular cross section varies jointly as the width and square of the depth and inversely as the length. A 1000-lb weight can safely be supported on a beam of a certain type of material that is 4 inches wide, 4 inches deep, and 6 feet long. How much weight could a beam of the same type of material support that is 6 inches wide, 2 inches deep, and 10 feet long?

Solve the following motion problems. Set up an equation and solve. (**10.4**)

18. A boy rode his bicycle into the country at the rate of 15 miles per hour and returned over the same route at 12 miles per hour. If he was bicycling for 4.5 hours, how far into the country did he ride?

19. A family moving from Iowa to the west coast rented a truck to move their furniture. When the truck was loaded, the father left with the truck traveling at a constant rate of 40 miles per hour. Two hours later, after taking care of some last-minute details, his wife followed him in the family car over the same route traveling at 55 miles per hour. How long did it take her to catch up with her husband?

20. A canoeist can travel upstream 15 miles in the same time she can travel downstream 25 miles. If she can canoe at the rate of 8 miles per hour in still water, how fast is the stream flowing?

11 Second-Degree Equations, Circles, and Parabolas

Contents

This chapter will present the graphs of some relations where either x or y or both are raised to the second power. The most common of these relations is the circle. After developing the formula for the distance between two points on a graph in Section one, it will use the distance formula in Section two to derive the equation of a circle. Section three will show you how to move the circle anywhere on the graph and still write its equation. Section four will show the graphs of quadratic equations which are parabolas. Finally this chapter will draw graphs of inequalities involving parabolas.

■ 11.1   Distance Formula

The Pythagorean Theorem from plane geometry relates the sides of a right triangle.

11.1A   Pythagorean Theorem

The **Pythagorean Theorem** tells us that in a right triangle the square of the hypotenuse is equal to the sum of the squares of the other two sides. In symbols we have

$$c^2 = a^2 + b^2$$

where a and b are legs of a right triangle and c is the hypotenuse.

In the following right triangle the side c, opposite the right angle, is called the **hypotenuse of the right triangle.** The hypotenuse is also the longest side of a right triangle.

■ Figure 11.1

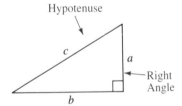

If two sides of a right triangle are known, the third side can be found by using the Pythagorean Theorem.

Example 1 □ Given that $a = 3$ and $b = 4$, find side x in the right triangle below.

Substitute in the formula.

■ Figure 11.2

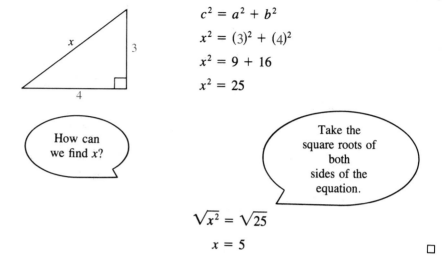

$$c^2 = a^2 + b^2$$
$$x^2 = (3)^2 + (4)^2$$
$$x^2 = 9 + 16$$
$$x^2 = 25$$

How can we find x?

Take the square roots of both sides of the equation.

$$\sqrt{x^2} = \sqrt{25}$$
$$x = 5$$

□

Example 2 □ Find x in the right triangle shown below.

Substitute in the formula.

■ Figure 11.3

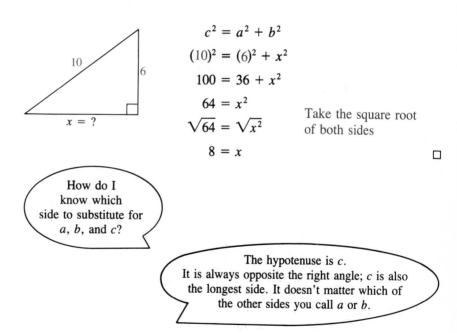

$$c^2 = a^2 + b^2$$
$$(10)^2 = (6)^2 + x^2$$
$$100 = 36 + x^2$$
$$64 = x^2$$
$$\sqrt{64} = \sqrt{x^2}$$

Take the square root of both sides

$$8 = x$$

□

How do I know which side to substitute for a, b, and c?

The hypotenuse is c. It is always opposite the right angle; c is also the longest side. It doesn't matter which of the other sides you call a or b.

The Pythagorean Theorem is immediately useful in finding the distance between two points on a graph.

Example 3 □ Find the distance between $(2, 3)$ and $(5, 7)$.

It's possible to view the distance between any two points on a graph as the hypotenuse of a right triangle. To illustrate, plot the two points on a graph.

■ Figure 11.4

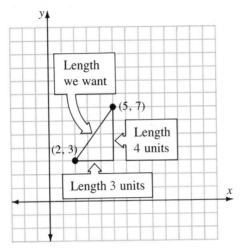

$$c^2 = a^2 + b^2$$

$$d^2 = (4)^2 + (3)^2$$

$$d^2 = 16 + 9$$

$$d^2 = 25$$

$$d = 5$$

We don't have to draw graphs and count squares to determine the lengths of the sides of the triangle. We can use the same procedure we used in determining slope. In Example 3 notice that the 4-unit length which we called side a is equal to the difference of the y-coordinates of the two points.

$$(7 - 3) = 4$$

Similarly, side b is the difference of the x-coordinates.

$$(5 - 2) = 3$$

The distance d is then

$$d = \sqrt{(5 - 2)^2 + (7 - 3)^2}$$
$$= \sqrt{3^2 + 4^2}$$
$$= \sqrt{25}$$
$$= 5$$

■ Figure 11.5

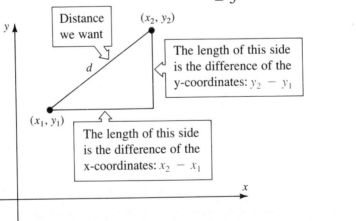

□

Chapter 11 ■ Second-Degree Equations, Circles, and Parabolas

11.1B Definition: Distance Between Two Points

The distance between any two points is an application of the Pythagorean Theorem. It is given by the following formula:

$$d = \sqrt{(x_2 - x_1)^2 + (y_2 - y_1)^2}$$

where (x_1, y_1) and (x_2, y_2) represent the coordinates of any two points on the graph.

Example 4 □ Find the distance between points $(2, 3)$ and $(8, -5)$.
Substitute $(2, 3)$ and $(8, -5)$ in the distance formula.

■ Figure 11.6

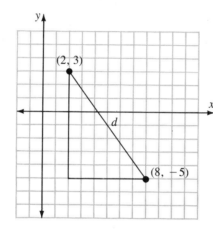

$$d = \sqrt{(x_2 - x_1)^2 + (y_2 - y_1)^2}$$
$$= \sqrt{(2 - 8)^2 + (3 - (-5))^2}$$
$$= \sqrt{(-6)^2 + 8^2}$$
$$= \sqrt{36 + 64}$$
$$= \sqrt{100}$$
$$= 10$$

□

Example 5 □ Find the distance between $(-4, 3)$ and $(-2, -3)$.

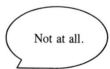

Does it matter which point I call (x_1, y_1)?

Not at all.

Consider $(-4, 3)$ as (x_1, y_1) and $(-2, -3)$ as (x_2, y_2).
Substitute in the formula.

■ Figure 11.7

2

$\sqrt{40}$

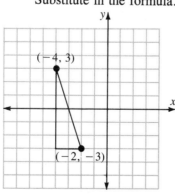

$$d = \sqrt{(x_2 - x_1)^2 + (y_2 - y_1)^2}$$
$$= \sqrt{[-2 - (-4)]^2 + [-3 - 3]^2}$$
$$= \sqrt{(\underline{})^2 + (-6)^2}$$
$$= \sqrt{4 + 36}$$
$$= \underline{}$$
$$= 2\sqrt{10}$$

□

The distance between two points won't always be a rational number.

Example 6 □ Find the perimeter of a triangle whose vertices are $(-5, -1)$, $(3, 5)$ and $(3, -1)$.

If we sketch the triangle, it's easier.

The perimeter is the distance around the outside of the triangle.

The distance from $(-5, -1)$ to $(3, 5)$ is

■ Figure 11.8

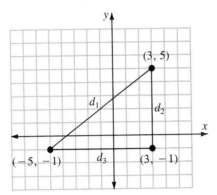

$$d_1 = \sqrt{(-5 - 3)^2 + (-1 - 5)^2}$$
$$= \sqrt{(-8)^2 + (-6)^2}$$
$$= \sqrt{64 + 36}$$
$$= \sqrt{100}$$
$$= 10$$

The distance from $(3, 5)$ to $(3, -1)$ is

$$d_2 = \sqrt{(3 - 3)^2 + (5 - (-1))^2}$$
$$= \sqrt{0^2 + 6^2}$$
$$= 6$$

> Since distance is always a positive number, there is no need to worry about the negative root.

The distance from $(-5, -1)$ to $(3, -1)$ is

$$d_3 = \sqrt{(-5 - 3)^2 + (-1 - (-1))^2}$$
$$= \sqrt{(-8)^2 + 0^2}$$
$$= 8$$

The perimeter is

$$p = d_1 + d_2 + d_3$$
$$= 10 + 6 + 8$$
$$= 24$$

□

Midpoint Formula

The x-coordinate of the midpoint between two points is halfway between the x-coordinates of points.

$$x_{\text{midpoint}} = \frac{x_1 + x_2}{2}$$

> That's an average x-coordinate.

The y-coordinate is halfway between the two y-coordinates.

$$y_{\text{midpoint}} = \frac{y_1 + y_2}{2}$$

□

Example 7 □ Find the midpoint between $(-5, 6)$ and $(3, 4)$.

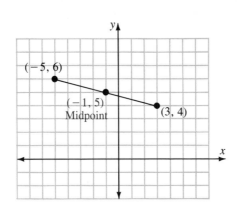

$$x_{mid} = \frac{-5 + 3}{2}$$

$$= \frac{-2}{2}$$

$$= -1$$

$$y_{mid} = \frac{6 + 4}{2}$$

$$= \frac{10}{2}$$

$$= 5$$

Coordinates of the midpoint are $(-1, 5)$.   □

Problem Set 11.1

(a) Find the distance between the following pairs of points. (b) Find the midpoint between the following pairs of points.

1. $(5, 2)$ and $(1, -1)$

2. $(3, -2)$ and $(-5, 4)$

3. $(-2, 4)$ and $(3, -8)$

4. $(4, 0)$ and $(0, 3)$

5. $(3, -4)$ and $(-2, 1)$

6. $(4, 3)$ $(-4, -3)$

7. $\left(3, 4\frac{1}{2}\right)$ and $\left(1\frac{1}{2}, 2\frac{1}{2}\right)$

8. $(-5, -3)$ and $(2, -1)$

9. $(-5, 6)$, $(3, 2)$

10. $(-3, -4)$, $(-9, 4)$

11. $(-1, 3)$, $(-6, 8)$

12. $(-6, -7)$, $(-2, -4)$

13. $(7, -2)$, $(1, -10)$

14. $\left(2\frac{1}{2}, 3\right)$, $\left(4\frac{1}{2}, 5\right)$

15. $\left(3\frac{1}{2}, 5\frac{1}{3}\right)$, $\left(-3\frac{1}{2}, -1\frac{1}{3}\right)$

16. $\left(-2\frac{3}{4}, -3\frac{2}{3}\right)$, $\left(-5\frac{3}{4}, 1\frac{2}{3}\right)$

Find the perimeter of the following triangles having the given vertices.

17. $(-4, 0)$, $(0, -3)$, $(-4, -3)$

18. $(1, 4)$, $(1, -2)$, $(5, -2)$

19. $(-3, 7)$, $(3, -1)$, $(-3, -7)$

20. $(3, -3)$, $(7, -3)$, $(7, 5)$

21. $(-2, 6)$, $(3, -4)$, $(-2, -4)$

22. $(-1, 5)$, $(6, -3)$, $(6, 5)$

23. $(-1, 2)$, $(2, 6)$, $(2, -3)$

24. $(-3, 4)$, $(-2, -2)$, $(6, 4)$

In the accompanying figures of right triangles, find the missing side. Simplify all radicals in your answer.

25.

26.

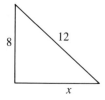

27.

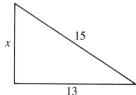

28.

29.

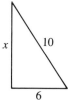

30.

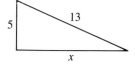

31.

In the following problems, leave your answer in simplest form.

32. A 25-foot ladder just reaches the top of a wall. If the foot of the ladder is 5 feet from the wall, how high is the wall?

33. A guy wire attached to the top of a 60-foot telephone pole is anchored 40 feet from the base of the pole. How long is the guy wire?

34. The top of an office desk is 5 feet by 3 feet. What is the length of a line running diagonally across the desk from one corner to another corner?

35. A plot of ground in the shape of a right triangle is to be fenced. The hypotenuse and one side are measured to be 80 feet and 60 feet, respectively. Determine the length of fencing needed to enclose the plot of ground.

36. An aerialist is to slide down a rope attached to the top of a 75-foot pole. If he has 90 feet of rope available to slide on, how far from the foot of the pole must the rope be attached to the floor?

37. A swimming pool is 40 feet by 15 feet. What is the length of a rope running diagonally across the top of the pool from one corner to the opposite corner?

Chapter 11 ■ Second-Degree Equations, Circles, and Parabolas

38. A 24-foot ladder leans against a wall. If the top of the ladder reaches a point on the wall 20-feet above the ground, how far is the foot of the ladder from the base of the wall?

39. A 4-foot by 5-foot wood gate was to be constructed. To keep the gate from sagging, a piece of lumber was attached running from one corner of the gate at the top to the opposite corner at the bottom. How long a piece of lumber was needed so it would reach exactly from the corner to the opposite corner?

40. On a camping trip, a tent was used for shelter. Each corner of the tent measured $5\frac{1}{2}$ feet above the ground. To prevent the tent from collapsing in the wind, a guy wire (rope) was cut so that 8 feet of rope was available to tie down the tent. How far from each corner of the tent should a stake be placed to keep the tent wall vertical?

■ **11.2**

Equation of a Circle

In geometry a circle is defined as the set of all points in a plane a given distance from a fixed point. The fixed point is called the **center of the circle** and the given distance is referred to as the **radius.**

Example 8 □ Write the equation of a circle centered at the origin with a radius of 4.

Any point (x, y) on the circle will always be four units from the origin. We can use the distance formula to express this.

■ Figure 11.10

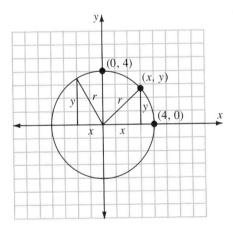

$$d = \sqrt{(x_2 - x_1)^2 + (y_2 - y_1)^2}$$

In this case,

$$4 = \sqrt{(x - 0)^2 + (y - 0)^2}$$

$$4 = \sqrt{x^2 + y^2}$$

Squaring both sides,

$16 = x^2 + y^2$ is the equation of a circle with a center at the origin and a radius of 4. □

Any point on the circle makes this equation true. Similarly any point off the circle makes the equation false. You can test this statement by substituting the coordinates of the points where the circle crosses the x- and the y-axis into the formula.

In general,

11.2A The Equation of a Circle

The equation of a circle centered at the origin with radius r is

$$x^2 + y^2 = r^2$$

Example 9 ☐ Write the equation of a circle centered at the origin with a radius of 7.

Using the equation of a circle,

$$x^2 + y^2 = r^2$$

$$x^2 + y^2 = (7)^2$$

$$x^2 + y^2 = 49 \qquad \text{is the equation of a circle with radius 7}$$
$$\text{centered at the origin.} \qquad ☐$$

Example 10 ☐ Graph $x^2 + y^2 = 25$.

This is a circle centered at the origin with radius of 5.

We use our compass to draw a circle with a radius of 5 centered at the origin.

 Figure 11.11

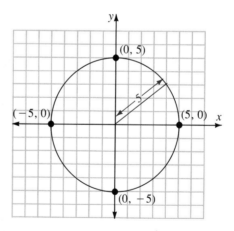

☐

Not all circles are centered at the origin, but the equation of any circle can be written using the distance formula.

Example 11 ☐ Write the equation of a circle with a center at $(3, 2)$ and a radius of 4.

By definition of a circle we are looking for the set of points (x, y) that are a distance of 4 units from the point $(3, 2)$.

Chapter 11 ■ Second-Degree Equations, Circles, and Parabolas

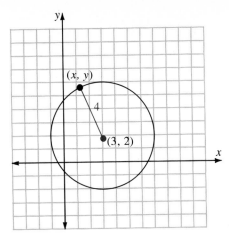

$$\sqrt{(x_2 - x_1)^2 + (y_2 - y_1)^2} = r$$

$$\sqrt{(x - 3)^2 + (y - 2)^2} = 4 \qquad \text{Substituting}$$

$$(x - 3)^2 + (y - 2)^2 = 16 \qquad \text{Squaring both sides}$$

This is the equation of a circle with center at $(3, 2)$ and a radius of 4.

□

We can write the equation of any circle whether or not its center is on the origin.

Example 12 □ Write the equation of a circle with center at (h, k) and radius r.

This is a direct application of the distance formula. We are looking for all points that are r units from the point (h, k).

■ Figure 11.13

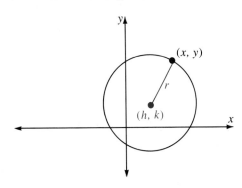

$$\sqrt{(x_2 - x_1)^2 + (y_2 - y_1)^2} = d$$

becomes

$$\sqrt{(x - h)^2 + (y - k)^2} = r$$

$$(x - h)^2 + (y - k)^2 = r^2 \qquad \text{Squaring both sides} \qquad □$$

11.2B Equation of a Circle

The equation of a circle with radius r centered at (h, k) is

$$(x - h)^2 + (y - k)^2 = r^2$$

Example 13 □ Write the equation of a circle with radius 5 and centered at $(-3, 1)$.

In this case, $h = -3$, $k = +1$, and $r = 5$.

The general equation of a circle

$$(x - h)^2 + (y - k)^2 = r^2$$

becomes

$$(x - (-3))^2 + (y - 1)^2 = 5^2$$
$$(x + 3)^2 + (y - 1)^2 = 25$$

If we were going to graph this circle, the form of the equation above would be most convenient because the length of the radius and location of the center can be read directly from the equation.

Frequently this equation will be "simplified" further.

$$x^2 + 6x + 9 + y^2 - 2y + 1 = 25$$
$$x^2 + 6x + y^2 - 2y - 15 = 0 \qquad \square$$

In the next example we complete the square of two trinomials.

Reminder:

How to complete the square.

We frequently use the technique called **completing the square** to rewrite equations.

For example, suppose we want to rewrite the equation $x^2 + 10x = 7$ with the left side as a perfect square of a binomial.

$x^2 + 10x \qquad = 7$ Rewrite with space after the first-degree term

$x^2 + 10x + 25 = 7 + 25$

Add the square of half the coefficient of the first-degree term ($10x$) to both sides

$(x + 5)(x + 5) = 7 + 25$ Factor the left side

$(x + 5)^2 = 32$

This is a perfect square of a binomial

Another reminder of completing the square:

Transform the following equation by completing the square.

$$x^2 + 3x = 6$$

Half of 3 is _____

$\dfrac{3}{2}$

The square of $\dfrac{3}{2}$ is _____

$\dfrac{9}{4}$

Add $\dfrac{9}{4}$ to both sides.

$$x^2 + 3x + \frac{9}{4} = 6 + \frac{9}{4}$$

$$\left(x + \frac{3}{2}\right)^2 = \frac{33}{4}$$

This is now a perfect square

Example 14 □ Graph $x^2 - 4x + y^2 + 8y = 5$.

In this form it's not obvious that this equation represents a circle. Now we will transform the equation into the form $(x - h)^2 + (y - k)^2 = r^2$ by completing the square on both the x- and y-terms.

First, separate the terms with x from the terms with y on the left.

Get all the constants on the right side of the equation

$$x^2 - 4x \qquad + y^2 + 8y \qquad = 5$$

Leave space here

4
16

(half of $-4)^2 =$ _____
(half of $\ 8)^2 =$ _____

Add $4 + 16$ to both sides.

$$x^2 - 4x + 4 + y^2 + 8y + 16$$
$$= 5 + 4 + 16$$

Write each trinomial as a perfect square.

$$(x - 2)^2 + (y + 4)^2 = 25$$

Now we can see this is the equation of a circle with

$$h = 2 \qquad k = -4 \qquad r = 5$$

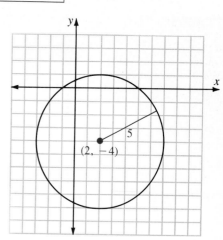

□

Example 15 □ Write the equation $x^2 + 6x + y^2 - 10y + 27 = 0$ in a form where its center and radius are apparent.

We will use completing the square to transform this equation.

$$x^2 + 6x \qquad + y^2 - 10y \qquad = -27 \qquad \text{Separate variables}$$

$$x^2 + 6x + 9 + y^2 - 10y + 25 = -27 + 9 + 25 \qquad \begin{array}{l}\text{Add constants to}\\ \text{complete the}\\ \text{squares}\end{array}$$

$$(x + 3)^2 + (y - 5)^2 = 7 \qquad \qquad \begin{array}{l}\text{Write in factored}\\ \text{form}\end{array}$$

The center of this circle is $(-3, 5)$; its radius is $\sqrt{7}$. □

Inequalities

A circle is defined as the set of points in a plane that are a given distance from a fixed point.

Although a center is needed to construct a circle, it is not part of the circle; neither is the radius or diameter part of the circle. Inequalities can refer to the points inside, on, or outside a circle.

Example 16 □ Sketch the inequality $x^2 + y^2 < 16$.

We want the points where the square of the first coordinate plus the square of the second coordinate is less than 16.

The circle $x^2 + y^2 = 16$ is the set of points where the square of the first coordinate plus the square of the second coordinate is 16. That circle will serve as a "fence" for this graph. First plot it with a dashed line. Then shade the points inside the circle.

■ Figure 11.15

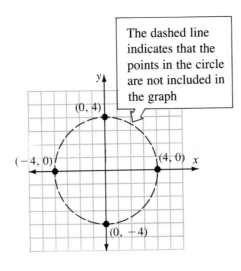

The dashed line indicates that the points in the circle are not included in the graph

$x^2 + y^2 = 16$

How do I know whether to shade the inside or outside of the circle?

Pick a test point [$(0, 0)$ is easy] and see if it satisfies the inequality. Since $0^2 + 0^2 < 16$ is true, we shade the points inside the circle.

Chapter 11 ■ Second-Degree Equations, Circles, and Parabolas

Notice that other points inside the circle such as $(1, 3)$, $(-2, -2)$ also satisfy the inequality. Points on the circle like $(0, 4)$ or points outside the circle like $(4, 5)$ do not satisfy this inequality.

We finish the graph by shading inside the circle.

■ Figure 11.16

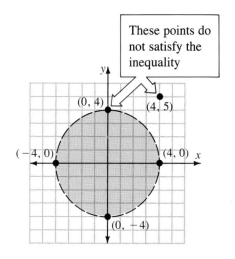

We finish the graph by shading inside the circle.

Problem Set 11.2

Find an equation of the circle having the center and radius as given.

1. Center $(0, 0)$, radius 8

2. Center $(0, 0)$, radius 2

3. Center $(-2, 3)$, radius 6

4. Center $(8, -7)$, radius 5

5. Center $(1, 0)$, radius 4

6. Center $(2, 4)$, radius 9

7. Center $(-5, -6)$, radius 3

8. Center $(0, -2)$, radius 10

9. Center $(3, -2)$, radius 5

10. Center $(-6, -5)$, radius 4

11. Center $(-4, -9)$, radius 2

12. Center $(-7, 8)$, radius 1

13. Center $(-8, 4)$, radius 4

14. Center $(9, 4)$, radius 6

15. Center $(0, 5)$, radius 3

16. Center $(-5, 0)$, radius 5

Find the center and radius of each circle. Sketch the graph.

17. $(x - 3)^2 + (y - 4)^2 = 9$

18. $(x + 4)^2 + y^2 = 4$

19. $(x + 2)^2 + (y - 3)^2 = 16$

20. $(x - 4)^2 + (y + 5)^2 = 1$

21. $x^2 + (y + 2)^2 = 25$

22. $x^2 + y^2 = 36$

23. $(x + 2)^2 + (y - 4)^2 = 4$

24. $(x - 3)^2 + (y + 3)^2 = 9$

25. $(x - 2)^2 + y^2 = 16$

26. $x^2 + (y - 3)^2 = 25$

27. $(x - 4)^2 + (y + 2)^2 = 25$

28. $(x + 1)^2 + (y - 3)^2 = 16$

Find the center and radius of each circle. Sketch the graph.

29. $x^2 + y^2 - 4x - 6y + 9 = 0$

30. $x^2 + y^2 + 8x - 4y + 4 = 0$

31. $x^2 + y^2 - 4x - 12 = 0$

32. $x^2 + y^2 + 10y + 9 = 0$

33. $x^2 + y^2 + 2x + 8y - 8 = 0$

34. $x^2 + y^2 - 8x + 4y + 19 = 0$

35. $x^2 + y^2 - 2x - 2y - 6 = 0$

36. $x^2 + y^2 - 6x + 8y + 21 = 0$

37. $x^2 + y^2 + 2x - 2y - 7 = 0$

38. $x^2 + y^2 + 4x + 6y + 1 = 0$

39. $x^2 + y^2 + 4x - 6y + 4 = 0$

40. $x^2 + y^2 - 8x - 2y + 1 = 0$

41. $x^2 + y^2 + 6x + 6y + 2 = 0$

42. $x^2 + y^2 + 8x - 10y + 37 = 0$

43. $x^2 + y^2 - 10x = 0$

44. $x^2 + y^2 + 8x = 0$

45. $x^2 + y^2 - 8x - 12y + 51 = 0$

46. $x^2 + y^2 + 2x + 6y - 15 = 0$

47. $x^2 + y^2 - 2x - 6y - 10 = 0$

48. $x^2 + y^2 + 4x - 2y - 7 = 0$

Sketch the following inequalities.

49. $x^2 + y^2 \geq 25$

50. $x^2 + y^2 < 9$

51. $(x - 2)^2 + y^2 < 36$

52. $x^2 + (y + 3)^2 \geq 4$

53. $x^2 + y^2 + 6x - 4y + 4 > 0$

54. $x^2 + y^2 + 2x + 8y + 1 \leq 0$

55. $x^2 + y^2 \leq 36$

56. $x^2 + y^2 \geq 16$

57. $x^2 + (y - 4)^2 > 16$

58. $(x + 2)^2 + y^2 < 25$

59. $x^2 + y^2 + 4x - 8y + 4 < 0$

60. $x^2 + y^2 - 8x + 2y + 8 > 0$

■ 11.3   Translations

Notice in the previous section that the equation of a circle with its center at the origin is much simpler than the equation of a circle with center at (h, k). To simplify problems, engineers draw secondary sets of axes chosen to make the calculation as simple as possible. Consider a circle centered at $(2, 4)$ with a radius of 5.

To make the equation as simple as possible, draw a new set of axes called the x', y' axes (pronounced "x prime," "y prime") through the center of the circle. In the (x', y') system, the equation of the circle is

$$x'^2 + y'^2 = 5^2$$

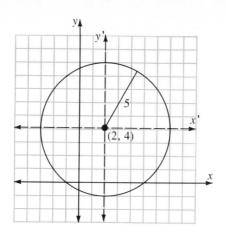

To write the equation in the x, y system consider the relationship between the coordinates of a point in the (x, y) system and the same point in the (x', y') system.

Start with the origin of the circle (h, k) in the (x, y) system. It is $(2, 4)$. In the (x', y') system it is $(0, 0)$. The figure below shows two sets of axes. Call the axes drawn with colored lines the (x', y') system. The origin of the (x', y') is at $(2, 4)$ in the original (x, y) system. In the (x', y') system that same point is $(0, 0)$.

Now examine the point $(3, 6)$ in the (x', y') system. As viewed from the (x, y) system, that point is much farther from the origin. It is $(5, 10)$.

■ Figure 11.18

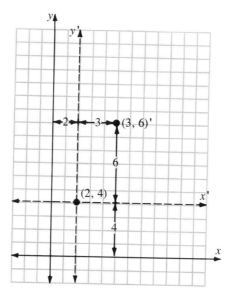

To get the y-coordinate of 10 in the old system, add the value of y' in the new system to the displacement of the new origin in the y-direction.

$$6 + 4 = 10$$

Next we will practice writing the coordinates of a point in the (x', y') system as a point in the (x, y) system.

Example 17 □ The system sketched has a new (x', y') set of axes with its origin of $(2, 4)$ in the original (x, y) system. The points labeled are in the new (x', y') system. Find the coordinates of the labeled points in the (x, y) system.

■ Figure 11.19

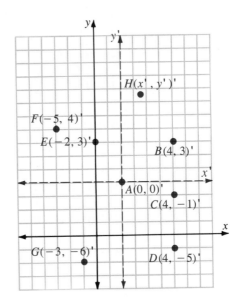

(6,7)

(6,3), (6, −1)

(0, 7), (−3, 8)

(−1, −2)

$(x' + 2, y' + 4)$

a. $(0, 0)' = (2, 4)$

c. $(4, −1)' = $ _____

e. $(−2, 3)' = $ _____

g. $(−3, −6)' = $ _____

b. $(4, 3)' = $ _____

d. $(4, −5)' = $ _____

f. $(−5, 4)' = $ _____

h. $(x', y')' = $ _____

□

The origin of the x', y' system above is at $(2, 4)$ in (x, y). In general the coordinates of any point in the (x, y) system are

$$x = x' + 2$$
$$y = y' + 4$$

And in the (x', y') system they are

$$x' = x - 2$$
$$y' = y - 4$$

Solve the equations for x' and y' to get these equations.

The two equations above are called **translation equations.** We can use translation equations to solve the problem at the beginning of this section.

Chapter 11 ■ Second-Degree Equations, Circles, and Parabolas

Example 18 □ What is the equation of a circle centered at $(2, 4)$ with a radius of 5?

First, we pick a new coordinate system with its origin at the center of the circle. The equation of the circle in the (x', y') system is

■ Figure 11.20

$$x'^2 + y'^2 = 25$$

The translation equations for this system are

$$x' = x - 2$$
$$y' = y - 4$$

Substituting the translation equations in the equation of the circle in the (x', y') system yields

$$(x - 2)^2 + (y - 4)^2 = 5^2$$

This is the equation of the circle in the (x, y) system.

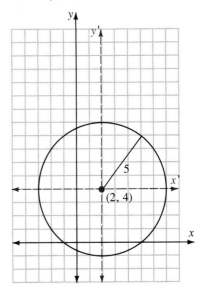

□

Next consider a system where the new origin is in the second quadrant.

Example 19 □ The system sketched has a new (x', y') set of axes with its origin at $(-2, 2)$ in the original (x, y) system. The points labeled are in the new (x', y') system. Find the coordinates of the labeled points in the (x, y) system

■ Figure 11.21

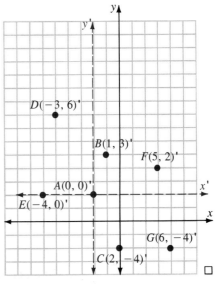

a. $(0, 0)' = (-2, 2)$

$(-1, 5)$ b. $(1, 3)' = $ _____

$(0, -2)$ c. $(2, -4)' = $ _____

$(-5, 8)$ d. $(-3, 6)' = $ _____

$(-6, 2)$ e. $(-4, 0)' = $ _____

$(3, 4)$ f. $(5, 2)' = $ _____

$(4, -2)$ g. $(6, -4)' = $ _____

_____

11.3 Translation Equations

The relationship between any point (x, y) in an (x, y) coordinate system, and that same point (x', y') in a translated system with origin at (h, k) is given by

$$x = x' + h \qquad \text{and} \qquad y = y' + k$$

$$x' = x - h \qquad\qquad\qquad y' = y - k$$

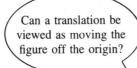

Can a translation be viewed as moving the figure off the origin?

Yes, but the movement is in the x- and y-directions only. Rotations are beyond this course.

Example 20 □ Write the equation of a circle with center at $(-5, 4)$ and radius 3.

First write the equation of a circle with a radius of 3 centered at the origin of a (x', y') system.

■ Figure 11.22

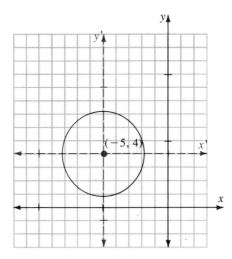

$$x'^2 + y'^2 = 3^2$$

The translation equations for the (x', y') system are

$$x' = x - h \qquad\qquad y' = y - 4$$
$$x' = x - (-5) \qquad y' = y - k$$

$$= x + 5$$

Substituting these into the equation in the (x', y') system

$$x'^2 + y'^2 = 3^2$$

becomes $\qquad (x + 5)^2 + (y - 4)^2 = 9.$

This is the equation of the circle in the (x, y) system. □

Example 21 ☐ Use a translation to graph $x^2 + 6x + y^2 - 10y + 27 = 0$.

We can write this equation in a form where the translation is apparent by completing the square of both variables.

$$x^2 + 6x \qquad + y^2 - 10y \qquad = -27 \qquad \text{Rewrite with space}$$

$$x^2 + 6x + 9 + y^2 - 10y + 25 = -27 + 9 + 25 \qquad \text{Add } 9 + 25 \text{ to both sides}$$

$$(x + 3)^2 + (y - 5)^2 = 7 \qquad \text{Rewrite each trinomial as the square of a binomial}$$

It is now apparent that if a pair of axes were chosen with an origin at $(-3, 5)$, the equation of the circle would be

■ Figure 11.23

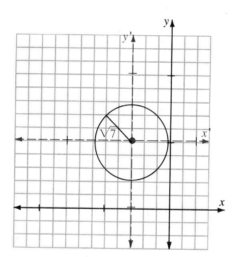

$$x'^2 + y'^2 = (\sqrt{7})^2$$

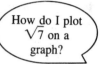
How do I plot $\sqrt{7}$ on a graph?

$\sqrt{7} \approx 2.646$. If you set your compass at a little over 2.5 units and draw a circle with center at $(-3, 5)$, it will be accurate enough for our purposes. ☐

Problem Set 11.3

In the following problems, first write the equation of a circle with the indicated radius in a (x', y') coordinate system with its center at the given coordinates. Then express the same equation in the original (x, y) coordinate system.

1. Radius 5, center $(3, 6)$

2. Radius $\sqrt{3}$, center $(-3, 4)$

3. Radius $\sqrt{5}$, center $(-6, -2)$

4. Radius 2, center $(4, -1)$

5. Radius 3, center $(-7, 4)$

6. Radius 6, center $(-6, -5)$

7. Radius 5, center $(8, 1)$

8. Radius 4, center $(3, -6)$

9. Radius $\sqrt{6}$, center $(2, 0)$

10. Radius $\sqrt{8}$, center $(0, -3)$

Use a translation to graph the following equations.

11. $x^2 + y^2 + 2x + 6y + 6 = 0$

12. $x^2 + y^2 + 2x - 8y + 8 = 0$

13. $x^2 + y^2 - 6x - 4y + 7 = 0$

14. $x^2 + y^2 - 12x + 6y + 40 = 0$

15. $x^2 + y^2 - 8x - 4y + 11 = 0$

16. $x^2 + y^2 + 6x + 4y - 12 = 0$

17. $x^2 + y^2 + 10x - 4y + 25 = 0$

18. $x^2 + y^2 - 6x + 10y + 33 = 0$

19. $x^2 + y^2 + 12x - 8y + 51 = 0$

20. $x^2 + y^2 - 2x + 2y - 34 = 0$

■ 11.4   **Parabolas**
_____

Circles are defined as the set of all points at a given distance from a fixed point. There is a similar definition for a parabola.

11.4A   Definition of a Parabola

A parabola is the set of all points in a plane that are equidistant from a fixed point called the **focus** and a fixed line called the **directrix.**

■ Figure 11.24

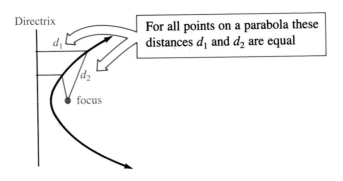

We will not derive the equation of a parabola; instead, we will start our study of parabolas by plotting the simplest case $y = x^2$.

Example 22 □ Graph $y = x^2$.

We will use a table of values.

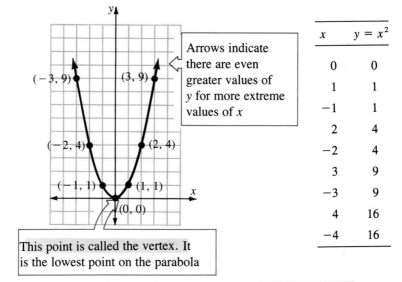

Arrows indicate there are even greater values of y for more extreme values of x

x	$y = x^2$
0	0
1	1
−1	1
2	4
−2	4
3	9
−3	9
4	16
−4	16

This point is called the vertex. It is the lowest point on the parabola

Axis of Symmetry

On the graph of $y = x^2$ the y-axis is the **axis of symmetry**. If you folded the graph in half along the y-axis, both sides would match.

Notice on the table of values that, since $y = x^2$, if you replace x with either $+a$ or $-a$ you will get the same value for y. □

Example 23 □ Graph $y = -x^2$.

Again we make a table of values.

This time the vertex is the highest point

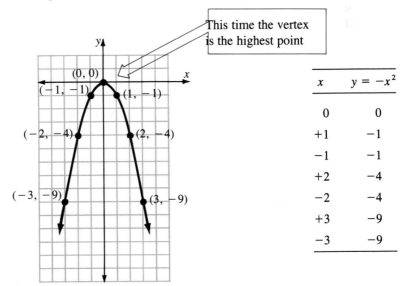

x	$y = -x^2$
0	0
+1	−1
−1	−1
+2	−4
−2	−4
+3	−9
−3	−9

This graph has exactly the same shape as $y = x^2$. The difference is that it opens down instead of upward. □

Example 24 □ On the same axis, plot $y = x^2$, $y = 2x^2$, $y = \frac{1}{2}x^2$.

Make a table of values.

■ Figure 11.27

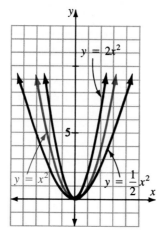

x	$y = x^2$	$y = 2x^2$	$y = \frac{1}{2}x^2$
0	0	0	0
1	1	2	$\frac{1}{2}$
−1	1	2	$\frac{1}{2}$
2	4	8	2
−2	4	8	2
+3	9	18	4.5
−3	9	18	4.5
4	16	32	8
−4	16	32	8

Things to notice about these graphs

a. These are graphs of $y = ax^2$ for $a = 1$, 2 and $\frac{1}{2}$.

b. Since a is positive, all three graphs open upward.

c. The larger the value of a, the narrower the graph.

d. All graphs have the same vertex $(0, 0)$.

e. All graphs have the same axis of symmetry.

We can consider $y = ax^2$, a sort of generic parabola. The effect of a is as follows.

Condition	Comment	Example
$|a| > 1$	graph will be narrower than $y = x^2$	$y = 3x^2$
$|a| < 1$	graph will be wider than $y = x^2$	$y = \frac{1}{2}x^2$
a is positive	graph will open up	$y = 2x^2$
a is negative	graph will open down	$y = -2x^2$

□

Example 25 □ Graph $y = x^2$, $y = x^2 + 3$ and $y = x^2 - 4$ on the same axis.

Using a table of values:

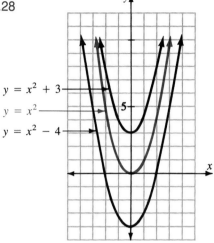

x	$y = x^2$	$y = x^2 + 3$	$y = x^2 - 4$
0	0	3	−4
1	1	4	−3
−1	1	4	−3
2	4	7	0
−2	4	7	0
3	9	12	5
−3	9	12	5

> These are all the same shape as $y = x^2$.

> The constant just moves the graph up or down.

In general the graph of $y = ax^2 + k$ looks just like the graph of $y = ax^2$ except it is moved up or down k units. □

> Are all parabolas symmetrical to the y-axis.

> No. Next we'll examine how to move the graph right or left.

One way to think of a parabola with its vertex away from the origin is as a translation.

$y = ax^2$ represents a parabola with its vertex at the origin.

> What if the vertex isn't on the y-axis?

> No problem. I'll just write a new coordinate system with its origin at the vertex. The equation of the parabola will be $y' = ax'^2$.

We want the equation of the parabola whose vertex is not at the origin in the old system. To get this, we use the translation equations

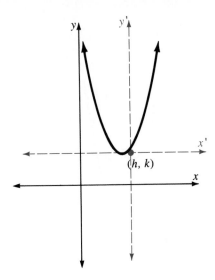

$$x' = x - h$$
$$y' = y - k$$

The equation in the (x', y') system is

$$y' = ax'^2$$

Substituting the translation equations in $y' = ax'^2$ we get

$$y - k = a(x - h)^2$$

The equation above is the general equation of a parabola with its vertex at (h, k).

Example 26 □ Sketch **a)** $y = x^2$ and
b) $y = (x - 3)^2 - 4$ on the same axis.

Rewrite Equation **b** in the same form as the general equation of a parabola.

b) $\qquad y = (x - 3)^2 - 4$

becomes $\qquad y + 4 = (x - 3)^2$

Equation **b** is now recognizable as Equation **a** in a system with its vertex at $(3, -4)$. Therefore, we need make a table of values only for Equation **a**.

■ Figure 11.30

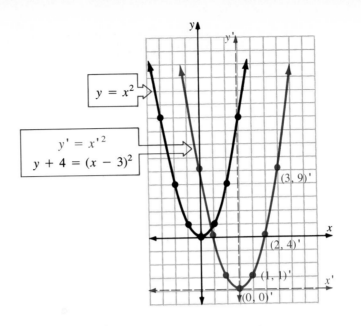

x	y
0	0
1	1
−1	1
2	4
−2	4
3	9
−3	9

□

Example 27 □ Graph $y - 2 = (x + 5)^2$.

Notice that this equation is the same form as

$$y - k = (x - h)^2 \quad \text{with} \quad k = +2 \quad \text{and} \quad h = -5.$$

Therefore, this is the parabola

$$y' = x'^2 \quad \text{with vertex at } (-5, 2).$$

Make a table of values for $y' = x'^2$ and draw the graph in the (x', y') system.

■ Figure 11.31

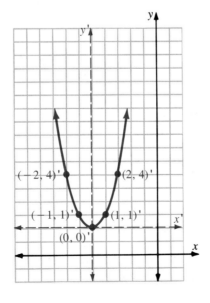

x'	y'
0	0
1	1
−1	1
2	4
−2	4
3	9
−3	9

Suppose I make a table of values in the (x, y) system. What will I get?

The same graph. Try plotting a few points like $(-4, 3)$ and $(-6, 7)$ on the (x, y) system graph to the left.

□

11.4B Equation of a Parabola Opening Parallel to the Y-Axis

The general equation of a parabola opening parallel to the y-axis is

$$y - k = a(x - h)^2$$

where

(h, k) is the vertex

$x = h$ is the axis of symmetry

a controls the rate of curvature

positive a—curve opens up

negative a—curve opens down

Example 28 □ Graph $y = -2(x + 3)^2 + 4$.

First rewrite the equation so we can see the translation.

$$y = -2(x + 3)^2 + 4$$
$$y - 4 = -2(x + 3)^2$$

Now we can see $a = -2$, $h = -3$, $k = 4$

This parabola opens down because a is less than zero.

Vertex is at $(-3, +4)$ in the original system, this is origin of the (x', y') system.

Axis of symmetry is $x = -3$ in the original system or $x' = 0$ in the translated system.

Equation in (x', y') system is $y' = -2x'^2$

Make a table of values for the equation in the (x', y') system.

■ Figure 11.32

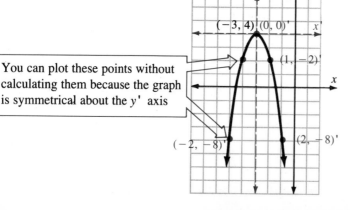

You can plot these points without calculating them because the graph is symmetrical about the y' axis

x'	$y' = -2x'^2$
0	0
1	−2
2	−8
3	−18
−2	−8

□

Chapter 11 ■ Second-Degree Equations, Circles, and Parabolas

Example 29 □ Graph $y = -\dfrac{1}{2}(x - 2)^2 - 5$.

Writing this equation so the translation is apparent gives

$$y + 5 = -\frac{1}{2}(x - 2)^2$$

We know that if we pick a coordinate system with its origin at $(2, -5)$, the vertex of this parabola will be at the origin of the new system. We also know the axis of symmetry goes through the vertex.

Since we know a lot about the shape of the graph from the equation, we will graph the equation without constructing an (x', y') system.

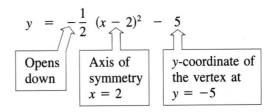

| Opens down | Axis of symmetry $x = 2$ | y-coordinate of the vertex at $y = -5$ |

Making an abbreviated table of values

■ Figure 11.33

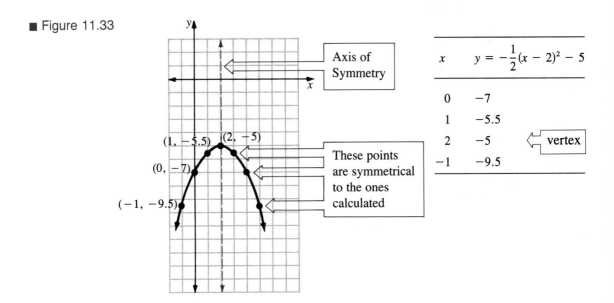

x	$y = -\dfrac{1}{2}(x - 2)^2 - 5$
0	−7
1	−5.5
2	−5 ⟵ vertex
−1	−9.5

Now work from the vertex and axis of symmetry. □

Example 30 □ Graph $y = 2x^2 + 8x + 9$.

First rewrite this equation into the form $y - k = a(x - h)^2$.

$$y = 2x^2 + 8x + 9$$

$$y - 9 = 2x^2 + 8x \qquad \text{Subtract 9}$$

$$y - 9 = 2(x^2 + 4x \qquad) \qquad \text{The coefficient of } x^2 \text{ must be 1}$$
to complete the square

$$y - 9 + 8 = 2(x^2 + 4x + 4) \qquad \text{Complete the square by}$$
adding 4 inside the parentheses, which is 8 outside the parentheses, to both sides

$$y - 1 = 2(x + 2)^2 \qquad \text{Write the trinomial as a square}$$

$$y = 2(x + 2)^2 + 1$$

Now in the original system identify the critical elements.

$(-2, 1)$ Vertex is _____

$x = -2$ Axis of symmetry _____

up   Opens _____

Calculating a few points

■ Figure 11.34

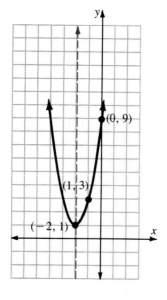

x	y
-2	1
-1	3
0	9

□

Parabolas may also open parallel to the x-axis.

Example 31 □ Graph $x = y^2$.

It is easiest to make a table of values for this equation if we pick values for y and calculate corresponding values for x.

■ Figure 11.35

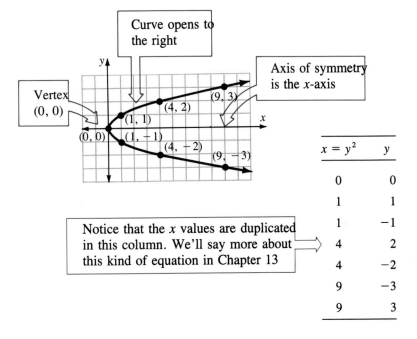

$x = y^2$	y
0	0
1	1
1	−1
4	2
4	−2
9	−3
9	3

□

Notice that in this equation y instead of x is squared. That's what makes the curve open parallel to the x-axis.

Example 32 □ Graph $x = 4 - y^2$.

Make a table of values by picking values for y and calculating x.

$$x = 4 - y^2$$

■ Figure 11.36

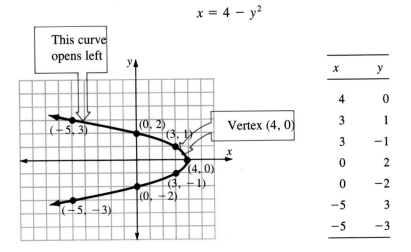

x	y
4	0
3	1
3	−1
0	2
0	−2
−5	3
−5	−3

□

> **Pointer about the Equations of a Parabola**

In all parabolas the value of the constant a determines the rate of curvature. $y - k = a(x - h)^2$ **is the equation of a parabola that opens parallel to the y-axis.**

Condition	Comment	Example
positive a	parabola opens up	$y - 2 = 3(x + 4)^2$
negative a	parabola opens down	$y - 2 = -3(x + 4)^2$
(h, k)	coordinates of vertex	$y - 2 = +3(x + 4)^2$ vertex at $(-4, 2)$
$x = h$	axis of symmetry	$y - 2 = 3(x + 4)^2$ axis of symmetry is $x = -4$

$x - h = a(y - k)^2$ **is the equation of a parabola that opens parallel to the x-axis.**

Condition	Comment	Example
positive a	parabola opens right	$x - 3 = 5(y + 4)^2$
negative a	parabola opens left	$x - 3 = -5(y + 4)^2$
(h, k)	coordinates of the vertex	$x - 3 = 5(y + 4)^2$ vertex at $(3, -4)$
$y = k$	axis of symmetry	$x - 3 = 5(y + 4)^2$ axis of symmetry is $y = -4$

Problem Set 11.4

Sketch the following graphs.

1. $y = 3x^2$

2. $y = -3x^2$

3. $y = \frac{1}{2}x^2$

4. $y = -\frac{1}{4}x^2$

5. $y = -4x^2$

6. $y = 4x^2$

7. $y = \frac{1}{3}x^2$

8. $y = -\frac{1}{2}x^2$

9. $y = -2x^2$

10. $y = -\frac{1}{3}x^2$

11. $y = \frac{1}{4}x^2$

12. $y = 5x^2$

Chapter 11 ■ Second-Degree Equations, Circles, and Parabolas

Sketch the graph, label the vertex, and draw the line of symmetry.

13. $y = x^2 + 4$

14. $y = x^2 - 4$

15. $y = -x^2 + 3$

16. $y = -2x^2 + 3$

17. $x = -2y^2 + 3$

18. $y = \frac{1}{2}x^2 - 4$

19. $y = -\frac{1}{2}x^2 + 4$

20. $x = 2(y - 3)^2$

21. $y = -2(x + 3)^2$

22. $y = -4(x + 6)^2$

23. $y = -x^2 - 2$

24. $y = x^2 - 5$

25. $y^2 = -2x^2 - 1$

26. $y = -\frac{1}{2}x^2 + 3$

27. $x = -\frac{1}{2}y^2 + 4$

28. $x = -2y^2 + 3$

29. $y = \frac{1}{2}(x - 4)^2$

30. $y = -2(x - 2)^2$

31. $x = 2(y + 1)^2$

32. $x = 3(y - 1)^2$

33. $y = 3(x - 1)^2$

34. $x = (y - 2)^2 - 3$

35. $y = -(x + 3)^2 + 2$

36. $y = x^2 - 4x$

37. $y - 3 = -(x + 3)^2$

38. $y - 5 = -2(x + 1)^2$

39. $y + 1 = 3(x - 2)^2$

40. $y + 2 = \frac{1}{3}(x - 2)^2$

41. $x - 5 = -2(y + 1)^2$

42. $x + 2 = 3(y - 1)^2$

43. $x - 3 = \frac{1}{2}(y + 2)^2$

44. $y = -(x - 2)^2 + 4$

45. $y = -(x + 3)^2 + 2$

46. $y + 3 = 2(x + 4)^2$

47. $y + 2 = 3(x - 2)^2$

48. $x = -2(y - 3)^2 + 3$

49. $x = -\frac{1}{2}(y + 1)^2 - 1$

50. $y - 4 = -2(x + 3)^2$

51. $y + 2 = \frac{1}{2}(x + 4)^2$

52. $x - 2 = \frac{1}{3}(y - 3)^2$

53. $x + 4 = 3(y + 1)^2$

54. $x + 3 = 2(y - 1)^2$

A parabola of the form $y = ax^2$ divides a plane into three regions: above the parabola, below the parabola, and on the parabola.

To graph parabolic inequalities, use the same technique as you used in graphing linear inequalities and circular inequalities.

1. Determine the boundary or "fence."
2. Sketch the boundary with a solid line if both an equality and inequality are indicated; otherwise use a dashed line.
3. Use a test point to decide which side of the boundary to shade.

Example 33 □ Sketch a graph of $y \geq x^2 + 2$.

Since this inequality uses a greater than or equal sign, draw the boundary with a solid line. $(0, 0)$ is a convenient test point.

■ Figure 11.37

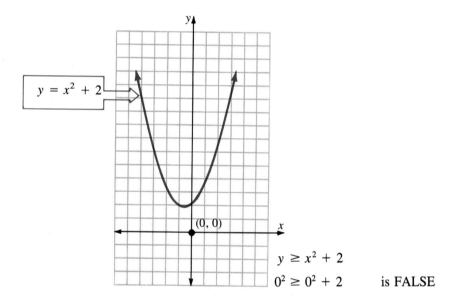

$y = x^2 + 2$

$(0, 0)$

$$y \geq x^2 + 2$$
$$0^2 \geq 0^2 + 2 \qquad \text{is FALSE}$$

■ Figure 11.38

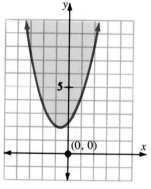

Therefore, shade the region that does not include $(0, 0)$. □

Circles divide a plane into three regions: inside the circle, on the circle and outside the circle.

Example 34 □ Sketch a graph of $x^2 + y^2 < 16$.

The boundary of this inequality is a circle. Since a less than sign is used, we draw the boundary with a dashed line.

■ Figure 11.39

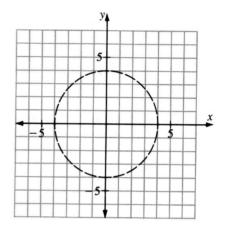

The dashed line represents the set of points where $x^2 + y^2$ is equal to 16. The points where $x^2 + y^2$ is *less than* 16 are inside the circle. $(0, 0)$ is a convenient test point.

$$x^2 + y^2 < 16$$

$$(0)^2 + (0) < 16 \qquad \text{is TRUE}$$

■ Figure 11.40

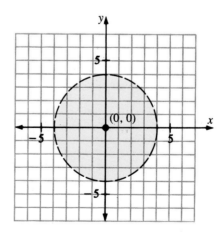

□

Example 35 □ Sketch a graph of $(x - 3)^2 + (y + 4)^2 \geq 25$.

This inequality involves a circle of radius 5 with its center at $(3, -4)$.

First, sketch the circle. Because the inequality uses the greater than or equal to sign, the points on the circle will satisfy the inequality. Therefore, draw the boundary with a solid line.

This inequality uses a greater than or equal to sign. Therefore, the points that satisfy the inequality will be outside the circle. Try testing this by using $(0, 0)$ as a test point in the inequality.

■ Figure 11.41

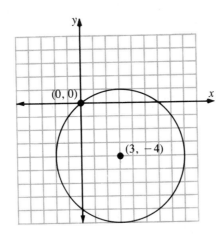

$$(x - 3)^2 + (y + 4)^2 \geq 25$$

Substituting $(0, 0)$

$$(0 - 3)^2 + (0 + 4)^2 \overset{?}{\geq} 25$$

$$9 + 16 \geq 25$$

is TRUE

This verifies that the inequality above is true for $(0, 0)$, but since $(0, 0)$ is a point on the circle, it does not help us confirm that the outside of the circle should be shaded. Pick a point that is clearly outside the circle, say $(4, 5)$.

Substitute this point in the inequality.

$$(4 - 3)^2 + (5 + 4)^2 \overset{?}{\geq} 25$$

$$1^2 + 9^2 \overset{?}{\geq} 25$$

$$1 + 81 \geq 25 \qquad \text{is TRUE}$$

■ Figure 11.42

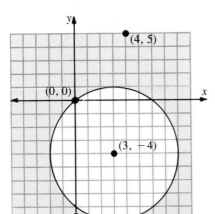

Therefore, we have verified that the points outside the circle should be shaded.

Therefore, shade the region that includes $(4, 5)$.

□

Chapter 11 ■ Second-Degree Equations, Circles, and Parabolas

Graph the following inequalities.

1. $y < x^2 + 1$

2. $y \geq x^2 - 2$

3. $y \geq (x - 2)^2$

4. $y < (x + 3)^2$

5. $y + 2 \leq (x + 1)^2$

6. $y - 1 > (x - 4)^2$

7. $x > y^2 - 1$

8. $x \leq y^2 + 2$

9. $y \geq 2(x + 1)^2$

10. $y \leq \frac{1}{2}(x - 3)^2$

11. $x^2 + y^2 > 4$

12. $x^2 + y^2 \geq 25$

13. $y + 3 < (x - 2)^2$

14. $y - 4 > (x + 2)^2$

15. $x^2 + y^2 \leq 9$

16. $x^2 + y^2 < 36$

17. $x + 2 > (y + 1)^2$

18. $x - 3 < (y - 2)^2$

19. $(x + 2)^2 + (y - 3)^2 \geq 4$

20. $(x - 4)^2 + (y - 2)^2 > 9$

21. $y - 2 \leq \frac{-1}{2}(x + 1)^2$

22. $y + 3 \geq -2(x - 1)^2$

23. $x^2 + y^2 + 6x - 2y - 6 < 0$

24. $x^2 + y^2 - 2x + 4y - 4 \leq 0$

Chapter 11 □ Key Ideas

11.1 **1.** The **Pythagorean Theorem**
$c^2 = a^2 + b^2$ where c is the hypotenuse of a right triangle.
  2. The **distance between two points** formula
$d = \sqrt{(x_2 - x_1)^2 + (y_2 - y_1)^2}$
  3. The **midpoint between two points**
$$x_{mid} = \frac{x_1 + x_2}{2} \qquad y_{mid} = \frac{y_1 + y_2}{2}$$

11.2 **1.** **Equation of a circle** with center at the origin and radius r
$x^2 + y^2 = r^2$
  2. **Equation of a circle** with center at point (h, k) and radius r
$(x - h)^2 + (y - k)^2 = r^2$
  3. Equation of form: $x^2 + y^2 - Dx + Ey + F = 0$ is a circle.
   a. Use the method of completing the square to put the equation in standard form.
$(x - h)^2 + (y - k)^2 = r^2$

11.3 The **translation equations**

$$x = x' + h \qquad \text{and} \qquad y = y' + k$$
$$x' = x - h \qquad\qquad\qquad y' = y - k$$

where

   (x, y) is a point in the original coordinate system
   (x', y') is a point in the translated coordinate system
   (h, k) is the center of the translated coordinate system.

11.4 1. The **equation of a parabola opening parallel** to the y-axis
   $$y - k = a(x - h)^2$$
   a. (h, k) is the vertex
   b. $x = h$ is the axis of symmetry
   c. $a > 0$, the curve opens upward
   d. $a < 0$, the curve opens downward.
2. The **equation of a parabola opening parallel** to the x-axis
   $$x - h = a(y - k)^2$$
   a. (h, k) is the vertex
   b. $y = k$ is the axis of symmetry
   c. $a > 0$, the curve opens to the right
   d. $a < 0$, the curve opens to the left.

11.5 1. **Graphs of inequalities** involving parabolas
   a. Sketch the boundary curve
   b. Use a solid line if both inequality and equality are indicated
   c. Use a dashed line if an inequality only
   d. Use a test point to determine which side of the boundary to shade.

Chapter 11   Review Test

Find the distance between the following pairs of points. **(11.1)**

   1. $(0, 2)$ and $(3, -1)$ **2.** $(-6, 4)$ and $(3, 4)$ **3.** $(-8, -2)$ and $(2, 4)$

   4. $(2, -5)$ and $(-6, 1)$

Find the midpoint between the following pairs of points. **(11.1)**

   5. $(8, -3)$ and $(-6, 5)$ **6.** $(-4, -3)$ and $(6, 11)$

Find the perimeter of the following triangles having the given vertices. **(11.1)**

   7. $(-2, -5), (-2, 3), (8, -5)$ **8.** $(-3, 5), (7, 5), (7, -3)$

In the accompanying right triangles, find the missing side. Simplify your answer. **(11.1)**

9.

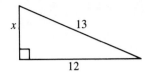

Solve the following problem. Leave your answer in simplest form.

10. A patio is in the form of a rectangle and has the measurements 28 feet long by 12 feet wide. What is the length of a line running diagonally across the patio from one corner to the other?

Find the equation of the circle having the given center and radius. **(11.2)**

11. center $(-4, 5)$, radius 3

12. center $(0, -3)$, radius 8

Find the center and radius of each circle. Sketch the graph. **(11.2)**

13. $(x - 1)^2 + (y + 2)^2 = 4$

14. $(x + 5)^2 + (y - 2)^2 = 8$

15. $x^2 + y^2 - 8x + 6y + 9 = 0$

16. $x^2 + y^2 + 10x - 2y + 17 = 0$

First write the equation of the indicated circle in the new (x', y') coordinate system with the given translation. Then express the same equation in the original (x, y) coordinate system. **(11.3)**

17. radius 4, center $(-5, 4)$

18. radius $\sqrt{7}$, center $(3, -6)$

Use a translation to graph the following equations.

19. $x^2 + y^2 - 6x + 8y + 9 = 0$

20. $x^2 + y^2 + 4x - 6y + 5 = 0$

Sketch the following graphs. **(11.4)**

21. $y = 3x^2$

22. $x = -\frac{1}{2}y^2$

23. $x = 2y^2 - 3$

24. $y = 2(x + 3)^2$

25. $y - 3 = \frac{1}{2}(x + 4)^2$

26. $y + 2 = -3(x - 4)^2$

Sketch the following inequalities.

27. $(x - 2)^2 + (y + 1)^2 \geq 9$

28. $x^2 + y^2 + 2x - 6y - 6 < 0$

29. $y - 1 < (x + 2)^2$

30. $x \geq y^2 - 2$

12 Second-Degree Equations, Ellipses, and Hyperbolas

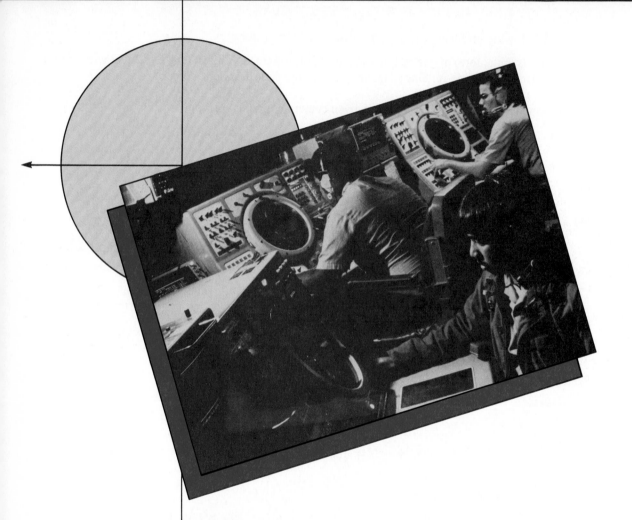

Contents

F our major kinds of curves are generated by second-degree equations. In the last chapter you learned of two of them—the circle and parabola.

In this chapter you will learn about the other two curves generated by second-degree equations—ellipses and hyperbolas. The path of a satellite in orbit is an ellipse. The path of some atomic particles fired at the nucleus of an atom is a hyperbola. Hyperbolas are also the basis of the primary system of long-range navigation used by ships and aircraft called LORAN.

The final sections of this chapter will show you how to identify each of the four second-degree equations. Then you learn how to find the solutions to systems of second-degree equations.

■ 12.1   Ellipses

Ellipse

In addition to circles and parabolas, there are two other types of second-degree equations. They are **ellipses** and **hyperbolas.**

In the last chapter, a circle was defined as the set of all points that are a given distance from a fixed point.

An ellipse can be defined in a similar manner.

Definition 12.1A   Ellipse

An ellipse is the set of all points in a plane, the sum of whose distances from two fixed points is a constant.

One way to draw an ellipse is to stick two tacks into a pad. Then tie a string loosely between the tacks. An ellipse is the figure you trace with a pencil as you keep the string tight.

■ Figure 12.1

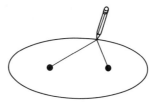

Foci

The thumb tacks are at the points called the **foci** (plural of focus) of the ellipse. Next we will derive the equation of an ellipse.

What does "derive an equation" mean?

It means to express a relationship between the x- and y-coordinates of each point on the ellipse in terms of some of the key features that define the ellipse.

Key Features of an Ellipse

The features that determine the shape of a particular ellipse are the location of its foci and the length of its major and minor axes.

An ellipse is the set of all points, the sum of whose distance from two fixed points is a constant. The two fixed points are called **foci.**

A line segment across the ellipse through the foci is called the **major axis.**

A line segment across the ellipse that is the perpendicular bisector of the major axis is called the **minor axis.**

Foci

Major Axis

Minor Axis

■ Figure 12.2

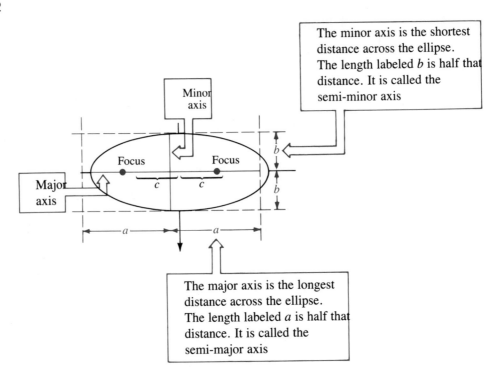

The minor axis is the shortest distance across the ellipse. The length labeled b is half that distance. It is called the semi-minor axis

The major axis is the longest distance across the ellipse. The length labeled a is half that distance. It is called the semi-major axis

To derive the equation of an ellipse with the x-axis along the major axis of the ellipse, we choose a coordinate system through the center of the ellipse.

■ Figure 12.3

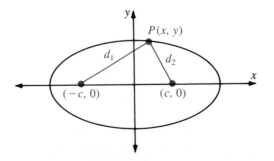

The two foci are at $(c, 0)$ and $(-c, 0)$.

For an ellipse, $d_1 + d_2$ is equal to a constant for all points $P(x, y)$ in the ellipse.

Chapter 12 ■ Second-Degree Equations, Ellipses, and Hyperbolas

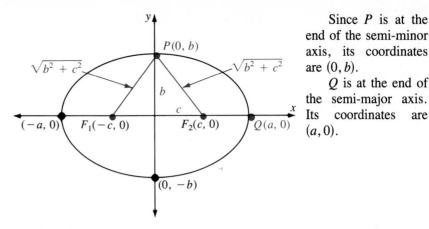

Since P is at the end of the semi-minor axis, its coordinates are $(0, b)$.

Q is at the end of the semi-major axis. Its coordinates are $(a, 0)$.

Since P and Q both lie on the ellipse, the sum of the distances from each of them to the foci will be the same.

Distance of P to F_1	+	Distance of P to F_2	=	Distance of Q to F_1	+	Distance of Q to F_2
$\sqrt{b^2 + c^2}$	+	$\sqrt{b^2 + c^2}$	=	$(a + c)$	+	$(a - c)$
		$2\sqrt{b^2 + c^2}$	=	$2a$		

This tells us that the sum of the distances from P to each of the foci is $2a$. Since an ellipse is the set of points whose sum of the distances to the foci is a constant, the sum of the distances from any point (x, y) on the ellipse must be $2a$.

■ Figure 12.5

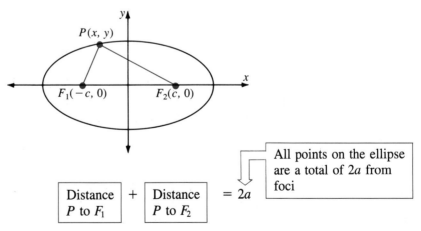

Distance P to F_1	+	Distance P to F_2	= $2a$

All points on the ellipse are a total of $2a$ from foci

Using the distance formula for PF_1 and PF_2

$$\sqrt{(x - (-c))^2 + (y - 0)^2} + \sqrt{(x - c)^2 + (y - 0)^2} = 2a$$
$$\sqrt{(x + c)^2 + y^2} + \sqrt{(x - c)^2 + y^2} = 2a$$

To solve the equation we eliminate the radicals.

$$\sqrt{(x + c)^2 + y^2} = 2a - \sqrt{(x - c)^2 + y^2} \qquad \text{Isolate the left radical}$$

$$(x + c)^2 + y^2 = 4a^2 - 4a\sqrt{(x - c)^2 + y^2} + (x - c)^2 + y^2$$

Square both sides

$$x^2 + 2cx + c^2 + y^2 = 4a^2 - 4a\sqrt{(x - c)^2 + y^2} + x^2 - 2cx + c^2 + y^2$$

Square binomials

$$2cx = 4a^2 - 4a\sqrt{(x - c)^2 + y^2} - 2cx$$

$$-4a^2 + 4cx = -4a\sqrt{(x - c)^2 + y^2} \qquad \text{Isolate the radical}$$

$$a - \frac{cx}{a} = \sqrt{(x - c)^2 + y^2} \qquad \text{Divide by } (-4a)$$

$$a^2 - \frac{2cx}{a}(a) + \frac{c^2x^2}{a^2} = (x - c)^2 + y^2 \qquad \text{Square both sides}$$

$$a^2 - 2cx + \frac{c^2x^2}{a^2} = x^2 - 2cx + c^2 + y^2 \qquad \text{Square binomial}$$

$$a^2 + \frac{c^2x^2}{a^2} - c^2 = x^2 + y^2 \qquad \text{Add } +2cx - c^2$$

What do I do with this mess?

Remember way back when we said the sum of the distance was $2a$?

$$2a = 2\sqrt{b^2 + c^2}$$

$$a = \sqrt{b^2 + c^2}$$

$$a^2 = b^2 + c^2$$

$$a^2 - c^2 = b^2 \text{ which we can substitute.}$$

$$b^2 + \frac{c^2x^2}{a^2} = x^2 + y^2$$

$$b^2 = x^2 - \frac{c^2x^2}{a^2} + y^2 \qquad \text{Subtract } \frac{c^2x^2}{a^2}$$

$$b^2 = \frac{a^2x^2 - c^2x^2}{a^2} + y^2$$

$$b^2 = \frac{x^2(a^2 - c^2)}{a^2} + y^2$$

I remember $a^2 - c^2 = b^2$.

Substitute it.

$$b^2 = \frac{x^2 b^2}{a^2} + y^2$$

$$1 = \frac{x^2}{a^2} + \frac{y^2}{b^2} \qquad \text{Multiply by } \frac{1}{b^2}$$

Chapter 12 ■ Second-Degree Equations, Ellipses, and Hyperbolas

12.1B Equation of an Ellipse at the Origin

$$\frac{x^2}{a^2} + \frac{y^2}{b^2} = 1$$

is the equation of an ellipse centered at the origin. The ellipse crosses the x-axis at $\pm a$ and crosses the y-axis at $\pm b$.
Note: The right member is always equal to one.

Example 1 □ Graph $\dfrac{x^2}{9} + \dfrac{y^2}{16} = 1$.

We know from the derivation of equation 12.1B that this is the equation of an ellipse.

We also know that the ellipse crosses the x-axis at $\pm\sqrt{9}$ and crosses the y-axis at $\pm\sqrt{16}$. We can verify the y-intercepts by substituting $x = 0$ into the equation.

$$\frac{x^2}{9} + \frac{y^2}{16} = 1$$

$$\frac{(0)^2}{9} + \frac{y^2}{16} = 1 \qquad \text{On the } y\text{-axis, } x = 0$$

$$\frac{y^2}{16} = 1$$

$$y^2 = 16$$

$$y = \pm 4$$

The equation crosses the y-axis at $(0, 4)$ and $(0, -4)$.
We can verify the x-intercepts by substituting $y = 0$ in the equation.
At the x-intercepts

$$\frac{x^2}{9} + \frac{y^2}{16} = 1$$

becomes

$$\frac{x^2}{9} + \frac{(0)^2}{16} = 1$$

$$\frac{x^2}{9} = 1$$

$$x^2 = 9$$

$$x = \pm 3$$

The x-intercepts are $(3, 0)$ and $(-3, 0)$.

Now we have four points to plot on the graph.

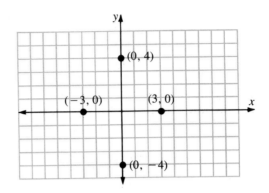

To draw the graph, we draw a smooth curve through these points. We can calculate other points by substituting additional values of x into the equation if we need more precise results.

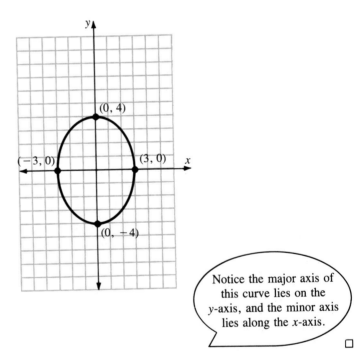

Notice the major axis of this curve lies on the y-axis, and the minor axis lies along the x-axis.

□

Example 2 □ Graph the equation $\dfrac{x^2}{25} + \dfrac{y^2}{4} = 1$.

We can read the intercepts directly from the equation.
The curve crosses the x-axis at ± 5 and crosses the y-axis at ± 2.

If we sketch in the lines $x = \pm 5$ and $y = \pm 2$, we get a box that neatly contains the ellipse.

■ Figure 12.8

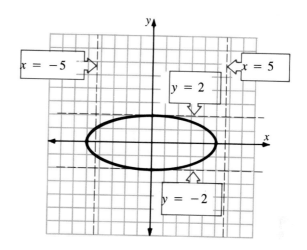

Example 3 □ Graph $4x^2 + 9y^2 = 36$.

This equation is not in a form recognizable for graphing. To get it into a more usable form, divide both sides of the equation by 36.

■ Figure 12.9

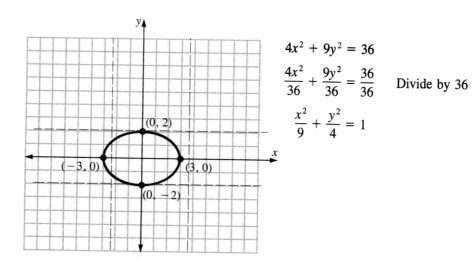

$$4x^2 + 9y^2 = 36$$

$$\frac{4x^2}{36} + \frac{9y^2}{36} = \frac{36}{36} \qquad \text{Divide by 36}$$

$$\frac{x^2}{9} + \frac{y^2}{4} = 1$$

The above equation is recognizable as an ellipse with x-intercepts of ± 3 and y-intercepts of ± 2. □

Graph each ellipse.

1. $\dfrac{x^2}{1} + \dfrac{y^2}{4} = 1$

2. $\dfrac{x^2}{4} + \dfrac{y^2}{1} = 1$

3. $\dfrac{x^2}{16} + \dfrac{y^2}{4} = 1$

4. $\dfrac{x^2}{4} + \dfrac{y^2}{36} = 1$

5. $\dfrac{x^2}{4} + \dfrac{y^2}{25} = 1$

6. $\dfrac{x^2}{9} + \dfrac{y^2}{1} = 1$

7. $\dfrac{x^2}{16} + \dfrac{y^2}{4} = 1$

8. $\dfrac{x^2}{1} + \dfrac{y^2}{36} = 1$

9. $x^2 + 4y^2 = 36$

10. $4x^2 + y^2 = 16$

11. $16x^2 + 25y^2 = 400$

12. $9x^2 + 4y^2 = 144$

13. $4x^2 + y^2 = 36$

14. $25x^2 + 4y^2 = 100$

15. $x^2 + 9y^2 = 36$

16. $x^2 + 9y^2 = 9$

Ellipses That Are Not at the Origin

To write the equation of an ellipse with its center off the origin, think of the ellipse as centered on an x'-, y'-coordinate system. The equation of the ellipse then becomes

$$\frac{x'^2}{a^2} + \frac{y'^2}{b^2} = 1$$

In which system are a and b?

a and b are the distances from the center of the ellipse to the x- and y-intercepts in either system. In a translation, the distance is independent of the coordinate system.

The center of the ellipse is at (h, k) in the original coordinate system. When the translation equations

$$x' = x - h$$
$$y' = y - k$$

are substituted in $\dfrac{x'^2}{a^2} + \dfrac{y'^2}{b^2} = 1,$

we get the equation of any ellipse as viewed from the original coordinate system.

$$\frac{(x-h)^2}{a^2} + \frac{(y-k)^2}{b^2} = 1$$

is the equation of an ellipse centered at (h, k), and the major axis is parallel to the x-axis if $a > b$.

In the translated ellipse, the lengths of the semi-major and semi-minor axes are still given by a and b.

Example 4 □ Graph $\dfrac{(x-3)^2}{16} + \dfrac{(y+2)^2}{25} = 1$.

The graph of this equation is equivalent to graphing $\dfrac{x'^2}{16} + \dfrac{y'^2}{25} = 1$ with the origin of the (x', y') system at $(3, -2)$.

Center $(3, -2)$

$a = 4$

$b = 5$ □

■ Figure 12.10

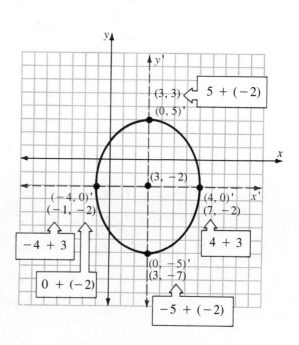

Example 5 □ Graph $\dfrac{(x + 5)^2}{9} + \dfrac{(y - 3)^2}{16} > 1$.

■ Figure 12.11

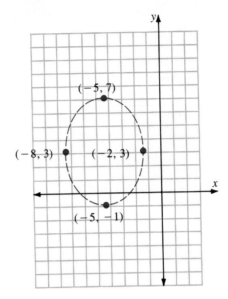

This time we are dealing with an inequality. First sketch the boundary with a dashed line.

Let's test several points inside and outside the ellipse.

Point	Inside or outside	Evaluation of left side	Is this value >1?
$(-5, 3)$	inside	$\dfrac{(-5 + 5)^2}{9} + \dfrac{(3 - 3)^2}{16} = 0$	NO
$(0, 0)$	outside	$\dfrac{(0 + 5)^2}{9} + \dfrac{(0 - 3)^2}{16} = \dfrac{25}{9} + \dfrac{9}{16}$	YES
$(-10, -10)$	outside	$\dfrac{(-10 + 5)^2}{9} + \dfrac{(-10 - 3)^2}{16} = \dfrac{25}{9} + \dfrac{169}{16}$	YES

□

Why pick those particular points?

I just wanted any points where I wouldn't have to work too hard. Notice I didn't even get common denominator for the fractions. All I needed to know was, "Is the sum greater than 1?"

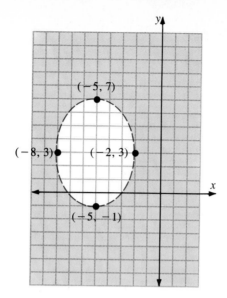

Since the points outside the ellipse satisfy the inequality, shade the entire region outside the ellipse.

To determine the part of the graph that must be shaded, pick test points that are obviously inside or outside the ellipse. □

Example 6 □ Graph $9x^2 + 4y^2 + 16y - 20 = 0$.

First we need to write the equation in standard form. Start by completing the square on y.

$$9x^2 + 4y^2 + 16y - 20 = 0$$

$$9x^2 + 4(y^2 + 4y \qquad) = 20$$

Factor 4 to get coefficient of $y^2 = 1$

$$9x^2 + 4(y^2 + 4y + 4) = 20 + 16$$

4 inside the parentheses is multiplied by 4 outside the parentheses to give 16 added to both sides

$$9x^2 + 4(y + 2)^2 = 36$$

Write as a binomial squared

$$\frac{x^2}{4} + \frac{(y + 2)^2}{9} = 1$$

Divide by 36

This is now recognizable as an ellipse with center at $(0, -2)$, with $a = 2$ and $b = 3$. □

■ Figure 12.13

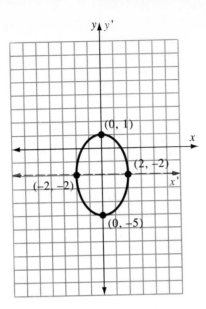

In general the ellipse extends b units above and b units below the center. It also extends a units to the right and a units to the left of the center.

Problem Set 12.1B

Graph each ellipse.

1. $\dfrac{(x+1)^2}{9} + \dfrac{(y-2)^2}{16} = 1$

2. $\dfrac{(x-3)^2}{16} + \dfrac{(y+2)^2}{9} = 1$

3. $\dfrac{(x-2)^2}{1} + \dfrac{(y-4)^2}{4} = 1$

4. $\dfrac{(x+4)^2}{4} + \dfrac{(y+3)^2}{9} = 1$

5. $\dfrac{x^2}{25} + \dfrac{(y+4)^2}{16} = 1$

6. $\dfrac{(x-4)^2}{9} + \dfrac{y^2}{25} = 1$

7. $\dfrac{(x-2)^2}{1} + \dfrac{(y+3)^2}{9} = 1$

8. $\dfrac{(x-3)^2}{4} + \dfrac{(y-3)^2}{1} = 1$

9. $\dfrac{(x-4)^2}{16} + \dfrac{y^2}{4} = 1$

10. $\dfrac{x^2}{9} + \dfrac{(y-1)^2}{25} = 1$

11. $\dfrac{(x+2)^2}{16} + \dfrac{(y-1)^2}{1} = 1$

12. $\dfrac{(x+1)^2}{16} + \dfrac{(y+1)^2}{25} = 1$

13. $4x^2 + 9y^2 - 18y - 27 = 0$

14. $x^2 + 4y^2 + 6x - 27 = 0$

Chapter 12 ■ Second-Degree Equations, Ellipses, and Hyperbolas

15. $4x^2 + y^2 + 16x - 6y + 9 = 0$

16. $x^2 + 9y^2 + 6x - 36y + 9 = 0$

17. $9x^2 + 4y^2 - 36x - 32y - 44 = 0$

18. $4x^2 + 25y^2 + 32x - 150y + 189 = 0$

19. $9x^2 + y^2 - 36x < 0$

20. $4x^2 + 9y^2 + 36y \leq 0$

21. $x^2 + 4y^2 + 6x - 32y + 69 \geq 0$

22. $16x^2 + 9y^2 + 128x - 36y + 148 > 0$

23. $9x^2 + 4y^2 + 36x + 16y + 16 \leq 0$

24. $9x^2 + 25y^2 - 18x - 150y + 9 < 0$

■ **12.2**   **Hyperbolas**

An ellipse is defined as the set of all points the sum of whose distances from two foci is a constant.

The definition of a hyperbola is similar to the definition of an ellipse.

Definition 12.2A Hyperbola

A hyperbola is defined as the set of all points in a plane where the difference of the distances from two fixed points (foci) is a constant.

The foci are at $(c, 0)$ and $(-c, 0)$. The vertices of the hyperbola are at distance a from the center. Their coordinates are $(a, 0)$ and $(-a, 0)$. For a hyperbola, $d_1 - d_2$ is constant.

■ Figure 12.14

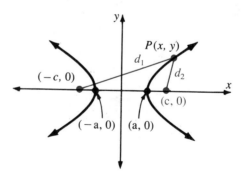

The equation of a hyperbola that opens parallel to the x-axis is

$$\frac{x^2}{a^2} - \frac{y^2}{b^2} = 1$$

If you believe us you may skip the derivation and continue with the bubble "What's b^2?"

Derivation of the
Equation of a
Hyperbola

A **hyperbola** is the set of points whose distance to the farther focus minus the distance to the nearer focus is a constant. To determine the constant, start with the simplest point on the hyperbola V_2.

■ Figure 12.15

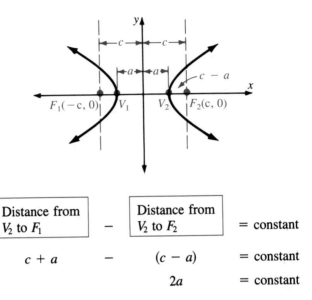

Distance from V_2 to F_1	−	Distance from V_2 to F_2	= constant
$c + a$	−	$(c - a)$	= constant
		$2a$	= constant

Thus for all points on the hyperbola

Distance from P to farther focus	−	Distance from P to closer focus	= $2a$

We will work with the right branch. Working with the left branch produces an identical result.

■ Figure 12.16

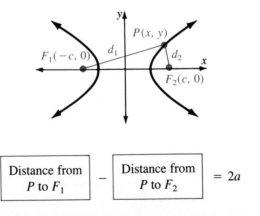

Distance from P to F_1	−	Distance from P to F_2	= $2a$

$$\sqrt{(x - (-c))^2 + (y - 0)^2} - \sqrt{(x - c)^2 + (y - 0)^2}$$
$$= 2a$$

$$\sqrt{(x + c)^2 + y^2} - \sqrt{(x - c)^2 + y^2}$$
$$= 2a$$

$$\sqrt{(x + c)^2 + y^2}$$
$$= 2a + \sqrt{(x - c)^2 + y^2} \qquad \text{Isolating one radical}$$

$$(x + c)^2 + y^2$$
$$= 4a^2 + 4a\sqrt{(x - c)^2 + y^2} + (x - c)^2 + y^2 \qquad \text{Squaring both}$$
sides

$$x^2 + 2cx + c^2 + y^2$$
$$= 4a^2 + 4a\sqrt{(x - c)^2 + y^2} + x^2 - 2cx + c^2 + y^2$$

$$4cx - 4a^2$$
$$= 4a\sqrt{(x - c)^2 + y^2} \qquad \text{Simplifying}$$

$$cx - a^2$$
$$= a\sqrt{(x - c)^2 + y^2} \qquad \text{Dividing by 4}$$

$$c^2x^2 - 2a^2cx + a^4$$
$$= a^2[(x - c)^2 + y^2] \qquad \text{Squaring both sides}$$

$$= a^2[x^2 - 2cx + c^2 + y^2] \qquad \text{Squaring the binomial}$$

$$c^2x^2 - 2a^2cx + a^4$$
$$= a^2x^2 - 2a^2cx + a^2c^2 + a^2y^2 \qquad \text{Distributive}$$

$$c^2x^2 + a^4$$
$$= a^2x^2 + a^2c^2 + a^2y^2 \qquad \text{Adding } 2a^2cx \text{ to both sides}$$

$$c^2x^2 - a^2x^2 - a^2y^2$$
$$= a^2c^2 - a^4 \qquad x^2 \text{ and } y^2 \text{ terms to the left}$$

$$(c^2 - a^2)x^2 - a^2y^2$$
$$= (c^2 - a^2)a^2 \qquad \text{Factoring}$$

Now if we let $\quad c^2 - a^2$
$$= b^2$$

A later bubble talks about b^2

$$b^2x^2 - a^2y^2$$
$$= b^2a^2 \qquad \text{Substituting}$$

$$\frac{x^2}{a^2} - \frac{y^2}{b^2}$$
$$= 1 \qquad \text{Dividing both sides by } b^2a^2$$

This is the equation of a hyperbola with center at the origin and vertices at $\pm a$ on the x-axis.

What's b^2?

Basically, a value we substituted for $c^2 - a^2$. However, it does turn out to be useful.

We can construct a box similar to the box we drew for an ellipse with sides $\pm a$ on the x-axis and $\pm b$ on the y-axis. The vertices of the hyperbola are where the box crosses the x-axis.

■ Figure 12.17

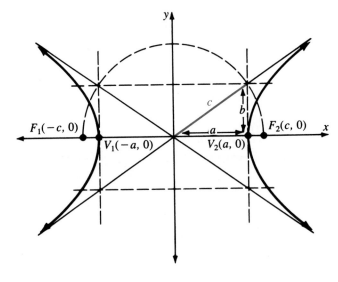

Asymptotes

The diagonals of the box are asymptotes (pronouned ass-m-totes) of the hyperbola.

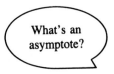

What's an asymptote?

It's a line that a curve gets closer and closer to but never touches.

The asymptotes pass through the origin. Their slope is $\pm \dfrac{b}{a}$ from the diagram. Therefore, the equations of the asymptotes are

$$y = +\frac{b}{a}x \qquad \text{and} \qquad y = -\frac{b}{a}x$$

More advanced courses prove that these lines are indeed asymptotes of the hyperbola.

Chapter 12 ■ Second-Degree Equations, Ellipses, and Hyperbolas

Example 7 □ Graph $\dfrac{x^2}{9} - \dfrac{y^2}{4} = 1$.

First draw the box with diagonals.

■ Figure 12.18

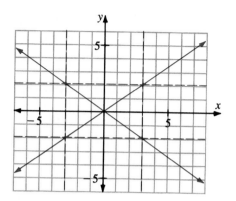

Next sketch in the graph using the points where the box crosses the x-axis for the vertices of the hyperbola.

■ Figure 12.19

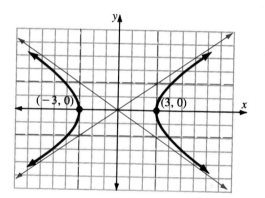

You can verify 2 points on the graph of $\dfrac{x^2}{9} - \dfrac{y^2}{4} = 1$ by substituting $y = 0$ in the equation and calculating x. Notice that the equation is undefined at $x = 0$.

$x = \pm 6$ isn't too hard to calculate if you want to test some additional points.

□

Hyperbolas can also have their foci on the y-axis. The equation of these hyperbolas is

$$\frac{y^2}{a^2} - \frac{x^2}{b^2} = 1$$

We use the same procedure to graph hyperbolas on the y-axis.

Example 8 □ Graph $\dfrac{y^2}{25} - \dfrac{x^2}{12} = 1$.

First draw the box with diagonals.

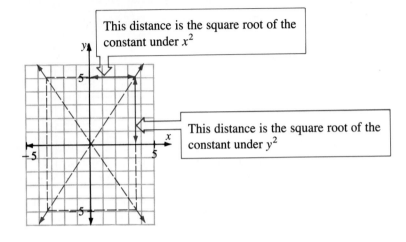

This distance is the square root of the constant under x^2

This distance is the square root of the constant under y^2

How do I get $\sqrt{12}$?

From $2\sqrt{3}$, or use your calculator. $\sqrt{12} \approx 3.46$. Eyeball the distance on the graph.

Where do I put the vertices of this hyperbola?

Look at the equation and read on.

Notice in

$$\dfrac{y^2}{25} - \dfrac{x^2}{12} = 1$$

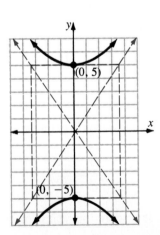

(0, 5)

(0, −5)

if $x = 0$, then y is defined. Therefore, the curve must cross the y-axis. However, if $y = 0$ you'll be trying to take the square root of a negative number. Therefore the curve can't cross the x-axis.

□

The vertices of the hyperbola lie on the y-axis when the plus sign is in front of the y^2 term.

And the vertices lie on the x-axis if the plus sign is in front of the x^2 term.

If the center of the hyperbola is not at the origin, we can view the hyperbola as the translation of a hyperbola centered at the origin of the (x', y') system. As in the case of a circle and an ellipse, the translation equations are

$$x' = x - h$$
$$y' = y - k$$

12.2B   Standard Form of the Equation of a Hyperbola

The equation of a hyperbola is

$$\frac{(x - h)^2}{a^2} - \frac{(y - k)^2}{b^2} = 1 \qquad \text{if the focal axis is parallel to the } x\text{-axis}$$

or

$$\frac{(y - k)^2}{a^2} - \frac{(x - h)^2}{b^2} = 1 \qquad \text{if the focal axis is parallel to the } y\text{-axis.}$$

In both cases the center of the hyperbola is at (h, k).

Example 9 $\square$ Graph $y^2 + 10y - 9x^2 + 18x = 20$.

First we'll change this into recognizable form by completing the square on x and y.

$$y^2 + 10y - 9x^2 + 18x = 20$$

$$y^2 + 10y + 25 - 9(x^2 - 2x + 1) = 20 + 25 - 9 \qquad \text{Complete the square}$$

$$(y + 5)^2 - 9(x - 1)^2 = 36$$

$$\frac{(y + 5)^2}{36} - \frac{(x - 1)^2}{4} = 1 \qquad \text{Divide by 36}$$

This is a hyperbola with center at $(1, -5)$ with vertices parallel to the y-axis

at $(1, -5 + 6)$

and $(1, -5 - 6)$

or $(1, 1)$

and $(1, -11)$

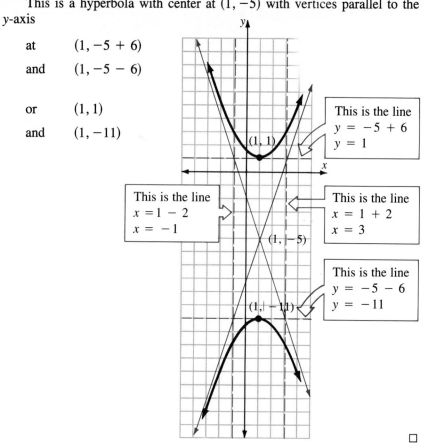

This is the line
$y = -5 + 6$
$y = 1$

This is the line
$x = 1 - 2$
$x = -1$

This is the line
$x = 1 + 2$
$x = 3$

This is the line
$y = -5 - 6$
$y = -11$

Example 10 □ Graph $\dfrac{(x + 5)^2}{9} - \dfrac{(y - 3)^2}{16} < 1$.

The boundary of this inequality is a hyperbola. Not counting the hyperbola itself, the plane is divided into three regions labeled A, B, and C. Test a point from each region to determine where to shade.

■ Figure 12.23

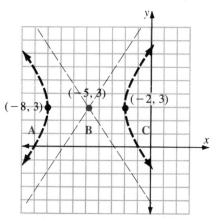

Chapter 12 ■ Second-Degree Equations, Ellipses, and Hyperbolas

Region	Point	Evaluation	Is this value < 1?
A	$(-9, 3)$	$\dfrac{(-9+5)^2}{9} - \dfrac{(3-3)^2}{16} = \dfrac{16}{9} - 0$	NO
B	$(-5, 3)$	$\dfrac{(-5+5)^2}{9} - \dfrac{(3-3)^2}{16} = \dfrac{0}{9} - \dfrac{0}{16}$	YES
C	$(0, 3)$	$\dfrac{(0+5)^2}{9} - \dfrac{(3-3)^2}{16} = \dfrac{25}{9} - \dfrac{0}{16}$	NO

■ Figure 12.24

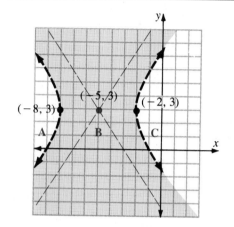

Region B is the only area that satisfies the inequality. Shade this region. To test the three regions, pick points that are obviously in each region. The center of the hyperbola is a good pick for a test point because it will make the arithmetic easier. The origin of the original coordinate system $(0, 0)$ is also usually a convenient test point. □

Problem Set 12.2

1. $\dfrac{x^2}{4} - \dfrac{y^2}{9} = 1$ **2.** $\dfrac{y^2}{36} - \dfrac{x^2}{9} = 1$ **3.** $\dfrac{y^2}{4} - \dfrac{x^2}{25} = 1$ **4.** $\dfrac{x^2}{16} - \dfrac{y^2}{25} = 1$

5. $\dfrac{x^2}{4} - \dfrac{y^2}{4} = 1$ **6.** $\dfrac{x^2}{9} - \dfrac{y^2}{9} = 1$ **7.** $\dfrac{x^2}{9} - \dfrac{y^2}{4} = 1$ **8.** $\dfrac{x^2}{16} - \dfrac{y^2}{9} = 1$

9. $\dfrac{y^2}{25} - \dfrac{x^2}{16} = 1$ **10.** $\dfrac{y^2}{4} - \dfrac{x^2}{4} = 1$ **11.** $\dfrac{y^2}{16} - \dfrac{x^2}{4} = 1$ **12.** $\dfrac{y^2}{9} - \dfrac{x^2}{1} = 1$

13. $x^2 - 4y^2 = 16$ **14.** $9y^2 - x^2 = 36$ **15.** $4x^2 - 9y^2 = 36$ **16.** $x^2 - 4y^2 = 36$

17. $\dfrac{(y-3)^2}{4} - \dfrac{(x-4)^2}{9} = 1$ **18.** $\dfrac{(y+4)^2}{36} - \dfrac{(x+4)^2}{16} = 1$

19. $\dfrac{(x-2)^2}{25} - \dfrac{(y+3)^2}{16} = 1$ **20.** $\dfrac{(x+1)^2}{1} - \dfrac{(y-2)^2}{8} = 1$ **21.** $\dfrac{(x+4)^2}{9} - \dfrac{(y-3)^2}{9} = 1$

22. $\dfrac{(y-4)^2}{16} - \dfrac{x^2}{25} = 1$ **23.** $\dfrac{(x+2)^2}{4} - \dfrac{y^2}{18} = 1$ **24.** $\dfrac{(y+2)^2}{9} - \dfrac{x^2}{12} = 1$

25. $\dfrac{(x+5)^2}{20} - \dfrac{(y+2)^2}{9} = 1$ **26.** $\dfrac{(x+1)^2}{16} - \dfrac{(y-2)^2}{9} = 1$ **27.** $\dfrac{(x-2)^2}{4} - \dfrac{(y+1)^2}{9} = 1$

28. $\dfrac{(x+3)^2}{16} - \dfrac{(y-1)^2}{4} = 1$ **29.** $\dfrac{(y-4)^2}{4} - \dfrac{(x-2)^2}{4} = 1$ **30.** $\dfrac{(y+5)^2}{9} - \dfrac{(x-2)^2}{9} = 1$

31. $\dfrac{(x-1)^2}{9} - \dfrac{y^2}{4} = 1$ **32.** $\dfrac{(x+3)^2}{9} - \dfrac{y^2}{16} = 1$ **33.** $\dfrac{(y+2)^2}{9} - \dfrac{x^2}{25} = 1$

34. $\dfrac{(y-3)^2}{16} - \dfrac{x^2}{25} = 1$ **35.** $\dfrac{(x+2)^2}{4} - \dfrac{(y+2)^2}{8} = 1$ **36.** $\dfrac{(x-3)^2}{8} - \dfrac{(y-3)^2}{9} = 1$

37. $4y^2 - 24y - x^2 - 4x - 4 = 0$

38. $y^2 - 4x^2 - 4y - 32 = 0$

39. $16x^2 - 25y^2 - 150y - 625 = 0$

40. $4y^2 - 9x^2 + 18x + 32y + 19 = 0$

41. $4x^2 - 3y^2 + 8x + 30y - 119 < 0$

42. $x^2 + 2x - 4y^2 - 16y - 31 \leq 0$

43. $9y^2 - 4x^2 - 8x - 54y + 41 \geq 0$

44. $16y^2 - 9x^2 + 54x + 32y - 209 > 0$

45. $9x^2 - 16y^2 + 36x - 32y - 124 \leq 0$

46. $x^2 - y^2 - 8x - 2y + 11 < 0$

47. $y^2 - x^2 + 8x + 2y - 24 > 0$

48. $4y^2 - x^2 - 6x + 16y - 9 \geq 0$

■ 12.3   Identifying Second-Degree Equations

Except for a few special cases, second-degree equations have graphs that are one of four forms: circle, ellipse, hyperbola, or parabola. Because these are the shapes formed if a cone is sliced by a plane, they are sometimes referred to as **conic sections.**

■ Figure 12.25

Circle

Ellipse

Parabola

Hyperbola

Chapter 12 ■ Second-Degree Equations, Ellipses, and Hyperbolas

12.3A General Second-Degree Equation

$$Ax^2 + Bxy + Cy^2 + Dx + Ey + F = 0$$

represents a second-degree equation. The specific equation is determined by the value assigned to each of the constants A, B, C, D, E, F. If A, B, C all equal zero, this becomes a first-degree equation.

A second-degree equation with B not equal to zero has a Bxy term. This term only occurs in equations that have been rotated. To develop rotations requires trigonometry, which is beyond the scope of this text. We will mention only one special case.

12.3B Equation of a Hyperbola with Axis of Symmetry $y = x$ or $y = -x$

$$xy = k$$

is the equation of a hyperbola that has been rotated 45°. Its axis of symmetry is the line $y = +x$ or $y = -x$. Its asymptotes are the x- and y-axes.

Example 11 □ Graph $xy = 3$.

We know the general shape of the curve from 12.3B. A brief table of values allows us to sketch the graph.

■ Figure 12.26

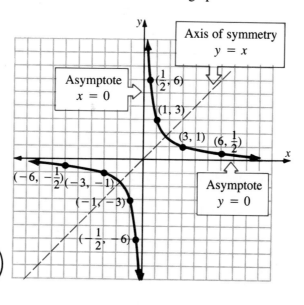

Axis of symmetry
$y = x$

Asymptote
$x = 0$

$(\frac{1}{2}, 6)$

$(1, 3)$

$(3, 1)$ $(6, \frac{1}{2})$

$(-6, -\frac{1}{2})$ $(-3, -1)$

$(-1, -3)$

Asymptote
$y = 0$

$(-\frac{1}{2}, -6)$

Notice the equation is undefined at $x = 0$ and $y = 0$.

$xy = 3$	
x	y
1	3
3	1
6	$\frac{1}{2}$
$\frac{1}{2}$	6
-1	-3
-3	-1
-6	$-\frac{1}{2}$
$-\frac{1}{2}$	-6

As long as there is no Bxy term in the equation, it is possible to identify the shape of the graph of a second-degree equation by examining the general form of the equation. Use the table below.

General Form $Ax^2 + Cy^2 + Dx + Ey + F = 0$

	Standard Form	Comments on General Form
Circle	$(x - h)^2 + (y - k)^2 = r^2$	A and C must be equal.
Ellipse	$\dfrac{(x - h)^2}{a^2} + \dfrac{(y - k)^2}{b^2} = 1$	A not equal to C, but A and C must have the same sign. Equation must be set equal to one in order to identify a and b.
Hyperbola	$\dfrac{(x - h)^2}{a^2} - \dfrac{(y - k)^2}{b^2} = 1$ or $\dfrac{(y - k)^2}{a^2} - \dfrac{(x - h)^2}{b^2} = 1$	A and C must have opposite signs. Equation must be set equal to one in order to identify a and b.
Parabola	$y = (x - h)^2 + k$ or $x = (y - k)^2 + h$	$C = 0$ $A = 0$
Straight line	$y = mx + b$	A and $C = 0$

□

Example 12 □ Identify the graph of each of the following equations.

a. $x^2 + 9y^2 + 2x - 18y + 1 = 0$ Ellipse

Hyperbola   b. $x^2 - 9y^2 - 4x + 18y - 14 = 0$ _____

Circle   c. $4x^2 + 4y^2 - 15 = 0$ _____

Hyperbola   d. $4x^2 - 4y^2 - 15 = 0$ _____

Parabola   e. $3x^2 + 5x + 6y - 5 = 0$ _____

Ellipse   f. $4x^2 + 25y^2 - 8x - 100y + 4 = 0$ _____

Straight Line   g. $6x + 6y - 1 = 0$ _____ □

Chapter 12 ■ Second-Degree Equations, Ellipses, and Hyperbolas

Identify each of the following equations. Do **not** graph.

1. $x^2 + y^2 + 2x - 6y + 5 = 0$

2. $x^2 + 2x + 4y - 6 = 0$

3. $4x^2 + y^2 - 16x + 4 = 0$

4. $y^2 - 9x^2 + 18x - 4y - 2 = 0$

5. $5x^2 - 5y^2 + 10x - 20y - 1 = 0$

6. $4x^2 + 4y^2 + 8x - 12y - 7 = 0$

7. $x^2 + 2y^2 + 6x - 12y - 3 = 0$

8. $y^2 + 2x - 3y - 6 = 0$

9. $5x^2 + 10x + 5y^2 - 30y - 6 = 0$

10. $6x^2 + 12x - 6y^2 - 24y - 4 = 0$

11. $4x^2 - 16x + 5y - 20 = 0$

12. $4x - 10y - 6 = 0$

13. $x^2 + y^2 - 3x + 4y - 6 = 0$

14. $x^2 - 2x + 5y + 4 = 0$

15. $4y^2 + 9x^2 - 8x + 18y - 6 = 0$

16. $9x^2 - 2y^2 + 18x - 2y - 4 = 0$

17. $x^2 - 4y^2 - 4x + 16y - 3 = 0$

18. $y^2 - 2x + 4y - 5 = 0$

19. $x - 3y + 4 = 0$

20. $4y^2 + x^2 + 2x + 2y - 1 = 0$

21. $y^2 + 2x - 3y + 2 = 0$

22. $x^2 + 3y^2 + 2x + 6y - 4 = 0$

23. $x^2 + 2y^2 + 4x + 4y - 3 = 0$

24. $2x - y + 4 = 0$

Sketch the graphs.

25. $xy = 6$

26. $4xy = 6$

27. $xy = -5$

28. $3xy = -2$

29. $2xy = -5$

30. $xy = -4$

31. $6xy = 9$

32. $2xy = 3$

Systems of Equations Involving a Line and a Second-Degree Equation

A system of two distinct straight lines can have at most one point in common. With a system that has a straight line and a second-degree equation there are more possibilities.

A Line and a Circle

These are possibilities for a system with a line and circle.

■ Figure 12.27

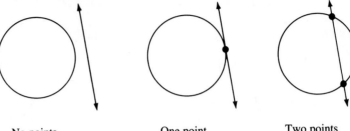

No points
in common
No solution

One point
in common
One solution

Two points
in common
Two solutions

Example 13 □ Solve the system.

$$x^2 + y^2 = 17$$
$$x + y = 5$$

This system consists of a circle and a straight line. Solve the straight line for y.

$$x + y = 5$$
$$y = 5 - x$$

Substitute in the circle

$$x^2 + y^2 = 17$$
$$x^2 + (5 - x)^2 = 17$$
$$x^2 + 25 - 10x + x^2 = 17$$
$$2x^2 - 10x + 8 = 0$$
$$x^2 - 5x + 4 = 0$$
$$(x - 4)(x - 1) = 0$$
$$x = 1 \qquad \text{or} \qquad x = 4$$

Substituting these values in the equation of the line gives

$$
\begin{array}{ll}
y = 5 - x \qquad \text{or} & y = 5 - x \\
\quad = 5 - 1 & \quad = 5 - 4 \\
y = 4 & y = 1 \\
(1, 4) & (4, 1)
\end{array}
$$

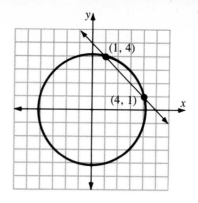

The line crosses the circle at $(1, 4)$ and $(4, 1)$. □

Example 14 □ Solve the system.

$$x^2 - 6x + y^2 + 4y - 3 = 0$$
$$y = x + 6$$

The line is already solved for y. Substitute it into the circle.

$$x^2 - 6x + y^2 + 4y - 3 = 0$$
$$x^2 - 6x + (x + 6)^2 + 4(x + 6) - 3 = 0$$
$$x^2 - 6x + x^2 + 12x + 36 + 4x + 24 - 3 = 0$$
$$2x^2 + 10x + 57 = 0$$

This is unfactorable.
By the quadratic formula

$$x = \frac{-10 \pm \sqrt{(10)^2 - 4(2)(57)}}{2(2)}$$

$$= \frac{-10 \pm \sqrt{100 - 456}}{4}$$

$$x = \frac{-10 \pm \sqrt{-356}}{4}$$

□

The discriminant is negative.

That means there is no real solution. The circle does not intersect the line.

Similar possibilities exist for systems of a line and an ellipse, hyperbola, or parabola.

Possible systems with a straight line:

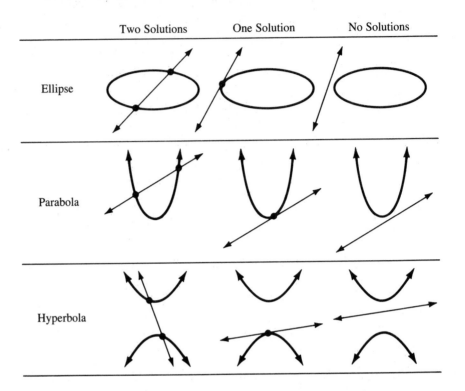

	Two Solutions	One Solution	No Solutions
Ellipse			
Parabola			
Hyperbola			

Can you have more than two solutions for a system with a hyperbola?

No, because the line would have to cross an asymptote more than once to intersect the curve more than twice.

Example 15 □ Solve the system.

$$y = x^2 + 2x - 8$$
$$y - 2x + 8 = 0$$

Solve the second equation for y.

$$y - 2x + 8 = 0$$
$$y = 2x - 8$$

Substitute in the first

$$y = x^2 + 2x - 8$$

$$2x - 8 = x^2 + 2x - 8$$

$$0 = x^2$$

$$0 = x$$

Substituting $x = 0$ in the second equation to find y

$$y = 2x - 8$$

$$y = \underline{\hspace{1.5cm}}$$

−8

 Figure 12.30

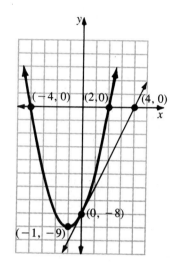

The solution is $(0, -8)$.

Only one
solution?

Yes. Notice
the graph.

□

Problem Set 12.3B

Using algebraic methods, find the points of intersection of the following systems of equations. A quick sketch will help you determine the number of solutions for each system. The sketch will not give you the exact values of all solutions.

1. $x^2 + y^2 = 25$
    $x + y = 7$

2. $y = x^2$
    $x - y + 2 = 0$

3. $3x^2 + 4y^2 = 19$
    $2x - y = 0$

4. $y^2 - x^2 = 4$
    $x + y = 1$

5. $y = x^2 + 2x$
    $y = -2x + 12$

6. $x^2 + y^2 = 36$
    $x - y = 6$

12.3 ■ Identifying Second-Degree Equations

7. $4x^2 + 9y^2 = 25$
$x - 2y = 0$

8. $x^2 + 4y^2 = 37$
$3x - y = 0$

9. $9x^2 - y^2 = 35$
$x - 2y = 0$

10. $x^2 + y^2 = 20$
$y - x = 2$

11. $y = x^2 + 2$
$x + y = 4$

12. $4y^2 - x^2 = 39$
$4x - 5y = 0$

13. $x^2 + y^2 = 4$
$4y + 5x - 20 = 0$

14. $y = x^2 - 4x - 5$
$2x + y + 6 = 0$

15. $xy = 12$
$y = x + 1$

16. $xy = -10$
$x - y + 7 = 0$

17. $4y^2 - x^2 = 20$
$x - 3y = 5$

18. $x^2 + y^2 - 4x - 8y + 20 = 0$
$3x - y = 12$

19. $2x^2 + y^2 = 6$
$y - x = 1$

20. $x^2 + 4y^2 = 8$
$y - 2x = 5$

21. $y^2 - 3x^2 = 1$
$2x - y = 0$

22. $4x^2 - y^2 = 7$
$3x - 2y = 0$

23. $y = x^2 + 3x - 5$
$4x - y = 3$

24. $y = x^2 - 2x + 2$
$y - x = 2$

25. $x^2 + y^2 = 13$
$x - y = -1$

26. $x^2 + y^2 = 17$
$x + y = 3$

27. $xy = 7$
$x - y = 6$

28. $xy = 4$
$x - y = -3$

29. $4x^2 + y^2 = 4$
$x + y = 4$

30. $4x^2 + y^2 = 17$
$x = 2y$

31. $4x^2 - 4y^2 = 9$
$4x - 5y = 0$

32. $x^2 - 9y^2 = 9$
$y = 2x + 1$

33. $y = x^2 - 6x + 8$
$y = -2x + 8$

34. $y = x^2 + 4x - 3$
$4x - y = 2$

35. $xy = -6$
$x - y = 5$

36. $xy = -8$
$x - y = 6$

■ 12.4 Solving Systems of Two Second-Degree Equations

A system of two second-degree equations can have up to four possible solutions. The following are graphs of some of the possibilities.

■ Figure 12.31

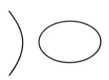
No solution

2 solutions

3 solutions

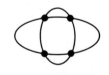
4 solutions

One solution

Chapter 12 ■ Second-Degree Equations, Ellipses, and Hyperbolas

Example 16 □ Solve the system.

$$x^2 + y^2 = 5$$
$$2x^2 + y^2 = 6$$

Multiply the first equation by -1 and use addition to eliminate y^2.

$$
\begin{aligned}
x^2 + y^2 &= 5 &\longrightarrow\quad -x^2 - y^2 &= -5 \\
2x^2 + y^2 &= 6 &\longrightarrow\quad \underline{2x^2 + y^2} &= \ \ \underline{6} \\
& & x^2\ \ \ \ \ \ &= 1 \\
& & x\ \ \ \ \ \ \ \ &= \pm 1
\end{aligned}
$$

Substituting into the first equation

for $x = +1$	for $x = -1$
$x^2 + y^2 = 5$	$x^2 + y^2 = 5$
$(+1)^2 + y^2 = 5$	$(-1)^2 + y^2 = 5$
$y^2 = 4$	$y^2 = 4$
$y = \pm 2$	$y = \pm 2$

The solutions are $(1, 2),\ (1, -2),\ (-1, 2),\ (-1, -2)$.

■ Figure 12.32

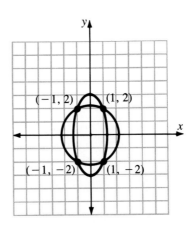

Example 17 □ Solve the system.

$$x^2 = 8y - 8$$
$$x^2 + 4y^2 = 4$$

The first equation will substitute directly into the second.

$$x^2 + 4y^2 = 4$$

becomes

$$8y - 8 + 4y^2 = 4$$
$$4y^2 + 8y - 12 = 0$$
$$y^2 + 2y - 3 = 0$$
$$(y + 3)(y - 1) = 0$$

Therefore, $y = -3$ or $y = +1$.

Substituting these values into the first equation

$$x^2 = 8y - 8 \qquad \text{and} \qquad x^2 = 8y - 8$$
$$x^2 = 8(-3) - 8 \qquad\qquad x^2 = 8(+1) - 8$$
$$x^2 = -24 - 8 \qquad\qquad x^2 = 0$$
$$x^2 = -32 \qquad\qquad x = 0$$

No real roots at $y = -3$.

$(0, 1)$ is a real root for the system.

■ Figure 12.33

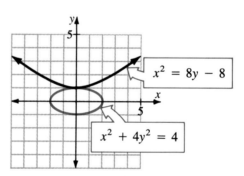

■ Problem Set 12.4A

When two second-degree equations are graphed, there may be zero through four points of intersection. Make a rough sketch similar to the ones before Example 16 in this chapter to indicate these possibilities for each of the following combinations.

1. Circle and Ellipse

2. Ellipse and Ellipse

3. Circle and Circle

4. Ellipse and Hyperbola

5. Circle and Hyperbola

6. Hyperbola and Hyperbola

7. Parabola and Ellipse

8. Parabola and Circle

9. Parabola and Hyperbola

10. Parabola and Parabola

Chapter 12 ■ Second-Degree Equations, Ellipses, and Hyperbolas

Example 18 □ Solve the system.

$$4x^2 + 25y^2 = 100$$
$$2x^2 - 9y^2 = 7$$

The easiest way to solve this system is by addition. Multiply the second equation by -2 and add.

$$4x^2 + 25y^2 = 100 \quad \longrightarrow \quad 4x^2 + 25y^2 = 100$$
$$2x^2 - 9y^2 = 7 \quad \longrightarrow \quad \underline{-4x^2 + 18y^2 = -14}$$
$$43y^2 = 86$$
$$y^2 = 2$$
$$y = \pm\sqrt{2}$$

Substitute the values for y into the second equation because it looks easier.

$$2x^2 - 9y^2 = 7$$
$$2x^2 - 9(+\sqrt{2})^2 = 7$$
$$2x^2 - 9 \cdot 2 = 7$$
$$2x^2 - 18 = 7$$
$$2x^2 = 25$$
$$x^2 = 12.5$$
$$x \approx \pm 3.536$$

That didn't work out evenly.

That's the way things happen in real life. Use your calculator.

The apparent solutions to the system, accurate to three places, are

$$(3.536, -1.414) \quad (3.536, +1.414)$$
$$(-3.536, -1.414) \quad (-3.536, +1.414)$$

Testing in the first equation

$$4x^2 + 25y^2 = 100$$
$$4(\sqrt{12.5})^2 + 25(\sqrt{2})^2 \stackrel{?}{=} 100$$
$$4(12.5) + 25(2) \stackrel{?}{=} 100$$
$$50 + 50 = 100$$

Why did you substitute $\sqrt{2}$ instead of 1.414?

Since I was going to square it anyway, $\sqrt{2}$ was less work and more accurate. Frequently, you can outthink a calculator.

Notice that we don't have to test all four cases because of the symmetry involved. Here is a graph of the system. □

■ Figure 12.34

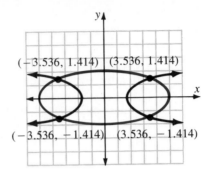

$(-3.536, 1.414)$ $(3.536, 1.414)$

$(-3.536, -1.414)$ $(3.536, -1.414)$

Example 19 □ Solve the system.

$$x^2 = 4y - 20$$
$$x^2 + 4y^2 = 4$$

Rewrite the system so we can eliminate the x^2 term by addition.

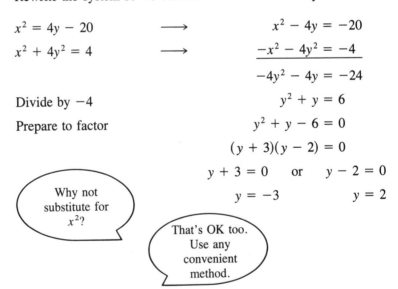

$x^2 = 4y - 20$ $\longrightarrow$ $x^2 - 4y = -20$

$x^2 + 4y^2 = 4$ $\longrightarrow$ $\underline{-x^2 - 4y^2 = -4}$

$-4y^2 - 4y = -24$

Divide by -4 $y^2 + y = 6$

Prepare to factor $y^2 + y - 6 = 0$

$(y + 3)(y - 2) = 0$

$y + 3 = 0$ or $y - 2 = 0$

$y = -3$ $y = 2$

Why not substitute for x^2?

That's OK too. Use any convenient method.

Substitute $y = -3$ and $y = 2$ in the first equation.

For $y = -3$ | For $y = 2$

$x^2 = 4y - 20$ | $x^2 = 4y - 20$

$x^2 = 4(-3) - 20$ | $x^2 = 4(2) - 20$

$x^2 = -12 - 20$ | $x^2 = 8 - 20$

$x^2 = -32$ | $x^2 = -12$

Chapter 12 ■ Second-Degree Equations, Ellipses, and Hyperbolas

Neither of these yields a real solution for x. A graph tells the story.

$\square$

■ Figure 12.35

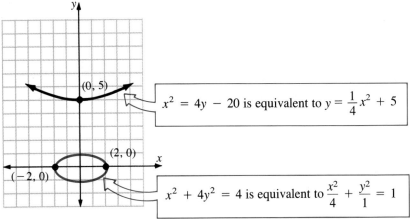

$x^2 = 4y - 20$ is equivalent to $y = \frac{1}{4}x^2 + 5$

$x^2 + 4y^2 = 4$ is equivalent to $\frac{x^2}{4} + \frac{y^2}{1} = 1$

A real solution for one variable is no guarantee of a real solution for the other.

A sketch also helps to visualize what is happening.

Problem Set 12.4B

Find the points of intersection for each of the following systems.

1. $x^2 + y^2 = 1$
$2x^2 + y^2 = 2$

2. $x^2 + y^2 = 8$
$3x^2 - y^2 = 8$

3. $x^2 + y^2 = 10$
$x^2 = -3y + 6$

4. $2x^2 + 3y^2 = 12$
$2x^2 - 3y^2 = 4$

5. $4x^2 + y^2 = 4$
$9x^2 - 4y^2 = 9$

6. $x^2 + 9y^2 = 36$
$x^2 + y^2 = 4$

7. $9x^2 + 4y^2 = 36$
$y = x^2 + 3$

8. $3x^2 + 2y^2 = 48$
$x^2 - y^2 = 16$

9. $x^2 + y^2 = 17$
$xy = 4$

10. $9y^2 - 4x^2 = 36$
$y = x^2 + 2$

11. $x^2 + y^2 = 9$
$4x^2 - y^2 = -4$

12. $4x^2 + y^2 = 21$
$y^2 = x^2 - 4$

13. $2x^2 + 3y^2 = 11$
$x^2 + 3y^2 = 7$

14. $xy = 6$
$2x^2 - y^2 = 14$

15. $x^2 + y^2 = 13$
$x^2 - y^2 = 5$

16. $x^2 + 2y^2 = 18$
$x^2 + y^2 = 9$

17. $y = 3x^2$
$y = 9x^2 - 2x - 4$

18. $y = 3x^2 + 2x$
$y = 2x^2 - 3x - 6$

19. $x^2 + y^2 = 3$
$3x^2 - y^2 = 1$

20. $x^2 + 4y^2 = 5$
$x^2 + y^2 = 2$

21. $x^2 + y^2 = 7$
$x^2 = 4y - 5$

22. $x^2 + y^2 = 4$
$x^2 + y^2 - 3x - 7 = 0$

23. $2x^2 + y^2 = 10$
$x^2 + y^2 = 6$

24. $3x^2 + y^2 = 8$
$2x^2 - y^2 = 2$

25. $4x^2 + 9y^2 = 36$
$x = y^2 + 3$

26. $3x^2 + y^2 = 15$
$y^2 = x^2 - 1$

27. $x^2 - y^2 = 1$
$y^2 = 2x + 2$

28. $xy = 4$
$x^2 + y^2 = 17$

29. $xy = -3$
$x^2 + 2y^2 = 9$

30. $xy = -6$
$y^2 = x^2 + 5$

31. $y = 4x^2 + 7x - 4$
$y = 2x^2$

32. $y = 4x^2 + 5x$
$y = 5x^2 + 6x - 6$

33. $xy = -12$
$x^2 - y^2 = 7$

34. $x^2 - 3y^2 = -9$
$2x^2 + y^2 = 10$

35. $2x^2 + 3y^2 = 12$
$y = x^2 - 2$

36. $x^2 + 4y^2 = 4$
$x^2 - y^2 = 9$

Chapter 12   □ Key Ideas

12.1 **1. Equation of an ellipse with center at the origin** in standard form is
$$\frac{x^2}{a^2} + \frac{y^2}{b^2} = 1$$
  a. The major axis is along the x-axis if $a > b$.
  b. The major axis is along the y-axis if $a < b$.
  c. To sketch an ellipse with center at the origin
   (1) determine the intercepts
   (2) draw a smooth curve through these points.
 2. The equation of an ellipse with center at (h, k) in standard form is
$$\frac{(x - h)^2}{a^2} + \frac{(y - k)^2}{b^2} = 1$$
  a. The major axis is parallel to the x-axis if $a > b$.
  b. The major axis is parallel to the y-axis if $a < b$.
  c. To sketch an ellipse with center at (h, k)
   (1) use translation equations
$$x' = x - h \qquad \text{and} \qquad y' = y - k$$
   to form the equation
$$\frac{x'^2}{a^2} + \frac{y'^2}{b^2} = 1 \qquad \text{with the center at } (h, k)$$
   (2) Sketch the graph in the (x', y') system.

Chapter 12 ■ Second-Degree Equations, Ellipses, and Hyperbolas

12.2 **1. Equation of hyperbola in standard form with center at the origin** is

 a. $\dfrac{x^2}{a^2} - \dfrac{y^2}{b^2} = 1$ if major axis is along the x-axis.

 b. $\dfrac{y^2}{a^2} - \dfrac{x^2}{b^2} = 1$ if major axis is along the y-axis.

 2. To sketch a hyperbola with center at the center:
   a. Determine a and b;
   b. Construct a box (a rectangle) with sides through $(\pm a, 0)$ and $(0, \pm b)$;
   c. Draw diagonals of rectangle. The extensions of these lines are the asymptotes of the hyperbola;
   d. Draw the hyperbola with vertices at $(a, 0)$ and $(-a, 0)$ between the asymptotes.

 3. Equation of a hyperbola in standard form with center at (h, k) is

 a. $\dfrac{(x - h)^2}{a^2} - \dfrac{(y - k)^2}{b^2} = 1$ if major axis is parallel to the x-axis.

 b. $\dfrac{(y - k)^2}{a^2} - \dfrac{(x - h)^2}{b^2} = 1$ if major axis is parallel to the y-axis.

 4. To sketch the hyperbola with center at (h, k)
   a. Use the translation equations
   $$x' = x - h \qquad \text{and} \qquad y' = y - k$$
   to form the equation
   $$\dfrac{x'^2}{a^2} - \dfrac{y'^2}{b^2} = 1 \qquad \text{or} \qquad \dfrac{y'^2}{a^2} - \dfrac{x'^2}{b^2} = 1$$
   with center at (h, k);
   b. Sketch the graph in the (x', y') system.

12.3 **1.** The **general second-degree equation** is
   $$Ax^2 + Bxy + Cy^2 + Dx + Ey + F = 0$$
   a. If $B = 0$
     $$Ax^2 + Cy^2 + Dx + Ey + F = 0 \qquad \text{is}$$
     1. a circle if $A = C$
     2. an ellipse if $A \neq C$ and $A, C > 0$
     3. a hyperbola if A and C have opposite signs
     4. a parabola if $A = 0$ and $C \neq 0$ or $A \neq 0$ and $C = 0$
     5. a straight line if $A, C = 0$.
   b. If $A, C, D, E = 0$, we have
     $$Bxy + F = 0$$
     The graph of this equation is a hyperbola with
     1. center at the origin
     2. x-axis and y-axis as the asymptotes.
 2. Systems of equations with one second-degree equation
   One equation is a line and the other one a second-degree equation.
   a. The graphs of a line and a second degree equation may have zero to two points of intersection resulting in 0, 1, or 2 solutions.
   b. Systems are solved by the methods of addition or substitution.

3. Systems of two second-degree equations

Both equations are second-degree equations.

a. The graphs of two second-degree equations may have zero to four points of intersection resulting in 0, 1, 2, 3, or 4 solutions.

b. Systems are solved by the methods of addition or substitution.

Chapter 12   Review Test

Sketch the graphs. **(12.1)**

1. $\dfrac{x^2}{16} + \dfrac{y^2}{1} = 1$

2. $9x^2 + y^2 = 36$

3. $4x^2 + 5y^2 = 80$

4. $\dfrac{(x+3)^2}{25} + \dfrac{(y-2)^2}{36} = 1$

5. $4x^2 + 9y^2 - 32x + 18y + 37 \geq 0$

6. $16x^2 + 9y^2 + 64x + 90y + 145 = 0$

Sketch the graphs. **(12.2)**

7. $\dfrac{x^2}{25} - \dfrac{y^2}{16} = 1$

8. $9y^2 - 4x^2 = 36$

9. $x^2 - 2y^2 = 8$

10. $4y^2 - x^2 - 2x - 8y - 33 = 0$

11. $\dfrac{(x+5)^2}{20} - \dfrac{(y-3)^2}{25} = 1$

12. $16x^2 - 9y^2 - 64x + 36y - 116 < 0$

Identify the following graphs. Do not sketch. **(12.3)**

13. $6x^2 + 9x + 6y^2 - 8y - 5 = 0$

14. $2x^2 + 4y^2 - 6x - 8y - 6 = 0$

15. $5x^2 + 10x - 4y^2 + 8y - 15 = 0$

16. $x^2 - 2x + 3y + 8 = 0$

Find the solution for each of the following. **(12.3)**

17. $x^2 + y^2 = 40$
$x + y = 4$

18. $2x^2 + 4y^2 = 12$
$2x - y = 3$

19. $2x^2 - y^2 = 1$
$x - y = 2$

20. $y = 4x^2 - 3x - 5$
$y = 5x + 7$

Find the solution for each of the following. **(12.4)**

21. $x^2 + y^2 = 48$
$x^2 - y^2 = 24$

22. $7x^2 - 2y^2 = 24$
$x^2 + 4y^2 = 12$

23. $x^2 + y^2 = 13$
$xy = 6$

24. $5x^2 + y^2 = 29$
$3y = x^2 + 5$

13 Functions, Inverses, and Logarithms

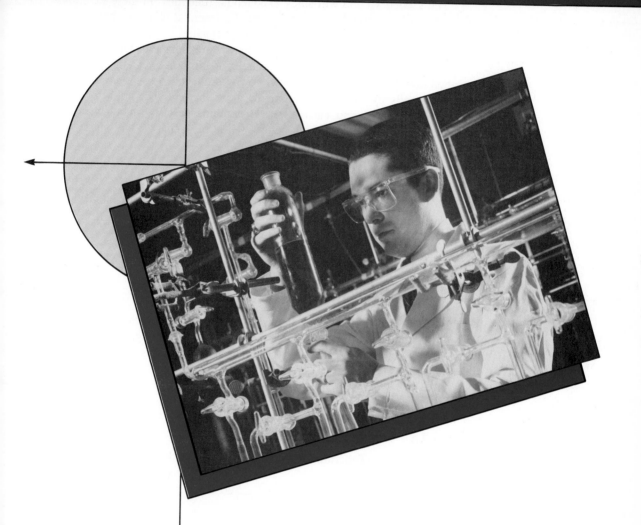

Contents

Preview

So far in this text in expressions that have had a variable raised to a power, the power has been a constant. This chapter will explore some of the possibilities when the exponent is a variable.

The first section talks about relations and functions. This section is concerned with defining the valid inputs to a relation and the possible outputs that can come from those inputs.

The second section will study exponential functions. In these functions the input is to a variable in the exponent. Exponential functions have many applications which we will study in the next chapter.

Logarithmic functions are closely related to exponential functions. Before calculators were commonly available, logarithms were the principal method of calculation for roots, powers, and products. Now, logarithms are used internally by calculators to provide these answers. The utility of logarithms goes far beyond calculation. Logarithms describe many chemical reactions and biological growth patterns. The next chapter will show you these and other applications of logarithms.

■ 13.1　Relations and Functions

Some quantities are related so directly that if you are given the value of one variable, you can supply the value of the second. The cost of gasoline at $1.50 per gallon is an example. If I tell you the number of gallons, you can tell me the cost.

13.1A　Definition: Function

y is said to be a **function** of x if for each value of x there is one and only one value for y.

The definition above only says that for each value of x there is a unique value of y. It does not tell you how to find the value of y.

One way to tell the value of y to be paired with each permissible value of x is to use a set of ordered pairs.

Ordered Pair

$(3, 5)$ is called an **ordered pair.** It is a pair because there are two numbers inside the parentheses. It is an ordered pair because $(3, 5)$ is different from $(5, 3)$. The order of the numbers is important since $(3, 5) \neq (5, 3)$.

Components

In the ordered pair $(3, 5)$, the numbers 3 and 5 are called **components** of the ordered pair; 3 is the first component and 5 is the second component.

Chapter 13 ■ Functions, Inverses, and Logarithms

Example 1 □ Ordered pairs can be used to show the price of gasoline.

Gallons	Cost	Ordered pair
1	1.50	(1, 1.50)
2	3.00	(2, 3.00)
5	7.50	(5, 7.50)
10	15.00	(10, 15.00)
20	30.00	(20, 30.00)

Functional Notation

If you were running a gas station, you would find the set of ordered pairs in Example 1 very incomplete. Another way to indicate the cost of gasoline is with **functional notation.**

$$P = f(n)$$

Read $P = f(n)$ as "P equals f of n."

The notation $P = f(n)$ says that the cost of gasoline represented by the variable P depends upon the number of gallons represented by the variable n. Exactly how the cost of filling your tank depends upon the number of gallons is explained by the equation

$$f(n) = 1.5 \cdot n$$

This function can be evaluated for each permissible replacement of n. $f(3)$ means the value of $f(n)$ when $n = 3$. □

Example 2 □ Find $f(3)$ if $f(n) = 1.50\ n$.

$$f(n) = 1.50 \cdot n$$
$$f(3) = 1.50 \cdot 3$$
$$= 4.50$$

The expression $f(x)$ is read "f of x." This expression in functional notation indicates that the value of some other quantity depends upon the replacement value selected for x. Suppose that the letter y is chosen to represent the quantity that is functionally related to x. Then we can write

$$y = f(x)$$

This expression tells us only that y is related to x. If we wish to specify exactly how y is related to x, we must specify $f(x)$. One possible way to specify $f(x)$ is with an equation like:

$$f(x) = 3x - 1$$

This is equivalent to saying $y = 3x - 1$. The function can be evaluated for each permissible replacement of x. $f(5)$ means the value of $f(x)$ when $x = 5$. We frequently write $y = f(x)$. This simply says that the value that depends upon x is called y. We may replace $f(x)$ with y.

The equation

$$f(x) = 3x - 1$$

now reads

$$y = 3x - 1$$

The value of y still depends upon the value used to replace x. □

Example 3 □ The value of y that is given by $f(x) = 3x - 1$ is shown for several values of x in the table below.

x	$f(x)$	$= 3x$	$- 1 = y$
1	$f(1)$	$= 3(1)$	$- 1 = 2$
2	$f(2)$	$= 3(2)$	$- 1 = 5$
0	$f(0)$	$= 3(0)$	$- 1 = -1$
-2	$f(-2)$	$= 3(-2)$	$- 1 = -7$
5	$f(5)$	$= 3(5)$	$- 1 = 14$

Value of x Resulting value of $f(x)$ or y

The table shows how each value of $y = f(x)$ is calculated for the various replacements for x. The important idea, however, is that for each x value that goes into the function exactly one $f(x) = y$ value comes out. We can use ordered pairs to express this information in shorter form.

From the table above

	Function	Ordered Pair
x	$f(x) = y$	(x, y)
1	2	$(1, 2)$
2	5	$(2, 5)$
0	-1	$(0, -1)$
-2	-7	$(-2, -7)$
5	14	$(5, 14)$

Value of x Resulting value of $f(x)$ or y

Chapter 13 ■ Functions, Inverses, and Logarithms

If $f(x) = 3x - 1$ is just another way to write $y = 3x - 1$, why bother?

Because $f(x)$ notation draws attention to the fact that a change in the value of x will cause a corresponding change in the value of y.

If the value of a quantity depends upon the value of a it might be expressed in this way

$$g(a) = 7a - 5$$

This is read "g of a equals seven a minus five." □

Example 4 □ Complete the following table for $g(a) = 7a - 5$.

a	$g(a)$	$= 7a$	$- 5$	
1	$g(1)$	$= 7(1)$	$- 5 = 2$	
3	$g(3)$	$= 7(3)$	$- 5 = 16$	
-3	$g(\_)$	$= 7(\_)$	$- 5 = -26$	$-3, -3$
0	$g(\_)$	$= 7(\_)$	$- 5 = -5$	$0, 0$
5	$g(\_)$	$= 7(\_)$	$- 5 = 30$	$5, 5$

□

A variable that may assume any one of a set of values is referred to as an **independent variable.** The quantity that changes as a result of a change in the independent variable is called the **dependent variable.** The value of the dependent variable depends upon the value of the independent variable. How the value of the dependent variable depends upon the value of the independent variable is specified by the equation.

In functional notation the independent variable is enclosed in parentheses.

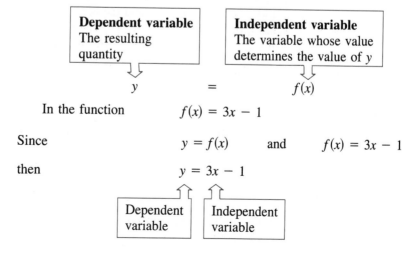

Dependent variable
The resulting quantity
y

Independent variable
The variable whose value determines the value of y
$f(x)$

In the function $f(x) = 3x - 1$

Since $y = f(x)$ and $f(x) = 3x - 1$

then $y = 3x - 1$

Dependent variable   Independent variable

When ordered pairs are used to express a function, the independent variable is always written first.

If $y = f(x)$, then $(x, f(x))$ and (x, y) are ways of identifying ordered pairs whose second components are determined by the choice of the first component in the ordered pair. Then $q = p(a)$ is a way to say the value of q depends in a predictable way upon the choice of a.

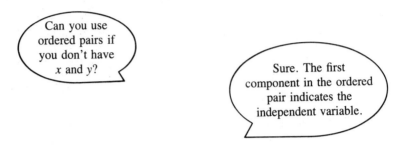

These three notations all express the same relationship.

$$q = p(a) \quad \longleftrightarrow \quad (a, p(a)) \quad \longleftrightarrow \quad (a, q)$$

The idea of functions that we have been developing started as early as 1694 with Leibniz who is best known as a contributor to the development of calculus. Lejeune Dirichlet (1805–1859) offered the definition of functions we have utilized. He said:

A variable is a symbol which represents any one of a set of numbers; if two variables x and y are so related that whenever a value is assigned to x, there is automatically assigned, by some rule or correspondence, a value to y, then we say y is a (single-valued) function of x. The variable x, to which values are assigned at will, is called the **independent variable,** and the variable y, whose values depend upon those of x, is called the **dependent variable.** The permissible values that x may assume constitute the domain of definition of the function, and the values taken on by y constitute the range of values of the function.*

Notice that the Dirichlet formulation of a function doesn't give any indication of how to find the value of y for a given value of x.

A classical way of indicating how one variable changes as a result of a change in another is a graph. On a graph we usually use the horizontal axis for the independent variable and the vertical axis for the dependent variable.

---

* Howard Eves, *An Introduction to the History of Mathematics* (New York: Holt, Rinehart, and Winston, 1976), p. 461.

Chapter 13 ■ Functions, Inverses, and Logarithms

Example 5 □ Without indicating the scale, draw a graph showing the following functional relationships.

    a. Cost of gasoline as a function of the number of gallons purchased.

■ Figure 13.1

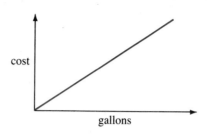

    b. Height of your foot above ground as you pedal a bicycle.

■ Figure 13.2

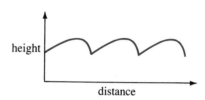

    c. Temperature of an oven as it warms up, cooks with the thermostat going on and off, then cools off.

■ Figure 13.3

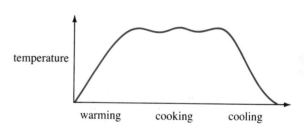

□

Problem Set 13.1A

Given $f(x) = 3x - 2$, find the following.

  1. $f(2)$ **2.** $f(3)$ **3.** $f(0)$ **4.** $f(5)$

Given $g(x) = -4x - 6$, find the following.

  5. $g(-3)$ **6.** $g(-1)$ **7.** $g(0)$ **8.** $g(2)$

Given $h(x) = \dfrac{3}{4}x - 3$, find the following.

  9. $h(-8)$ **10.** $h(-4)$ **11.** $h(4)$ **12.** $h(0)$

Given $p(x) = -\frac{2}{3}x + 5$, find the following.

13. $p(-6)$ **14.** $p(0)$ **15.** $p(-3)$ **16.** $p(3)$

Given $f(x) = 3 - \frac{1}{2}x$, find the following.

17. $f(2)$ **18.** $f(0)$ **19.** $f(-6)$ **20.** $f(10)$

Given $t(x) = 8 + \frac{2}{3}x$, find the following.

21. $t(3)$ **22.** $t(0)$ **23.** $t(-3)$ **24.** $t(4)$

Given $v(x) = -6 + \frac{3}{5}x$, find the following.

25. $v(5)$ **26.** $v(-10)$ **27.** $v(0)$ **28.** $v(4)$

Given $g(x) = -4 - \frac{3}{4}x$, find the following.

29. $g(-8)$ **30.** $g(4)$ **31.** $g(6)$ **32.** $g(-2)$

Given $f(x) = 3x + 5$, find the following.

33. $f(a)$ **34.** $f(a + h)$ **35.** $f(a + h) - f(a)$ **36.** $\dfrac{f(a + h) - f(a)}{h}$

Given $f(x) = \frac{4}{3}x + 7$, find the following.

37. $f(a)$ **38.** $f(a + h)$ **39.** $f(a + h) - f(a)$ **40.** $\dfrac{f(a + h) - f(a)}{h}$

Complete the following tables.

41.

x	$f(x) = 4x + 2 = y$	(x, y)
3	$f(3) = 4(3) + 2 = 14$	$(3, 14)$
-4		
0		
-6		
2		
1		

42.

x	$f(x) = -2x + 5 = y$	(x, y)
-3		
4		
2		
0		
-1		
6		

43.

x	$g(x) = \frac{1}{2}x - 4 = y$	(x, y)
-6		
-4		
0		
2		
8		

44.

x	$g(x) = -\frac{2}{5}x + 6 = y$	(x, y)
-10		
-5		
0		
5		
6		

Without indicating the scale, draw a graph showing the following functional relationships.

45. The more long-playing records you buy, the more it costs.

46. The height of your head from the ground as you ride a ferris wheel is a function of the time since you got on.

47. The larger the diameter of the pan is, the more pizza you get.

48. The amount of money in your savings account when the interest becomes a part of the principal is a function of how long you leave your money in the account.

49. The chances of acquiring cancer of the lungs is a function of the number of cigarettes smoked during a day.

50. The grade received in a course is a function of the time spent studying for the course.

51. The height of the valve stem on a tire of a car changes as the car moves forward.

52. The volume of a right circular cylinder is a function of the radius if the height remains constant.

The key idea is that a function implies predictability.
$y = f(x)$ means that the value of y depends in some predictable way upon x. The notation $y = f(x)$ does not tell us how to determine y given the value of x. It only says that y can be determined.

Sometimes we define a function as a set of ordered pairs with no two first components alike.

Why can't the two first components be equal?

That would destroy predictability. If $(2, 3)$ and $(2, 4)$ were both allowed in a function, then when I said 2 was the first component of an ordered pair you couldn't tell the second component.

Vertical Line Test

■ Figure 13.4

Another test for a function is the **vertical line test.** Any vertical line will cross the graph of a function one time at most.

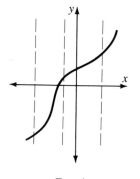

Function

Not a function

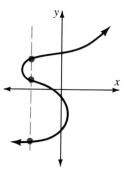

Not a function

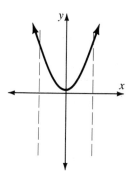

Function

13.1 ■ Relations and Functions

There are obviously many important equations that are not functions. Functions are a subset of relations.

13.1B Definition: Relation

A relation is any set of ordered pairs.

13.1C Domain and Range of a Relation

The **domain** of a relation is the set of possible replacements for the first component of the ordered pairs of the relation.

The **range** of a relation is the set of possible replacements for the second component of the ordered pairs of the relation.

13.1D Definition: Function

A function is a relation in which each first component has a unique second component. Note: This definition is equivalent to Definition 13.1A.

All functions are relations, but not all relations are functions.

If the function is defined as a set of **ordered pairs,** the first component in each ordered pair is a member of the domain, and the second component is a member of the range.

Example 6 □ List the elements of the domain and range of the function $\{(1, 2), (3, 4), (-5, -6), (-4, -6), (-3, -6)\}$.

Domain is $\{1, 3, -3, -4, -5\}$
Range is $\{2, 4, -6\}$

The range has fewer elements than the domain.

That's okay; but notice that no elements of the domain are repeated.

□

In the example on the previous page there is only a finite number of ordered pairs in the function. The idea of a function is dependence, not quantity.

► **A Pointer About Graphs**

On a graph we generally use the horizontal axis for the independent variable or domain and we plot the range along the vertical axis.

When we use a smooth graph to represent a function we are saying that the coordinates of every one of the infinite number of ordered pairs on the curve is a member of the function.

Example 7 □ Examine the graph below. Notice there are no scales on the axis; however, we can tell several things about this function just from looking at the graph.

The arrows on the curve indicate that the graph continues without limit in the direction of the arrows.

Domain and Range on a Graph

■ Figure 13.5

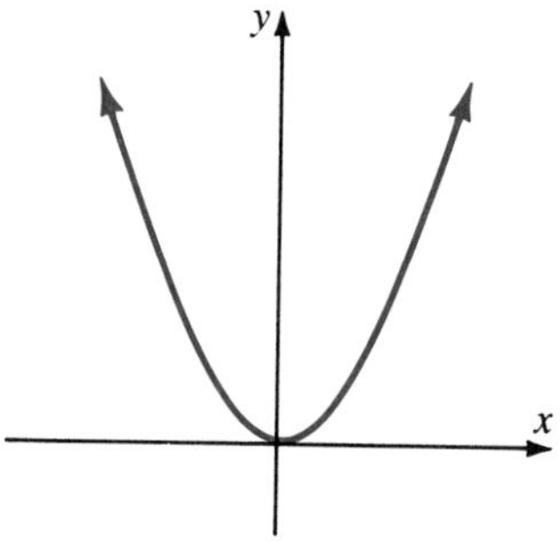

Yes, we just don't know exactly where.

Is there a point where the x-coordinate is 2? _____

Yes

Is there a point where the x-coordinate is −10? _____

Yes

Is there a point where the x-coordinate is −1000? _____

One

How many points have an x-coordinate of +1500? _____

Yes

Is there a point where the y-coordinate is +5? _____

2

How many points have a y-coordinate of +329? _____

0

How many points have a y-coordinate of −2? _____

There is no maximum x value.

What is the maximum x-coordinate? _____

All real numbers.

What is the domain of x? _____

The curve appears to touch the x-axis. Therefore the minimum y value is zero.

What is the minimum y-value? _____

There is no maximum y value.

What is the maximum y-value? _____

$y \geq 0$

What is the range of y? _____

□

13.1E   Definition: One-to-One Function

A **one-to-one function** is a function such that for each second component there is a unique first component.

In a one-to-one function, if you know an element of the range, it's possible to identify the corresponding element of the domain.

If a graph can pass both a vertical and horizontal line test, it is a one-to-one function.

■ Figure 13.6

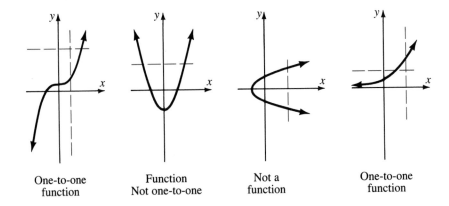

| One-to-one function | Function Not one-to-one | Not a function | One-to-one function |

Example 8 □ Give the domain and range of the relations below. If the relation is a function tell if it is one-to-one.

a. $\{(0, 1), (2, 3)\}$
Domain: $\{0, 2\}$
Range: $\{1, 3\}$
Function: yes
One-to-one: yes

0, −1, −2, −3, −4
Yes
Yes

b. $\{(0, 0), (1, -1), (2, -2), (3, -3), (4, -4)\}$
Domain: $\{0, 1, 2, 3, 4\}$
Range: $\{$————$\}$
Function: ————
One-to-one: ————

> Since each element in the domain is paired with exactly one element in the range, this is a function.
>
> Also, each element in the range is paired with only one element in the domain so this is a one-to-one function.

1, 2
0, 2, 3, 4
No
No

c. $\{(1, 0), (1, 2), (2, 3), (2, 4)\}$
Domain: $\{$————$\}$
Range: $\{$————$\}$
Function: ————
One-to-one: ————

d. $\{(1, 1), (2, 1), (3, 1), (-5, 2), (-6, 2)\}$
Domain: $\{$————$\}$
Range: $\{$————$\}$
Function: ————
One-to-one: ————   □

1, 2, 3, −5, −6
1, 2
Yes
No

Chapter 13 ■ Functions, Inverses, and Logarithms

Example 9 □ Give the domain and range for the following equations. Also tell if they represent one-to-one functions.

a. $y = x^2$

Domain: {all real numbers}

> How did you know that?

> The first component in each ordered pair is assumed to be from the domain. Unless specified otherwise, the domain of any relation is assumed to be all real numbers.

Range: {all reals ≥ 0}

> Why not all reals?

> When you square a negative you get a positive. This relation isn't defined for y values less than zero.

Yes, No

Function?_____ One-to-one?_____

b. $\dfrac{x^2}{4} + \dfrac{y^2}{25} = 1$

You need to recognize this as the equation of an ellipse centered at the origin.

Domain: $-2 \leq x \leq 2$

■ Figure 13.7

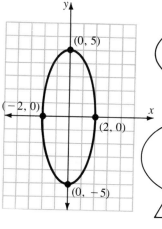

> Why not a domain of all real numbers?

> We consider all reals as possible replacements for x. But a quick sketch of the graph of this equation tells us it isn't defined for $x > 2$ or $x < -2$. Domain and range tell us where the equation is defined.

$-5 \leq y \leq 5$

Range:_____

No

Function:_____

No

One-to-one:_____

□

13.1 ■ Relations and Functions

Inverse of a Relation

If we interchange the components of each ordered pair of a relation we produce another relation called its **inverse**.

Example 10 □ Find the inverse relation of the relation below and graph it.

■ Figure 13.8

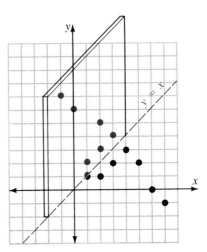

Relation	Inverse
$(1, 1)$	$(1, 1)$
$(2, 1)$	$(1, 2)$
$(2, 3)$	$(3, 2)$
$(4, 3)$	$(3, 4)$
$(5, 2)$	$(2, 5)$
$(6, 0)$	$(0, 6)$
$(7, -1)$	$(-1, 7)$

Notice that if you placed a mirror on the line $y = x$, the points of the inverse relation would be the reflections of the points of the original relation.

It may help you to see why the inverse of a relation is a reflection about the line $y = x$ if you consider just one point (P, Q).

Notice that reversing the order of the components of the ordered pair (P, Q) to produce (Q, P) places (Q, P) as far over as (P, Q) is up and as far up as (P, Q) is over.

■ Figure 13.9

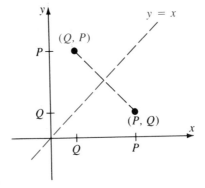

In general the inverse of any relation is a reflection of that relation across the line $y = x$. You find the inverse of a relation by interchanging the coordinates (x, y) of each point to get (y, x). Therefore, a way to write the equation of the inverse of a relation is to interchange x and y in the original equation.

Chapter 13 ■ Functions, Inverses, and Logarithms

Example 11 □ Write the inverse of the relation $y = 2x - 5$.
To write the inverse of a relation:

interchange y and x

relation $y = 2x - 5$

inverse $x = 2y - 5$

Now solve the inverse for y.

$$x = 2y - 5$$

$$x + 5 = 2y$$

$$\frac{x + 5}{2} = y$$

Inverse: $$y = \frac{x + 5}{2}$$

■ Figure 13.10

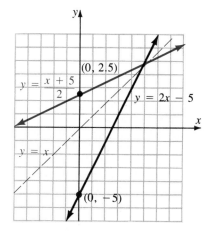

Which is the
inverse?

Taken together
each equation is the
inverse of the other
equation.

Inverse is a symmetric relationship. That is, if relation P is the inverse of relation R, then R is the inverse of P.

Like who is
the spouse in a
marriage.

You can't have
just one
spouse.

Example 12 □ Write the equation of the inverse of $y = x^3$.

Original equation: $y = x^3$
Inverse: $\qquad$ $x = y^3$
$\qquad\qquad$ $y = \sqrt[3]{x}$ $\qquad$ Solving for y

Yes

Is the original equation a function? _____

■ Figure 13.11

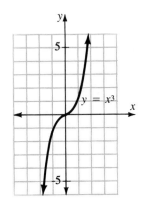

How do you tell by looking at an equation if it's a function?

Normally, you'll have to sketch the graph and apply the vertical line test.

$y = x^3$

□

Yes

Is the inverse a function? _____

■ Figure 13.12

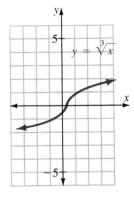

Is the inverse of a function always a function?

No. Look at the next example.

$y = \sqrt[3]{x}$

■ Figure 13.13

Example 13 □ Write the equation of the inverse of the relation $y = x^2 + 1$.

To write the equation of the inverse simply interchange x and y.

Relation: $y = x^2 + 1$
Inverse: $\ x = y^2 + 1$

$x - 1 = y^2$
$y = \pm\sqrt{x - 1}$

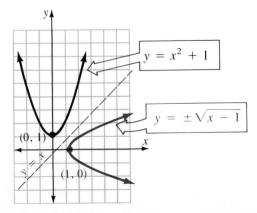

$y = x^2 + 1$

$y = \pm\sqrt{x - 1}$

$(0, 1)$

$(1, 0)$

$y = x$

Chapter 13 ■ Functions, Inverses, and Logarithms

Yes

Is the original equation $y = x^2 + 1$ a function? _____

■ Figure 13.14

Notice that $y = x^2 + 1$ passes a vertical line test.
For each value of x, there is only one value of y.

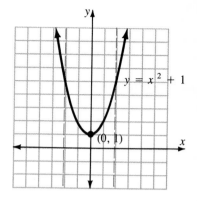

No

Is the inverse of $y = x^2 + 1$ a function? _____

■ Figure 13.15

Because for each value of x there are two values of y, $y = \pm\sqrt{x - 1}$ does not pass a vertical line test.

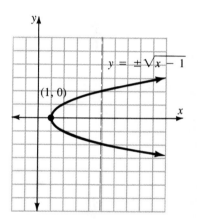

The inverse of a function may or may not be a function. The inverse of a one-to-one function, however, is always a function. □

Problem Set 13.1B

(a) List the elements of the domain and range of the following relations. (b) Determine if the relation is a function. (c) Determine if the relation is one-to-one.

1. $\{(1, 2), (3, -6), (5, -4), (-2, 1)\}$

2. $\{(-1, -1), (3, 2), (6, -1), (5, 4)\}$

3. $\{(2, -1), (5, 6), (4, 0), (-2, -2)\}$

4. $\{(-2, 1), (3, 4), (5, -2), (-2, 5), (0, 1)\}$

5. $\{(-3, -4), (-2, -3), (-1, -2)\}$

6. $\{(0, 1), (1, 2), (-1, 3), (0, -1), (5, -6)\}$

7. $\{(-2, 3), (-5, 6), (2, 3), (3, 4), (6, -2)\}$

8. $\{(-5, -6), (-4, -5), (0, 2), (-2, -3)\}$

9. $\{(-6, 4), (-2, 8), (0, 4), (2, 7), (8, 1)\}$

10. $\{(2, 6), (5, 4), (-2, 3), (2, 8), (-3, 6)\}$

11. $\{(5, 9), (3, 2), (0, 7), (-1, -4)\}$

12. $\{(-6, 1), (7, 2), (9, 0), (-5, 4), (-7, 2)\}$

13. $\{(3, 6), (0, 0), (5, 4), (-3, 2)\}$

14. $\{(8, -6), (7, -5), (0, 9)\}$

15. $\{(-1, 2), (-1, 3), (2, 4), (2, 8)\}$

16. $\{(7, 4), (-1, 2), (3, 5), (9, 0), (3, 2)\}$

(a) Determine if the graph is a function. (b) Determine if the relation is one-to-one.

17.
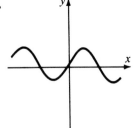

18.
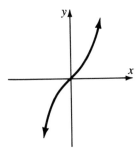

19.
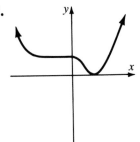

20.
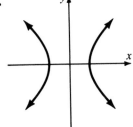

21.
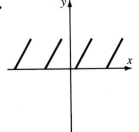

22.
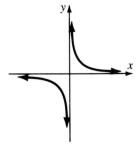

23.
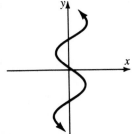

24.
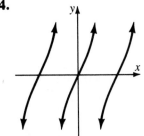

25.
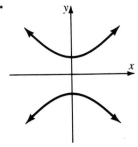

Chapter 13 ■ Functions, Inverses, and Logarithms

26.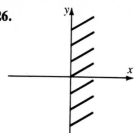

27.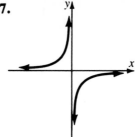

28.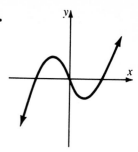

(a) Determine the domain and range of the following graphs. (b) Determine if the relation is one-to-one.

29.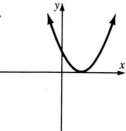

30.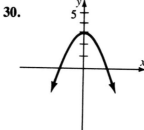

31.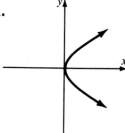

32.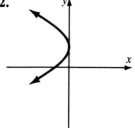

33.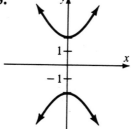

34.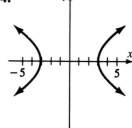

35.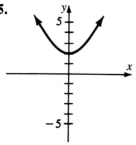

36.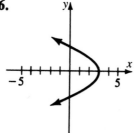

37.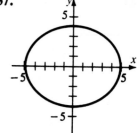

38.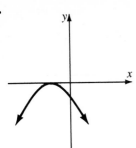

39.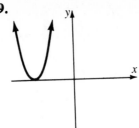

40.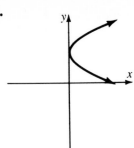

(a) Determine the domain and range of the following equations. (b) Determine if the equations represent one-to-one functions.

41. $y = x^2 + 2$

42. $y = x^3$

43. $\dfrac{x^2}{16} - \dfrac{y^2}{12} = 1$

44. $x^2 + y^2 = 36$

45. $x = y^2$

46. $x - 3y = 8$

47. $y^2 = x + 1$

48. $y = x^2 - 3$

49. $2x - 3y + 4 = 0$

50. $3x - 4y = 5$

51. $y = x^3 - 6$

52. $x = y^2 + 2$

(a) Write the inverse of the following equations. Solve for y in terms of x. (b) Determine if the original equation is a function. (c) Determine if the inverse is a function.

53. $2x + 3y = 8$

54. $y = x^2 + 3$

55. $y = x^3 - 3$

56. $x^2 + y^2 = 25$

57. $x^2 + 4y^2 = 1$

58. $x^2 - 9y^2 = 1$

59. $4x = 3y - 1$

60. $y = 4x - 6$

61. $x^2 = y + 4$

62. $y = x^2 - 8$

63. $y = x^3 + 6$

64. $x^3 = y - 5$

Write the inverse of the following equations. Sketch the graph of the original equation and its inverse on the same axis.

65. $y = 2x + 3$

66. $y^2 = x$

67. $y = x^2 + 2$

68. $3x + 2y = 6$

69. $4x - 6y = 7$

70. $5y - 6x = 2$

71. $x^2 = y - 3$

72. $y = x^2 + 4$

Exponential Functions

Equations with variables as exponents are referred to as **exponential equations.**

$$y = 2^x \qquad \text{is an exponential function.}$$

Let us begin by making a table of values for this function.

x	$y = 2^x$
0	1
1	2
2	4
3	8
$\dfrac{1}{2}$	1.414
-1	$\dfrac{1}{2}$
-2	$\dfrac{1}{4}$
-3	$\dfrac{1}{8}$

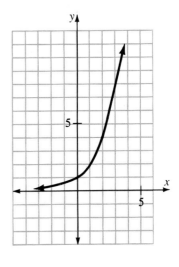

It is very tempting, and in fact possible, to draw a smooth curve connecting these points. However, you should be aware that when you draw the curve you are saying that 2^x is defined for all real numbers. Up to now, we have established that x can represent any rational exponent such as b^2, b^3, b^{-2}, $b^{\frac{1}{2}}$ or $b^{\frac{2}{3}}$. We have not dealt with irrational exponents, like $b^{\sqrt{2}}$ and $b^{\sqrt{3}}$. Proving that $b^{\sqrt{3}}$ exists and has a value that fits on a smooth curve between b^1 and b^2 is part of a calculus course. At this time we merely state that the laws of exponents are valid for all real numbers.

13.2A   Definition: Exponential Function

The general exponential function is

$$y = a^{kx}$$

where a, k are constants, $a > 0$ and $k \neq 0$.

Example 14 □ Graph the functions $y = 2^x$ and $y = 3^x$ on the same axis.
Setting up a table of values

x	2^x	3^x
0	1	1
1	2	3
2	4	9
$\dfrac{1}{2}$	1.414	1.732
-1	$\dfrac{1}{2}$	$\dfrac{1}{.3}$
-2	$\dfrac{1}{4}$	$\dfrac{1}{9}$

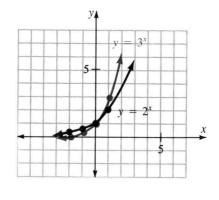

□

By studying Example 14 we can make several statements about exponential functions of the form $y = a^x$.

Observations about $y = a^x$ when a is greater than one:

Effect of $a > 1$ on the
exponential curve

a. In all cases the curve passes through $(0, 1)$ because $a^0 = 1$ for all $a \neq 0$.

b. If $a > 1$ the curve climbs steeply to the right. The steepness of the curve depends on the value of a. As a increases, the curve climbs more steeply.

c. For values of $x < 0$, the curve approaches the $-x$-axis asymptotically. The larger the value of a is, the faster the curve approaches the $-x$-axis.

Example 15 □ Graph $y = \left(\dfrac{1}{2}\right)^x$ and $y = \left(\dfrac{1}{3}\right)^x$ on the same axis.
Setting up a table of values

x	$\left(\dfrac{1}{2}\right)^x$	$\left(\dfrac{1}{3}\right)^x$
0	1	1
1	$\dfrac{1}{2}$	$\dfrac{1}{3}$
2	$\dfrac{1}{4}$	$\dfrac{1}{9}$
-1	2	3
-2	4	9

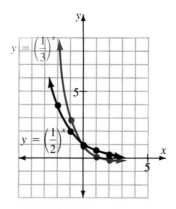

□

Observations about $y = a^x$ **when** a **is less than one but greater than zero:**

Effect of $0 < a < 1$ on exponential curve

a. The curve falls to the right.

b. For all values of a the curve still passes through $(0, 1)$.

c. The curve asymptotically approaches the $+x$-axis.

d. The smaller the value of a, the steeper the curve. ☐

Example 16 ☐ Graphs of typical exponential functions.

■ Figure 13.19

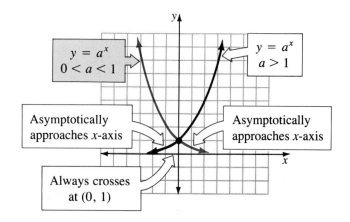

Exponential functions play a very important role in chemistry, physics, finance, biology, probability, and statistics. Some things that happen exponentially are

—Electrical buildup or decay in a transient circuit

—Decay in radioactivity

—Nuclear fission

—Growth of capital invested to earn compound interest

—Inflation

Functions where $a > 1$ are increasing exponential functions.

Functions where $a < 1$ are decreasing exponential functions.

Functions where $a = 1$ produce a straight line parallel to the x-axis because $y = 1^x$ is equivalent to $y = 1$.

Exponential functions grow very rapidly. The y^x key of your calculator can be used to evaluate exponentials.

Using Your Calculator

The y^x Key

Scientific calculators have a key that will raise positive values to any positive exponent.

Let's start with 2^3.

Press 2 $\boxed{y^x}$ 3 $\boxed{=}$

The display should be 8.

In this case y was 2 and $x = 3$.

The $\boxed{y^x}$ key also works for roots to find $(16)^{\frac{1}{2}}$.

Press 16 $\boxed{y^x}$ .5 $\boxed{=}$

The display shows 4.

To find a cube root, you must evaluate the exponent. There are two ways to evaluate the exponent.

1. Evaluate the exponent and store it in memory before you use the $\boxed{y^x}$ key. To find $\sqrt[3]{8}$ this way:

   Press: 1 $\boxed{\div}$ 3 $\boxed{=}$ $\boxed{STO}$ or $\boxed{x \to m}$.

   This places an approximation of $\frac{1}{3}$ in memory.

   Now press: 8 $\boxed{y^x}$ $\boxed{RCL}$ or $\boxed{RM}$ $\boxed{=}$.

   The display should be 2.0

2. Evaluate the exponent in parentheses. To find $\sqrt[3]{8}$ this way:

   Press: 8 $\boxed{y^x}$ $\boxed{(}$ 1 $\boxed{\div}$ 3 $\boxed{)}$ $\boxed{=}$.

   The display should be 2.0.

$\sqrt[3]{-8} = -2$?
Will the calculator find that?

Probably not. For reasons that can be explained after the next chapter, your calculator restricts the values of y and x on the $\boxed{y^x}$ key to positive values.

To find powers greater than one, use the same process.

Example 17 □ Find 3^4.

Press: 3 $\boxed{y^x}$ 4 $\boxed{=}$.

The display will show 81.

Find $2.08^{5.4}$.

Press: 2.08 $\boxed{y^x}$ 5.4 $\boxed{=}$.

The display will show 52.184558.   □

Example 18 □ Use your calculator to calculate each of the following.

$$2^{10} = 1024$$

59,049

$$3^{10} = \underline{\hspace{2cm}}$$

1,048,576

$$2^{20} = \underline{\hspace{2cm}}$$

Computers use powers of 2 internally. 1K memory means 1024 bytes.

□

Example 19 □ Compound interest on a savings account is usually paid quarterly. The interest is then credited to the account. The next quarter, interest is paid on the principal and on the previous interest. Let's examine an account with a starting balance of $1000. If the interest rate is 8% per year, the interest earned at the end of the first quarter is

$$I = PRT$$

$$I = (1000)(0.08)\left(\frac{1}{4}\right)$$

$$I = 20$$

Now depositing the interest in the account, the earnings for the second quarter are

$$I = PRT$$

$$I = (1000 + 20)(0.08)\frac{1}{4}$$

$$I = 20.40$$

□

Compound Interest Formula

A formula to express the balance in the account after n compounding periods is

$$B = P\left(1 + \frac{r}{n}\right)^{nt}$$

where B = ending balance

P = starting principal

r = annual interest rate

t = number of years

n = number of times per year interest compounded

Example 20 □ Use the interest formula to find the balance of a savings account where \$1000 was left at 8% interest, compounded quarterly for 9 years.

$$B = P\left(1 + \frac{r}{n}\right)^{nt}$$

$$B = 1000\left(1 + \frac{0.08}{4}\right)^{(4)(9)}$$

$$= 1000(1.02)^{36}$$

$$\approx 1000(2.03989)$$

$$\approx 2039.89 \qquad \qquad \Box$$

Exponential Equations

It is possible to solve some equations with variables as exponents by using the following property of equality.

13.2B Property of Equality for Exponentials

If $a^x = a^y$ then $x = y$ for all $a > 0$ and $a \neq 1$.

Example 21 □ Solve $2^x = 8$.
    Write 8 as a power of 2.

$$2^x = 8$$

$$2^x = 2^3 \qquad \text{Rewrite 8 as a power of 2}$$

$$x = 3 \qquad \text{Property 13.2B} \qquad \Box$$

Example 22 □ Solve $3^{2x} = 81$.
    When you apply property 13.2B, the bases on both sides of the equation must be equal.

$$3^{2x} = 81$$

$$3^{2x} = 3^4 \qquad \text{Rewrite 81 as a power of 3}$$

$$2x = 4 \qquad \text{Property 13.2B}$$

$$x = 2 \qquad \Box$$

Example 23 □ Solve $2^{3x-5} = \dfrac{1}{16}$.

Write $\dfrac{1}{16}$ as a power of 2.

$$2^{3x-5} = 2^{-4}$$
$$3x - 5 = -4 \qquad \text{Property 13.2B}$$
$$3x = 1$$
$$x = \dfrac{1}{3}$$

□

Example 24 □ Solve $9^x = 27^2$.

Rewrite both sides of this equation to get a common base.

$$9^x = 27^2$$
$$(3^2)^x = (3^3)^2$$
$$3^{2x} = 3^6 \qquad \text{Multiplication law of exponents}$$
$$2x = 6 \qquad \text{Property 13.2B}$$
$$x = 3$$

□

Problem Set 13.2

Make a table of values and graph each of the following exponential functions.

1. $y = 4^x$ **2.** $y = \left(\dfrac{1}{4}\right)^x$ **3.** $y = \left(\dfrac{3}{4}\right)^x$ **4.** $y = \left(\dfrac{2}{3}\right)^x$ **5.** $y = 2^{2x}$

6. $y = 2^{-x}$ **7.** $y = 5^x$ **8.** $y = 6^x$ **9.** $y = \left(\dfrac{2}{5}\right)^x$ **10.** $y = \left(\dfrac{3}{5}\right)^x$

11. $y = 3^{2x}$ **12.** $y = 3^{-x}$ **13.** $y = 2^x + 2$ **14.** $y = 3^x - 3$ **15.** $y = 2^{x-1}$

16. $y = 2^{x+1}$ **17.** $y = 2^{2x-1}$ **18.** $y = 3^{-x-1}$ **19.** $y = 2^x - 3$ **20.** $y = 2^x + 4$

21. $y = 2^{x-2}$ **22.** $y = 2^{x+2}$ **23.** $y = 3^{-x+2}$ **24.** $y = 2^{2x+1}$

Solve each of the following equations.

25. $2^x = 16$ **26.** $3^x = 27$ **27.** $2^{2x} = 64$ **28.** $2^{2x+1} = 32$

29. $3^{3x-1} = 81$ **30.** $2^{4x-3} = \dfrac{1}{32}$ **31.** $3^{2x+5} = \dfrac{1}{27}$ **32.** $4^x = 8$

13.2 ■ Exponential Functions

33. $9^{\frac{x}{2}} = \dfrac{1}{81}$

34. $3^{2x+5} + 5 = 32$

35. $2^{3x-4} - 3 = 29$

36. $2^{3x+5} = \dfrac{1}{16}$

37. $3^{x+2} = 27$

38. $3^{x-4} = 81$

39. $8^x = 16$

40. $9^x = 27$

41. $4^{3x+1} = 64$

42. $8^{2x-1} = 16$

43. $8^{\frac{x}{3}} = \dfrac{1}{16}$

44. $27^{\frac{x}{3}} = \dfrac{1}{9}$

45. $2^{3x-4} - 4 = 28$

46. $2^{2x+3} + 4 = 36$

47. $3^{2x-1} + 3 = 30$

48. $3^{3x-2} + 4 = 85$

49. $\left(\dfrac{1}{3}\right)^x = \dfrac{1}{81}$

50. $\left(\dfrac{1}{2}\right)^{2x} = \dfrac{1}{64}$

51. $\left(\dfrac{2}{3}\right)^{3x+1} = \dfrac{16}{81}$

52. $\left(\dfrac{1}{4}\right)^x = \dfrac{1}{32}$

53. $\left(\dfrac{4}{9}\right)^x = \dfrac{8}{27}$

54. $\left(\dfrac{3}{4}\right)^{3x+1} = \dfrac{81}{256}$

55. $\left(\dfrac{1}{2}\right)^{2x+1} = \dfrac{1}{32}$

56. $\left(\dfrac{1}{3}\right)^{2x-3} = \dfrac{1}{27}$

57. $\left(\dfrac{3}{4}\right)^{2x-1} = \dfrac{27}{64}$

58. $\left(\dfrac{2}{3}\right)^{3x-2} = \dfrac{16}{81}$

59. $\left(\dfrac{2}{3}\right)^{2x+3} = \dfrac{27}{8}$

60. $\left(\dfrac{3}{4}\right)^{3x+4} = \dfrac{16}{9}$

Use the formula $B = P\left(1 + \dfrac{r}{n}\right)^{nt}$ to solve the following problems. Find the balance (B) for each of the following investments.

61. $800 for 5 years at 14% compounded semi-annually.

62. $1250 for 6 years at 12% compounded semi-annually.

63. $1800 for 3 years at 16% compounded quarterly.

64. $1500 for 4 years at 12% compounded quarterly.

65. $2500 for 3 years at 15% compounded monthly.

66. $4000 for 6 years at 18% compounded monthly.

67. $1600 for 6 years at 8% compounded annually.

68. $2250 for 4 years at 9% compounded annually.

69. $3200 for 5 years at 7% compounded quarterly.

70. $3600 for 3 years at 10% compounded quarterly.

71. $2750 for 2 years at 8% compounded monthly.

72. $3500 for 5 years at 16% compounded monthly.

■ 13.3   **Logarithmic Functions**

We have been dealing with exponential equations of the form

$$y = a^x$$

Since this equation represents a one-to-one function, if we interchange x and y, we get the equation of the inverse function.

$x = a^y$ is the inverse of $y = a^x$. One way to read the inverse is to say y is the exponent on a to produce x or shorter yet

$$y = \text{Exponent}_a(x)$$

Instead of using the word **exponent,** for historical reasons we use the word **log,** which is short for logarithm.

$$y = \log_a(x)$$

13.3A   Definition: Logarithm

$y = \log_a(x)$ is the logarithmic function. It is read "y is equal to the log base a of x." It means that y is the exponent on a that will produce x.

$y = \log_a(x)$ is equivalent to $a^y = x$.

The most important thing to notice is that logarithms are exponents, and as such they obey all the laws of exponents.

Let's graph $y = \log_a(x)$.

a is a constant, so for our first graph we choose $a = 2$ and graph $y = \log_2(x)$ in the next example.

Example 25 □ Graph $y = \log_2(x)$.

This is equivalent to graphing $2^y = x$. It is inconvenient to build a table of values for $2^y = x$ by selecting values for x and calculating y. Therefore make a table of values for its inverse function $y = 2^x$, and interchange the x and y entries of that table to produce a table for $2^y = x$.

■ Figure 13.20

Inverse		Function	
$y = 2^x$		$x = 2^y$	
Exponential		Logarithm: $y = \log_2 x$	
x	y	x	y
0	1	1	0
1	2	2	1
2	4	4	2
3	8	8	3
-1	$\dfrac{1}{2}$	$\dfrac{1}{2}$	-1
-2	$\dfrac{1}{4}$	$\dfrac{1}{4}$	-2
-3	$\dfrac{1}{8}$	$\dfrac{1}{8}$	-3
-5	$\dfrac{1}{32}$	$\dfrac{1}{32}$	-5

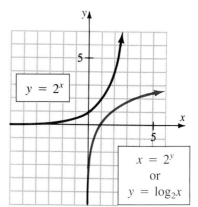

□

Things to Notice About This Curve:

$y = \log_2(x)$ and $y = 2^x$ are inverse functions.

$\log_2(x)$ is only defined for positive x.

For values of $0 < x < 1$, $\log_2(x)$ is negative.

For $x > 1$, $\log_2(x)$ is positive.

As x becomes a very large number, the value of $y = \log_2(x)$ increases at a much smaller rate. However, there is no upper bound on the value of $\log_2(x)$; that is, $\log_2 x$ will eventually get to any value you name. As the value of x approaches zero, the value of $\log_2(x)$ becomes a very large negative number.

There is nothing magical about using 2 as a base for logarithms. Any number greater than one will produce the same shape curve. Later we will use bases of 10 and $e \approx 2.71828$. First, let's examine the properties of logarithms using a base of 2. Here the calculations are easy.

Chapter 13 ■ Functions, Inverses, and Logarithms

Changing from Exponential to Logarithmic Form

Since a logarithm is just another way to write an exponential function, we expect that we can readily convert from one form of notation to the other.

Example 26 □ Change $8 = 2^3$ to logarithmic form.
$8 = 2^3$ is equivalent to $\log_2(8) = 3$.

Notice these are equivalent equations.

Right, they are the same statement written in two different notations.

□

Example 27 □ Write $\log_{10}(100) = 2$ in exponential form.
$\log_{10}(100) = 2$ is equivalent to $10^2 = 100$.

□

Example 28 □ Write $\log_{10}(0.001) = -3$ in exponential form.
$\log_{10}(0.001) = -3$ is equivalent to $10^{-3} = 0.001$.

The $\log_{10}$ of 0.001 is negative.

□

Example 29 □ If $\log_2 x = 4$, what is x?
Remembering that a logarithm is an exponent, write $\log_2 x = 4$ in exponential form.

$$2^4 = x$$

$$16 = x$$

$\text{Log}_2 x = 4$ says that 4 is the power of 2 that yields 16.

□

Example 30 □ What are the following logarithms?

3

2

5

    a. $\log_2 8 = $ _____

    b. $\log_2 4 = $ _____

    c. $\log_2 32 = $ _____

$\log_2 8 = 3$ because $2^3 = 8$.

□

Example 31 □ If $\log_{10}(0.01) = y$, what is y?

Write $\log_{10}(0.01) = y$ in exponential form.

> We are working to write both sides of the equation as 10 to some power.

$$10^y = 0.01$$

or $\quad 10^y = \dfrac{1}{100} \qquad$ Change 0.01 to a fraction

$$10^y = \dfrac{1}{10^2}$$

$$10^y = 10^{-2}$$

$$y = -2 \qquad\qquad \square$$

Example 32 □ If $\log_8 x = \dfrac{4}{3}$, what is x?

Write $\log_8 x = \dfrac{4}{3}$ in exponential form.

$$8^{\frac{4}{3}} = x$$

$$\left(8^{\frac{1}{3}}\right)^4 = x \qquad \text{Law of exponents}$$

$$2^4 = x$$

$$16 = x \qquad\qquad \square$$

Example 33 □ If $\log_a 4 = \dfrac{2}{3}$, what is a?

Write $\log_a 4 = \dfrac{2}{3}$ in exponential form.

$$a^{\frac{2}{3}} = 4$$

$$\left(a^{\frac{2}{3}}\right)^{\frac{3}{2}} = 4^{\frac{3}{2}} \qquad \text{Raise both sides to power } \dfrac{3}{2} \text{ to solve for } a$$

$$a = 4^{\frac{3}{2}}$$

$$a = \left(4^{\frac{1}{2}}\right)^3 \qquad \text{Law of exponents}$$

$$a = 2^3$$

$$a = 8$$

> Why did you raise $a^{\frac{2}{3}}$ to the $\dfrac{3}{2}$ power, that is, $\left(a^{\frac{2}{3}}\right)^{\frac{3}{2}}$?

> Since we are solving for a, a must have an exponent of one. Raising a power to a power, we multiplied $\dfrac{2}{3}$ by its reciprocal.
> $$\dfrac{2}{3} \cdot \dfrac{3}{2} = 1.$$

□

Chapter 13 ■ Functions, Inverses, and Logarithms

Write each of the following in logarithmic form.

1. $16 = 2^4$

2. $243 = 3^5$

3. $\dfrac{1}{32} = 2^{-5}$

4. $0.0001 = 10^{-4}$

5. $\dfrac{27}{8} = \left(\dfrac{3}{2}\right)^3$

6. $\dfrac{16}{81} = \left(\dfrac{3}{2}\right)^{-4}$

7. $0.001 = 10^{-3}$

8. $\dfrac{27}{8} = \left(\dfrac{2}{3}\right)^{-3}$

9. $\dfrac{1}{81} = 3^{-4}$

10. $\dfrac{81}{16} = \left(\dfrac{3}{2}\right)^4$

11. $\dfrac{4}{9} = \left(\dfrac{3}{2}\right)^{-2}$

12. $\dfrac{32}{243} = \left(\dfrac{3}{2}\right)^{-5}$

Write each of the following in exponential form.

13. $\log_2 128 = 7$

14. $\log_4 256 = 4$

15. $\log_{10} 0.01 = -2$

16. $\log_3 \dfrac{1}{81} = -4$

17. $\log_5 \dfrac{1}{125} = -3$

18. $\log_{10} 0.00001 = -5$

19. $\log_2 64 = 6$

20. $\log_3 243 = 5$

21. $\log_{10} 0.0001 = -4$

22. $\log_2 \dfrac{1}{32} = -5$

23. $\log_3 \dfrac{1}{81} = -4$

24. $\log_{10} 0.1 = -1$

Sketch the graph of each of the following and its inverse on the same axis.

25. $y = \log_3 x$ **26.** $y = \log_5 x$ **27.** $y = \log_{10} x$ **28.** $y = \log_4 x$ **29.** $y = \log_6 x$

30. $y = \log_7 x$ **31.** $y = \log_8 x$

Evaluate each of the following.

32. $\log_7 49$

33. $\log_2 \dfrac{1}{4}$

34. $\log_6 \dfrac{1}{36}$

35. $\log_3 81$

36. $\log_{10} 1000$

37. $\log_{10} 0.001$

38. $\log_{10} \dfrac{1}{10}$

39. $\log_2 2^{10}$

40. $\log_3 3^{25}$

41. $\log_3 \dfrac{1}{27}$

42. $\log_5 \dfrac{1}{25}$

43. $\log_6 \dfrac{1}{216}$

44. $\log_7 \dfrac{1}{49}$

45. $\log_5 125$

46. $\log_{10} \dfrac{1}{1000}$

47. $\log_{10} 0.0001$

48. $\log_{10} \dfrac{1}{10}$

Solve each of the following equations.

49. $\log_2 \dfrac{1}{16} = x$ **50.** $\log_3 x = 4$ **51.** $\log_x 64 = 3$ **52.** $\log_{\frac{1}{2}} x = -4$

53. $\log_x \dfrac{1}{27} = -3$ **54.** $\log_{\frac{1}{3}} 81 = x$ **55.** $\log_a 3 = \dfrac{1}{3}$ **56.** $\log_4 x = 0$

57. $\log_9 x = \dfrac{3}{2}$ **58.** $\log_{10} 10 = x$ **59.** $\log_a 4 = -2$ **60.** $\log_8 x = \dfrac{4}{3}$

61. $\log_2 x = -4$ **62.** $\log_3 x = -2$ **63.** $\log_a \dfrac{1}{25} = -2$ **64.** $\log_a \dfrac{1}{125} = -3$

65. $\log_2 \dfrac{1}{32} = x$ **66.** $\log_3 \dfrac{1}{81} = x$ **67.** $\log_{\frac{2}{3}} x = -3$ **68.** $\log_{\frac{3}{4}} x = -2$

69. $\log_a \dfrac{1}{64} = -3$ **70.** $\log_x \dfrac{81}{16} = -4$ **71.** $\log_x 36 = -2$ **72.** $\log_x 125 = -3$

Properties of Logarithms

Since each logarithmic equation has an equivalent exponential equation, we expect the properties of exponents to apply to logarithms.

Logarithm of a Product

$$\text{Let} \quad P = \log_a x \longleftrightarrow a^P = x$$
$$Q = \log_a y \longleftrightarrow a^Q = y$$

The product of x and y is

$$x \cdot y = a^P \cdot a^Q$$
$$x \cdot y = a^{P+Q} \qquad \text{Simplifying using the Law of exponents}$$

Writing this statement in logarithmic form

$$\log_a(x \cdot y) = P + Q$$
$$\log_a(x \cdot y) = \log_a x + \log_a y \qquad \text{Substituting } P = \log_a x, \, Q = \log_a y$$

13.3B Logarithm of a Product

$$\log_a(x \cdot y) = \log_a x + \log_a y$$

The logarithm of a product is the sum of the logarithms of each factor.

Corresponding Property of Exponents

$$a^x \cdot a^y = a^{x+y}$$

Example 34 □ Find the logarithm base 2 of $8 \cdot 4$.

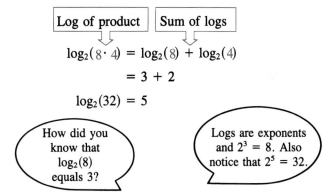

Log of product	Sum of logs

$$\log_2(8 \cdot 4) = \log_2(8) + \log_2(4)$$
$$= 3 + 2$$
$$\log_2(32) = 5$$

How did you know that $\log_2(8)$ equals 3?

Logs are exponents and $2^3 = 8$. Also notice that $2^5 = 32$.

The logarithm of a quotient also works as you would expect.

$$\text{Let} \quad P = \log_a x \longleftrightarrow a^P = x$$
$$Q = \log_a y \longleftrightarrow a^Q = y$$

Then

$$\frac{x}{y} = \frac{a^P}{a^Q} \qquad \text{Using the exponential form}$$

$$\frac{x}{y} = a^{P-Q} \qquad \text{Laws of exponents}$$

Writing the above in logarithmic form

$$\log_a \frac{x}{y} = P - Q$$

or $\quad \log_a \dfrac{x}{y} = \log_a x - \log_a y \qquad \text{Substituting} \qquad P = \log_a x$

$$Q = \log_a y \qquad \square$$

13.3C Logarithm of a Quotient

$$\log_a\left(\frac{x}{y}\right) = \log_a x - \log_a y$$

The logarithm of a quotient is the difference of the logs.

Corresponding Property of Exponents

$$\frac{a^x}{a^y} = a^{x-y}$$

Example 35 □ Find the $\log_2$ of $(32 \div 4)$.

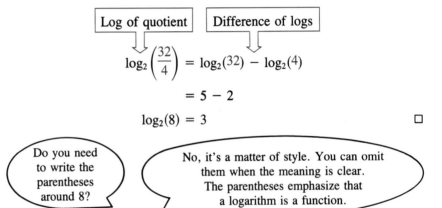

| Log of quotient | Difference of logs |

$$\log_2\left(\frac{32}{4}\right) = \log_2(32) - \log_2(4)$$

$$= 5 - 2$$

$$\log_2(8) = 3 \qquad\qquad □$$

Do you need to write the parentheses around 8?

No, it's a matter of style. You can omit them when the meaning is clear. The parentheses emphasize that a logarithm is a function.

Logarithms of a Power

$$\text{Let} \quad P = \log_a(x) \longleftrightarrow a^P = x$$

Let's examine x^n.

$$\text{If} \quad x = a^P$$

$$x^n = (a^P)^n$$

$$x^n = a^{n \cdot P}$$

Using the definition of logarithms, rewrite this exponential equation in logarithmic form.

$$\log_a(x^n) = n \cdot P$$

$$= n \log_a(x) \qquad \text{by substituting} \quad P = \log_a(x)$$

13.3D Logarithm of a Power

$$\log_a(x^n) = n \log_a(x)$$

The logarithm of x to the nth power is equal to n times the logarithm of x.

Corresponding Property of Exponents

$$(a^x)^n = a^{nx}$$

Example 36 □ What is the $\log_2$ of 8^2?

$$\log_2(8^2) = 2 \log_2(8)$$
$$= 2(3)$$
$$\log_2(64) = 6$$

which says

$$2^6 = 64 \qquad \square$$

Other Properties of Logarithms

A few special cases of the previous properties deserve mention.

13.3E Important Special Cases of Logarithms

I. $\log_a(a) = 1$ Because $a^1 = a$

II. $\log_a(1) = 0$ Because $a^0 = 1$

III. $\log_a\left(\dfrac{1}{x}\right) = -\log_a(x)$ View this as the log of a quotient;

$$\log_a\left(\frac{1}{x}\right) = \log_a(1) - \log_a(x)$$
$$= 0 - \log_a(x)$$
$$= -\log_a(x)$$

IV. $a^{\log_a x} = x$ $\log x$ is the power to which you raise a to yield x

V. $\log_a a^x = x$ x is the power to which a is raised to give a^x

Example 37 □ Express $2 \log_a(x) + \log_a(y)$ as a single logarithm.

$$2 \log_a(x) + \log_a(y) =$$
$$\log_a(x^2) + \log_a(y) = \log_a(x^2 y) \qquad \text{Property 13.3B} \qquad \square$$

Example 38 □ Express $\frac{1}{2}\log_a(x) + \log_a(y) - 3\log_a(z)$ as a single logarithm.

$$\frac{1}{2}\log_a(x) + \log_a(y) - 3\log_a(z)$$

$$= \log_a(x^{\frac{1}{2}}) + \log_a(y) - \log_a(z^3) \quad \text{Logarithm of a power}$$

$$= \log_a\left(x^{\frac{1}{2}}y\right) - \log_a(z^3) \qquad\qquad \text{Logarithm of a product}$$

$$= \log_a\left(\frac{x^{\frac{1}{2}}y}{z^3}\right) \qquad\qquad\qquad \text{Logarithm of a quotient} \qquad □$$

Example 39 □ Use the properties of logs to rewrite $\log_a\left(\sqrt[3]{\frac{x^2}{b}}\right)$ in terms of logarithms of x and b.

$$\log_a\left(\sqrt[3]{\frac{x^2}{b}}\right) = \frac{1}{3}\log_a\frac{x^2}{b}$$

$$= \frac{1}{3}[\log_a x^2 - \log_a b]$$

$$= \frac{1}{3}[2\log_a x - \log_a b] \qquad\qquad □$$

Example 40 □ Express $\log_a\left(\sqrt[4]{\frac{x^3}{y^2 z}}\right)$ in terms of logarithms of x, y, and z.

$$\log_a\left(\sqrt[4]{\frac{x^3}{y^2 z}}\right) = \frac{1}{4}\log_a\left(\frac{x^3}{y^2 z}\right)$$

$$= \frac{1}{4}[\log_a x^3 - \log_a(y^2 z)]$$

$$= \frac{1}{4}[3\log_a x - (\log_a(y^2) + \log_a(z))]$$

$$= \frac{1}{4}[3\log_a x - (2\log_a y + \log_a z)]$$

$$= \frac{1}{4}[3\log_a x - 2\log_a y - \log_a z] \qquad\qquad □$$

Use the properties of logarithms to evaluate the following. Do not use your calculator.

1. $\log_2(32 \cdot 8)$

2. $\log_2 8^2$

3. $\log_2(4^3 \cdot 8)$

4. $\log_2(\sqrt[3]{8} \cdot 4)$

5. $\log_2 \dfrac{64}{16}$

6. $\log_2 \dfrac{32^2}{8}$

7. $\log_3(81 \cdot 27)$

8. $\log_3(81^2)$

9. $\log_3[(27)^3 \cdot 9]$

10. $\log_3(\sqrt{81} \cdot 9)$

11. $\log_3 \left[\dfrac{(27)^2}{81} \right]$

12. $\log_3 \left[\dfrac{(81)^3}{27} \right]$

13. $\log_2 \sqrt{\dfrac{128}{8}}$

14. $\log_2 \sqrt[3]{\dfrac{32 \cdot 16}{8}}$

15. $\log_3 \dfrac{27 \cdot 81}{9}$

16. $\log_3(27)^3 \cdot \sqrt{81}$

17. $\log_3 \sqrt{\dfrac{27 \cdot 81}{3}}$

18. $\log_3 \sqrt{\dfrac{243 \cdot 27}{9}}$

19. $\log_2 \sqrt[3]{\dfrac{32 \cdot 8}{4}}$

20. $\log_2 \sqrt[4]{\dfrac{16 \cdot 64}{4}}$

21. $\log_2(32)^3\sqrt{16}$

22. $\log_2(16)^4 \cdot \sqrt[3]{64}$

23. $\log_3 \sqrt{\dfrac{(9)^3 \cdot 81}{(27)^2}}$

24. $\log_3 \sqrt{\dfrac{(27)^4 \cdot 81}{(243)^2}}$

Express each of the following as the sums or difference of logarithms of x, y, and z.

25. $\log_a 3xy$

26. $\log_a 5x^2 y$

27. $\log_a \dfrac{8xy^2}{z}$

28. $\log_a 24x^2\sqrt{y}$

29. $\log_a 7x^2 y$

30. $\log_a \dfrac{32x^3 y}{z^2}$

31. $\log_a \dfrac{18x^2 \sqrt[3]{y}}{z^2}$

32. $\log_a \dfrac{15\sqrt[3]{x}y^3}{z^2}$

33. $\log_a \dfrac{12\sqrt{x}\,y}{5z^3}$

34. $\log_a \dfrac{x^{\frac{1}{2}}y^{\frac{1}{3}}}{z^{\frac{1}{4}}}$

35. $\log_b \sqrt{\dfrac{x^2 y^3}{z^4}}$

36. $\log_b \sqrt[3]{\dfrac{6x^3 y^4}{z^2}}$

37. $\log_b \sqrt[4]{\dfrac{7x^5 y^4}{6z^2}}$

38. $\log_a \dfrac{14\sqrt[3]{x}\,y^2}{9z^2}$

39. $\log_b \dfrac{16\sqrt{x}\,y^3}{7\sqrt[3]{z}}$

40. $\log_b \dfrac{x^{\frac{2}{3}}y^{\frac{3}{4}}}{z^2}$

41. $\log_b \dfrac{x^{\frac{3}{4}}y^{\frac{2}{5}}}{z^{\frac{1}{3}}}$

42. $\log_b \sqrt{\dfrac{3x^2 y^3}{z^4}}$

43. $\log_b \sqrt[3]{\dfrac{5x^3 y^7}{2z^2}}$

Express each of the following as a single logarithm and simplify if possible.

44. $2 \log_a 3x + 3 \log_a y$

45. $\dfrac{2}{3} \log_b x - \dfrac{1}{2} \log_b y^2$

46. $4 \log_a x^2 + 6 \log_a y^{\frac{1}{2}}$

47. $4 \log_a x - 2 \log_a y$

48. $3 \log_a 4x - 2 \log_a y^2$

49. $3 \log_a x^3 + 4 \log_a y^{\frac{1}{2}}$

50. $6 \log_a x^2 + 3 \log_a y$ **51.** $4 \log_a 2x^{\frac{1}{2}} - 3 \log_a 3y^2$ **52.** $\frac{1}{2}(\log_a x^2 + \log_a y)$

53. $\frac{1}{3}(\log_a x^2 - \log_a y + \log_a z^3)$ **54.** $\frac{2}{3}(\log_a x^2 - \log_a \sqrt{x})$

55. $\frac{1}{4}(\log_a x^3 y + \log_a x^2 \sqrt{y})$ **56.** $\log_a \sqrt{xy^3} - \log_a \dfrac{x}{\sqrt{y}} + \log_a \dfrac{\sqrt{x}}{y}$

57. $\frac{1}{3}(\log_a x^3 - \log_a y^2)$ **58.** $\frac{1}{4}(\log_a x^6 - \log_a y^3)$

59. $\frac{2}{3}(\log_a x^3 y + \log_a x^2 \sqrt{y})$ **60.** $\frac{3}{4}(\log_a x^3 y^6 + \log_a x \sqrt[3]{y})$

■ 13.4   Calculations with Logarithms

The concept of logarithm was first published before that of exponents. Logs were used in 1614 by John Napier in Edinburgh, Scotland. He originally published a paper about logarithms he developed on comparing the relative motion of two points. The term **logarithm** comes from the Latin and means **ratio number.** By 1624, Henry Briggs of London, after working with Napier, published a table of logarithms with a base of 10 that was accurate to 20 places.* Because it is easier to add than multiply, Napier's discovery was rapidly adopted as an aid to numerical calculations and remained the principal method of scientific calculation until the advent of economical computers in the 1960s.

During this long period, calculations with logarithms became a common topic of algebra texts primarily due to their great utility.

Today we no longer use logarithms for calculations. We are more interested in logarithms as a means of describing many natural phenomenon. A brief treatment of how logs can be used to calculate is included here for historical interest. This will also give you an opportunity to improve your understanding of the properties of logarithms. In the next unit, these properties will be used extensively to solve equations.

 Using Your Calculator

Finding Logarithms

Your scientific calculator can find logarithms in both base 10 and base e. A later section will deal with the LN key which finds logarithms in base e.

First we will see how to find logarithms using base 10. To find logarithms in base 10, locate the key marked LOG. This will probably be a primary key on your calculator but on some calculators it may be a 2nd FN or INV key.

* Ibid., p. 249.

Example 41 □ Find $\log_{10} 1000$.

Press: 1000 LOG .

The display should be 3, because $10^3 = 1000$. □

Example 42 □ Find $\log_{10} 70.1$.

Press: 70.1 LOG .

The display is 1.845718. □

Example 43 □ Find $\log_{10} .002$.

Press: .002 LOG .

The display is -2.69897. □

Your calculator will also tell you which number has a given logarithm. The problem of finding the number with a particular logarithm is the inverse operation of finding a logarithm, so most calculators use INV LOG or 2nd FN LOG to find the number with a particular logarithm.

Since, by definition x is the base 10 logarithm of y, if $10^x = y$, many calculators have the key for the inverse logarithm labeled 10^x .

Example 44 □ Find the number with a base 10 logarithm of 3.

Press: 3 INV LOG .

The display is 1000, because $10^3 = 1000$. □

Example 45 □ Find the number with a base 10 logarithm of 0.5.

Press: .5 INV LOG .

The display is 3.1622777. □

From this point on in the text, if no base is given, the base is understood to be 10.

Getting Started

Example 46 □ Use your calculator to find the following logs.

			because	
	$\log(10)$	$= 1$	because	$10^1 = 10$
0	$\log(1)$	$=$ _____	because	$10^0 = 1$
2	$\log(100)$	$=$ _____	because	$10^2 = 100$
-2	$\log(.01)$	$=$ _____	because	$10^{-2} = .01$
0.5	$\log(3.16227)$	$=$ _____	because	$10^{.5} = \sqrt{10}$ □

The following will illustrate how the properties of logarithms work with some actual calculations.

In the next examples we will:

a. Rewrite the problem using the properties of logarithms.
b. Use a calculator to find the logarithms.
c. Do the arithmetic to produce a single logarithm.
d. Find the number that corresponds to the logarithm (INV LOG).

Now let's try some products.

Recall that $\log(A \cdot B) = \log A + \log B$. To illustrate this, here is an arithmetic example.

Example 47 □ Find (50)(6) using logarithms.

First write a logarithm equation.

$$\log(50)(6) = \log 50 + \log 6$$

$$\approx 1.69897 + .778151$$

$$\approx 2.47712$$

That's nice, but $50 \cdot 6$ isn't 2.47712.

True. 2.47712 is the log of $(50 \cdot 6)$. Use the INV LOG or 10^x key to find the number with a log of 2.47712.
$10^{2.47712} = 299.99913$.

$$\text{INV LOG}(2.47712) \approx 299.99913$$

That's still not 300.

No, but it's close. Even though a calculator has many decimal places, it doesn't give the exact log of a number.

□

You could find (50)(6) by simply using your calculator to multiply. We are trying to illustrate here how the laws of logarithms operate.

Example 48 □ Find (3456.7)(87.654).

$$\log(3456.7)(87.654) = \log(3456.7) + \log(87.654)$$

1.942772

$$\approx 3.538666 + \underline{\hspace{2cm}}$$

$$\approx 5.481433$$

$$\text{INV LOG}(5.481433) \approx 302993.58 \qquad \square$$

Chapter 13 ■ Functions, Inverses, and Logarithms

The INV LOG(x) is simply the number that has x as its logarithm. Sometimes INV LOG(x) is called **antilog(x).**

If your calculator doesn't have an INV LOG key, look for a 10^x key. Remember $10^{\log x} = x$.

We can verify this answer by multiplying $(3456.7)(87.654)$ on the calculator.

Moving On to Quotients

$$\log\left(\frac{A}{B}\right) = \log A - \log B$$

Example 49 □ Find $200 \div 0.02$.

$$\log\left(\frac{200}{0.02}\right) = \log(200) - \log(0.02)$$

-1.698970

$$\approx 2.30103 - \underline{\qquad}$$

> Note that $\log(0.02)$ is negative. Be careful when you subtract.

4.0

$$\approx \underline{\qquad}$$

INV LOG$(4.0) = 10000$ □

Example 50 □ Use logarithms to find $\dfrac{(0.6334)(21.78)}{56.45}$.

Writing a logarithm equation

$$\log\left(\frac{(0.6334)(21.78)}{56.45}\right) = \log(0.6334) + \log(21.78) - \log(56.45)$$

1.751664

$$\approx -0.198322 + 1.338058 - \underline{\qquad}$$

1.139736

$$\approx \underline{\qquad} - 1.751664$$

-0.611928

$$\approx \underline{\qquad}$$

INV LOG$(-0.611928) \approx 0.244383$ □

Logarithm of a Power

$$\text{Log}(A^n) = n \log A$$

Example 51 □ Find $(3.14)^{\frac{2}{3}}$ using logs.

Writing an equation

$$\log(3.14)^{\frac{2}{3}} = \frac{2}{3}\log(3.14)$$

0.496930

$$\approx \frac{2}{3}(\underline{\qquad})$$

INV LOG$(0.33128) \approx 2.1444$ □

13.4 ■ Calculations with Logarithms

Example 52 □ Find $\sqrt[4]{\dfrac{(23.1)^3(19.4)}{179}}$ using logs.
The equation is

$$\log \sqrt[4]{\dfrac{(23.1)^3(19.4)}{179}} = \dfrac{1}{4}(3 \log(23.1) + \log(19.4) - \log 179)$$

$$\approx \dfrac{1}{4}(3(1.363612) + 1.287802 - 2.252853)$$

$$\approx \dfrac{1}{4}(4.090836 + 1.287802 - 2.252853)$$

$$\approx \dfrac{1}{4}(3.125785)$$

$$\approx 0.781446$$

6.045692
$$10^{0.781446} \approx \underline{\hspace{1.5cm}}$$
□

Problem Set 13.4

Use logarithms to evaluate the following.

1. $(200)(10)$

2. $(230)(0.0694)$

3. $\dfrac{0.8674}{1.246}$

4. $\dfrac{38.92}{0.01867}$

5. $(0.5869)^2$

6. $(3.4829)^3$

7. $(0.9068)^{2.5}$

8. $(30.009)^{4.6}$

9. $\sqrt{0.86592}$

10. $\sqrt[3]{67.843}$

11. $(1.862)^4$

12. $(4.6394)^3$

13. $(0.9851)^{3.4}$

14. $(0.4186)^{5.2}$

15. $\sqrt[4]{0.7294}$

16. $\sqrt[5]{10.1754}$

17. $\dfrac{0.9164}{0.03485}$

18. $\dfrac{25.163}{5.0192}$

19. $\sqrt{(2.8164)^3}$

20. $\sqrt{(12.6931)^2}$

21. $\dfrac{(8205)(9.426)}{68.845}$

22. $\dfrac{(3.26)^2(9.6486)}{18.924}$

23. $\dfrac{(302.9)\sqrt{0.003294}}{8.792}$

24. $\dfrac{(99.86)(3.045)^2}{\sqrt{221.94}}$

25. $\dfrac{(59.6)\sqrt[3]{0.9438}}{\sqrt{578.42}}$

26. $\dfrac{(396.01)^2\sqrt{0.0062}}{(79.86)^3}$

27. $\sqrt{\dfrac{(38.569)^2(0.0976)}{(2.8769)^3}}$

28. $\sqrt[3]{\dfrac{(7.4921)^4(86.9)^2}{(967.48)^3}}$

29. $\dfrac{(51.68)(9.421)^2}{324.93}$

30. $\dfrac{(6.1549)^3(27.943)}{(33.416)}$

31. $\dfrac{(810.64)\sqrt{51.468}}{394.62}$

32. $\dfrac{(0.5869)\sqrt[3]{1.8427}}{0.02483}$

33. $\dfrac{(6.792)^2\sqrt{0.5142}}{\sqrt[3]{14.562}}$

34. $\dfrac{(2.1396)^3\sqrt{38.974}}{(7.1692)^2}$

35. $\sqrt[3]{\dfrac{(24.862)^2(3.1682)^3}{75.921}}$

36. $\sqrt{\dfrac{(6.7854)^2(3.4319)^3}{(2.8534)^4}}$

Chapter 13 □ Key Ideas

13.1   **1.** *Y* is said to be a **function** of *x* if for each value of *x* there is one and only one value of *y*.
 2. **Ordered pairs** are used to show how the value of *y* is paired with a permissible value of *x*.
 3. The number in the ordered pairs are called **components.**
 4. **Functional notation**
   $y = f(x)$ shows that *y* is related to *x*.
   a. independent variable—the value of *x* or of the first component in an ordered pair;
   b. dependent variable—the value of *y* or the second component in an ordered pair.
 5. A **relation** is any set of ordered pairs.
   a. Domain—set of the first components of the ordered pairs of a relation.
   b. Range—set of the second components of the ordered pairs of a relation.
 6. A **function** is a relation in which the first component of each ordered pair has a unique second component.
   In a function defined as a set of **ordered pairs,** the first component in each ordered pair is a member of the **domain** and the second component is a member of the **range.**
 7. A **one-to-one function** is a function in which the second component of each ordered pair has a unique first component.
 8. **Inverse of a relation** is a relation produced by interchanging the components in each ordered pair of the given relation.

13.2   **1.** The **general exponential function**
   $y = a^{kx}$ where $a > 0$, *a* and *k* are constant and $k \neq 0$.
   a. $a > 1$—the curve climbs steeply to the right:
    1. The curve approaches the negative *x*-axis asymptotically;
    2. The curve passes through the point $(0, 1)$.
   b. $0 < a < 1$—the curve falls to the right:
    1. The curve approaches the positive *x*-axis asymptotically;
    2. The curve passes through the point $(0, 1)$.
 2. **Property of Equality for Exponentials**
   a. If $a^x = a^y$ then $x = y$ for all $a > 0$ and $a \neq 1$.

13.3 **1.** The **logarithm function**

$y = \log_a x$

$y = \log_a x$ is equivalent to $a^y = x$.

2. The **graph of a logarithm** to some base a is the graph of the inverse of the exponential function $y = a^x$.

3. Properties of logarithms

a. $\log_a(x \cdot y) = \log_a x + \log_a y$

b. $\log_a \dfrac{x}{y} = \log_a x - \log_a y$

c. $\log_a(x^n) = n \log_a x$

4. Important special cases of logarithms

a. $\log_a a = 1$

b. $\log_a(1) = 0$

c. $\log_a\left(\dfrac{1}{x}\right) = -\log_a x$

d. $a^{\log_a x} = x$

e. $\log_a a^x = x$

13.4 Logarithms can be used to find products and quotients, raise to powers, extract roots, or any combination of these.

Chapter 13 Review Test

(a) List the elements of the domain and range of the following relations. (b) Determine if the relation is a function. (c) Determine if the relation is one-to-one. **(13.1)**

1. $\{(-3, 2), (-2, 6), (5, 4), (3, 2)\}$

2. $\{(0, 6), (-2, 0), (0, 7), (8, 2), (9, 3)\}$

(a) Determine if the graph is a function. (b) Determine if the graphs represent one-to-one functions. **(13.1)**

3.

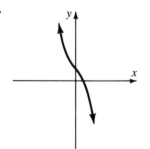

4.

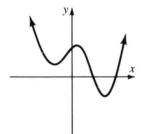

5.

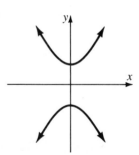

(a) Determine the domain and range of the following graphs. (b) Determine if the graphs represent one-to-one functions. **(13.1)**

6.

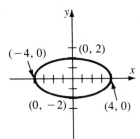

7.

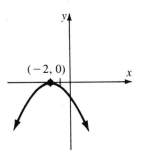

8.

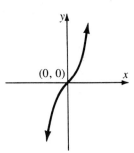

(a) Determine the domain and range of the following equations. (b) Determine if the equations represent one-to-one functions. **(13.1)**

9. $\dfrac{x^2}{25} - \dfrac{y^2}{12} = 1$

10. $4x^2 + y^2 = 16$

11. $y = x^2 + 4$

(a) Write the inverse of the following equations. Solve for y. (b) Determine if the original equation is a function. (c) Determine if the inverse is a function. **(13.1)**

12. $3x - 2y = 8$

13. $y = x^3 - 2$

14. $9x^2 + 4y^2 = 36$

(a) Write the inverse of the following equations. (b) Sketch the graph of the original equation and its inverse on the same axis. **(13.1)**

15. $y = \dfrac{1}{2}x + 2$

16. $y = x^2 + 4$

Graph each of the following exponential functions. **(13.2)**

17. $y = 2^{3x}$

18. $y = 2^{2x+3}$

19. $y = 2^{2x} + 3$

20. $y = 3^{-2x}$

Solve the following exponential equations. **(13.2)**

21. $2^x = 32$

22. $4^x = 16$

23. $2^{3x-1} = 256$

24. $4^{3x} - 3 = 29$

Given the formula $B = P\left(1 + \dfrac{r}{n}\right)^{nt}$, find the balance (B) for each investment. **(13.2)**

25. $600 for 4 years at 10% compounded semi-annually.

26. $1250 for 5 years at 12% compounded quarterly.

27. $1800 for 3 years at 18% compounded monthly.

Write in logarithmic form. **(13.3)**

28. $81 = 3^4$

29. $0.001 = 10^{-3}$

Write in exponential form. (**13.3**)

30. $\log_2 \dfrac{1}{8} = -3$

31. $\log_{10} 0.01 = -2$

Sketch the graph. (**13.3**)

32. $y = \log_4 x$

Evaluate each of the following. (**13.3**)

33. $\log_6 36$

34. $\log_2 \dfrac{1}{32}$

35. $2^{\log_2 100}$

Solve the following for the variable. (**13.3**)

36. $\log_8 32 = x$

37. $\log_a \dfrac{1}{32} = -5$

38. $\log_3 x = -4$

Use the properties of logarithms to evaluate. (**13.3**)

39. $\log_2 \dfrac{16 \cdot 8}{32}$

40. $\log_2 \sqrt{\dfrac{8 \cdot 2^5}{4}}$

41. $\log_3 9^2 \sqrt{81}$

Express each as a sum or difference of simpler logarithmic quantities. (**13.3**)

42. $\log_a 15\sqrt{x}\, y^3$

43. $\log_a \sqrt[3]{\dfrac{6x^2 y}{4z^3}}$

Express each as a single logarithm and simplify. (**13.3**)

44. $\dfrac{1}{3}\log_a x - \dfrac{2}{3}\log_a y^2$

45. $\dfrac{1}{4}(\log_a x^2 \sqrt{y} + \log_a \sqrt{x}\, y^2 - \log_a \sqrt{xy})$

Use logarithms to evaluate the following. (**13.4**)

46. $\dfrac{(8.964)(0.028691)}{0.002146}$

47. $\dfrac{(0.6792)^3 \sqrt{64.869}}{(2.6894)^2}$

48. $\sqrt[3]{\dfrac{(46.945)^2(0.009684)}{(3.9634)^3}}$

14 Applications of Logarithms and Exponential Functions

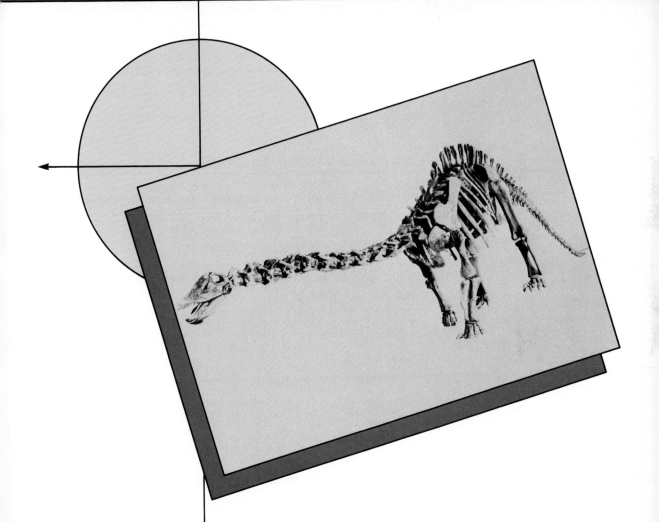

Contents

n the last chapter you learned about logarithms and exponential functions. This chapter will show you some of the many uses for these functions.

Section one will show you how to use logarithms to solve equations where the variable is an exponent. Section two will show you the natural number **e**. It is called the natural number because it is the base of exponential expressions that occur in many fields using **e**. The methods of variation you learned in Chapter 10 can be extended to develop mathematical expressions for interest growth in banking, atmospheric pressure, half-life of radioactive material, and population growth.

In the final section of this chapter you will learn about applications of logarithms which include carbon dating of ancient material like dinosaur bones and measurement of earthquakes.

■ 14.1 Using Logarithms to Solve Equations

Chapter 13 showed how to solve some equations with the variable as part of the exponent. The technique used was based on the property of exponents.

$$\text{if} \qquad x^a = x^b \qquad \text{then } a = b.$$

This method requires us to write both sides of the equation so that the same base is raised to a given power.

Example 1 □ Solve $9^x = 243$.

$$9^x = 243$$
$$3^{2x} = 3^5$$
$$2x = 5$$
$$x = \frac{5}{2}$$

This requires a lot of insight or luck to be able to write both sides as a power of 3.

□

Logarithms provide a more general method to solve such equations.

Example 2 □ Solve $9^x = 243$ using logs.

$$9^x = 243$$

$\log(9^x) = \log(243)$ Take the logarithm of both sides

$x \log(9) = \log(243)$ Log of a power

$$x = \frac{\log(243)}{\log(9)}$$

Note: $\log(243)$ is a number and $\log(9)$ is a number. Therefore, this is the quotient of two numbers not the logarithm of a quotient

$$x \approx \frac{2.38560}{0.95424}$$ Use calculator to find logs

$$x \approx 2.49999$$

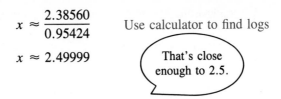
That's close enough to 2.5.

□

The general way to solve equations where the variable is an exponent is to take the logarithm of both sides of the equation.

Example 3 □ Solve $13^{2x} = 200$ for x.

$$13^{2x} = 200$$
$$\log(13^{2x}) = \log(200) \qquad \text{Take log of both sides}$$
$$2x \log(13) = \log(200) \qquad \text{Log of a power}$$
$$2x = \frac{\log(200)}{\log(13)}$$
$$2x \approx \frac{2.30103}{1.11394} \qquad \text{Logs by calculator}$$
$$2x \approx 2.06566$$
$$x \approx 1.03283$$

□

How can I check this result?

Use the y^x key of your calculator to raise 13 to the 2(1.03283) power.

I get 199.999. How come my answers are never exactly right?

Even though your calculator gives 8 or 9 places, that is still only an approximation.

Example 4 □ Solve $25^{x^2+2x} = 5^{-x}$.

$$25^{x^2+2x} = 5^{-x}$$
$$\log(25^{x^2+2x}) = \log(5^{-x})$$
$$(x^2 + 2x)\log(25) = -x(\log 5)$$
$$(x^2 + 2x)(1.39794) \approx -x(0.69897)$$
$$x^2 + 2x = -x(0.5) \qquad \text{Divide by 1.39794}$$
$$x^2 + 2.5x = 0$$
$$x(x + 2.5) = 0$$
$$x = 0 \qquad \text{or} \qquad x = -2.5$$

Checking the answer

$$25^{(-2.5)^2+2(-2.5)} \overset{?}{=} 5^{-(-2.5)}$$

$$25^{6.25-5} \overset{?}{=} 5^{2.5}$$

$$25^{1.25} \overset{?}{=} 5^{2.5}$$

$$55.9016 = 55.9016 \qquad\qquad \square$$

▶ **Pointers About Example 4**

Since the bases of the exponential equation in Example 4 were both powers of 5, it would have been easier to work Example 4 if we used logarithms with a base of 5.

Below, Example 4 is repeated using base 5 logarithms.

$$25^{x^2+2x} = 5^{-x}$$

$$\log_5(25^{x^2+2x}) = \log_5(5^{-x})$$

$$(x^2 + 2x)(\log_5 25) = -x(\log_5 5)$$

$$(x^2 + 2x)(2) = -x(1)$$

$$2x^2 + 4x = -x$$

$$2x^2 + 5x = 0$$

$$x(2x + 5) = 0$$

Therefore

$$x = 0 \qquad \text{or} \qquad x = -\frac{5}{2}$$

Why didn't you use base five logarithms in the first place?

Calculators don't have base five logarithms. So this second approach is useful only when the bases of the exponential equation are both powers of the base of the logarithm.

Logarithms with a base of ten or **e** are available on scientific calculators and can be used to simplify exponential equations with any combination of bases. In checking the solution we might have observed

$$25^{1.25} \overset{?}{=} 5^{2.5}$$

$$(5^2)^{1.25} \overset{?}{=} 5^{2.5}$$

$$5^{2.5} = 5^{2.5}$$

Again we chose the more general approach of using the calculator to evaluate both powers.

However, as a student you should observe that, even though you're not as fast as a calculator, you are smarter. You can certainly approach problems with more flexibility. Frequently human beings can use insight to solve problems that are unsolvable by cookbook methods.

Chapter 14 ■ Applications of Logarithms and Exponential Functions

Example 5 □ Solve $\log(x + 5) - \log(x - 1) = \log 3$.

$$\log(x + 5) - \log(x - 1) = \log 3$$

$$\log\left(\frac{x + 5}{x - 1}\right) = \log 3 \qquad \text{Log of a quotient}$$

$$\frac{x + 5}{x - 1} = 3 \qquad \text{If } \log a = \log b, \text{ then } a = b$$

$$x + 5 = 3(x - 1)$$

$$x + 5 = 3x - 3$$

$$8 = 2x$$

$$4 = x$$

Why can't I divide both sides by Log in the first step?

The first step is similar to $\log a - \log b = \log\left(\frac{a}{b}\right)$. If you tried to divide both sides by log, you'd get $a - b = \frac{a}{b}$, which is false. Log represents an operation or function, not a number.

□

▶ A Pointer About the Importance of One-to-One Functions

Consider the following inferences:

$$\sqrt{x} = \sqrt{y} \qquad \text{implies} \qquad x = y$$

$$2^x = 2^y \qquad \text{implies} \qquad x = y$$

$$\log x = \log y \qquad \text{implies} \qquad x = y$$

But

$$x^2 = y^2 \qquad \text{does not imply} \qquad x = y$$

Why not?

Suppose $x = 2$ and $y = -2$. Then $x \neq y$ but $x^2 = y^2$.

The question to be resolved for each process is, Is there more than one input that produces a specific output?

In general we are asking, if the ZOT(a) = ZOT(b), does a = b?

What's a ZOT(a)?

I don't know. I just made it up to represent any function with an input of a.

The answer is, not necessarily. We can determine the input from the output only if we are dealing with a one-to-one function.

In more formal language we are saying

For a one-to-one function,

$$\text{if } a = b \qquad \text{then} \qquad f(a) = f(b)$$

AND

$$\text{if } f(a) = f(b) \qquad \text{then} \qquad a = b$$

Since the logarithm is a one-to-one function, we can properly make the step

$$\text{from} \qquad \log(a) = \log(b)$$
$$\text{to} \qquad a = b.$$

Then in solving logarithmic equations the equation must be reduced to something like $\log(a) = \log(b)$.

Right. Use the laws of logarithms to rewrite log expressions so that the word **log** appears only once on each side.

Example 6 □ Solve $\log(x + 12) + \log(x + 3) = 1$.

$$\log a + \log b = \log(a \cdot b)$$
$$\log(x + 12) \cdot (x + 3) = 1$$

Before taking antilogs of both sides, write 1 as log 10. (Assume the use of logarithms to base 10.)

$$\log[(x + 12) \cdot (x + 3)] = \log 10$$
$$(x + 12)(x + 3) = 10 \quad \text{Taking the inverse log of both sides}$$
$$x^2 + 15x + 36 = 10$$
$$x^2 + 15x + 26 = 0$$
$$(x + 13)(x + 2) = 0$$
$$x = -13 \qquad \text{or} \qquad x = -2$$

Check: $x = -13$

$$\log(x + 12) + \log(x + 3) = 1$$
$$\log(-13 + 12) + \log(-13 + 3) \overset{?}{=} 1$$
$$\log(-1) \cdot \log(-10) \overset{?}{=} 1$$

-13 is not a solution because $\log(-1)$ is undefined.

Check: $x = -2$

$$\log(x + 12) + \log(x + 3) = 1$$

$$\log(-2 + 12) + \log(-2 + 3) \stackrel{?}{=} 1$$

$$\log(10) + \log 1 \stackrel{?}{=} 1$$

$$\log 10 + 0 \stackrel{?}{=} 1$$

It checks:

$$1 = 1 \qquad \square$$

Do you reject all negative solutions?

No; reject only those values of x that require taking the logarithm of negative numbers or zero because logarithms of negative numbers and zero are not defined.

Example 7 □ Solve $\log(3x + 4) + \log(4 - x) = \log(8 - 2x)$.
First we will rewrite each side of this equation as a single logarithm.

$$\log(3x + 4) \cdot (4 - x) = \log(8 - 2x) \qquad \text{Log of a product}$$

$$(3x + 4)(4 - x) = 8 - 2x \qquad \text{Taking the inverse log of both sides}$$

$$(3x + 4)(-x + 4) = 8 - 2x \qquad \text{Rewrite to make multiplication easier}$$

$$-3x^2 + 8x + 16 = 8 - 2x$$

$$-3x^2 + 10x + 8 = 0$$

$$3x^2 - 10x - 8 = 0 \qquad \text{Multiply both sides by } -1$$

$$(x - 4)(3x + 2) = 0 \qquad \text{to make factoring easier}$$

$$x - 4 = 0 \qquad 3x + 2 = 0$$

$$x = 4 \qquad 3x = -2$$

$$x = -\frac{2}{3}$$

Checking the first solution: $x = 4$

$$\log(3x + 4) + \log(4 - x) = \log(8 - 2x)$$

$$\log[3(4) + 4] + \log(4 - 4) \stackrel{?}{=} \log[8 - 2(4)]$$

$$\log 14 + \log 0 \stackrel{?}{=} \log 0$$

$\log 0$ is undefined, hence reject $x = 4$

Why is log 0 undefined?

There isn't any power to which you can raise a base in order to get zero.

Testing the other solution, $x = -\dfrac{2}{3}$

$$\log(3x + 4) + \log(4 - x) = \log(8 - 2x)$$

$$\log\left[3\left(-\frac{2}{3}\right) + 4\right] + \log\left[4 - \left(-\frac{2}{3}\right)\right] \overset{?}{=} \log\left[8 - 2\left(-\frac{2}{3}\right)\right]$$

$$\log(-2 + 4) + \log\left(4 + \frac{2}{3}\right) \overset{?}{=} \log\left(8 + \frac{4}{3}\right)$$

$$\log 2 + \log 4\frac{2}{3} \overset{?}{=} \log 9\frac{1}{3}$$

$$\log 2 \cdot 4\frac{2}{3} \overset{?}{=} \log 9\frac{1}{3}$$

$$\log 9\frac{1}{3} = \log 9\frac{1}{3} \qquad \square$$

I thought all positive values of x would work and all negative values had to be rejected.

No. We reject only those values of x that require us to take logarithms of negative numbers or zero. Logarithms are not defined for negative numbers or zero.

Problem Set 14.1

Solve using logarithms.

1. $2^x = 16$ **2.** $2^x = 12$ **3.** $2^x = 35$ **4.** $12^{2x} = 240$

5. $5^{4x-9} = 125$ **6.** $7^{3x-1} = 260$ **7.** $4^x = 9$ **8.** $6^{2x} = 225$

9. $3^x = 18$ **10.** $3^x = 24$ **11.** $6^x = 450$ **12.** $9^x = 360$

13. $8^{3x} = 126$ **14.** $4^{4x} = 69$ **15.** $7^{2x+3} = 730$ **16.** $15^{3x-1} = 425$

17. $9^x = 750$ **18.** $5^{x^2+2x} = 125$ **19.** $6^{x^2+5x} = \dfrac{1}{36^2}$ **20.** $2^{9x} \cdot 4^{x^2} = \dfrac{1}{16}$

21. $27^{x^2+2x} = 3^{2x}$ **22.** $81^{x^2+2x} = 36^{6+3x}$ **23.** $4^{x^2+3x} = 2^{x+12}$ **24.** $3^{x^2+2x} = 81^2$

25. $5^{x^2-x} = 25^3$ **26.** $4^{x^2+7x} = \dfrac{1}{16^3}$ **27.** $8^{x^2+5x} = \dfrac{1}{8^2}$ **28.** $3^{5x} \cdot 9^{x^2} = \dfrac{1}{27}$

29. $2^{7x} \cdot 4^{x^2} = 16$ **30.** $8^{x^2+2x} = 2^{8+4x}$ **31.** $25^{x^2+2x} = 5^{12+9x}$

Solve each of the following equations.

32. $\log(x + 2) = \log(2x - 3)$

33. $\log(x^2 + 2) = \log(6 - 3x)$

34. $\log x = \log(1 - 2x)$

35. $\log_5(2x + 3) - \log_5 x = \log_5 6$

36. $\log_2(2x + 5) - \log_2(x - 1) = \log_2 5$

37. $\log_3(3x + 1) + \log_3(x + 1) = \log_3 5$

38. $\log(2x - 4) = \log(5x - 8)$

39. $\log(3 - 2x) = \log(1 + 6x)$

40. $\log(x^2 - 4) = \log(8 - x)$

41. $\log(x^2 + 6) = \log(7x - 4)$

42. $\log_3(3x + 6) - \log_3(x - 2) = \log_3 6$

43. $\log_2(6x - 4) - \log_2(2x - 5) = \log_2 4$

44. $\log(x + 5) - \log(x - 1) = 1$

45. $\log x + \log(x + 3) = 1$

46. $\log x = 2 - \log(x + 15)$

47. $\log x = 2 - \log(x - 21)$

48. $\log x = 1 - \log(x + 9)$

49. $\log x + \log(5x - 8) - \log 10 - \log(2x + 15) = -1$

50. $\log x + \log(x + 2) + \log 10 - \log(3x + 12) = 1$

■ **14.2**   **Exponents and Logarithms to the Base e**

The irrational number e ≈ 2.71828 occurs so frequently in physics, mathematics, chemistry, biology, and even finance, that it is referred to as **the natural number.** A proper definition of **e** requires calculus, but an application from finance can give you an idea of how **e** arises.

Earlier we used a formula for compound interest.

$$A = P\left(1 + \frac{r}{n}\right)^{nt}$$ where A = final amount in the account

P = principal at the start of the interest period

r = annual interest rate

n = number of compounding periods per year

t = time in years

Banks advertise that they compound interest quarterly, monthly, or daily. In fact some banks advertise continuous compounding of interest. From this advertising you would think that increasing the frequency of compounding was of significant advantage to the consumer. This is true to a point.

The table below shows the interest earned on $1000 at 12% for various compounding periods.

Period	Amount in account after 1 year
year	$1120.00
quarter	1125.51
month	1126.83
day	1127.47
hour	1127.49
minute	1127.49

Not much difference after monthly compounding.

Let's examine the formula to see what's happening here.

$$A = P\left(1 + \frac{r}{n}\right)^{nt}$$

Notice that all P does is multiply the result by the initial principal.

In our examination of the frequency of compounding, we held r (rate) constant. Let's pay ourselves an interest rate of 100% per year. Thus $r = 1$. For $r = 1$ and $t = 1$, $\left(1 + \frac{r}{n}\right)^{nt}$, the factor affected by frequency of compounding becomes $\left(1 + \frac{1}{n}\right)^{n}$. Let's look at this factor for increasing values of n.

n	$\left(1 + \frac{1}{n}\right)^{n}$
1	2
2	2.25
4	2.4414063
8	2.5657845
16	2.6379285
32	2.6769901
64	2.6973450
128	2.7077391
256	2.7129908
5000	2.7180087
50000	2.7182682
100000	2.7182818

In fact, no matter how large we make n, the value of $\left(1 + \dfrac{1}{n}\right)^n$ gets closer and closer to a number commonly called **e**. In calculus, **e** is defined as the limit of this expression as n approaches infinity. Below is a formal definition of **e**. However, all you need to know at this point is that **e** is an irrational number approximately equal to 2.718.

14.2A   Definition of e

$$\mathbf{e} = \lim_{n \to \infty}\left(1 + \frac{1}{n}\right)^n \approx 2.7182818284. \ldots$$

Since powers of **e** occur naturally in biology, chemistry, physics, business, and mathematics, logarithms with a base **e** are frequently used. These logarithms are called natural logs or ln.

14.2B   Definition: $\ln(x)$

$$y = \ln(x) \qquad \text{if} \qquad \mathbf{e}^y = x$$

y is the natural logarithm of x if y is the power to which you would raise **e** in order to have it equal x.

$y = \ln(x)$ is equivalent to $y = \log_e(x)$. All the rules of operation for logarithms apply to logarithms with a base of **e**.

 Using Your Calculator

Finding Natural Logs

Your calculator will provide logarithms with a base of **e** similar to the way it provides base 10 logarithms.

Example 8 □ Use your calculator to find ln 2.7182818.

Press: 2.7182818 ⬚LN⬚ .

The display will be .99999999.   □

Shouldn't the ln of **e** be 1?

It is. You found ln of a number slightly less than **e**.

When we say ln(**e**) = 1 we are saying that the power to which you would raise **e** in order to get **e** is 1.

Example 9 □ Find ln 100.

Press: 100 [LN].
The display will be 4.6051702.   □

> Log 100 was 2. This appears to be irrational.

> 100 is not an integer power of **e**. $e^{4.651702} = 100$.

 Using Your Calculator

Finding Powers of e

Since the natural logarithm of a number is the power to which you raise **e** in order to get that number, you would expect that there is a way to raise **e** to a power. Your calculator probably provides two ways to find a power of **e**.
  The first way to find a power of **e** is the y^x key.
  First let's learn how to use the y^x key.

Example 10 □ Find 2^3.
  To find 2^3 using the $\boxed{y^x}$ key, do the following:

a. Enter 2.
b. Press the $\boxed{y^x}$ key.
c. Enter 3.
d. Press $\boxed{=}$.

You should get $2^3 = 8$.   □

Example 11 □ Find e^2.
  To find e^2 using the $\boxed{y^x}$ key:

a. Enter 2.7182818.
b. Press the $\boxed{y^x}$ key.
c. Enter 2.
d. Press $\boxed{=}$.

You should get $e^2 \approx 7.3890$.
The second way to find a power of **e** is the $\boxed{e^x}$ key.   □

> My calculator doesn't have an e^x key.

> Then press $\boxed{INV}$ followed by $\boxed{LN}$ for e^x.

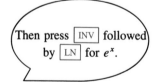

To find e^2 using the e^x key:

a. Enter 2.
b. Press $\boxed{e^x}$ or $\boxed{INV}$ $\boxed{LN}$, depending on your calculator.

You should get $e^2 \approx 7.3890$.   □

Example 12 □ Find $\ln(e^{13})$.

Depending on your calculator, do either A or B.

A	B	
a. Enter 13	**a.** Enter 13	
b. Press $\boxed{e^x}$	**b.** Press $\boxed{\text{INV}}$	This says find the inverse of the natural log of x which is e^x
c. Press $\boxed{\text{LN}}$	**c.** Press $\boxed{\text{LN}}$	
	d. Press $\boxed{\text{LN}}$	

You should get $\ln(e^{13}) = 13$, because the power to which you would raise e in order to get e^{13} is 13. □

Example 13 □ Find the number whose natural log is 1.

Since logs are exponents, finding the number whose natural log is 1 is equivalent to finding e^1. Depending upon your calculator, do either A or B.

A	B
a. Enter 1	**a.** Enter 1
b. Press $\boxed{e^x}$ key	**b.** Press $\boxed{\text{INV}}$
	c. Press $\boxed{\text{LN}}$

to get 2.7182818 which is e. □

Atmospheric Pressure

One of the many things that decreases exponentially is atmospheric pressure. Up to about 30,000 feet the pressure of air is approximated by the formula below.

14.2C Atmospheric Pressure Equation

$$P = 14.7e^{-0.037a} \ \frac{\text{lbs}}{\text{in}^2} \qquad \text{where } a = \text{altitude in thousands of feet.}$$

Example 14 □ Determine standard atmospheric pressure at sea level. Since sea level is zero-feet elevation, evaluate the formula.

$$P = 14.7e^{-0.037a}\,\frac{\text{lbs}}{\text{in}^2} \qquad \text{at } a = 0 \text{ feet}$$

$$P = 14.7(e^0)\,\frac{\text{lbs}}{\text{in}^2}$$

$$P = 14.7\,\frac{\text{lbs}}{\text{in}^2}$$

Physicists refer to the value of P when $a = 0$ as P_0 (say "P sub zero") or the initial condition. □

Example 15 □ What is atmospheric pressure at 18,000 feet?

$$P = 14.7e^{-(0.037)a}\,\frac{\text{lbs}}{\text{in}^2}$$

$$= 14.7e^{-(0.037)(18)}\,\frac{\text{lbs}}{\text{in}^2}$$

$$= 14.7e^{-(0.666)}\,\frac{\text{lbs}}{\text{in}^2}$$

Enter .666, press the $\boxed{+/-}$ key to get $-.666$, and then use the $\boxed{e^x}$ key or the $\boxed{\text{INV}}$ $\boxed{\text{LN}}$ keys on the calculator for $e^{-0.666} = 0.513759$.

$$P \approx 14.7(0.514)\,\frac{\text{lbs}}{\text{in}^2}$$

Round off to 0.514 since 14.7 is 3-place accuracy.

$$\approx 7.5\,\frac{\text{lbs}}{\text{in}^2}$$

About half of the earth's atmosphere is below 18,000 feet. □

Example 16 □ Find the altitude where atmospheric pressure is 0.7 of the pressure at sea level.
Use the formula $P = P_0e^{-0.037a}$
We are looking for the value of a

$$\text{where} \qquad P = 0.7P_0$$

$$0.7P_0 = P_0e^{-0.037a}$$

$$0.7 = e^{-0.037a}$$

P_0 is read "P sub zero." It refers to the initial or starting value of pressure.

I can use logs base 10 to solve this.

True, but, since we want a power of e, ln works much nicer than using base 10 logs.

Take the natural logarithm of both sides.

$$\ln(0.7) = \ln(e^{-0.037a})$$

$$-0.3566 \approx -0.037a \qquad \ln(e^x) = x$$

$$9.6 \approx a \qquad \text{Divide by } -0.037$$

Recall that a is in thousands of feet.

Therefore at 9600 feet the atmospheric pressure is about seven tenths of sea level pressure. □

Half-Life of Radioactive Material

Some materials such as uranium emit particles naturally. The period of time required for half the radioactive mass of material to decay is referred to as the half-life of the material. After one half-life, one half of the original material would remain. After two half-lives, one fourth of the original material remains and so on. At the time of a half-life, the decay equation is given by the formula below.

14.2D   Half-Life

$$\frac{1}{2} = e^{-\lambda t} \qquad \text{where } \lambda \text{ is the decay constant}$$
$$t \text{ is time in the same units as the decay constant}$$

λ is pronounced "Lambda."

Example 17 □ Uranium 238 has a decay constant of 1.54×10^{-10}/year. Find its half-life.

This requires solving

$$\frac{1}{2} = e^{-\lambda t} \qquad \text{for } t.$$

Taking the ln of both sides

$$\ln\left(\frac{1}{2}\right) = \ln(e^{-\lambda t})$$

$$-0.693 = -\lambda t$$

$$-0.693 = -(1.54 \times 10^{-10})t \qquad \text{Substituting for } \lambda$$

$$\frac{-0.693}{-(1.54 \times 10^{-10})} = t$$

$$4.5 \times 10^9 \text{ years} \approx t$$

This means that it takes 4,500,000,000 years for half of the mass of a quantity of uranium to go away. This long half-life is a big part of the problem of radioactive waste. The stuff just doesn't decay at the same rate as normal garbage. □

Population Growth

14.2E Population Growth

The formula for population growth over a short period of time is
$P = P_0 e^{kt}$

where P_0 is the initial population

k is the growth constant

P is the population after t years

Chapter 14 ■ Applications of Logarithms and Exponential Functions

Example 18 □ What is the growth constant for a city whose population doubles every nine years?

When the original population doubles $P = 2P_0$.

Then the formula above

$$P = P_0e^{kt}$$

becomes

$$2P_0 = P_0e^{kt}$$

Since this doubling took 9 years, we can replace t with 9.

$$2P_0 = P_0e^{k(9)}$$

$$2 = e^{9k} \qquad \text{Dividing both sides by } P_0$$

$$\ln 2 = \ln e^{9k} \qquad \text{Take the ln of both sides}$$

$$0.6931 \approx 9k$$

$$0.0770 \approx k$$

This is the growth constant k. □

Example 19 □ If the population of the town in Example 18 was 20,000 in 1988, what would you expect its population to be in 1996?

$$P = P_0e^{kt}$$

From Example 18, $k = 0.0770$. In 8 years, the population will be

$$P = 20000e^{0.0770(8)}$$

$$= 20000e^{0.6161}$$

$$= 20000(1.852)$$

$$P \approx 37,000$$

My calculator gives more accuracy than that.

This is an estimate. 3-digit accuracy over 8 years is pretty optimistic.

We expect the population to be 37,000 in 1996. □

Example 20 □ Find the rate of growth for the town in Example 18.

The rate of growth is how much the town grows in one year expressed as a percent. From Example 18, $k = 0.0770$.

$$P = P_0 e^{kt}$$

$$= P_0 e^{0.0770(1)} \qquad \text{1 year}$$

$$= P_0 e^{0.0770}$$

$$\approx P_0(1.08)$$

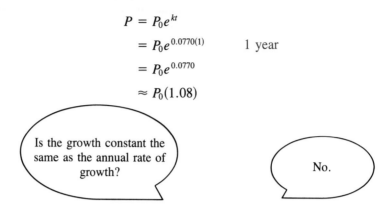

Is the growth constant the same as the annual rate of growth?

No.

Each year the population will be 1.08 times the previous year's population. Hence the rate of growth is 8% per year. In this example, the growth constant is $k = 0.0770$. □

Example 21 □ If a town grows 15% in one year, what is its growth constant?

At the end of one year the population P is $1.15P_0$. The time t is 1. Substituting in the formula $P = P_0 e^{kt}$

$$1.15P_0 = P_0 e^{k(1)}$$

$$1.15 = e^k$$

$$\ln(1.15) = k$$

$$0.13976 \approx k \qquad\qquad □$$

Example 22 □ If the town in Example 21 has a population of 25,000 this year, what would you expect its population to be in 3 years?

$$P = P_0 e^{kt}$$

From Example 21, $k = 0.13976$.

$$P = 25000e^{.13976(3)}$$

$$= 25000e^{.41928}$$

$$\approx 25000(1.520)$$

$$P \approx 38,000$$

Chapter 14 ■ Applications of Logarithms and Exponential Functions

We have mentioned exponential functions because so many things naturally follow an exponential pattern.

The general pattern for exponential growth or decay is:

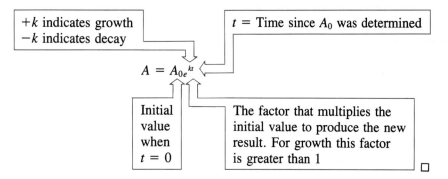

$+k$ indicates growth
$-k$ indicates decay

$t =$ Time since A_0 was determined

$A = A_0 e^{kt}$

Initial value when $t = 0$

The factor that multiplies the initial value to produce the new result. For growth this factor is greater than 1

Problem Set 14.2

Solve the following problems.

The formula for population growth over a short period of time is $P = P_0 e^{kt}$ where P_0 is the initial population.

k is the growth constant

P is the population after t years

1. In Jamesville the population doubled during a 5-year period (May, 1980–May, 1985). Determine the growth constant and the annual rate of growth.

2. If the population of Jamesville was 650 in May, 1980, how many people lived in Jamesville in May, 1982?

3. How many years would it take for Jamesville to quadruple its population if it had 650 people initially?

4. Due to a fast growing industry, Johnsville tripled its population in 6 years. Find the growth constant. Find the annual rate of growth.

5. Johnsville tripled its population in 6 years (May, 1982–May, 1988). How long did it take to double its population?

6. If Johnsville continues to grow at its present rate, what would be its population in May, 1995, if its population in May, 1985, was 1250?

7. The world population in 1980 was approximately 4 billion and growing at the rate of 2% per year. Assuming continuous growth, estimate the population in the year 2000.

8. During the gold rush in the 19th century, a certain town tripled its population in 2 years. Determine the growth constant.

9. If the population of the town in problem 8 was 540 people just prior to the gold rush, what was its population 3 years later?

10. Determine the annual rate of growth for the town in problem 8.

11. If a certain town grows 12% in one year, what is its growth constant?

12. If the town in problem 11 had an original population (at its first census) of 1450 people, what was its population 5 years later?

13. If a town grows 18% per year, how long would it take to double its population?

For problems 14–25, use the formula $P = 14.7\ e^{-0.037a}$
where $\qquad P =$ air pressure in pounds per square inch
$\qquad\qquad a =$ altitude in thousands of feet

14. What is the atmospheric pressure at 16,000 feet?

15. What is the atmospheric pressure at 12,000 feet?

16. What is the atmospheric pressure at 8500 feet?

17. What is the atmospheric pressure at 3500 feet?

18. Denver, Colorado, is called The Mile-High City. What is its atmospheric pressure?

19. Big Bear Mountain in California is approximately 6700 feet high. What is the atmospheric pressure at Big Bear Mountain?

20. Mt. McKinley is considered the highest mountain in the United States (approximately 20,000 feet). What is its atmospheric pressure?

21. Mt. Everest is considered the highest mountain in the world (approximately 29,000 feet). What is the atmospheric pressure at the top of Mt. Everest?

22. Federal aviation regulations require pilots to have supplemental oxygen above 14,000 feet. What part of the air pressure at sea level is available to the pilot at this altitude?

23. Sam Jones, who has mild heart problems has difficulty breathing at an elevation of 6000 feet. What part of the air pressure at sea level is available to Sam at 6000 feet?

24. What is the atmospheric pressure outside of a jet airplane flying at 35,000 feet?

25. If an airplane flying at 18,000 feet lost its ability to pressurize its cabin, what part of the air pressure at sea level would then be available?

The formula for finding the amount of an investment after compounding for a period of time is given by $A = P\left(1 + \dfrac{r}{n}\right)^{nt}$

The formula for finding the amount of an investment after compounding continuously for a period of time is given by $A = Pe^{rt}$

where A = amount after a period of time t = time in years
               P = initial investment r = rate of interest per year

Find the amount of an investment of $2000 at 15%

26. Compounded yearly for 4 years

27. Compounded semi-annually for 4 years

28. Compounded quarterly for 4 years

29. Compounded monthly for 4 years

30. Compounded weekly for 4 years

31. Compounded daily for 4 years

32. Compounded hourly for 4 years

33. Compounded continuously for 4 years

Suppose you inherited $10,000 today. You might consider investing your inheritance for your retirement. If you were able to invest at 12% compounded continuously, how long would it take to:

34. Double your investment

35. Triple your investment

36. Quadruple your investment

37. Have $100,000 in your account

Use the general exponential formula $A = A_0 e^{kt}$ for the following problems.

38. A certain type of bacterium weighs 10^{-12} grams. The mass of the earth is 6×10^{24} grams. If the mass of a bacteria culture doubles every half hour, how long would it take a single bacterium to grow to a culture with a mass equivalent to the earth's mass?

39. If a population of bacteria has a generation time of 40 minutes, how long will it take a population to grow from 5×10^4 cells to 10^9 cells?

40. If the number of bacteria in a culture doubles every half hour, how long would it take a culture of 2500 bacteria to grow to 1,000,000?

41. A particular bacterium divides every 45 minutes. If we start with a culture of 5000 bacteria, after t hours how many bacteria will we have? How long will it take to have a culture of 50,000 bacteria?

■ **14.3**   **Other Applications of Logarithms**

A measure of the acidity of a solution is pH. Chemists define pH as follows:

14.3A   Definition: pH

$pH = -\log_{10}(H^+)$ where H^+ = hydrogen ion concentration in moles/liter

---

Most solutions in a chemical laboratory have a pH between 0 and 14. Water with a pH of 7 is considered neutral. A pH of 0 is an extremely strong acid.

Example 23 □ A sample of tomato juice was found to have a hydrogen ion concentration of 10^{-4}. What is the pH of this drink?

$$\text{pH} = -\log_{10}[H^+]$$
$$= -\log_{10}[10^{-4}]$$
$$= -(-4)$$
$$\text{pH} = 4 \qquad\qquad\qquad\qquad □$$

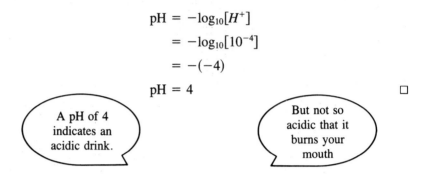

A pH of 4 indicates an acidic drink.

But not so acidic that it burns your mouth

Example 24 □ A sample of orange juice has a hydrogen ion concentration of 2.7×10^{-4}. What is the pH of the orange juice?

$$\text{pH} = -\log_{10}[H^+]$$
$$= -\log_{10}[2.7 \times 10^{-4}]$$
$$\approx -(-3.568)$$
$$\approx 3.6 \qquad\qquad\qquad\qquad □$$

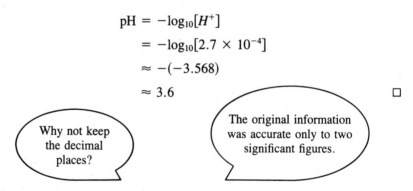

Why not keep the decimal places?

The original information was accurate only to two significant figures.

Example 25 □ A certain brand of coffee had a pH measurement of 5. What is its hydrogen ion concentration?

$$\text{pH} = -\log_{10}[H^+]$$
$$5 = -\log_{10}[H^+]$$
$$-5 = \log_{10}[H^+]$$
$$10^{-5} = H^+ \qquad\qquad \text{Taking inverse log} \qquad □$$

Radiocarbon Dating

Radiocarbon Dating

CARBON-14 is an isotope of carbon produced by cosmic ray bombardment of nitrogen atoms in the atmosphere. Hence there is a small portion of C-14 in the air. Plants which get their carbon from the air also contain a small portion of C-14.

When a plant dies, it no longer takes in C-14 from the air. The C-14 already in the plant begins to decay with a half-life of 5730 years. Thus by measuring the portion of C-14 in a piece of wood, archeologists can determine how long ago the tree was cut.

Example 26 □ How old is a piece of wood that has one fourth the radioactivity of a similar sample of modern wood?

The equation for half-life from Section 14.2 is

$$\frac{1}{2} = e^{-\lambda t}$$

As mentioned in the introduction, we know the half life of C-14. It is 5730 years. We must find λ.

$$\frac{1}{2} = e^{-\lambda(5730 \text{ years})}$$

$$\ln(0.5) = -\lambda(5730 \text{ yrs}) \qquad \text{Taking ln of both sides}$$

$$-0.6931 \approx -\lambda(5730)$$

$$0.00012 \approx \lambda$$

Substituting this value of λ in the decay equation

$$A = A_0 e^{-\lambda t}$$

$$A = A_0 e^{-0.00012t}$$

$$\frac{A}{A_0} = e^{-0.00012t} \qquad \text{Dividing by } A_0$$

We are looking for the value of t that makes the ratio $\dfrac{A}{A_0} = \dfrac{1}{4}$.

$$\frac{1}{4} = e^{-0.00012t}$$

$$\ln\left(\frac{1}{4}\right) = -0.00012t \qquad \text{Taking ln of both sides}$$

$$-1.3863 \approx -0.00012t$$

$$11552 \approx t$$

The wood is approximately 11,500 years old. □

Why approximate?

There are several assumptions connected to this method of dating.

One of the main assumptions is that the earth's magnetic field, and consequently the level of cosmic rays in the upper atmosphere, has been constant for the period in question. Using more sophisticated techniques, adjustments can be made for known variations.

Measurement of Earthquakes

A seismograph is used to measure the amount of earth movement due to an earthquake.

The amplitude of this motion can vary from barely perceptible to quite large. Therefore, the scale used to measure earthquakes uses logarithms. In order to have some basic standard to compare earthquakes a "zero" earthquake was defined by Dr. C. F. Richter and Dr. Beno Gutenberg of California Institute of Technology. The magnitude of an earthquake on the Richter scale is the difference in motion between the earthquake and a zero earthquake.

14.3B   Richter Scale

The magnitude of an earthquake is given by

$$m = \log A - \log A_0$$ where A_0 = Amplitude of a zero level earthquake

   A = Amplitude of this earthquake

This can be rewritten as

$$m = \log\left(\frac{A}{A_0}\right)$$

Chapter 14 ■ Applications of Logarithms and Exponential Functions

The chart below gives the magnitude of some modern earthquakes. These are by no means all the earthquakes during this time period. Since 1918, there have been nine earthquakes with a magnitude of 8.6 or greater. If we could detect all the magnitude zero earthquakes, we would probably record several million earthquakes a year.

Modern Earthquakes

Date	Location	Lives lost	Richter magnitude
1906	Columbia, Ecuador	Unknown	8.9
1933	Japan	3000	8.9
1964	Alaska	114	8.6
1985	Mexico City	7000	8.1
1906	San Francisco	700	8.0
1952	Kern County, CA	14	7.7
1971	San Fernando, CA	65	6.6
1983	Coalinga, CA	0	6.5
1933	Long Beach, CA	115	6.3
1987	Whittier, CA	5	6.1

Example 27 □ Calculate the magnitude of an earthquake that is 100 times the size of a magnitude A earthquake.

An earthquake 100 times as large as an earthquake with magnitude A has magnitude $100A$. Using the definition of magnitude of an earthquake

$$m = \log\left(\frac{A}{A_0}\right)$$

We wish to compare two values of A. Therefore we need to solve this equation for A.

Taking the inverse log of both sides yields

$$10^m = \frac{A}{A_0}$$

OR
$$A = A_0 10^m \quad \longleftarrow \boxed{\text{Amplitude of magnitude } m \text{ earthquake}}$$

$$100A = 100A_0 10^m \longleftarrow \boxed{\text{Amplitude of an earthquake 100 times as large}}$$

$$= 10^2 A_0 10^m \qquad 100 = 10^2$$

$$= A_0 10^{m+2} \qquad \text{To multiply, add exponents}$$

An increase of 2 on the Richter scale means the size of the earthquake has increased 100 times. Therefore, a magnitude 7 earthquake is 100 times the strength of a magnitude 5 earthquake. □

Example 28 □ How many times larger is the amplitude of a magnitude 7.5 earthquake than the amplitude of a magnitude 5.2 earthquake?

The amplitude of an earthquake is given by

$$A = A_0 10^m$$

We can use this equation to find the amplitude of a magnitude 7.5 earthquake.

$$A_{7.5} = A_0 10^{7.5}$$

Similarly for a magnitude 5.2 earthquake

$$A_{5.2} = A_0 10^{5.2}$$

The ratio of the two amplitudes is

$$\frac{A_{7.5}}{A_{5.2}} = \frac{A_0 10^{7.5}}{A_0 10^{5.2}}$$

Amplitude of a 7.5 earthquake

Amplitude of a 5.2 earthquake

$$= \frac{10^{7.5}}{10^{5.2}}$$

$$= 10^{2.3} \qquad \text{Subtract exponents to divide}$$

Since $10^{2.3} \approx 199.5$, a 7.5 earthquake is approximately 200 times as powerful as a 5.2 earthquake. □

Energy Released by an Earthquake

The energy released by an earthquake is not proportional to the amount of earth movement.

The energy released by an earthquake is given by the equation

$$\log E = 11.4 + 1.5m$$

where $\qquad E = $ energy in ergs

$\qquad\qquad m = $ magnitude on the Richter scale

Example 29 □ Calculate the energy released by a magnitude 8.0 earthquake.

$$\log E = 11.4 + 1.5m$$

$$\log E = 11.4 + 1.5(8)$$

$$\log E = 11.4 + 12$$

$$\log E = 23.4$$

$$E = 10^{23.4}$$

$$E \approx 2.5 \times 10^{23} \text{ ergs}$$

To visualize this amount of energy it helps to know that one foot-pound is the energy required to raise one pound a distance of one foot.

$$1 \text{ ft-lb} = 1.356 \times 10^7 \text{ ergs}$$

Therefore the energy in this example is

$$E = \frac{2.5 \times 10^{23} \text{ ergs}}{1.356 \times 10^7 \text{ ergs/ft-lb}}$$

$$\approx 1.84 \times 10^{16} \text{ ft/lbs}$$

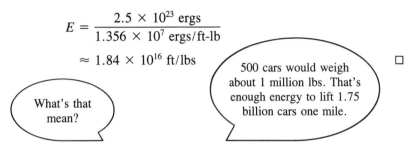

What's that mean?

500 cars would weigh about 1 million lbs. That's enough energy to lift 1.75 billion cars one mile.

It is estimated that if an earthquake of magnitude 8.0 occurred in Los Angeles, between five and fifteen thousand people would be killed.

Measurement of Sound

Measurement of sound levels is an important consideration in protecting the environment.

The level of sound is determined by the pressure amplitude of a sound wave and is measured in watts per square meter. To give you some idea of the levels of energy involved we repeat an illustration from *University Physics* by Sears, Zemansky, and Young.

If the entire population of New York City (8 million people) were to speak at once in a normal tone, about 8 watts of power would be generated. That's enough power to light a medium-size electric bulb. Sears, Zemansky, and Young go on to indicate that the power required to fill a large auditorium with sound is 2500 watts.

Why this wide discrepancy in levels of power? The answer is that the response of the human ear is logarithmic. This permits us to hear very soft sounds and tolerate extremely loud ones.

The volume, V, of sound is measured in decibels and given by the equation

$$V = 10 \log_{10} \frac{I}{I_0}$$

Where I equals the intensity of sound in watts/m^2.

Where I_0 is assumed to be the energy level of the faintest sound that can be heard, like a pin dropping (about 10^{-12} watts/m^2).

Decibels are a comparison of two levels.

Which levels are being compared?

Normally, we compare the level of the sound in question to the level of the quietest sound that can be heard.

Example 30 □ Find the decibel level of the faintest sound that can be heard.

$$V = 10 \log_{10}\left(\frac{I}{I_0}\right)$$

The intensity of the faintest sound that can be heard is defined to be I_0. The equation then becomes

$$V = 10 \log_{10}\left(\frac{I_0}{I_0}\right)$$

$$V = 10 \log 1$$

$$V = 10(0) \qquad \text{Since } \log 1 = 0$$

$$V = 0 \qquad\qquad\qquad □$$

The faintest sound that can be heard is at a volume level of 0 decibels. The maximum level that the human ear can tolerate is about 120 decibels.

Table 14.3C

Common Sound Levels

The levels of some common sounds are

Whisper	20 db
Normal conversation	65 db
Small airplane 1 mile away	65 db
Traffic sounds on a sidewalk alongside a busy street	70 db
Wide-body jet 1 mile away	92 db
Riveter 20 feet away	95 db
Loud rock concert	120 db

Example 31 □ How much higher than a normal conversation is the energy level of a rock concert?

This example asks for the ratio of two energy levels. It is similar to Example 28 where two earthquakes were compared.

We need to solve the decibel equation for the energy level I.

$$V = 10 \log_{10}\left(\frac{I}{I_0}\right)$$

$$\frac{V}{10} = \log_{10}\left(\frac{I}{I_0}\right) \qquad \text{Divide by 10}$$

$$10^{\frac{V}{10}} = \frac{I}{I_0} \qquad \text{Take the inverse log of both sides}$$

$$I_0 10^{\frac{V}{10}} = I \qquad \text{Multiply both sides by } I_0$$

Chapter 14 ■ Applications of Logarithms and Exponential Functions

This formula can be used to find the ratio of the energy level of a rock concert to the energy level of normal conversation.

$$\frac{I_{\text{rock convert}}}{I_{\text{conversation}}} = \frac{I_0 10^{\frac{120}{10}}}{I_0 10^{\frac{65}{10}}}$$

Table 14.3C gives energy of a rock concert as 120 db

Table 14.3C gives the energy of conversation as 65 db

$$= \frac{10^{12}}{10^{6.5}} \qquad \text{Divide out } \frac{I_0}{I_0}; \text{ simplify exponents}$$

$$= 10^{5.5}$$

$$\approx 316{,}228$$

The energy level of the rock concert is over 300 thousand times higher than normal conversation. □

How come rock fans aren't deaf?

Because your ear detects sounds logarithmically. However, many rock musicians show a hearing loss at an early age.

Example 32 □ Determine the energy level in watts per square meter of the sound along a busy street where the intensity level of the sound is 70 db. Recall $I_0 = 10^{-12}$ watts/square meter.

$$V = 10 \log_{10}\left(\frac{I}{I_0}\right)$$

$$70 = 10 \log\left(\frac{I}{10^{-12}}\right)$$

$$7 = \log\left(\frac{I}{10^{-12}}\right)$$

$$10^7 = \frac{1}{10^{-12}} \qquad \text{Taking inverse log}$$

$$10^{-5} \frac{\text{watts}}{\text{meter}^2} = I \qquad \text{Multiplying by } 10^{-12}$$

□

Problem Set 14.3

Solve the following problems

1. A solution of drain cleaner was found to have a hydrogen ion concentration of 2.5×10^{-13}. What is the pH?

2. At a sidewalk stand run by children lemonade was sold. The neighborhood child chemist tested the lemonade and found the hydrogen ion concentration to be about 10^{-3}. What was the pH?

3. Checking the detergent used by the family, the father, who is a chemist, found the pH to be approximately 9.4. What was the hydrogen ion concentration?

4. The family chemist also checked the soap used by the family and found the pH to be approximately 7.6. What was the hydrogen ion concentration?

5. A sample of vinegar has a hydrogen ion concentration of 1.03×10^{-3}. What is its pH?

6. A sample of black coffee has a hydrogen ion concentration of 1.16×10^{-4}. What is its pH?

7. How old is a piece of wood that has three-fourths of the radioactivity of a similar sample of modern wood? (Hint: Use the formula for C-14 with a half-life of 5730 years.)

8. An unstable isotope of cobalt CO-60 is widely used in medicine. It has a half-life of 5.3 years. Determine its decay constant. If the original amount of cobalt is 5 mg (milligrams), how much will be left after 12 years?

9. The half-life of a certain radioactive substance is 10 minutes. If a laboratory has 5 grams of the substance, how much will be left after 20 minutes? After 30 minutes?

10. A certain isotope of radium decomposes at $A = A_0 e^{-0.038t}$, where t is in centuries. What is its half-life? If the original amount has 10 grams, how much remains after 500 years?

11. How old is a piece of wood that has one fifth of the radioactivity (C-14) of a similar piece of modern wood? One-tenth of the radioactivity?

12. The decay rate of a certain radioactive substance is expressed by $A = A_0 e^{-0.0175t}$. If you begin with 50 grams of the substance how much will remain after 5 years? After 10 years? After 20 years?

13. How much greater than the energy level of a normal conversation is the energy level developed by the sound of a riveter?

    Normal conversation: $I = 65$ db

    Riveter: $I = 95$ db

14. How much higher is the energy level developed by the sound of a small airplane one mile away than the energy level of the sound inside a quiet automobile?

    Small airplane: $V = 65$ db

    Quiet automobile: $V = 50$ db

15. Find the noise level (L) in decibels of a quiet radio if the intensity (I) in watts/meter^2 is 10^{-8}.

16. Find in decibels the noise level (L) of an average whisper if the intensity (I) in watts/meter^2 is 10^{-10}

17. Calculate the energy released by an earthquake with magnitude of 4.2 on the Richter scale. Give your answer in ft-lbs.

18. Calculate the energy released by an earthquake with magnitude of 5.8 on the Richter scale. Give your answer in ft-lbs.

19. Calculate the energy released by the 1983 earthquake in Coalinga, Ca. The earthquake read 6.5 on the Richter scale. Give your answer in ft-lbs.

20. What is the magnitude on the Richter scale of an earthquake that is $10^{2.4}$ times larger than the 1933 earthquake in Long Beach, California, whose Richter scale reading was 6.3?

21. What is the reading on the Richter scale of an earthquake that is 1000 times larger than the San Fernando, California, earthquake in 1971? Richter scale reading shows a magnitude of 6.6.

22. What is the magnitude for the Richter scale of an earthquake that is $10^{5.8}$ times larger than the zero earthquake?

23. How many times larger is a magnitude 8 earthquake than a magnitude 6.5 earthquake?

24. How many times larger was the 1933 Yokohama, Japan, earthquake with magnitude 8.9 than the 1933 Long Beach, California, earthquake with magnitude of 6.3?

Chapter 14 □ Key Ideas

14.1   1. **Logarithms** can be used to solve a variety of exponential equations and logarithmic equations.

      a. Solving equations when the variable is an exponent.

      b. Solving equations involving logarithmic functions.

14.2 **1. Logarithms and exponents** to the base **e** are used frequently in physics, mathematics, chemistry, biology, finance, and many other scientific and economic fields.

    a. Compound interest is given by:

$$A = P\left(1 + \frac{r}{n}\right)^{nt} \qquad \text{for a fixed number}$$

    b. Atmospheric pressure:

$$P = 14.7e^{-0.037a} \frac{\text{lbs}}{\text{in}^2}$$

    c. Half-life of radioactive substance:

$$\frac{1}{2} = e^{-\lambda t}$$

    d. Decay:

$$A = A_0 e^{kt} \qquad \text{where } k < 0$$

    e. Growth:

$$A = A_0 e^{kt} \qquad \text{where } k > 0$$

    f. Population growth:

$$P = P_0 e^{kt}$$

14.3 **1. Other applications**

    a. From chemistry, to find the pH:

$$\text{pH} = -\log_{10}(\text{H}^+)$$

    b. Radio carbon dating
        1. Half-life: $\dfrac{1}{2} = e^{-\lambda t}$
        2. $A = A_0 e^{kt}$

    c. Measurement of earthquakes
        1. Magnitude of an earthquake:

$$m = \log_{10}\left(\frac{A}{A_0}\right)$$

        2. Energy released by an earthquake:

$$\log_{10} E = 11.4 + 1.5m$$
E is measured in ergs.

d. Measurement of sound. The volume (V) of sound is given by:

$$V = 10 \log_{10}\left(\frac{I}{I_0}\right)$$

Volumes of sound are measured in decibels.

Chapter 14   Review Test

Solve the following equations, using logarithms.

1. $2^x = 14$ **2.** $5^{2x} = 650$ **3.** $3^{9x} \cdot 9^{x^2} = \dfrac{1}{81}$ **4.** $16^{x^2+2x} = 2^{7x+3}$

Solve each of the following equations.

5. $\log(3x - 4) = \log(5x - 8)$ **6.** $\log(x^2 + 4) = \log(4x + 9)$

7. $\log_2(3x + 4) + \log_2(x + 3) = \log_2(4 - x)$ **8.** $\log(2x - 1) = 1 - \log x$

The basic formula for exponential growth is $A = A_0 e^{kt}$.

9. If the population of Jerriville doubled during a 6-year period, 1976–1981, determine the growth constant. Find the annual rate of growth.

10. Estimate the population in Jerriville at the end of 1977 if the population at the beginning of 1976 was 1250.

11. The growth constant in a city is 0.05. If the population at the beginning of 1970 was 200,000, what was the population at the beginning of 1980? At the beginning of 2000?

12. A bacteria culture has 10,000 bacteria initially. If the rate of growth is 10% each day, how many bacteria will there be in 10 days? In 30 days?

Atmospheric pressure is given by the formula $P = 14.7e^{-0.037a}$.
P is the atmospheric pressure in lb/in²
a is the altitude in 1000 ft.

13. What is the atmospheric pressure at 15,000 feet?

14. Mt. McKinley in Alaska is approximately 20,000 feet high. What is the atmospheric pressure at its peak?

15. Mammoth Ski Resort in California is about 8000 feet. What part of air pressure at sea level is available?

16. What would be the amount of an $8000 investment at 15% annual interest rate compounded continuously for 5 years? For 10 years?

17. Find the amount of a $6000 investment at 12% annual rate compounded quarterly for 5 years? What is the amount if compounded continuously at the same rate for the same length of time?

Chemistry

18. A sample of cow's milk has a pH of 6.4. What is its hydrogen ion concentration?

19. A sample of sea water has a hydrogen ion concentration of 1.2×10. What is its pH?

20. The half-life of a certain radioactive substance is 2 hours. If you have 10 grams of the substance, how much will remain after one half hour? After one hour? After five hours?

21. An isotope of radium decomposes at $A = A_0 e^{-0.0125t}$ where t is in years. What is its half-life? If the original amount was 20 grams, how much remains after one half year? After two years? After 20 years?

22. The half-life of C-14 is 5730 years. How old is a piece of wood that has five-sixths of the radioactivity of a similar piece of modern wood?

23. How much higher than normal conversation ($V = 65$ db) is the energy level of a wide-body jet a mile away ($V = 92$ db)?

24. Find the noise level of an automobile with an intensity in watts per meter squared of 10^{-6}.

25. Calculate the energy released by an earthquake with a magnitude of 4.8 on the Richter scale. Give your answer in ft-lbs.

26. What is the magnitude on the Richter scale of an earthquake that is $10^{1.8}$ times stronger than an earthquake with a Richter scale reading of 5.2?

27. How many times stronger is a magnitude 8.2 earthquake than a magnitude 4.8 earthquake?

15 Matrices and Determinants

Contents

This chapter will show you methods for dealing with rectangular arrays of numbers. These arrays are called **matrices.** Matrices are used in many areas where there are large numbers of conditions to be considered simultaneously. Some of these areas are computer graphics, business modeling, aeronautical engineering, traffic flow, and electrical network analysis. In all of these areas, systems of up to several hundred equations need to be solved. This chapter will show you how Cramer's Rule can be used with matrices to solve these systems of equations.

Before computers, the tediousness of computation limited the applications of matrices. Now the availability of computers makes applications of matrix mathematics a growing field.

■ 15.1 Matrices

The branch of mathematics that deals with matrices is called **matrix algebra.** This section will show you a few of the basic ideas. These ideas are applied in many fields. Here is partial list:

Business   -Input output analysis

-Transportation problems

-Demand forecasting

-Linear programming

Computers -Data base management

-Translations, rotations and scalings in computer graphics

Engineering-Network analysis

-Aerodynamics

15.1A   Definition: Matrix

A matrix is a rectangular array of numbers.

Some examples of matrices are:

	Preferred	Reserved	General
Adults	30.00	21.50	7.50
Juniors	20.00	14.75	3.75

		% of voters			
Candidate	Support	Favor	Neutral	Opposed	Undecided
Jones	13	30	1	22	34
Smith	6	40	30	2	22
Johnson	2	10	40	10	38

Other examples in formal mathematical notation are:

$$A = \begin{bmatrix} 3 & -2 & 7 \\ 4 & 1 & 10 \end{bmatrix}$$

column 1   column 2   column 3

$$B = \begin{bmatrix} -1 & 4 \\ 7 & 13 \\ 10 & 5 \\ 11 & 19 \end{bmatrix}$$

⟵ row 1
⟵ row 2
⟵ row 3
⟵ row 4

Matrix A has 2 rows and 3 columns. Matrix B has 4 rows and 2 columns. Each number in a matrix is specified by a subscript representing the row and column where it appears.

In matrix A

$$A_{(1,2)} = -2$$
$$A_{(1,3)} = 7$$
$$A_{(2,1)} = 4$$
$$A_{(2,2)} = \underline{\hspace{1cm}}$$

The first subscript tells which row.

The second subscript tells which column.

In matrix B

$$B_{(3,1)} = 10 \qquad B_{(3,2)} = \underline{\hspace{1cm}}$$
$$B_{(1,1)} = \underline{\hspace{1cm}} \qquad B_{(4,2)} = \underline{\hspace{1cm}}$$

Remember: columns are vertical; rows are horizontal.

Like the vertical columns that hold up a building.

A matrix with the same number of rows as columns is called a **square matrix**.

$$A = \begin{bmatrix} a & b \\ c & d \end{bmatrix}$$

A square matrix of order 2

Each square matrix has a number associated with it called its **determinant**.

1

5

−1, 19

15.1B Definition: Determinant of a Matrix of Order 2

The determinant of the matrix $\begin{bmatrix} a & b \\ c & d \end{bmatrix}$

is $\qquad \begin{vmatrix} a & b \\ c & d \end{vmatrix} = ad - bc$

Only square matrices have determinants.

A matrix is enclosed in brackets [ ]. A matrix is an array of numbers. It cannot be expressed as a single number.

A determinant is enclosed in bars | |. A determinant evaluates to a single number.

Example 1 □ Evaluate the determinant of the matrix $\begin{bmatrix} 2 & 3 \\ 4 & 5 \end{bmatrix}$.

$$\begin{vmatrix} 2 & 3 \\ 4 & 5 \end{vmatrix} = (2 \times 5) - (3 \times 4)$$

$$= 10 - 12$$

$$= -2 \qquad\qquad □$$

Example 2 □ Evaluate the determinant of the matrix $\begin{bmatrix} 4 & 2 \\ -3 & 5 \end{bmatrix}$.

$$\begin{vmatrix} 4 & 2 \\ -3 & 5 \end{vmatrix} = (4 \times 5) - (2)(-3)$$

$$= 20 - (-6)$$

$$= 26 \qquad\qquad □$$

Problem Set 15.1

Evaluate the following determinants.

1. $\begin{vmatrix} 2 & 1 \\ 1 & 3 \end{vmatrix}$
   2. $\begin{vmatrix} -5 & 4 \\ 2 & 6 \end{vmatrix}$
   3. $\begin{vmatrix} 3 & 2 \\ -5 & 4 \end{vmatrix}$
   4. $\begin{vmatrix} 1 & 6 \\ 3 & -2 \end{vmatrix}$

5. $\begin{vmatrix} -5 & 0 \\ 6 & 3 \end{vmatrix}$

6. $\begin{vmatrix} -5 & 3 \\ -4 & 0 \end{vmatrix}$

7. $\begin{vmatrix} -4 & 0 \\ -3 & -7 \end{vmatrix}$

8. $\begin{vmatrix} -5 & 2 \\ -3 & -4 \end{vmatrix}$

9. $\begin{vmatrix} 0 & -7 \\ -12 & -3 \end{vmatrix}$

10. $\begin{vmatrix} -5 & 7 \\ -6 & 0 \end{vmatrix}$

11. $\begin{vmatrix} -2 & 4 \\ 3 & -1 \end{vmatrix}$

12. $\begin{vmatrix} -1 & -3 \\ 5 & 2 \end{vmatrix}$

13. $\begin{vmatrix} 5 & -3 \\ 6 & 4 \end{vmatrix}$

14. $\begin{vmatrix} -2 & 5 \\ 0 & 4 \end{vmatrix}$

15. $\begin{vmatrix} 7 & 0 \\ 1 & 2 \end{vmatrix}$

16. $\begin{vmatrix} -6 & 2 \\ 5 & 0 \end{vmatrix}$

17. $\begin{vmatrix} -4 & 4 \\ 3 & -3 \end{vmatrix}$

18. $\begin{vmatrix} 2 & 4 \\ -3 & -5 \end{vmatrix}$

19. $\begin{vmatrix} -1 & 5 \\ 0 & 3 \end{vmatrix}$

20. $\begin{vmatrix} 9 & -2 \\ 7 & 3 \end{vmatrix}$

21. $\begin{vmatrix} \dfrac{1}{2} & -2 \\ 6 & -2 \end{vmatrix}$

22. $\begin{vmatrix} \dfrac{2}{3} & \dfrac{3}{4} \\ -12 & 15 \end{vmatrix}$

23. $\begin{vmatrix} \dfrac{1}{2} & -\dfrac{5}{6} \\ \dfrac{7}{5} & \dfrac{7}{3} \end{vmatrix}$

24. $\begin{vmatrix} 0.5 & 0.6 \\ 4 & -7 \end{vmatrix}$

25. $\begin{vmatrix} 0.08 & 4 \\ 0.007 & 0.9 \end{vmatrix}$

26. $\begin{vmatrix} 0.9 & -0.8 \\ 1.06 & 1.2 \end{vmatrix}$

27. $\begin{vmatrix} \dfrac{2}{3} & \dfrac{3}{4} \\ 0.20 & -0.012 \end{vmatrix}$

28. $\begin{vmatrix} \dfrac{5}{6} & -0.2 \\ -\dfrac{3}{5} & -0.42 \end{vmatrix}$

29. $\begin{vmatrix} -\dfrac{3}{4} & 9 \\ \dfrac{2}{3} & -8 \end{vmatrix}$

30. $\begin{vmatrix} \dfrac{3}{4} & -\dfrac{5}{6} \\ 12 & 4 \end{vmatrix}$

31. $\begin{vmatrix} \dfrac{5}{6} & -\dfrac{1}{2} \\ \dfrac{4}{3} & -\dfrac{16}{5} \end{vmatrix}$

32. $\begin{vmatrix} \dfrac{7}{4} & \dfrac{1}{2} \\ -\dfrac{1}{6} & \dfrac{5}{3} \end{vmatrix}$

33. $\begin{vmatrix} \dfrac{1}{4} & \dfrac{3}{8} \\ \dfrac{1}{3} & \dfrac{5}{6} \end{vmatrix}$

34. $\begin{vmatrix} 0.4 & 0.5 \\ 5 & -8 \end{vmatrix}$

35. $\begin{vmatrix} 0.04 & 0.008 \\ 6 & 7 \end{vmatrix}$

36. $\begin{vmatrix} \dfrac{2}{3} & \dfrac{3}{4} \\ 0.16 & 0.15 \end{vmatrix}$

■ 15.2   **Cramer's Rule**

One application of determinants is the solution of systems of equations. Consider a general system of two equations.

$$ax + by = p$$
$$cx + dy = q$$

This system can be solved by addition using the techniques given in Chapter 4.

$$ax + by = p \implies \quad adx + bdy = \quad pd \qquad \text{Multiply by } d$$

$$cx + dy = q \implies \quad \underline{-bcx - bdy = \quad -bq} \qquad \text{Multiply by } -b$$

$$adx - bcx = pd - bq \qquad \text{Adding}$$

$$(ad - bc)x = pd - bq \qquad \text{Factor } x$$

$$x = \frac{pd - bq}{ad - bc} \qquad \text{Divide}$$

Using a similar procedure,

$$y = \frac{aq - pc}{ad - bc}$$

15.2A The General Solution of a System of Two Linear Equations

$$ax + by = p$$
$$cx + dy = q$$

is

$$x = \frac{pd - bq}{ad - bc} \qquad\qquad y = \frac{aq - pc}{ad - bc}$$

provided $ad - bc \neq 0$

Notice, if you write the coefficients only of the system

$$ax + by = p$$
$$cx + dy = q$$

you get the matrix $\begin{bmatrix} a & b \\ c & d \end{bmatrix}$.

The matrix of the constant terms is $\begin{bmatrix} p \\ q \end{bmatrix}$.

Calculating the determinants of the coefficient matrix gives

$$\text{determinant of } \begin{bmatrix} a & b \\ c & d \end{bmatrix} = \begin{vmatrix} a & b \\ c & d \end{vmatrix} = ad - bc$$

Notice, this is the denominator of the general solution of a system of the two linear equations given in 15.2A.

Using determinant notation we can rewrite 15.2A.

15.2B Cramer's Rule: The Solution of a System of Two Equations

$$ax + by = p$$
$$cx + dy = q$$

is

$$x = \frac{\begin{vmatrix} p & b \\ q & d \end{vmatrix}}{\begin{vmatrix} a & b \\ c & d \end{vmatrix}} \qquad\qquad y = \frac{\begin{vmatrix} a & p \\ c & q \end{vmatrix}}{\begin{vmatrix} a & b \\ c & d \end{vmatrix}}$$

provided $\begin{vmatrix} a & b \\ c & d \end{vmatrix} \neq 0$

This looks difficult to remember.

Notice, both denominators are alike. They are the determinants of the coefficient matrix.

To remember the numerators in Cramer's rule, notice that they are the determinants of the coefficient matrix with the constant matrix replacing the coefficients of the variable we want.

Use the symbols D, D_x and D_y to help remember. (D_x is read "D sub x.")

$$D_x = \begin{vmatrix} p & b \\ q & d \end{vmatrix} \longleftarrow \boxed{\text{Replace the coefficients of } x \text{ with the matrix of the constant terms}}$$

replaces ↓

$$D = \begin{vmatrix} a & b \\ c & d \end{vmatrix} \longleftarrow \boxed{\text{This is the determinant of the coefficient matrix}}$$

↑ replaces

$$D_y = \begin{vmatrix} a & p \\ c & q \end{vmatrix} \longleftarrow \boxed{\text{Replace the coefficients of } y \text{ with the matrix of the constant terms}}$$

In shorter form Cramer's Rule is

$$x = \frac{D_x}{D} \qquad\qquad y = \frac{D_y}{D}$$

Example 3 □ Use Cramer's Rule to find the solution of the system.

$$3x - 2y = -1$$
$$2x + y = 4$$

First we find the determinant of the coefficient matrix

$$D = \begin{vmatrix} 3 & -2 \\ 2 & 1 \end{vmatrix} = (3)(1) - (-2)(2)$$

$$= 7$$

Since D is not zero we may proceed

$$x = \frac{D_x}{D} \qquad\qquad y = \frac{D_y}{D}$$

$$x = \frac{\begin{vmatrix} -1 & -2 \\ 4 & 1 \end{vmatrix}}{7} \qquad\qquad y = \frac{\begin{vmatrix} 3 & -1 \\ 2 & 4 \end{vmatrix}}{7}$$

$$x = \frac{(-1)(1) - (-2)(4)}{7} \qquad y = \frac{(3)(4) - (-1)(2)}{7}$$

$$x = \frac{7}{7} \qquad\qquad y = \frac{14}{7}$$

$$x = 1 \qquad\qquad y = 2$$

Testing this solution in the original system

$$
\begin{array}{ll}
3x - 2y = -1 & 2x + y = 4 \\
3(1) - 2(2) = ? & 2(1) + (2) = ? \\
3 - 4 = -1 & 2 + 2 = 4
\end{array}
$$

□

Example 4 □ Solve the system.

$$8x + 12y = 24$$
$$-2x - 3y = -6$$

First determine D.

$$D = \begin{vmatrix} 8 & 12 \\ -2 & -3 \end{vmatrix} = (8)(-3) - (12)(-2) = 0$$

Recall $x = \dfrac{D_x}{D}$ and $y = \dfrac{D_y}{D}$, but in this case $D = 0$.

Because you can't divide by zero, there is no solution to this system of equations. □

If D, D_x, and D_y are all zero, the system of equations is dependent.
If $D = 0$ and either D_x or D_y is nonzero, then the system is inconsistent.

Example 5 □ Use Cramer's Rule to solve the system.

$$\frac{1}{2}x + \frac{1}{3}y = \frac{7}{3}$$

$$\frac{3}{2}x - \frac{1}{2}y = 2$$

You could apply Cramer's Rule directly, but to simplify calculation, first clear the fractions from this system.

$$m6\!\mid \qquad \frac{1}{2}x + \frac{1}{3}y = \frac{7}{3} \longrightarrow 3x + 2y = 14$$

$$m2\!\mid \qquad \frac{3}{2}x - \frac{1}{2}y = 2 \longrightarrow 3x - y = 4$$

$$x = \frac{D_x}{D} \qquad\qquad y = \frac{D_y}{D}$$

$$x = \frac{\begin{vmatrix} 14 & 2 \\ 4 & -1 \end{vmatrix}}{\begin{vmatrix} 3 & 2 \\ 3 & -1 \end{vmatrix}} \qquad\qquad y = \frac{\begin{vmatrix} 3 & 14 \\ 3 & 4 \end{vmatrix}}{D}$$

$$x = \frac{(14)(-1) - (4)(2)}{(3)(-1) - (2)(3)} \qquad\qquad y = \frac{(3)(4) - (3)(14)}{D}$$

$$x = \frac{-14 - 8}{-3 - 6} \qquad\qquad y = \frac{12 - 42}{D}$$

$$x = \frac{-22}{-9} \qquad\qquad y = \frac{-30}{-9}$$

$$x = \frac{22}{9} \qquad\qquad y = \frac{10}{3}$$

□

Problem Set 15.2

Use Cramer's Rule to find the solution set.

1. $2x - y = -4$
$-x + 2y = 5$

2. $x - 2y = 11$
$2x + 3y = -6$

3. $3x + y = -11$
$4x - 3y = 7$

4. $4x - 3y = -7$
$-6x + 4y = 9$

5. $y = -\dfrac{1}{4}x + 4$
$3x + 2y = -2$
Hint: Clear fractions first.

6. $5x + 3y = 6$
$-4x - y = 5$

7. $3x - 4y = 9$
$\dfrac{3}{2}x - 2y = 4$

8. $6x = 12y + 11$
$4x + 2y = -1$

9. $3x - 4y = 4$
$-6x + 2y = -5$

10. $x + 8y = 1$
$2x - 4y = -13$

11. $2x - 3y = 12$
$3x + 4y = 1$

12. $4x + 5y = -11$
$3x - 2y = -14$

13. $x + \dfrac{3}{2}y = -4$
$2x - \dfrac{1}{2}y = 13$

14. $3x - 2y = -11$
$x + \dfrac{3}{2}y = 5$

15. $4x - 9y = -1$
$2x + 3y = 2$

16. $3x - 2y = 6$
$-9x + 6y = -5$

17. $3x + 8y = 4$
$6x - 4y = 3$

18. $6x + 4y = 3$
$12x + 2y = 11$

19. $2x - 3y = 5$
$-4x + 6y = 7$

20. $3x + 4y = 1$
$3x - 2y = -2$

21. $6x - 4y = 9$
$y + \dfrac{9}{4} = \dfrac{3}{2}x$

22. $\dfrac{1}{2}x + \dfrac{1}{3}y = \dfrac{2}{3}$
$-\dfrac{2}{3}x + \dfrac{1}{2}y = \dfrac{23}{6}$

23. $4x - \dfrac{9}{2}y = -1$
$-\dfrac{4}{3}x + 2y = 1$

24. $-x + 2y = \dfrac{9}{2}$
$\dfrac{3}{4}x - \dfrac{3}{2}y = -3$

25. $x + \dfrac{3}{2}y + \dfrac{5}{2} = 0$
$\dfrac{1}{3}x + y + \dfrac{4}{3} = 0$

26. $\dfrac{3}{4}x + y - \dfrac{13}{4} = 0$
$\dfrac{1}{2}y - x - 3 = 0$

27. $\dfrac{1}{3}x - \dfrac{1}{2}y = -1$
$\dfrac{1}{3}x - \dfrac{1}{4}y = 2$

28. $\dfrac{5}{3}x + y = 2$
$\dfrac{7}{4}x - \dfrac{1}{2}y = -1$

29. $4x - y = 7$
$2y = 8x + 6$

30. $x + \dfrac{3}{4}y = 1$
$\dfrac{2}{3}x - \dfrac{4}{3}y = -3$

31. $2x + 3y = -5$
    $x - 6y = -5$

32. $3x - 2y = 8$
    $6x = 4y + 11$

33. The difference of two numbers is 8. The larger number is three times the smaller number. Find the two numbers.

34. Suong has been saving dimes and quarters. After some time he finds he has 55 coins for a total of $8.95. How many of each kind of coin does he have?

35. The difference of two numbers is 3. Three times the smaller number diminished by the larger number is 7. Find the two numbers.

36. A youngster has a collection of 45 coins. If he has twice as many dimes as quarters, how many of each kind of coin does he have?

■ 15.3   Determinants of Higher Order Matrices

Cramer's Rule applies to systems of more than two equations. However, to use it we need to evaluate the determinants of higher order matrices. To do that we need a few definitions.

The common notation for a specific element of a matrix is $A_{(i,j)}$

A indicates that the element is from matrix A.

i indicates the row number.

j indicates the column number.

$B_{2,3}$ is the element in the second row and third column of matrix B.

15.3A   Definition: Minor of an Element

The minor of an element in a square matrix is the determinant of the matrix that remains after deleting the row and column in which the element appears.

Example 6 □ Find the minor of $A_{3,1}$ in $A = \begin{bmatrix} 3 & -2 & 1 \\ 7 & 4 & 5 \\ 9 & -6 & -3 \end{bmatrix}$.

$A_{3,1}$ is the element in the third row and the first column. Deleting this row and column gives

$$\begin{bmatrix} 3 & -2 & 1 \\ 7 & 4 & 5 \\ 9 & -6 & -3 \end{bmatrix} \qquad \text{minor}(A_{3,1}) = \begin{vmatrix} -2 & 1 \\ 4 & 5 \end{vmatrix}.$$

$$= (-2)(5) - (1)(4)$$

$$= -14 \qquad \qquad □$$

Example 7 □ Using the matrix A of Example 6, find the minor of $A_{2,2}$.
Delete the second row and the second column.

$$\begin{bmatrix} 3 & -2 & 1 \\ 7 & 4 & 5 \\ 9 & -6 & -3 \end{bmatrix} \qquad \text{minor } (A_{2,2}) = \begin{vmatrix} 3 & 1 \\ 9 & -3 \end{vmatrix}$$

$$= (3)(-3) - (1)(9)$$
$$= -18 \qquad \qquad \square$$

Does each element
in a matrix have
a minor?

Yes. Notice the
minor is a
determinant, not
a matrix.

Example 8 □ Using matrix A in Example 6, find the following minors.

a. $\text{minor}(A_{1,2}) = \begin{vmatrix} 7 & 5 \\ 9 & -3 \end{vmatrix}$

$$= (7)(-3) - (5)(9)$$
$$= \underline{\qquad}$$

b. $\text{minor}(A_{2,3}) = \begin{vmatrix} 3 & \underline{\qquad} \\ 9 & \underline{\qquad} \end{vmatrix}$

$$= (3)(-6) - \underline{\qquad}$$
$$= \underline{\qquad} \qquad \qquad \square$$

−66
−2
−6
(−2)(9)
0

15.3B Definition: Cofactor of an Element

If A is a square matrix, the cofactor of A_{ij} is the minor of A_{ij} multiplied by $(-1)^{i+j}$.

Example 9 □ Find the cofactor of $A_{1,2}$ if $A = \begin{bmatrix} -2 & 3 & 4 \\ 7 & -8 & 5 \\ 6 & 9 & 1 \end{bmatrix}$.

$$\text{cofactor of} \qquad (A_{1,2}) = (-1)^{1+2} \begin{vmatrix} 7 & 5 \\ 6 & 1 \end{vmatrix}$$

$$= (-1)[(7)(1) - (5)(6)]$$
$$= (-1)[-23]$$
$$= 23$$

$1 + 2 = 3$, an odd
number. Therefore
$(-1)^{i+j} = -1$.

□

Example 10 □ Find the cofactor of -8 in matrix A of Example 9.

$$-8 \text{ is } A_{2,2}$$

$$\text{cofactor of} \qquad A_{2,2} = (-1)^{2+2} \begin{vmatrix} -2 & 4 \\ 6 & 1 \end{vmatrix}$$

$$= (+1)[(-2)(1) - (4)(6)]$$

$$= (+1)[-26]$$

$$= -26 \qquad \qquad □$$

The determinant of a 3×3 matrix is found by the method called **expansion by cofactors.**

15.3C Expansion by Cofactors

The determinant of a square matrix is found by forming the product of each element in any row (or column) with its cofactor, then taking the sum of these products.

One way to find the determinant of

$$\begin{bmatrix} a_{1,1} & a_{1,2} & a_{1,3} \\ a_{2,1} & a_{2,2} & a_{2,3} \\ a_{3,1} & a_{3,2} & a_{3,3} \end{bmatrix}$$

is to expand about the first column

$$(+1)(a_{1,1})[(a_{2,2})(a_{3,3}) - (a_{3,2})(a_{2,3})]$$

$$+ (-1)(a_{2,1})[(a_{1,2})(a_{3,3}) - (a_{3,2})(a_{1,3})]$$

$$+ (+1)(a_{3,1})[(a_{1,2})(a_{2,3}) - (a_{2,2})(a_{1,3})]$$

We could also expand about any other row or column.

Example 11 □ Find the determinant of $B = \begin{bmatrix} 7 & 3 & 2 \\ 0 & 7 & -4 \\ 2 & 1 & 3 \end{bmatrix}$.

You may expand about any row or column. However, if you pick the second row, since it involves a zero, computation will be reduced.

$$\text{Det}(B) = (0)(-1)[\text{minor of } B_{2,1}] + (7)(+1)[(7)(3) - (2)(2)]$$

$$+ (-4)(-1)[(7)(1) - (2)(3)]$$

$$= 0 + (7)[17] + (4)[1]$$

$$= 119 + 4$$

$$= 123 \qquad \qquad □$$

You could also have taken advantage of the zero by expanding about the first column in the example on the previous page.

The method in 15.3C works for any square matrix. For example in a 4×4 matrix you multiply each element of any row or column by the cofactor of that element which involves applying the method again to evaluate four 3×3 determinants. This tactic is particularly well suited to a powerful computer programming technique called **recursion.**

Problem Set 15.3

Evaluate the following determinants by expansion about any row or column.

1. $\begin{vmatrix} 1 & 2 & 2 \\ -2 & 1 & -1 \\ 2 & 1 & 0 \end{vmatrix}$ **2.** $\begin{vmatrix} 1 & 5 & -2 \\ 0 & 1 & 0 \\ 3 & -1 & 2 \end{vmatrix}$ **3.** $\begin{vmatrix} 2 & -1 & 2 \\ -1 & 0 & 3 \\ 6 & 2 & 7 \end{vmatrix}$ **4.** $\begin{vmatrix} 1 & 5 & -5 \\ -3 & 3 & 2 \\ 2 & 0 & 0 \end{vmatrix}$ **5.** $\begin{vmatrix} 1 & 4 & 3 \\ 0 & -1 & -2 \\ -1 & 10 & 20 \end{vmatrix}$

6. $\begin{vmatrix} 1 & -4 & 1 \\ 2 & 0 & -1 \\ 1 & 4 & 1 \end{vmatrix}$ **7.** $\begin{vmatrix} 1 & 3 & 3 \\ -2 & 2 & 0 \\ 8 & -4 & 6 \end{vmatrix}$ **8.** $\begin{vmatrix} 2 & 0 & -3 \\ 1 & -4 & 1 \\ 2 & 1 & -3 \end{vmatrix}$ **9.** $\begin{vmatrix} 1 & 6 & -2 \\ 0 & -1 & 4 \\ 2 & 10 & 20 \end{vmatrix}$ **10.** $\begin{vmatrix} 0 & 2 & 2 \\ 0 & 4 & -3 \\ -1 & 0 & 2 \end{vmatrix}$

11. $\begin{vmatrix} 1 & 1 & 3 \\ 5 & 5 & 6 \\ 3 & 3 & 2 \end{vmatrix}$ **12.** $\begin{vmatrix} 2 & 1 & -4 \\ 3 & 3 & 2 \\ 6 & -5 & 0 \end{vmatrix}$ **13.** $\begin{vmatrix} 2 & -1 & 5 & -2 \\ -1 & 4 & -3 & 3 \\ 3 & -2 & 4 & -5 \\ -2 & 1 & -7 & 4 \end{vmatrix}$ **14.** $\begin{vmatrix} 4 & 2 & -4 & -6 \\ 3 & -1 & 4 & 2 \\ 1 & -3 & 11 & -4 \\ -5 & 4 & -13 & 2 \end{vmatrix}$

15. $\begin{vmatrix} 6 & -5 & 2 & 4 \\ 1 & 2 & -1 & 3 \\ -1 & 4 & -2 & 5 \\ -4 & -7 & 3 & -5 \end{vmatrix}$ **16.** $\begin{vmatrix} 3 & -1 & 5 \\ -2 & 0 & -1 \\ 4 & 2 & 1 \end{vmatrix}$ **17.** $\begin{vmatrix} 1 & -5 & 2 \\ -2 & 1 & -4 \\ 3 & 4 & 6 \end{vmatrix}$ **18.** $\begin{vmatrix} -4 & -2 & -1 \\ 2 & 4 & 3 \\ -1 & -5 & 6 \end{vmatrix}$

19. $\begin{vmatrix} 2 & 6 & 1 \\ -4 & 2 & 3 \\ -2 & -1 & 4 \end{vmatrix}$ **20.** $\begin{vmatrix} 1 & 3 & 5 \\ 2 & 4 & 6 \\ -3 & 1 & 2 \end{vmatrix}$ **21.** $\begin{vmatrix} 4 & -2 & 3 \\ -3 & 2 & 1 \\ -4 & 2 & -3 \end{vmatrix}$ **22.** $\begin{vmatrix} -3 & 1 & 2 \\ 4 & 0 & 0 \\ 6 & 4 & 3 \end{vmatrix}$ **23.** $\begin{vmatrix} 6 & 2 & 3 \\ -2 & 0 & 2 \\ 5 & 1 & -1 \end{vmatrix}$

24. $\begin{vmatrix} 3 & 4 & -3 \\ -6 & -2 & 6 \\ 5 & 1 & -5 \end{vmatrix}$ **25.** $\begin{vmatrix} 2 & -1 & 5 & -2 \\ -1 & 4 & -3 & 3 \\ 3 & -2 & 4 & -5 \\ -2 & 1 & -7 & 4 \end{vmatrix}$ **26.** $\begin{vmatrix} 4 & 2 & -4 & -6 \\ 3 & -1 & 4 & 2 \\ 1 & -3 & 11 & -4 \\ -5 & 4 & -13 & 2 \end{vmatrix}$ **27.** $\begin{vmatrix} 6 & -5 & 2 & 4 \\ 1 & 2 & -1 & 3 \\ -1 & 4 & -2 & 5 \\ -4 & -7 & 3 & -5 \end{vmatrix}$

28. $\begin{vmatrix} 3 & -3 & 1 & 5 \\ -1 & 0 & -2 & 3 \\ 4 & 2 & 3 & -5 \\ 2 & 5 & -4 & 2 \end{vmatrix}$ **29.** $\begin{vmatrix} 6 & -2 & 4 & 5 \\ 2 & 0 & -3 & -2 \\ -3 & 1 & -1 & 1 \\ 1 & -2 & 0 & 4 \end{vmatrix}$ **30.** $\begin{vmatrix} 5 & -1 & 0 & -5 \\ -3 & 4 & 1 & 2 \\ 1 & 6 & -3 & 3 \\ -2 & 0 & -4 & 1 \end{vmatrix}$

Cramer's Rule will provide the solution to higher order square systems of equations. For a system of three equations Cramer's Rule is

$$x = \frac{D_x}{D} \qquad\qquad y = \frac{D_y}{D} \qquad\qquad z = \frac{D_z}{D}$$

Use Cramer's Rule to solve the system.

$$x + 2y + z = 8$$
$$x + y + 2z = 7$$
$$2x + y + z = 9$$

First evaluate D by cofactors of the first column.

$$D = \begin{vmatrix} 1 & 2 & 1 \\ 1 & 1 & 2 \\ 2 & 1 & 1 \end{vmatrix}$$

$$= 1(-1)^{1+1}\begin{vmatrix} 1 & 2 \\ 1 & 1 \end{vmatrix} + 1(-1)^{2+1}\begin{vmatrix} 2 & 1 \\ 1 & 1 \end{vmatrix} + 2(1)^{3+1}\begin{vmatrix} 2 & 1 \\ 1 & 2 \end{vmatrix}$$

$$= (1)(-1)^{1+1}[(1)(1) - (1)(2)] + (1)(-1)^{2+1}[(2)(1) - (1)(1)]$$
$$+ (2)(-1)^{3+1}[(2)(2) - (1)(1)]$$

$$= [1 - 2] - 1[2 - 1] + 2[4 - 1]$$

$$= -1 - 1 + 6$$

$$= 4$$

Since $D \neq 0$ you may proceed.

$$D_x = \begin{vmatrix} 8 & 2 & 1 \\ 7 & 1 & 2 \\ 9 & 1 & 1 \end{vmatrix}$$

Evaluating D_x by cofactors about the first column,

$$= 8[1 - 2] - 7[2 - 1] + 9[4 - 1]$$

$$= -8 - 7 + 27$$

$$= 12$$

$$x = \frac{D_x}{D} = \frac{12}{4} = 3$$

$$D_y = \begin{vmatrix} 1 & 8 & 1 \\ 1 & 7 & 2 \\ 2 & 9 & 1 \end{vmatrix}$$

Evaluating D_y by cofactors about the first row,

Why use the first row?

To illustrate that it doesn't matter which row or column we use. The first row also has two ones in it.

$$D_y = (1)(-1)^{1+1}[7 - 18] + 8(-1)^{1+2}[1 - 4] + 1(-1)^{1+3}[9 - 14]$$
$$= [-11] - 8[-3] + [-5]$$
$$= 8$$

$$y = \frac{D_y}{D} = \frac{8}{4} = 2$$

$$D_z = \begin{vmatrix} 1 & 2 & 8 \\ 1 & 1 & 7 \\ 2 & 1 & 9 \end{vmatrix}$$

$$D_z = 1(-1)^2(9 - 7) + 1(-1)^3(18 - 8) + 2(1)^4(14 - 8)$$
$$= 1(2) - 1(10) + 2(6)$$
$$= 4$$

$$z = \frac{D_z}{D} = \frac{4}{4} = 1$$

Problem Set 15.4

Use Cramer's Rule to find the solution sets.

1. $x + y + z = 4$
    $2x - y - 2z = -1$
    $3x + 4y - 3z = -7$

2. $3x + y + 3z = 1$
    $2x - y - 4z = -2$
    $4x + 2y + z = -3$

3. $3x + 2y + z = 0$
    $-3y + 5z = -7$
    $2x + 4z = -2$

4. $2x - 2y - 4z = -1$
    $4x + 2y + 3z = 2$
    $-6x + 3y + 4z = -4$

5. $3x + 6y - 2z = -2$
    $6x + 4y - z = 0$
    $9x - 2y = 4$

6. $x + 3y + z = -2$
    $2x + 3z = -5$
    $-x + 6y - 2z = 4$

7. $x + 3y + 2z = 0$
    $\frac{1}{2}x + 2y = -1$
    $x - y - 4z = 1$

8. $x + 2y - 3z = -4$
    $-x + 4y = 3$
    $2x + 3z = -5$

9. $-x + 2y + z = 0$
    $2x + y - 2z = -2$
    $-x + 3y + z = 4$

Chapter 15 ■ Matrices and Determinants

10.
$$2x + y - z = 1$$
$$x + 2z = -2$$
$$-x + 2y - 3z = -1$$

11.
$$2x + z = -4$$
$$-4x + y = 4$$
$$3y + 4z = -8$$

12.
$$2x - y - 2z = 3$$
$$-x + 2y + z = -2$$
$$x + 3y - z = 5$$

13.
$$x + y + 2z = 4$$
$$2x + 3y - z = -1$$
$$3x + 4y - z = -1$$

14.
$$3x - 4y + 3z = -3$$
$$2x - 5y + z = 1$$
$$4x + 4y + z = 5$$

15.
$$3x + y + 3z = -2$$
$$2x - y = -8$$
$$y + 5z = 4$$

16.
$$2x + 3y + 4z = -1$$
$$3x + y = 3$$
$$2y + 7z = 1$$

17.
$$2x + y + 3z = 1$$
$$5x - 2y - 4z = -4$$
$$6x + y - 5z = 9$$

18.
$$\frac{1}{2}x + 4y - 2z = 5$$
$$2x - 6y = 1$$
$$3x + 2y - 3z = 10$$

19.
$$\frac{1}{3}x + 3y + 4z = 1$$
$$2x + 6y + 5z = -2$$
$$\frac{2}{3}x + 9y + 3z = 4$$

20.
$$2x + 5y + 3z = -4$$
$$-3x + 3y - 4z = 7$$
$$x + 7y + 3z = 1$$

21.
$$4x + 4y - z = 5$$
$$x - 6y - 2z = 3$$
$$3x + 8y + z = 1$$

22.
$$2x - 3y + z = 4$$
$$-2x + y - 3z = 2$$
$$-4x + 6y - 2z = 5$$

23.
$$2x + 3y + 4z = 4$$
$$x - 5y - 6z = 6$$
$$7y - 8z = 4$$

24.
$$3x + 2y + z = -3$$
$$6x - 3y - 4z = 8$$
$$-3x + 3y + 5z = -7$$

25. The sum of three numbers is 18. If the first is two-thirds as large as the second, and the sum of the first two numbers is two more than the third number, find the numbers.

26. A college student carrying some coins in his pocket had a total of $1.80 consisting of nickels, dimes, and quarters. If he had a total of twelve coins and there were two more quarters than nickels, how many of each kind of coin did he have?

27. Jose divided a total of $18,000 into three investments. He invested twice as much at 12% as he did at 8%. The remaining amount was invested at 10%. If his total annual interest income was $1880, how much was invested at each rate?

28. At a local football game 1220 tickets were sold. Adult tickets sold for $4.00, student tickets for $2.00, and tickets for children aged 5 years to 12 years for $1.00. Fifteen times as much money was collected from the adults as from the children aged 5 years to 12 years. If the total gate receipts were $3180, how many tickets of each kind were sold?

15.1   1. A **matrix** is a rectangular array of numbers.
2. A **determinant** is a number associated with a square matrix.

    a. The determinant of the matrix $\begin{vmatrix} a & b \\ c & d \end{vmatrix}$

    is: $\begin{vmatrix} a & b \\ c & d \end{vmatrix} = ad - bc$

    b. Only square matrices have determinants.

15.2   1. **Cramer's Rule** is used to find the solution set of any system of equations where there are the same number of equations as there are variables.
2. The general solution of a system of two linear equations

$$ax + by = p$$
$$cx + dy = q$$

    is $x = \dfrac{pd - bq}{ab - cd}$, $y = \dfrac{aq - pc}{ab - cd}$

    provided $ab - cd \neq 0$

15.3   1. The **minor** of an element in a matrix is the determinant that remains after deleting the row and column in which the element appears.
2. The **cofactor** of an element in the ith row and j^{th} column labeled as A_{ij} is the minor of A_{ij} multiplied by $(-1)^{i+j}$.
3. The value of any determinant is equal to the sum of the products of each element in any row or column and its corresponding cofactor.

15.4   1. **Cramer's Rule** for three equations in three unknowns is given by:

$$x = \frac{D_x}{D}, \, y = \frac{D_y}{D}, \, z = \frac{D_z}{D}$$

    provided: $D \neq 0$

Chapter 15   Review Test

Evaluate the following determinants. **(15.1)**

1. $\begin{vmatrix} -3 & 8 \\ -5 & 6 \end{vmatrix}$
           2. $\begin{vmatrix} -6 & -9 \\ 4 & 6 \end{vmatrix}$
           3. $\begin{vmatrix} 5 & -7 \\ -6 & 8 \end{vmatrix}$

Use Cramer's Rule to evaluate the following equations. **(15.2)**

4. $x + 2y = 4$
           5. $3x - 6y = 10$
           6. $x + \dfrac{3}{4}y = -\dfrac{7}{4}$

    $2x - 3y = -27$
                $-4x + 8y = 7$
                $-2x - \dfrac{4}{3}y = 3$

7. $0.2x - 0.9y = 2.4$
$0.25x + 0.4y = -0.05$

Evaluate the following determinants. **(15.3)**

8. $\begin{vmatrix} 3 & 2 & -1 \\ 4 & 5 & -3 \\ -2 & 6 & 1 \end{vmatrix}$

9. $\begin{vmatrix} 5 & 4 & 6 \\ -2 & 6 & 9 \\ 3 & -8 & -12 \end{vmatrix}$

10. $\begin{vmatrix} 2 & 6 & 5 \\ 3 & -6 & 0 \\ -7 & 1 & -6 \end{vmatrix}$

Use Cramer's Rule to solve the following equations. **(15.4)**

11. $6x - 2y + 4z = -20$

$4x + 2y - 5z = 12$

$5x - 3y - 6z = -2$

12. $4x + 3y - 8z = 5$

$-6x + 9y - 4z = 1$

$8x - 6y + 4z = 1$

13. $\dfrac{3}{2}x + y - \dfrac{1}{2}z = \dfrac{1}{2}$

$-\dfrac{2}{3}x + z = 2$

$x + \dfrac{5}{4}y = \dfrac{1}{2}$

Set up the appropriate equations and solve, using Cramer's Rule. **(15.2)**

14. Twice the length of a rectangle is 9 feet greater than three times its width. Find the dimensions of the rectangle if the perimeter is 44 feet.

15. A cookie jar contains $7.25 in nickels, dimes and quarters. There are 53 coins in all, with five more dimes than nickels. How many coins of each kind are there?

16 Sequences, Series, and the Binomial Expansion

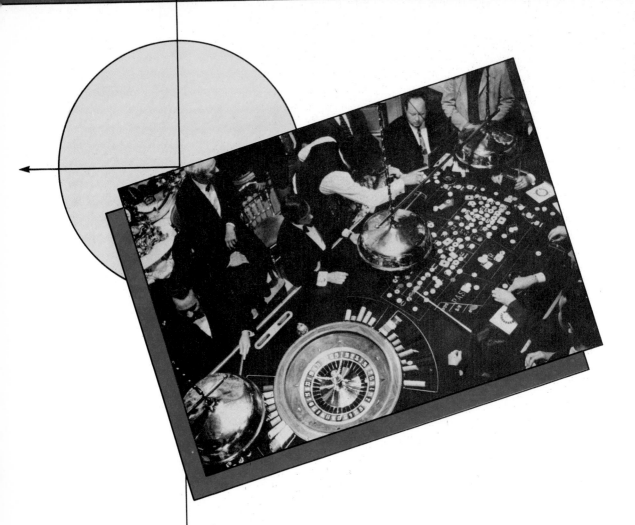

Contents

T his final chapter of the book will deal with lists of numbers written in a predetermined order. Sequences can describe the timing of naturally occurring events, like phases of the moon and locust plagues. Some sequences describe how the branches of a tree divide. The sum of a sequence is called a **series.** The chapter on logarithms illustrated how a series can be used to find the logarithm of a number. Other series are used in the study of trigonometry, electronics, and probability.

The final section of the chapter will show you a convenient way to expand a binomial. The binomial expansion is frequently used in the study of probability where it helps count the number of possible combinations of events. Another application is Poisson's Distribution which helps predict random events like the distribution of telephone calls at a switchboard or the arrival of skiers at a lift line.

■ 16.1   Sequences

A sequence is a function whose domain is restricted to a subset of the counting numbers. Consider making a table of values for: $f(x) = 2x - 1$ if the domain of x is limited to the counting numbers from 1 to 5.

x	$f(x) = 2x - 1$
1	1
2	3
3	5
4	7
5	9

We could express this information as the set of ordered pairs

$$(1, 1), (2, 3), (3, 5), (4, 7), (5, 9)$$

Since the domain of the function is readily apparent, we could write the values of $f(x)$ for the first five counting numbers without writing the corresponding values of x. They are:

$$1, 3, 5, 7, 9 \longrightarrow \text{This is an example of a sequence.}$$

16.1 Definition: Sequence

A **sequence** is a list of numbers that is written in a predetermined order. Each member of the list is called a **term** of the sequence.

By the definition above, your phone number is a sequence with seven terms. The first digit is the first term, the second digit is the second term, and so on through the seven digits. The sequences that are most interesting mathematically are those where we can determine the value of a term in a sequence from its position in the sequence. This chapter will restrict our attention to sequences where the value of each term is a function of its position in the sequence.

Consider the sequence ⟶ 2, 4, 6, 8, 10, 12

We will use it to define some vocabulary to describe sequences.

Terms of a sequence are elements or members of the sequence. In this case the terms of the sequence are:

2, 4, 6, 8, 10, 12

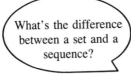

What's the difference between a set and a sequence?

In a sequence, the order of the numbers is important. In a set, order is unimportant. Also members of a set cannot be duplicated while terms of a sequence can be duplicated.

Usually letters are used to denote a sequence.

$$a_1 = \text{first term of sequence } a$$

$$a_2 = \text{second term of sequence } a$$

$$a_3 = \text{third term of sequence } a$$

$$\vdots$$

$$a_n = n\text{th term of sequence } a$$

If we call the example above sequence a, then:

$$a = 2, 4, 6, 8, 10, 12$$

and

$$a_1 = 2 \quad a_2 = 4 \quad a_3 = 6 \quad a_4 = 8 \quad a_5 = 10 \quad a_6 = 12$$

What's a_n?

That's the nth or general term of the sequence.

The value of term number n of the sequence above is: $a_n = 2n$.

Chapter 16 ■ Sequences, Series, and the Binomial Expansion

The **general term** of a sequence is an expression that gives the value of any term in the sequence as a function of its term number.

Example 1 □ Find the first seven terms of the sequence where $a_n = n^2 - 2$.
Substituting 1 through 7 for n in $a_n = n^2 - 2$ yields:

$$a_1 = (1)^2 - 2 = -1$$
$$a_2 = (2)^2 - 2 = 2$$
$$a_3 = (3)^2 - 2 = 7$$
$$a_4 = (4)^2 - 2 = 14$$
$$a_5 = (5)^2 - 2 = 23$$
$$a_6 = (6)^2 - 2 = 34$$
$$a_7 = (7)^2 - 2 = 47$$

□

You asked for "the first seven terms" in the sequence above. Are there more terms?

That depends on the sequence. Sequences can have either a finite or an infinite number of terms.

An **infinite sequence** is a function whose domain is the set of counting numbers.

A **finite sequence** is a function whose domain is the first n counting numbers.

Example 2 □ List the first four terms of the sequence b if $b_n = \dfrac{1}{n + 1}$.
Substituting 1 . . . 4 for n yields:

$$b_1 = \frac{1}{1 + 1} = \frac{1}{2}$$

$$b_2 = \frac{1}{2 + 1} = \frac{1}{3}$$

$$b_3 = \frac{1}{3 + 1} = \frac{1}{4}$$

$$b_4 = \frac{1}{4 + 1} = \frac{1}{5}$$

□

Example 3 □ Find the first four terms of the sequence with a general term of $\frac{1}{24}(50 - 35n + 10n^2 - n^3)$.

Substituting 1 . . . 4 for n in $a_n = \frac{1}{24}(50 - 35n + 10n^2 - n^3)$ yields:

$$a_1 = \frac{1}{24}[50 - 35(1) + 10(1)^2 - (1)^3]$$

$$= \frac{1}{24}(50 - 35 + 10 - 1)$$

$$= \frac{1}{24}(24)$$

$$= 1$$

$$a_2 = \frac{1}{24}[50 - 35(2) + 10(2)^2 - (2)^3]$$

$$= \frac{1}{24}(50 - 70 + 40 - 8)$$

$$= \frac{1}{24}(12)$$

$$= \frac{1}{2}$$

$$a_3 = \frac{1}{24}[50 - 35(3) + 10(3)^2 - (3)^3]$$

$$= \frac{1}{24}(50 - 105 + 90 - 27)$$

$$= \frac{1}{24}(8)$$

$$= \frac{1}{3}$$

$$a_4 = \frac{1}{24}[50 - 35(4) + 10(4)^2 - (4)^3]$$

$$= \frac{1}{24}(50 - 140 + 160 - 64)$$

$$= \frac{1}{24}(6)$$

$$= \frac{1}{4}$$

It looks as if the next term is $\frac{1}{5}$ and the general term is $\frac{1}{n}$.

The first four terms of the sequence are:

$$1, \frac{1}{2}, \frac{1}{3}, \frac{1}{4}.$$

That is tempting. Look at the next example to see what happens when we try evaluating a_5 for the general term given in Example 3.

□

Chapter 16 ■ Sequences, Series, and the Binomial Expansion

Example 4 □ Find a_5 for $a_n = \dfrac{1}{24}[50 - 35(n) + 10(n)^2 - n^3]$.

$$a_5 = \frac{1}{24}[50 - 35(5) + 10(5)^2 - 5^3]$$

$$= \frac{1}{24}(50 - 175 + 250 - 125)$$

$$= \frac{1}{24}(0)$$

$$= 0 \qquad\qquad □$$

That's not the $\dfrac{1}{5}$ I expected.

Trying to find the general term of a sequence from a few terms always involves guessing. Without extra terms to verify our guess, we can't be certain the guess is correct.

Finding the general term of a sequence involves making a conjecture, testing the conjecture against the known terms, and revising the conjecture on the basis of how well it fits the known facts.

That sounds like guessing to me.

It is. The difference is we don't guess blindly. Read the following pointers for a few suggestions on how to find the general term of a sequence.

Pointers on How to Find the General Term of a Sequence

1. Study each term of the sequence as it compares to its term number. Ask yourself these questions:
   a. Is it a multiple of the term number?
   b. Is it a multiple of the square or cube of the term number? If each term is a multiple of the term number, there will be a common factor.
2. Examine the sequence. Does it increase or decrease?
   a. If it increases slowly, consider expressions that involve the term number plus or minus a constant like: $n + 2$ or $n - 3$.
   b. If it increases moderately, think about multiples of the term number plus or minus a constant like: $2n$ or $3n - 1$.
   c. If the sequence increases very rapidly, try powers of the term number like: n^2 or $n^3 + 1$.
3. If the sequence consists of fractions, examine how the denominator and numerator change as separate sequences.

   For example: $n = \dfrac{n + 1}{n^2}$ yields: $\dfrac{2}{1}, \dfrac{3}{4}, \dfrac{4}{9}, \dfrac{5}{16}, \dfrac{6}{25}, \cdots$

Example 5 ☐ Find the general term for the sequence: $x = 0, 3, 8, 15, 24,$ $35, \ldots$

Examine the given terms of the sequence to observe that:

a. There is no common factor; therefore, the general term is not a multiple of the term number.

b. The terms grow fairly rapidly.

Since the terms grow fairly rapidly, let's compare the sequence to the first terms of

$$a_n = n^2$$

$$a = 1, 4, 9, 16, 25, 36, 49, \ldots$$

It appears that $x_n = n^2 - 1$.   ☐

Example 6 ☐ Find the general term of the sequence: $x = 0, 2, 6, 12,$ $20, \ldots$

Since the sequence grows fairly rapidly, compare it to $a_n = 2n$: $2, 4, 6,$ $8, 10$. It grows more rapidly than: $a_n = 2n$ so compare it to $a_n = 3n$: $3, 6,$ $9, 12, 15$.

This doesn't seem to work either. Try a power of n.

Compare x_n to $a_n = n^2$: $a = 1, 4, 9, 16, 25, 36$.

This is growing more rapidly than x_n.

Let's look at the difference between a_n and x_n.

$$a = 1, 4, 9, 16, 25, 36, \ldots$$

$$x = 0, 2, 6, 12, 20$$

$$a - x = 1, 2, 3, 4, 5$$

The difference is $a_n = n$.

It appears that: $x_n = n^2 - n$.

It is true that

$$x_n = n^2 - n \quad \text{gives the sequence:} \quad x = 0, 2, 6, 12, 20, 30, \ldots \quad ☐$$

Problem Set 16.1

Find the first five terms for each of the sequences below where the value of a_n is given.

1. $a_n = n + 2$ **2.** $a_n = 2n - 1$ **3.** $a_n = \dfrac{n}{2}$ **4.** $a_n = -\dfrac{n}{3}$

5. $a_n = 2n^2 + 1$ **6.** $a_n = 2n^2 - 3$ **7.** $a_n = 15 - 2n$ **8.** $a_n = 20 - 3n$

9. $a_n = \dfrac{1}{n}$ **10.** $a_n = \dfrac{1}{n^2}$ **11.** $a_n = \dfrac{1}{2(n + 1)}$ **12.** $a_n = \dfrac{n}{n + 1}$

Chapter 16 ■ Sequences, Series, and the Binomial Expansion

13. $a_n = \dfrac{n^2 + 1}{2n - 1}$ **14.** $a_n = (-1)^n$ **15.** $a_n = n(-1)^n$ **16.** $a_n = \dfrac{2n}{(-1)^n}$

Find the general term for each of the following sequences.

17. $1, 3, 5, 7, 9, \ldots$

18. $1, \dfrac{1}{2}, \dfrac{1}{3}, \dfrac{1}{4}, \dfrac{1}{5}, \ldots$

19. $1, \dfrac{1}{4}, \dfrac{1}{9}, \dfrac{1}{16}, \dfrac{1}{25}, \ldots$

20. $2, \dfrac{1}{2}, \dfrac{2}{9}, \dfrac{1}{8}, \dfrac{2}{25}, \ldots$

21. $1, \dfrac{2}{3}, \dfrac{1}{2}, \dfrac{2}{5}, \dfrac{1}{3}, \ldots$

22. $1, 4, 9, 16, 25, \ldots$

23. $2, 8, 18, 32, 50, \ldots$

24. $3, 12, 27, 48, 75, \ldots$

25. $2, 5, 10, 17, 26, \ldots$

26. $3, 9, 19, 33, 51, \ldots$

27. $2, 6, 12, 20, 30, \ldots$

28. $0, 2, 6, 12, 20, \ldots$

■ 16.2 Arithmetic Progressions

Some events occur in a regular pattern where each occurrence can be found by adding a constant to the previous value. For example, consider the weight that each floor of a five-story building must support. The first floor must support the four floors above it. The second floor has to support three floors above it, and so on. If each floor weighs 50 tons, the weights to be supported are 200, 150, 100, 50, 0 tons.

Each term of the sequence above can be found by adding -50 tons to the preceding term. Such a sequence is an example of an arithmetic progression. Leap years and locust plagues are examples of arithmetic progressions.

16.2A Definition: Arithmetic Progression

An arithmetic progression is a sequence where each term can be found by adding a constant to the previous term.

To discuss arithmetic progressions we will define some terms.

Vocabulary of Arithmetic Progressions

By historical convention the following letters refer to arithmetic progressions:

a_1 = first term of the arithmetic progression

l = last term of the arithmetic progression

n = number of terms in the arithmetic progression

d = common difference between any two terms of an arithmetic progression

Election years are: 1980, 1984, 1988, 1992.
The common difference between any two election years is 4 years.
In general:

$$d = a_n - a_{n-1}$$

Example 7 □ What is the common difference for the arithmetic progression: 3, 10, 17, 24, 31?

Taking the difference of the first two terms,

$$d = a_2 - a_1$$
$$d = 10 - 3$$
$$d = 7$$

Testing with the difference of the fourth and fifth terms,

$$d = a_5 - a_4$$
$$d = 31 - 24$$
$$d = 7 \qquad □$$

It will help us to write an expression for the nth term of an arithmetic progression if we examine how we can write the arithmetic progression in Example 7 in terms of the first term, the term numbers, and the common difference.

$$a = 3,\ 3 + (2 - 1)7,\ 3 + (3 - 1)7,\ 3 + (4 - 1)7,\ 3 + (5 - 1)7 \ldots$$
$$3, \qquad 10 \quad , \qquad 17 \quad , \qquad 24 \quad , \qquad 31 \ldots$$

If we generalize this result we can write an expression for the nth term of an arithmetic progression.

Chapter 16 ■ Sequences, Series, and the Binomial Expansion

16.2B *n*th Term of an Arithmetic Progression

The *n*th term of an arithmetic progression is the first term of the arithmetic progression plus $(n - 1)$ times the common difference.

$$a_n = a_1 + (n - 1)d$$

Example 8 □ Find the 10th term of the arithmetic progression: $-6, -4, -2, \ldots$.

In this case: $a_1 = -6$.

The common difference is:

$$d = a_n - a_{n-1}$$
$$= -4 - (-6)$$
$$d = +2$$

The tenth term is then:

$$a_{10} = a_1 + (10 - 1)d$$
$$= -6 + 9(2)$$
$$= -6 + 18$$
$$= 12 \qquad \square$$

Example 9 □ Find the general term of the arithmetic progression: 15, 12, 9, 6, 3.

First find the common difference by subtracting the first term from the second term.

$$12 - 15 = -3$$

That's a negative number. Why not say you are subtracting +3 from each term to get the next term?

To be consistent with the formula $a_n = a_1 + (n - 1)d$, we think of it as adding a negative number to get the next term.

Now that we know the first term a_1 is 15 and d is -3, we can write an expression for a_n.

$$a_n = 15 + (n - 1)(-3)$$
$$= 15 - 3(n - 1)$$
$$= 15 - 3n + 3$$
$$= 18 - 3n \qquad \square$$

Example 10 □ Find the 101st term of an arithmetic progression where $a_1 = 4$ and $d = 6$.

Applying the formula: $a_n = a_1 + (n - 1)d$

$$d_{101} = 4 + (101 - 1)6$$

$$= 4 + 600$$

$$= 604$$

The 101st term is 604. □

Example 11 □ If the first term of an arithmetic progression is -10 and the sixth term $+10$, find the twenty-first term.

Since you know the first term and know the sixth term, you can apply the formula:

$$a_n = a_1 + (n - 1)d$$

With $n = 6$ to find d

$$a_6 = a_1 + (6 - 1)d$$

$$10 = -10 + (5)d$$

$$20 = 5d$$

$$4 = d$$

Knowing $a_1 = -10$, $d = 4$, you can compute a_{21}.

$$a_{21} = a_1 + (21 - 1)d$$

$$= -10 + (20)4$$

$$= -10 + 80$$

$$a_{21} = 70$$ □

Example 12 □ Find the tenth term of an arithmetic progression where the 90th term is 1300 and the 95th term is 1400.

Between the 90th and 95th term, d has been added five times.

$$5d = 1400 - 1300$$

$$= 100$$

$$d = 20$$

If we knew a_1, we could solve the general term for the tenth term.

We can find a by substituting the information we have for the 95th term and d into the formula for the general term.

Chapter 16 ■ Sequences, Series, and the Binomial Expansion

$$a_n = a_1 + (n - 1)d$$

$$1400 = a_1 + (95 - 1)20 \qquad \text{For the 95th term}$$

$$1400 = a_1 + 1880$$

$$-480 = a_1$$

Now that we know a and d, we can write the general term and calculate the 10th term.

$$a_n = -480 + (n - 1)20$$

$$a_{10} = -480 + (9)20$$

$$= -480 + 180$$

$$= -300 \qquad\qquad \square$$

The Sum of an Arithmetic Progression

Legend has it that the famous mathematician Carl Friedrich Gauss (1777–1855) was always finishing his school work early and irritating his teacher. To keep Carl occupied, the teacher asked him to find the sum of the first 100 whole numbers. Carl returned with the answer in an amazingly short period of time, so his teacher gave him another similar problem. Again he returned shortly with the answer. Carl Gauss had derived the following formula:

16.2C   Formula for the Sum of an Arithmetic Progression

$$s_n = \frac{n}{2}(a_1 + l) \text{ where } s_n = \text{sum of the first } n \text{ terms of an arithmetic progression.}$$

a_1 = first term

l = last term

n = number of terms

One way to derive this formula is to think of Carl Gauss' problem:

$$1 + 2 + 3 + 4 + \cdots + 97 + 98 + 99 + 100$$

$$+ 101$$
$$+ 101$$
$$+ 101$$
$$+ 101$$

We will be able to make 50 or $\frac{n}{2}$ pairings from this progression. Each pairing is equal to the sum of the first plus the last term of the progression. You can also visualize the formula as:

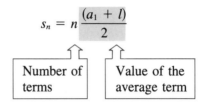

$$S_n = n\frac{(a_1 + l)}{2}$$

Number of terms

Value of the average term

A more formal proof of the formula can be made by writing an expanded form of the sum of n terms of an arithmetic progression.

First write the sum by writing the first term and adding the appropriate multiple of the difference to express each term.

$$S_n = a_1 + [a_1 + d] + [a_1 + 2d] + \cdots + [a_1 + (n - 1)d]$$

First term

Second term

Last term

Next write the progression backwards by starting with l and subtracting d each time.

$$S_n = l + [l - d] + [l - 2d] + [l - 3d] \cdots [l - (n - 1)d]$$

Last term

Last term minus one difference

First term

Now add the two expressions for the sum term by term.

$$S_n = a_1 + [a_1 + d] + [a_1 + 2d] + [a_1 + 3d] + \cdots + [a_1 + (n - 1)d]$$
$$\underline{+ S_n = l + [l - d] + [l - 2d] + [l - 3d] + \cdots + [l - (n - 1)d]}$$
$$2S_n \quad (a_1 + l) + (a_1 + l) + (a_1 + l) + (a_1 + l) + \cdots (a_1 + l)$$

There will be n terms of $(a_1 + l)$ above which we write as:

$$2S_n = n(a_1 + l)$$

$$S_n = \frac{n}{2}(a_1 + l) \quad \text{Dividing both sides by 2}$$

Chapter 16 ■ Sequences, Series, and the Binomial Expansion

Example 13 □ Find the sum of the whole numbers from 1 to 100.
In this case, $n = 100$, $d = 1$, $a_1 = 1$, $l = 100$.

$$S_n = \frac{n}{2}(a_1 + l)$$

$$S_{100} = \frac{100}{2}(1 + 100)$$

$$= 50(101)$$

$$= 5050 \qquad \qquad \square$$

Example 14 □ Find the sum of the first 20 terms of the arithmetic progression: 4, 7, 10, 13, ...
In this example we know $a = 4$ and can observe $d = 3$. We must compute a_{20} before we can find the sum.

$$a_n = a_1 + (n - 1)d$$

$$a_{20} = a_1 + (20 - 1)3$$

$$= 4 + (19)3$$

$$= 4 + 57$$

$$= 61$$

Now we can compute the sum of the first 20 terms.

$$S_n = \frac{n}{2}(a_1 + l)$$

$$= \frac{20}{2}(4 + 61)$$

$$= 10(65)$$

$$S_n = 650 \qquad \qquad \square$$

Problem Set 16.2

Find the common difference for each of the following arithmetic progressions.

1. 2, 4, 6, 8, . . .

2. 3, 6, 9, 12, . . .

3. 10, 6, 2, −2, . . .

4. −20, −24, −28, −32, . . .

Write the formula for the general term of each of the following arithmetic progressions.

5. 1, 3, 5, 7, . . .

6. 10, 14, 18, 22, . . .

7. 15, 10, 5, 0, . . .

8. −16, −12, −8, −4, . . .

9. Find the tenth term of the arithmetic progression with general term of $a_n = 3 + (n - 1)7$.

10. Find the eighth term of the arithmetic progression with general term of $a_n = -5 + (n - 1)7$.

11. Find the 101st term of the arithmetic progression with general term of $a_n = 0 + (n - 1)7$.

12. Find the 101st term of the arithmetic progression with general term of $a_n = 1000 + (n - 1)(-3)$.

Given the indicated terms of an arithmetic progression, find the requested term.

13. $a_1 = 5$, $a_{10} = 50$, $a_{100} = $ _____

14. $a_1 = 6$, $a_9 = 38$, $a_{20} = $ _____

15. $a_1 = -10$, $a_5 = 0$, $a_{15} = $ _____

16. $a_1 = 10$, $a_5 = 0$, $a_{15} = $ _____

17. $a_7 = 10$, $a_{17} = 64$, $a_1 = $ _____

18. $a_{10} = 96$, $a_{30} = 56$, $a_5 = $ _____

19. $a_{100} = 1000$, $a_{900} = 600$, $a_{200} = $ _____

20. $a_{50} = 64$, $a_{34} = 60$, $a_{60} = $ _____

21. Find the sum of the first 20 terms of the arithmetic progression: 13, 16, 19, 22, . . .

22. Find the sum of the first 10 terms of the arithmetic progression: 10, 14, 18, 22, . . .

23. Find the sum of the first 40 terms of the arithmetic progression: $-6, -4, -2, 0, \ldots$

24. Find the sum of the first 50 terms of the arithmetic progression: $-9, -6, -3, 0, \ldots$

25. Find the sum of the first 10 terms of the arithmetic progression: 10, 7, 4, 1, . . .

26. Find the sum of the first 200 terms of the arithmetic progression: $6, 0, -6, -12, \ldots$

27. Find the sum of the first 20 terms of the arithmetic progression: 10, 10.5, 11, 11.5, . . .

28. Find the sum of the first 20 terms of the arithmetic progression: 10, 9.25, 8.5, 7.75, . . .

The previous section studied sequences where each term could be found by adding a constant term to the previous term. This section will examine the sequences that can be obtained by multiplying each term by a constant term. Such a sequence is called a **geometric progression.**

16.3A   Geometric Progression

A geometric progression is a sequence where each term a_n is equal to a constant times the previous term of the sequence.

$$a_n = a_{n-1} \cdot r \qquad \text{where } a_n = n\text{th term of the sequence}$$
$$r = \text{common ratio}$$

Example 15 □ Write the geometric progression where each term is obtained by multiplying the previous term by 2 if the first term is one.

The first term is 1

The second term is $1 \cdot 2 = 2$

The third term is $2 \cdot 2 = 4$

The fourth term is $4 \cdot 2 = 8$

The sequence is 1, 2, 4, 8, 16, 32, . . .   □

Notice in the sequence above that if you divide any term by its predecessor, you obtain the common ratio between two terms.

$$\frac{32}{16} = 2, \qquad \frac{8}{4} = 2, \qquad \frac{4}{2} = 2$$

We usually label the common ratio between two terms in a geometric progression r.

Since by definition $a_n = a_{n-1} \cdot r$, we can solve for r by dividing both sides of the equation by a_{n-1}.

$$\frac{a_n}{a_{n-1}} = \frac{a_{n-1}}{a_{n-1}} \cdot r = r$$

This is the common ratio between any two consecutive terms in a geometric sequence

Example 16 □ Write the first five terms of a sequence with a first term of 2 and a common ratio of 3.

Multiply each term by 3 to get its successor:

$$2, 6, 18, 54, 162$$

The sequence above also can be written in the factored form:

$$2, \quad 2 \cdot 3, \quad 2 \cdot 3 \cdot 3, \quad 2 \cdot 3 \cdot 3 \cdot 3, \quad 2 \cdot 3 \cdot 3 \cdot 3 \cdot 3$$

or in exponential form:

$$2 \cdot 3^0, \quad 2 \cdot 3^1, \quad 2 \cdot 3^2, \quad 2 \cdot 3^3, \quad 2 \cdot 3^4.$$

With a representing the first term and r representing the common ratio, the sequence above can be written:

$$ar^0, ar^1, ar^2, ar^3, ar^4 \qquad\qquad □$$

Based on the sequence above, the following is the expression for the nth term of a geometric progression.

16.3B General Term of a Geometric Progression

The nth term of a geometric progression is given by:

$$a_n = a \cdot r^{n-1}$$

where: n = term number

a = first term

r = common ratio

Example 17 □ Find the general term of the geometric progression: $1, 5, 25, 125, 625$.

We find r by dividing any term by its predecessor.
Taking the third term 25 and dividing it by its predecessor 5 yields:

$$\frac{25}{5} = 5$$

Verifying with the fifth and fourth terms:

$$\frac{\text{fifth term}}{\text{fourth term}} = \frac{625}{125}$$

$$r = 5$$

Chapter 16 ■ Sequences, Series, and the Binomial Expansion

Since you know that $r = 5$ and the first term is $a = 1$, you can write an expression for the general term.

$$a_n = 1 \cdot 5^{n-1}$$ ☐

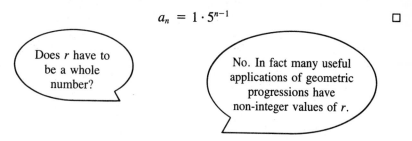

Does r have to be a whole number?

No. In fact many useful applications of geometric progressions have non-integer values of r.

If r is a fraction, each term of the progression is smaller than the term before it.

Example 18 ☐ Find the first 6 terms of the geometric progression where: $a = 4$ and $r = \dfrac{1}{2}$.

The first six terms are:

$$4\left(\frac{1}{2}\right)^0, \ 4\left(\frac{1}{2}\right)^1, \ 4\left(\frac{1}{2}\right)^2, \ 4\left(\frac{1}{2}\right)^3, \ 4\left(\frac{1}{2}\right)^4, \ 4\left(\frac{1}{2}\right)^5$$

$$4, \ 2, \ 1, \ \frac{1}{2}, \ \frac{1}{4}, \ \frac{1}{8}$$ ☐

Example 19 ☐ Each year the value of a refrigerator is approximately 80% of its value the previous year. What would the value of a refrigerator that costs $600 new be after 9 years?

This is equivalent to asking for the tenth term of a geometric sequence with $a_1 = 600$ and $r = 0.8$.

Why the tenth term? It was nine years.

It was worth $600 at the beginning of the first year. At the end of the first year the value of the refrigerator was given by the second term in the sequence.

Since $a_n = a \cdot r^{n-1}$

$$a_{10} = 600(0.8)^9$$

How do I find 0.8^9?

The easiest way is by using the $\boxed{y^x}$ key on your calculator. See the calculator help on the next page.

Calculator Help

To find (0.8^9) on your calculator, do the following:

Enter .8.

Press $\boxed{y^x}$.

Press 9.

Press $\boxed{=}$.

The value 0.1342 should appear on the display.

The value of the refrigerator after 10 years then is:

$$a_{10} = 600(0.1342)$$

$$= \$80 \qquad \qquad \square$$

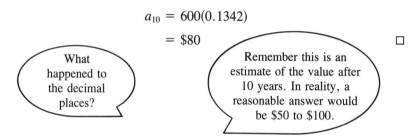

What happened to the decimal places?

Remember this is an estimate of the value after 10 years. In reality, a reasonable answer would be \$50 to \$100.

Example 20 $\square$ Find the sixth term of a geometric progression if the third term is 10.3680 and the fourth term is 12.4416.

First find r. Since r is the ratio of two successive terms, you can divide the fourth term by the third term to find r.

$$r = \frac{\text{value of fourth term}}{\text{value of third term}}$$

$$= \frac{12.4416}{10.3680}$$

$$r = 1.2$$

To find the sixth term evaluate $a_n = a \cdot r^{(n-1)}$

Before you can find a_6 you must find a.

If you substitute the values you know for the third term in the expression for the general term, you can solve for a.

$$a_n = ar^{(n-1)}$$

$$a_3 = ar^2$$

$$10.3680 = a(1.2)^2$$

$$10.3680 = a(1.44)$$

$$7.2 = a$$

Now that you know a, r, and n, you can evaluate the expression for the general term for the value of the sixth term.

$$a_n = ar^{(n-1)}$$

$$a_6 = 7.2(1.2)^5$$

$$= 7.2(2.48832)$$

$$a_6 = 17.915904 \qquad \square$$

Why so many places this time?

To let you know you are doing the right thing with your calculator. This is also a theoretical result so we can express the answer exactly. See Appendix A for more information on accuracy.

Example 21 $\square$ Find the first term of a geometric progression if the second term is 3 and the fifth term is 0.081.

This time:

$$a_2 = ar^1 \qquad \text{and} \qquad a_5 = ar^4$$

Substituting:

$$3 = ar^1 \qquad \text{and} \qquad 0.081 = ar^4$$

Dividing:

$$\frac{a_5}{a_2} = \frac{0.081}{3.0} = \frac{ar^4}{ar^1}$$

$$0.027 = r^3$$

$$\sqrt[3]{0.027} = r$$

To find $\sqrt[3]{0.027}$ if you don't recognize it, raise 0.027 to the (1/3) power on your calculator.

How do I do that?

First STORE $1 \div 3$ in memory. Then raise 0.027 to the $\boxed{\text{RCL}}$ using the $\boxed{y^x}$ key.

$$(0.027)^{\frac{1}{3}} = r$$

$$0.3 = r$$

The common ratio is 0.3. To find the first term we substituted $r = 0.3$ into the expression for either a_2 or a_5.

Using $a_2 = ar^1$:

$$3 = a(0.3)$$

$$\frac{3}{0.3} = a$$

$$10 = a \qquad \square$$

The Sum of a Geometric Progression

To find the sum of the first n terms of a geometric progression start by writing out the sum of n terms of a geometric progression in general terms.

$$S_n = a + ar^1 + ar^2 + ar^3 + \ldots ar^{n-2} + ar^{n-1}$$

Now multiply both sides by r.

$$rs_n = ar + ar^1 \cdot r + ar^2 \cdot r + \ldots ar^{n-2} \cdot r + ar^{n-1} \cdot r$$

$$= ar^1 + ar^2 + ar^3 + \ldots ar^{n-1} + ar^n$$

Next subtract the second equation from the first.

$$S_n = a + ar^1 + ar^2 + ar^3 + \ldots ar^{n-2} + ar^{n-1}$$

$$-rS_n = \quad ar^1 + ar^2 + ar^3 + \ldots \quad + ar^{n-1} + ar^n$$

$$S_n - rS_n = a \qquad\qquad\qquad\qquad - ar^n$$

$$S_n - rS_n = a - ar^n$$

$$S_n(1 - r) = a - ar^n \qquad \text{factor } S_n$$

$$S_n = \frac{a - ar^n}{1 - r}$$

$$= \frac{a(1 - r^n)}{1 - r}$$

16.3C   Sum of n Terms of a Geometric Progression

The sum of n terms of a geometric progression is given by:

$$S_n = \frac{a(1 - r^n)}{1 - r}$$

Where: a = first term of the geometric progression

r = common ratio

Example 22 □ Find the sum of the first 6 terms of the sequence 3, 6, 12, 24, ...

The common ratio is $\frac{6}{3} = 2$.

The sum of 6 terms is given by:

$$S_6 = \frac{3(1 - 2^6)}{1 - 2}$$

$$= \frac{3(1 - 64)}{-1}$$

$$= \frac{3(-63)}{-1}$$

$$= 189 \qquad\qquad □$$

Chapter 16 ■ Sequences, Series, and the Binomial Expansion

In a computer, numbers are stored in memory locations where spots are magnetized or not magnetized. These spots can be interpreted as zero or one to represent numbers in a base-two number system. A base-two number system uses the same principle of positional notation as a base-ten number system.

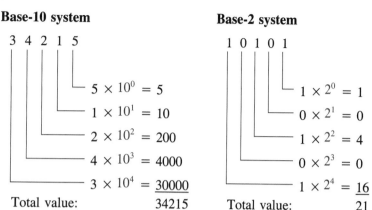

Base-10 system

3 4 2 1 5

$5 \times 10^0 = 5$
$1 \times 10^1 = 10$
$2 \times 10^2 = 200$
$4 \times 10^3 = 4000$
$3 \times 10^4 = \underline{30000}$
Total value: 34215

Base-2 system

1 0 1 0 1

$1 \times 2^0 = 1$
$0 \times 2^1 = 0$
$1 \times 2^2 = 4$
$0 \times 2^3 = 0$
$1 \times 2^4 = \underline{16}$
Total value: 21

A single one or zero in a base-two number system is called a *bit* which is short for binary digit. Designating a computer as an 8-bit or 16-bit computer indicates the number of bits that can be stored in a single memory location. The value of each bit read from right to left across the number is given by the corresponding term of a geometric progression with a common ratio of two.

Example 23 □ What is the value of the fourth bit from the right in the binary number 101111?

This is a geometric progression with $a = 1$ and $r = 2$. We are asking for the value of a_4.

$$a_n = ar^{n-1} \qquad a_4 = 1 \cdot 2^{4-1}$$
$$= 2^3$$
$$= 8 \qquad\qquad\qquad □$$

Example 24 □ What is the largest number that can be stored in an eight-bit binary number?

This is equivalent to asking for the sum of the geometric progression of eight terms with $a = 1$ and $r = 2$.

$$S_n = \frac{a(1 - r^n)}{1 - r} \qquad S_8 = \frac{1(1 - 2^8)}{1 - 2}$$
$$= \frac{1(1 - 256)}{-1}$$
$$= \frac{-255}{-1}$$

$S_8 = 255$ is the largest base-ten number that can be stored in eight bits. □

Example 25 □ Find the sum of 10 terms of the progression: $1, \dfrac{1}{2}, \dfrac{1}{4}, \ldots$

In this case: $a = 1$, $r = \dfrac{1}{2}$, $n = 10$.

$$S_n = \frac{a(1 - r^n)}{1 - r}$$

$$S_{10} = \frac{1\left[1 - \left(\dfrac{1}{2}\right)^{10}\right]}{1 - \dfrac{1}{2}}$$

$$= \frac{\left[1 - \dfrac{1}{1024}\right]}{\dfrac{1}{2}}$$

$$= \left[\frac{1024}{1024} - \frac{1}{1024}\right] \cdot \frac{2}{1}$$

$$= \frac{1023}{1024} \cdot 2$$

$$S_{10} \approx 1.998 \qquad\qquad \square$$

Problem Set 16.3

1. Write the first 5 terms of a geometric progression with: $a = 2$, $r = 3$.

2. Write the first 4 terms of a geometric progression with: $a = 3$, $r = 2$.

3. Write the first 7 terms of a geometric progression with: $a = 10$, $r = \dfrac{1}{2}$.

4. Write the first 6 terms of a geometric progression with: $a = 32$, $r = \dfrac{1}{4}$.

5. Write the first 5 terms of a geometric progression with: $a = 3$, $r = -2$.

6. Write the first 6 terms of a geometric progression with: $a = 2$, $r = -3$.

Find the common ratio for each of the following geometric progressions.

7. 10, 20, 40, . . .

8. 60, 90, 135, . . .

9. 4, 2, 1, . . .

10. 81, 27, 9, . . .

11. 5, −10, 20, −40, . . .

12. 1, −4, 16, −64, . . .

Find the general term of the following geometric progressions.

13. 1, 5, 25, 125, . . . **14.** 4, 16, 64, 256, . . . **15.** 64, 32, 16, 8, . . . **16.** 1000, 100, 10, 1, . . .

17. Find the sum of the first 8 terms of the geometric progression: 2, 4, 8, 16, . . .

18. If you worked at a job where you were paid one cent for the first day and your pay was doubled for each of the next 30 days, how much money would you earn on the 30th day?

19. What is the value of the largest binary number that can be stored in a 16-bit computer?

20. What is the sum of the first 3 terms of the geometric progression: $1, \dfrac{1}{2}, \dfrac{1}{4}, \dfrac{1}{8}, \ldots$?

21. What is the sum of the first 10 terms of: $1, \dfrac{1}{2}, \dfrac{1}{4}, \dfrac{1}{8}, \ldots$?

22. What is the sum of the first 30 terms of: $1, \dfrac{1}{2}, \dfrac{1}{4}, \dfrac{1}{8}, \ldots$?

23. Assuming a generation lasts 20 years, what is the total number of ancestors you had in the 200 years before you were born or 10 generations?

24. Find the sum of the first 7 terms of the geometric series: 3, 9, 27, 81, ...

25. Find the sum of the first 4 terms of the geometric series: 1, 10, 100, 1000, ...

26. Find the sum of the first 6 terms of the geometric progression in problem 25.

■ 16.4　　　The Binomial Expansion

An earlier chapter found the square of a binomial by multiplying it by itself, as in

$$(a + b)^2 = a^2 + 2ab + b^2$$

This section will show you a convenient way to raise a binomial to any power. However, to start you must be familiar with **factorial notation.**

16.4A　*n* Factorial–*n*!

n factorial is written *n*! It means:

$$n! = n \cdot (n - 1)(n - 2) \ldots [n - (n - 1)]$$

By definition: $0! = 1$

Example 26 ☐ Find 4!

$$4! \text{ means } 4 \cdot 3 \cdot 2 \cdot 1$$

Therefore,
$$4! = 4 \cdot 3 \cdot 2 \cdot 1$$
$$= 24 \qquad \square$$

Example 27 ☐ Find 5!

$$5! = 5 \cdot 4 \cdot 3 \cdot 2 \cdot 1$$
$$= 120$$

A generalization of the fact that

$$6! = 6 \cdot 5!$$

is
$$n! = n(n-1)! \qquad \square$$

This relationship can be used to simplify quotients of two factorials.

Example 28 ☐ Find $\dfrac{5!}{3!}$

$$\frac{5!}{3!} = \frac{5 \cdot 4 \cdot 3 \cdot 2 \cdot 1}{3 \cdot 2 \cdot 1}$$
$$= 5 \cdot 4$$
$$= 20 \qquad \square$$

Example 29 ☐ Find $\dfrac{100!}{95!}$

Think of the numerator as a product that contains the denominator which is 95!. Your work is then reduced.

$$\frac{100!}{95!} = \frac{100 \cdot 99 \cdot 98 \cdot 97 \cdot 96 \cdot 95!}{95!}$$

That is a big number.

$$= 100 \cdot 99 \cdot 98 \cdot 97 \cdot 96$$
$$= 9034502400$$

Factorials grow very rapidly. $\qquad \square$

Example 30 ☐ Write $8 \cdot 7 \cdot 6 \cdot 5$ using factorial notation.

Since 8! is the product of the whole numbers from 8 down to 1, you need a way to divide out the extra factors in 8!.

$$8 \cdot 7 \cdot 6 \cdot 5 = \frac{8 \cdot 7 \cdot 6 \cdot 5 \cdot 4 \cdot 3 \cdot 2 \cdot 1}{4 \cdot 3 \cdot 2 \cdot 1}$$
$$= \frac{8!}{4!} \qquad \square$$

Chapter 16 ■ Sequences, Series, and the Binomial Expansion

Example 31 □ Write $301 \cdot 300 \cdot 299$ using factorial notation.

$$301 \cdot 300 \cdot 299 = \frac{301!}{298!}$$

□

Example 32 □ Write $n \cdot (n - 1)(n - 2)(n - 3)$ using factorial notation.

$$n(n - 1)(n - 2)(n - 3) = \frac{n!}{(n - 4)!}$$

□

Where did the $n - 4$ come from?

$n!$ includes all the whole numbers from n down to one. Divide out the factors following $(n - 3)$. They are $(n - 4)!$

Problem Set 16.4A

Find the following products.

1. $3!$ **2.** $4!$ **3.** $2!$ **4.** $5!$ **5.** $1!$ **6.** $0!$ **7.** $9!$

Find the following quotients.

8. $\dfrac{10!}{8!}$ **9.** $\dfrac{7!}{5!}$ **10.** $\dfrac{5!}{2!}$ **11.** $\dfrac{12!}{8! \cdot 5!}$ **12.** $\dfrac{21! \cdot 5!}{15! \cdot 9!}$ **13.** $\dfrac{24!}{19! \cdot 6!}$

Write each of the following using factorial notation.

14. $9 \cdot 8 \cdot 7$ **15.** $22 \cdot 21 \cdot 20 \cdot 19$

16. $100 \cdot 99 \cdot 98 \ldots 50$ **17.** $200 \cdot 199 \cdot 198 \ldots 10$

18. $n \cdot (n - 1)(n - 2)$ **19.** $n \cdot (n - 1)(n - 2)(n - 3)(n - 4)$

Pascal's Triangle

Now that you can use factorial notation, here is a convenient way to raise a binomial to a higher power. Before we generalize a formula for $(a + b)^n$, find $(a + b)^3$ by multiplication.

$$(a + b)^3 = (a + b)^2(a + b)$$
$$= (a^2 + 2ab + b^2)(a + b)$$

At this stage we will use a form of multiplication similar to multiplication of whole numbers. We write each partial product with like terms in a column to make addition easier.

$$a^2 + 2ab + b^2$$
$$a + b$$

$$a^3 + 2a^2b + \ ab^2 \quad \longleftarrow \boxed{\text{Product of } a \text{ times } (a^2 + 2ab + b^2)}$$
$$a^2b + 2ab^2 + b^3 \ \longleftarrow \boxed{\text{Product of } b \text{ times } (a^2 + 2ab + b^2)}$$

$$(a + b)^3 =$$
$$a^3 + 3a^2b + 3ab^2 + b^3 \quad \longleftarrow \boxed{\text{Sum of the partial products}}$$

You could find $(a + b)^4$ by multiplying $(a + b)^3$ by $(a + b)$.

If you were to repeat this process for the first few powers of $(a + b)^n$, you would get the following results:

$$(a + b)^0 = \qquad\qquad\qquad 1$$
$$(a + b)^1 = \qquad\qquad\qquad a + b$$
$$(a + b)^2 = \qquad\qquad a^2 + 2ab + b^2$$
$$(a + b)^3 = \qquad\quad a^3 + 3a^2b + 3ab^2 + b^3$$
$$(a + b)^4 = \qquad a^4 + 4a^3b^1 + 6a^2b^2 + 4a^1b^3 + b^4$$
$$(a + b)^5 = \quad a^5 + 5a^4b^1 + 10a^3b^2 + 10a^2b^3 + 5a^1b^4 + b^5$$
$$(a + b)^6 = a^6 + 6a^5b^1 + 15a^4b^2 + 20a^3b^3 + 15a^2b^4 + 6a^1b^5 + b^6$$

By referring to the expansions above we can make several observations about $(a + b)^n$ at least as far as $n = 6$.

Observations:

1. The first term is always a^n.

2. The last term is always b^n.

3. The power of a decreases by one in each successive term.

4. The power of b increases by one in each successive term.

5. The sum of the exponents of a and b is always n.

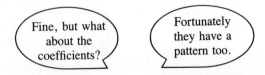

Fine, but what about the coefficients?

Fortunately they have a pattern too.

Chapter 16 ■ Sequences, Series, and the Binomial Expansion

The pattern of the coefficients is easier to see if you write the coefficients alone.

$$
\begin{array}{lc}
n = 0 & 1 \\
n = 1 & 1\ 1 \\
n = 2 & 1\ 2\ 1 \\
n = 3 & 1\ 3\ 3\ 1 \\
n = 4 & 1\ 4\ 6\ 4\ 1 \\
n = 5 & 1\ 5\ 10\ 10\ 5\ 1 \\
n = 6 & 1\ 6\ 15\ 20\ 15\ 6\ 1
\end{array}
$$

Notice in the triangular pattern above, any coefficient is the sum of the two coefficients directly above it. This relationship was first observed by Chinese writers some time before 1300. Later, Blaise Pascal (1623–1662) independently developed many applications of the triangle including the coefficients of binomials and the laws of probability. Since his work, rather than the early Chinese work, was published in the European Press, the pattern is called *Pascal's Triangle*.

Use the observations above and Pascal's triangle to expand $(a + b)$ to any whole number power.

Example 33 □ Expand $(a + b)^5$ using Pascal's triangle.
The pattern of the variable is:

$$(a + b)^5 = ?a^5b^0 + ?a^4b^1 + ?a^3b^2 + ?a^2b^3 + ?a^1b^4 + ?a^0b^5$$

To get the coefficients replace the question marks with the line of Pascal's triangle for $n = 5$.

$$(a + b)^5 = 1a^5 + 5a^4b^1 + 10a^3b^2 + 10a^2b^3 + 5a^1b^4 + 1b^5 \qquad □$$

Example 34 □ Use Pascal's triangle to expand $(2x - 3)^5$.
This is almost the same as the example above if you let $a = 2x$ and $b = -3$.

$$(a + b)^5 = a^5 + 5a^4b^1 + 10a^3b^2 + 10a^2b^3 + 5a^1b^4 + b^5$$

Substituting $a = 2x$ and $b = -3$ yields

$$(2x - 3)^5 = (2x)^5 + 5(2x)^4(-3)^1 + 10(2x)^3(-3)^2$$
$$+ 10(2x)^2(-3)^3 + 5(2x)(-3)^4 + (-3)^5$$
$$= 32x^5 + 5(16x^4)(-3) + 10(8x^3)(9)$$
$$+ 10(4x^2)(-27) + 5(2x)(81) + (-243)$$
$$= 32x^5 - 240x^4 + 720x^3 - 1080x^2 + 810x - 243 \qquad □$$

There is another way to expand a binomial without having to write out Pascal's triangle.

Let's look at $(a + b)^5$ to see how we can write any term in the expansion from the previous term.

Notice that the sum of the exponents is always 5

$$(a + b)^5 = 1a^5 + 5a^4b^1 + 10a^3b^2 + 10a^2b^3 + 5a^1b^4 + b^5$$

Observation 6. If we multiply the coefficient of any term by the power of a in that term and divide by the term number, we get the coefficient of the next term.

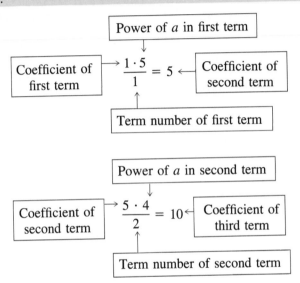

Example 35 □ Use Observation 6 to expand $(a + b)^4$.

$(a + b)^4 = 1a^4b^0 +$

Coefficient of first term $\dfrac{1 \cdot 4}{1} = 4$ ⇒ $4a^3b^1 +$

Coefficient of second term $\dfrac{4 \cdot 3}{1 \cdot 2} = 6$ ⇒ $6a^2b^2 +$

Coefficient of third term $\dfrac{4 \cdot 3 \cdot 2}{1 \cdot 2 \cdot 3} = 4$ ⇒ $4ab^3 +$

Coefficient of fourth term $\dfrac{4 \cdot 3 \cdot 2 \cdot 1}{1 \cdot 2 \cdot 3 \cdot 4} = 1$ ⇒ $1a^0b^4$

$$= a^4 + 4a^3b + 6a^2b^2 + 4ab^3 + b^4$$

In the expansion of $(a + b)^n$ there is always one more term than the value of n.

Next let r represent the exponent of the b value in any term, then r will be one less than the term number. That is, for the fourth term in an expansion, r will equal 3.

Now rewrite the expansion of $(a + b)^4$ above with the coefficients written as indicated products.

First Term $r = 0$	Second Term $r = 1$	Third Term $r = 2$	Fourth Term $r = 3$	Fifth Term $r = 4$

$$(a + b)^4 = a^4 + \frac{4 \cdot a^3 b^1}{1} + \frac{4 \cdot 3a^2 b^2}{2 \cdot 1} + \frac{4 \cdot 3 \cdot 2a^1 b^3}{3 \cdot 2 \cdot 1} + \frac{4 \cdot 3 \cdot 2 \cdot 1a^0 b^4}{4 \cdot 3 \cdot 2 \cdot 1}$$

Next rewrite the expression above replacing the exponents with expressions involving n, the power of the expansion, and r, where r is one less than the term number.

For $n = 4$:

$$(a + b)^4 =$$

$r = 0$	$r = 1$	$r = 2$	$r = 3$	$r = 4$

$$\frac{a^{n-r} b^r}{1} + \frac{4 \cdot a^{n-r} b^r}{1} + \frac{4 \cdot 3a^{n-r} b^r}{2 \cdot 1} + \frac{4 \cdot 3 \cdot 2a^{n-r} b^r}{3 \cdot 2 \cdot 1} + \frac{4 \cdot 3 \cdot 2 \cdot 1a^{n-r} b^r}{4 \cdot 3 \cdot 2 \cdot 1}$$

Next notice that the denominator of each coefficient is $r!$

$$(a + b)^4 = \frac{a^{n-r} b^r}{r!} + \frac{4 \cdot a^{n-r} b^r}{r!} + \frac{4 \cdot 3a^{n-r} b^r}{r!}$$

$$+ \frac{4 \cdot 3 \cdot 2a^{n-r} b^r}{r!} + \frac{4 \cdot 3 \cdot 2 \cdot 1a^{n-r} b^r}{r!}$$

Next notice that the numerator of each coefficient is $\dfrac{n!}{(n - r)!}$, that is, for the third term:

$$4 \cdot 3 = \frac{4 \cdot 3 \cdot 2 \cdot 1}{(4 - 2)!}$$

$$(a + b)^4 = \frac{n!a^{n-r} b^r}{(n - r)!r!} + \frac{n!a^{n-r} b^r}{(n - r)!r!} + \frac{n!a^{n-r} b^r}{(n - r)!r!}$$

$$+ \frac{n!a^{n-r} b^r}{(n - r)!r!} + \frac{n!a^{n-r} b^r}{(n - r)!r!} \quad \square$$

Generalize the observations above to get the following expression for any term of a binomial expansion.

16.4B   A Term of a Binomial Expansion

To find a term of $(a + b)^n$, let r be one less than the term number, then:

$$a_{r+1} = \frac{n!}{(n - r)!r!} a^{n-r}b^r$$

Example 36 □ Find the third term of $(a + b)^5$.
For the third term of $(a + b)^5$, $r = 2$ and $n = 5$

$$a_{r+1} = \frac{n!}{(n - r)!r!} a^{n-r}b^r$$

$$a_{2+1} = \frac{5!}{(5 - 2)!2!} a^{5-2}r^2$$

$$= \frac{5!}{3!2!} a^3b^2$$

$$= \frac{5 \cdot 4 \cdot 3 \cdot 1}{3 \cdot 2 \cdot 1} a^3b^2$$

$$a_3 = 10a^3b^2 \qquad \text{Third term of } (a + b)^5$$

Is there an easy way to remember the formula?

Yes. Focus on the exponent of b for each term. Notice that it is r, which is always one less than the term number.

For the fourth term of $(a + b)^5$ $n = 5$ and $r = 4$

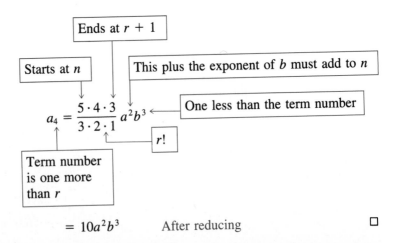

$$= 10a^2b^3 \qquad \text{After reducing} \qquad □$$

Chapter 16 ■ Sequences, Series, and the Binomial Expansion

Example 37 ▢ Find the coefficient of the third term of $(a + b)^4$.

In this case: $n = 4$, and since you are looking for the third term you want $r + 1 = 3$ or $r = 2$.

$$\text{Coefficient of } a_{r+1} = \frac{n!}{(n - r)!r!}$$

$$\text{Coefficient of } a_{2+1} = \frac{4!}{(4 - 2)!2!}$$

$$= \frac{4!}{2! \cdot 2!}$$

$$= \frac{4 \cdot 3 \cdot 2!}{2 \cdot 1 \cdot 2!}$$

$$\text{Coefficient of } a_3 = 6 \qquad \text{for } n = 4 \qquad \qquad ▢$$

Example 38 ▢ Find the coefficient of the third term of $(a + b)^6$.

$$\text{Coefficient of } a_{2+1} = \frac{n!}{(n - r)!r!}$$

$$= \frac{6!}{(6 - 2)!2!} = \frac{6!}{4! \cdot 2!}$$

$$= \frac{6 \cdot 5}{2!}$$

$$= 15 \qquad \qquad ▢$$

Example 39 ▢ Find the fourth term of $(2x - 4)^7$.

In this case $n = 7$, $r = 3$, $a = 2x$ and $b = -4$.

First find the coefficient.

$$\text{Coefficient of } a_{r+1} = \frac{n!}{(n - r)!r!}$$

$$\text{Coefficient of } a_{3+1} = \frac{7!}{(7 - 3)!3!} = \frac{7!}{4! \cdot 3!}$$

$$= \frac{7 \cdot 6 \cdot 5 \cdot 4!}{4! \cdot 3 \cdot 2 \cdot 1}$$

$$= 35$$

Now that you know the coefficient, you can write the fourth term of $(a + b)^7$ and substitute the appropriate values for a and b.

$$a_4 = 35a^{7-3}b^3$$

$$= 35a^4b^3$$

$$= 35(2x)^4(-y)^3$$

$$= 35(16x^4)(-y^3)$$

$$= -560x^4y^3 \qquad \qquad ▢$$

Example 40 □ Find the sixth term of $(3x + 2y)^7$.

In this case $r = 5$ and $n = 7$.

Let $a = 3x$ and $b = 2y$.

$$a_{5+1} = \frac{7 \cdot 6 \cdot 5 \cdot 4 \cdot 3 \cdot a^2 b^5}{5 \cdot 4 \cdot 3 \cdot 2 \cdot 1}$$

$$= 21a^2 b^5$$

$$= 21(3x)^2(2y)^5 \qquad \text{Substituting for } a \text{ and } b$$

$$= 21(9x^2)(32y^5)$$

$$a_6 = 6048x^2 y^5 \qquad\qquad\qquad\qquad □$$

Problem Set 16.4B

1. Using Pascal's triangle, write the coefficients for the expansion of $(a + b)^7$.

2. Using Pascal's triangle, write the coefficients for the expansion of $(a + b)^9$.

Use Pascal's triangle to expand the following:

3. $(a + b)^3$ **4.** $(x + y)^4$ **5.** $(p - q)^4$ **6.** $(r + 2s)^5$

Use the formula to find a term of a binomial expansion to expand the following binomials.

7. $(a + b)^4$ **8.** $(a + b)^7$ **9.** $(x - y)^5$ **10.** $(2x + y)^5$

Find the requested term.

11. Fourth term of $(a + b)^9$.

12. Fifth term of $(x + y)^9$.

13. First term of $(3x - 10y)^5$.

14. Sixth term of $(8p - 2q)^5$.

15. Fourth term of $\left(2x - \dfrac{y}{2}\right)^6$.

16. Third term of $\left(3x + \dfrac{y}{3}\right)^7$.

17. Fourth term of $(3x + y)^7$.

18. Seventh term of $(x - 2y)^9$.

Write the requested term as an indicated product of factors in lowest terms. Do not multiply out the factors.

19. Ninth term of $(7x + 4y)^{20}$.

20. Eleventh term of $(5x - 3y)^{15}$.

Chapter 16 ■ Sequences, Series, and the Binomial Expansion

16.1 **1.** A **sequence** is a list of numbers written in a predetermined order. Each element of the list is called a **term of the sequence.**

2. A **sequence** can be either infinite or finite.

3. The **general term of a sequence** is an expression that gives the value of a term in the sequence as a function of the term number.

16.2 **1.** An **Arithmetic Progression** is a sequence where each term of the sequence can be found by adding a constant to the previous term.

2. The following **variables are usually used with arithmetic progressions:**

a_1 = first term
l = last term
n = number of terms
d = difference between any two terms
S_n = sum of the first n terms

3. The **formula for the nth term of an arithmetic progression** is: $a_n = a_1 + (n - 1)d$

4. The **formula for the sum of the first n terms of an arithmetic progression** is:

$$S_n = \frac{n}{2}(a_1 + l)$$

16.3 **1.** A **Geometric Progression** is a sequence where each term a_n is equal to a constant times the previous term of the sequence. The constant is called the **common ratio between any two successive terms.**

2. The following **variables are usually used for a geometric progression:**

a = first term
n = term number
r = common ratio
S_n = sum of the first n terms of a geometric progression

3. The **general term of a Geometric Progression** is given by: $a_n = a \cdot r^{n-1}$

4. The **sum of n terms of a Geometric Progression** is given by: $S_n = \dfrac{a(1 - r^n)}{1 - r}$

16.4 **1.** n **factorial** − $n!$ means the product of each counting number from 1 to n.
$n! = n \cdot (n - 1)(n - 2) \ldots [n - (n - 1)]$
by definition: $0! = 1$

2. **Pascal's triangle** can be used to find the coefficients of binomial expansions.

3. In the **expansion of $(a + b)^n$** the following are always true:

  1. The first term is always a^n.
  2. The last term is always b^n.
  3. The power of a decreases by one in each successive term.
  4. The power of b increases by one in each successive term.
  5. The sum of the exponents of a and b is always n.
  6. If we multiply the coefficient of any term by the power of a in that term and divide by the term number, we get the coefficient of the next term.

4. Any **term of a binomial expansion** is given by:

$$a_{r+1} = \frac{n!}{(n-r)!r!}a^{n-r}b^r$$

where $r = $ one less than the term number

Review Test   Chapter 16

Find the first five terms for each of the following sequences where the value of a_n is given. **(16.1)**

1. $a_n = \dfrac{n}{n^2 + 1}$

2. $a_n = \dfrac{(-1)^n \cdot n}{n + 2}$

Find the general term for each of the following sequences. **(16.1)**

3. 0, 3, 8, 15, 24, . . .

4. $-\dfrac{1}{4}, \dfrac{1}{6}, -\dfrac{1}{8}, \dfrac{1}{10}, -\dfrac{1}{12}, \ldots$

Find the general term for each of the following arithmetic progressions. **(16.2)**

5. 1, 4, 7, 10, 13, . . .

6. $-13, -9, -5, -1, 2, \ldots$

7. Find the eleventh term of the arithmetic progression where the first term is 9 and the common difference is -2. **(16.2)**

Find the requested term, given the indicated terms of an arithmetic progression. **(16.2)**

8. $a_1 = 11$, $a_8 = -10$, $a_{15} = $ _____

9. $a_7 = 24$, $a_{16} = 69$, $a_3 = $ _____

10. Find the sum of the first 11 terms of the arithmetic progression: 24, 20, 26, 12, ... **(16.2)**

11. Write the first 6 terms of a geometric progression with $a = 5$, $r = -2$. **(16.3)**

12. Write the first 5 terms of a geometric progression with $a = 27$, $r = \dfrac{2}{3}$. **(16.3)**

Find the common ratio for each of the following geometric progressions. **(16.3)**

13. 3, -6, 12, -24, . . .

14. 80, 60, 45, . . .

15. Find the general term of the following geometric progression: 3, 9, 27, 81, ... **(16.3)**

16. Find the sum of the first 8 terms of the progression: 1, 3, 9, 27, ... **(16.3)**

17. What is the sum of the first 5 terms of the geometric progression: $1, \dfrac{1}{2}, \dfrac{1}{4}, \dfrac{1}{16}, \ldots$ . . . . **(16.3)**

Chapter 16 ■ Sequences, Series, and the Binomial Expansion

18. A man is offered a job with a starting salary of $18,000 annually, with yearly increases of 6%. What would be his annual salary during the fifth year? **(16.3)**

19. Referring to problem 18, what would be his accumulated earnings during the five-year period? **(16.3)**

Find the following products. **(16.4)**

20. 6!

21. 8!

Find the following quotients. **(16.4)**

22. $\dfrac{12!}{9!}$

23. $\dfrac{22!}{17! \cdot 7!}$

24. Use the formula for any term of a binomial expansion to find all the terms in the expansion of $(x + 2y)^5$. **(16.4)**

Find the requested term. **(16.4)**

25. Fourth term of $(2x - 3y)^6$

26. Sixth term of $(x + 2y)^8$

Cumulative Review   Chapters 1–16

Simplify the following.

1. $(-45)\left(-\dfrac{2}{5}\right) + 6^2 - 24 \div 6$ **(1.1)**

2. $(-8)^{\frac{1}{3}}$ **(8.1)**

3. $-\sqrt[5]{-32}$ **(8.2)**

4. $\dfrac{\sqrt{32a^5b^3}}{\sqrt{8ab}}$ **(8.2)**

Simplify. Write without negative exponents.

5. $(-3x^{-3}y^2)^2(2x^{-2}y)^3$ **(5.1)**

6. $\left(\dfrac{a^3b^{-2}}{2a^{-3}b^0}\right)^{-2}$ **(5.1)**

7. $\dfrac{x^{\frac{2}{3}}y^{-3}}{x^{\frac{1}{6}}y^{\frac{3}{4}}}$ **(8.1)**

Evaluate the following.

8. $\dfrac{2}{3}x - 15ax + 18xy^3$ if $a = \dfrac{1}{5}$, $x = -6$, $y = \dfrac{2}{3}$ **(1.3)**

9. $\sqrt{-24}$ **(9.1)**

10. i^{11} **(9.1)**

Perform the indicated operations and simplify.

11. $(4a^2 - 3a + 5)(3a^2 + 6a - 4)$ **(5.3)**

12. $(3a - 2b)(3a + 2b)$ **(5.3)**

13. $(15x^4 + 4x^3 - 4x^2 - 10x + 1) \div (5x - 2)$ **(5.4)**

14. $\dfrac{x^2 - 9}{3x^2 + 8x - 3} \cdot \dfrac{9x^2 - 3x}{x^3 - 27}$ **(5.4)**

15. $\dfrac{2a^2 + ab - 6a - 3b}{2a^2 - 9a + 9} \div \dfrac{4a^2 + 4ab + b^2}{2a^2 - ab - b^2}$ **(7.1)**

16. $\dfrac{y}{y^2 - 16} - \dfrac{y + 1}{y^2 - 5y + 4}$ **(7.2)**

17. $\dfrac{a + 2 - \dfrac{12}{a + 3}}{a - 5 + \dfrac{16}{a + 3}}$ **(7.3)**

18. $\dfrac{x - 2}{1 - \dfrac{4}{x^2}}$ **(7.3)**

19. $\sqrt{27x} - x\sqrt{12x} + 4\sqrt{3x}$ **(8.3)**

20. $(\sqrt{2x} + 4)(\sqrt{2x} - 6)$ **(8.4)**

21. $\dfrac{\sqrt{6}}{\sqrt{3} - \sqrt{2}}$ **(8.4)**

22. $(-2i + 4) - (5 - 6i)$ **(9.1)**

23. $(-5 + \sqrt{-3})(3 + 2\sqrt{-3})$ **(9.1)**

24. $\dfrac{3 - 2i}{3 + 4i}$ **(9.1)**

Factor the following.

25. $8x^2 + 14x - 15$ **(6.2)**

26. $25x^2 + 20xy + 4y^2$ **(6.3)**

27. $16x^8 - y^4$ **(6.3)**

28. $x^6 - 8y^3$ **(6.3)**

29. $x^4 - 18x^2 + 81$ **(6.4)**

30. $(x^2 - 2x)^2 - 11(x^2 - 2x) + 24$ **(6.4)**

31. $2xy + 12x - 5y - 30$ **(6.4)**

Solve the following equations or inequalities.

32. $6x - 4(x - 3) - 4 = 3x - (2 - 4x)$ **(2.1)**

33. $\dfrac{x - 3}{5} + \dfrac{x}{6} = \dfrac{1}{2}$ **(2.1)**

34. $N = \dfrac{a}{2}(c + d)$ for d **(2.2)**

35. $|x + 3| \le 6$ **(2.4)**

36. $|3x - 8| + 2 \ge 6$ **(2.4)**

37. $\dfrac{1}{a - 1} + \dfrac{2}{3a - 3} = -\dfrac{5}{12}$ **(7.4)**

38. $3x^2 + 10x - 8 = 0$ by factoring **(9.2)**

39. $2x^2 - x - 15 = 0$ by completing the square **(9.3)**

40. $4x^2 - 2x + 3 = 0$ by quadratic formula **(9.4)**

41. $\sqrt{2x - 1} + x = 8$ **(10.1)**

42. $x^4 - 7x^2 + 12 = 0$ **(10.2)**

43. $3x^{-2} + 10x^{-1} - 8 = 0$ **(10.2)**

44. $2^{3x+1} = 128$ **(13.2)**

45. $\log_a \dfrac{1}{8} = -3$ **(13.3)**

46. $3^{2x} = 40$ (Use logarithms, 5-digit accuracy.) **(14.1)**

47. $\log_2(3x - 4) + \log_2(x - 5) = \log_2(4x - 10)$ **(14.1)**

Solve the following systems.

48. $2x + 4y = -2$ **(4.1)**

$\quad x - 2y = 7$

49. $2x + y - z = 3$ **(4.3)**

$\quad 3x - 2y - 3z = 2$

$\quad 4x + 3y + 2z = -4$

50. $x^2 + y^2 = 100$ **(12.3)**

$\quad y - x = 2$

51. $2x^2 + 5y^2 = 22$ **(12.4)**

$\quad 3x^2 - y^2 = -1$

52. Find the equation of the line with slope of $-\dfrac{4}{5}$ passing through the point $(5, -6)$. **(3.2)**

53. Find the equation of the line passing through the points $(-5, 3)$ and $(4, -9)$. **(3.2)**

54. Find the equation of the line perpendicular to the line $2x + 5y = 8$, passing through the point $(4, -3)$. **(3.3)**

55. Use scientific notation to evaluate. **(5.2)**

$\dfrac{0.000027}{7200}$

56. Use synthetic division to divide. **(5.5)**

$(2x^4 + 5x^3 - 7x^2 + 16) \div (x + 3)$

57. Find the distance between the points $(-3, 6)$ and $(7, -2)$. **(11.1)**

58. Find the midpoint of the points $(-3, 6)$ and $(7, -1)$. **(11.1)**

59. Find the center and radius of the circle: $x^2 + y^2 - 6x + 10y - 2 = 0$ **(11.2)**

60. Express the following equations in an (x', y') system with the origin at the given point in the original system. **(11.3)**

Radius 6, center: $(3, -5)$

61. Express as a sum or difference of simpler logarithmic quantities.

$$\log_a \sqrt[4]{\dfrac{8x^3 y}{z^2}}$$

62. Use logarithms to evaluate. Correct to five significant digits. **(13.4)**

$$\dfrac{(0.5869)^2 \sqrt{36.924}}{(3.654)^3}$$

Solve and graph the following inequalities. **(2.3)**

63. $5(1 - x) + 3 \geq 3x - (4x + 5)$

64. $5x + 6 \leq 1$ or $2x - 3 > -1$

Graph the following.

65. $3x - 4y = 12$ **(3.3)**

66. $y < -\dfrac{2}{3}x + 4$ **(3.4)**

67. $y - 2 = -2(x + 3)^2$ **(11.4)**

68. $9x^2 + 4y^2 - 36x + 8y + 4 = 0$ **(12.1)**

69. $\dfrac{(x - 1)^2}{16} - \dfrac{(y + 2)^2}{9} = 1$ **(12.2)**

Set up the appropriate number of equations and solve.

70. The difference of two numbers is 3. If seven times the smaller is three more than four times the larger, find the numbers. **(4.2)**

71. A collection of coins consists of nickels, dimes, and quarters valued at $3.75. If there are twice as many dimes as nickels and three fewer quarters than dimes, find the number of each kind of coin in the collection. **(4.3)**

72. A typist can complete a typing job in 4 hours. A less-experienced typist can complete the same job in 6 hours. How long would it take them to complete the job if they worked together? **(7.5)**

73. The resistance (R) of an electrical circuit varies directly as the length (l) and inversely as the square of the diameter (d). If a certain type of wire 30 feet long with a diameter of $\dfrac{1}{16}$ inch has a resistance of 20 ohms, how much resistance would 150 feet of the same type of wire have if its diameter was $\dfrac{1}{32}$ inch? **(10.3)**

74. A fisherman can travel 18 miles upstream on a river in the same time he can travel 30 miles downstream. If his boat can travel 12 miles per hour in still water, what is the speed of the current of the river? **(10.4)**

75. Find the amount of a $5000 investment at 12% for 5 years if compounded:
   a. annually
   b. semi-annually
   c. quarterly
   d. weekly
   e. daily
   f. continuously   **(14.2 or 14.3)**

76. A sample of black coffee has a hydrogen ion concentration of 1.05×10^{-4}. What is its pH? **(14.3)**

77. Determine the atmospheric pressure at the Tioga Pass in Yosemite. The Tioga Pass has an elevation of 9941 feet. **(14.2)**

78. A desert town of 1200 people doubled its population in 6 years. Find the rate of growth. Assuming a continuous rate of growth, how long would it take for the town to triple its population? **(14.2)**

79. If the number of bacteria in a culture doubles every 30 minutes, how long would it take a culture of 1.5×10^3 to grow to a culture of 4.5×10^9? **(14.2)**

80. If a certain type of radioactive substance has a decay constant of $1.35 + 10^{-9}$ per year, what is its half-life? **(14.2)**

81. How old is a piece of wood that has two-fifths of the activity (decay of C-14) of a similar piece of modern wood? **(14.3)**

82. If a non-living substance has 3 grams of carbon-14, how much will be present 500 years later? (Half-life of C-14 is 5730 years) **(14.3)**

83. How much louder is a busy street traffic sound than a whisper? **(14.3)**

84. Determine the energy level in watts per square meter of the sound of a normal conversation where the intensity level of the sound is 65 db. **(14.3)**

85. How many times more powerful is an earthquake of 6.8 magnitude than one of 4.8 magnitude? **(14.3)**

Evaluate the following determinants

86. $\begin{vmatrix} -8 & -9 \\ 5 & 7 \end{vmatrix}$ **(15.1)**

87. $\begin{vmatrix} 5 & 3 & -8 \\ -4 & -1 & 2 \\ -7 & -2 & 4 \end{vmatrix}$ **(15.3)**

Use Cramer's Rule to solve the following equations.

88. $x - 3y = 7$ **(15.2)**

$$\frac{1}{2}x - \frac{1}{4}y = -\frac{3}{2}$$

89. $\frac{1}{2}x - y + \frac{3}{2}z = \frac{1}{2}$ **(15.3)**

$$x + \frac{1}{3}y - \frac{2}{3}z = 2$$

$$\frac{3}{2}x + 2y - z = 6$$

Set up two equations and solve using Cramer's Rule. **(15.2)**

90. A sum of money is invested, part at 12% and the remainder at 10%. If two-thirds as much is invested at 10% as at 12% and the total income from both investments is $224, find the amount invested at each rate.

Set up three equations and solve using Cramer's Rule.

91. The sum of three numbers is eight. Three times the first plus twice the second added to the third is zero, and the second minus the third is equal to the first. Find the three numbers. **(15.4)**

92. Find the first five terms of the following sequence where the value of a_n is given. **(16.1)**

$$a_n = \frac{(-1)^n n^2}{n + 2}$$

93. Find the general term of the following arithmetic progression: $-22, -18, -14, -10, \ldots$ **(16.2)**

Find the indicated term of the arithmetic progression. **(16.2)**

94. $a_1 = -5$, $a_7 = 19$, $a_{15} = $ _____

95. $a_5 = 17$, $a_{27} = -27$, $a_{19} = $ _____

96. Find the sum of the first 12 terms of the arithmetic progression: $-16, -13, -11, -9, \ldots$ **(16.3)**

97. Write the first six terms of a geometric progression with $a = 81$, $r = -\dfrac{2}{3}$. **(16.3)**

98. Find the general term of the following geometric progression: $16, -8, 4, -2, \ldots$ **(16.3)**

99. Find the sum of the first six terms of the geometric progression: $-4, 2, -1, -\dfrac{1}{2}, \ldots$ **(16.3)**

100. Find the following quotient: $\dfrac{21!}{15!\,7!}$ **(16.4)**

101. Use the formula $(r + 1)$st term of a binomial expansion to expand the binomial: $(2x - y)^5$ **(16.4)**

102. Find the fourth term of $(3x - 2y)^6$. **(16.4)**

Selected Answers

■ Chapter 1

Problem Set 1.1A

1. 13 **3.** 0 **5.** 13 **7.** 56 **9.** 46 **11.** 17 **13.** 25 **15.** 12 **17.** 42
19. 1001 **21.** 750 **23.** 125

Problem Set 1.1B

1. 14 **3.** 4 **5.** -2 **7.** -14 **9.** 9 **11.** -9 **13.** 19 **15.** 10
17. 16 **19.** -11 **21.** 13 **23.** 8 **25.** 1 **27.** -55 **29.** 0 **31.** 36
33. 25 **35.** 34 **37.** 0 **39.** -44 **41.** 44 **43.** -20 **45.** 13

Problem Set 1.2A

1. $5 \cdot 3 + 5 \cdot 6$ **3.** $8 \cdot 3 + 8 \cdot 1$ **5.** $8 \cdot a + 8 \cdot b$ **7.** $4 \cdot x + 4 \cdot y$ **9.** $a \cdot b + a \cdot 2$
11. $x \cdot a + x \cdot b$

Problem Set 1.2B

1. $x + 5a$ **3.** $(a + 4) \cdot 6$ **5.** $6 \cdot a + 6 \cdot 4$ **7.** $a \cdot 2 + 2 \cdot 4$ or $2 \cdot a + 4 \cdot 2$
9. $(6 + 8) + c$ **11.** $(0 + 4) \cdot 6$ **13.** $(a + 6) \cdot 0$ **15.** $(5 \cdot a) \cdot 4$
17. $9 \cdot (3 + x)$ **19.** $3 \cdot a + 3 \cdot b$
21. Associative Prop. Mult. **23.** Associative Prop. Addition
25. Commutative Prop. of Addition **27.** Multiplication by zero
29. Multiplication by zero **31.** Distributive Prop.
33. Commutative Prop. of Addition **35.** Distributive Prop.
37. Commutative Prop. of Mult. **39.** Commutative Prop. of Mult.
41. Distributive Prop. **43.** Distributive Prop.
45. Multiplication by zero **47.** Commutative Prop. of Mult.
49. Commutative Prop. of Addition **51.** Commutative Prop. of Addition

Problem Set 1.3

1. $\dfrac{1}{4}$ **3.** 3   **5.** $-\dfrac{7}{12}$ **7.** -3 **9.** $-\dfrac{1}{b}$ **11.** -9 **13.** 2.4

15. -54 **17.** -30 **19.** 70   **21.** 0.024   **23.** 0.72   **25.** 8   **27.** -16

29. -16 **31.** 12   **33.** 30   **35.** -17 **37.** 12   **39.** -33 **41.** 7

43. -12 **45.** $\dfrac{-3}{4}$ **47.** $\dfrac{3}{2}$ **49.** -1.2 **51.** $-1\dfrac{1}{9}$ **53.** $\dfrac{-8y^2}{15}$ **55.** $\dfrac{16x^3}{5}$

57. $\dfrac{21y^3}{20}$ **59.** 6.45   **61.** $\dfrac{2x^2}{25}$ **63.** $20x$ **65.** $9x^2$ **67.** -1 **69.** 24

71. 0.72   **73.** 20   **75.** -28 **77.** -2 **79.** $-\dfrac{4}{3}$ **81.** -17 **83.** 10

Problem Set 1.4

1. $\dfrac{-85a - 32}{60}$ **3.** $\dfrac{-22xy + 21}{36}$ **5.** $\dfrac{25ax^2}{56}$ **7.** $\dfrac{23x}{24a}$

9. $\dfrac{-5bcy + 13aby}{ac}$ **11.** $\dfrac{-7b - 9a}{abx}$ **13.** $\dfrac{-12x^2 + 7y}{a^2xy}$ **15.** $\dfrac{-95a^2 + 68b^2}{60aby}$

17. $\dfrac{-35x^4 + 27a^2y^2}{42abx^2y}$ **19.** $\dfrac{35xb^2 + 42a^2y}{90x^2y^2ab}$ **21.** $\dfrac{-12a^2 - 6b^2}{abx^2}$ **23.** $\dfrac{-33ax + 26bc}{24aby}$

25. $\dfrac{32ay - 81ax}{72xy}$ **27.** $\dfrac{-9ax^2 + 22ay^2}{24bxy}$ **29.** $\dfrac{-49y^4 + 33a^2x^2}{42abxy^2}$ **31.** $\dfrac{33a^2y + 22xb^2}{36x^2y^2ab}$

33. $\dfrac{11b^2y + 3a^2y}{9abx^2}$ **35.** $\dfrac{66abx + 55aby}{90x^2y^2c}$

Problem Set 1.5

1. $16\dfrac{5}{64}$ sq in   **3.** $14\dfrac{1}{6}$ sq ft   **5.** $15\dfrac{2}{3}$ ft   **7.** 141.3 in

9. 1.37 sq ft   **11.** 7.3 cu ft   **13.** 17.8 cu ft   **15.** 1.73 gal

17. 47.1 in   **19.** 63.6 sq in   **21.** 94.2 sq in   **23.** 1017.4 sq ft

25. 9216 sq ft   **27.** 37.7 sq in   **29.** 2880 sq in   **31.** 5400 cu in

33. 1728 sq in   **35.** 40176 cu in   **37.** \$605   **39.** 432 sq ft

41. 10.1 or 11 rolls   **43.** 25434 sq mi   **45.** 2370 sq in   **47.** 3555 cu in

49. 9 sq ft   **51.** 116.875 cu in   **53.** 7680 sq ft   **55.** 274 ft

57. 1644 blocks

Chapter 1   Review Test

1. 35   **2.** 0   **3.** 3   **4.** 3

5. 21   **6.** -15 **7.** 0   **8.** 3

9. -32 **10.** 1   **11.** -20 **12.** Assoc. Prop. Add.

13. Comm. Prop. Mult.   **14.** Mult. by zero   **15.** $p \cdot 2 + p \cdot q$ **16.** $x + 0$

17. $1(x)$ **18.** $(x + 0) \cdot 1$ **19.** $1(0 + x)$ **20.** 24

21. -4 **22.** 21 **23.** -9 **24.** -24

25. $\dfrac{7}{5}$ **26.** -0.06 **27.** 18 **28.** $\dfrac{-20x^2y}{9}$

29. $\dfrac{21z^3}{y}$ **30.** $4n^2$ **31.** $\dfrac{24n^3}{5}$ **32.** $\dfrac{8x^4}{3}$

33. $\dfrac{4x^3}{9}$ **34.** $\dfrac{x}{36a}$ **35.** $\dfrac{-33ay - 14x}{24bxy}$ **36.** $\dfrac{-25aby + 16abx}{30x^2y^2}$

37. 10.05 cu in   **38.** 754 yds   **39.** $1733\dfrac{1}{3}$ yds   **40.** a) 282.6 cu in
                                                                                                         b) 1.2 gal

41. 1.1 gal   **42.** a) 1962.5 sq ft
                                b) 1537.5 sq ft

■ Chapter 2

Problem Set 2.1A

1. 3 **3.** 3 **5.** 5 **7.** $\dfrac{2}{3}$ **9.** -3 **11.** 2

13. 4 **15.** 2 **17.** $\dfrac{-7}{2}$ **19.** $\dfrac{5}{2}$ **21.** $\dfrac{5}{2}$ **23.** $\dfrac{15}{7}$

Problem Set 2.1B

1. 7 **3.** $\dfrac{28}{3}$ **5.** 6 **7.** $\dfrac{36}{11}$ **9.** $-\dfrac{21}{5}$ **11.** 3

13. 6 **15.** $\dfrac{1}{3}$ **17.** $\dfrac{-2}{3}$ **19.** 5 **21.** $\dfrac{4}{5}$ **23.** 1

25. All real numbers   **27.** $\dfrac{-1}{2}$ **29.** All real numbers   **31.** -10 **33.** $\dfrac{-9}{2}$ **35.** 1

37. 3 **39.** No solution   **41.** $\dfrac{-4}{3}$ **43.** 3 **45.** -10 **47.** -6

49. 40 **51.** 2 **53.** 0 **55.** -3 **57.** -1 **59.** 2

Problem Set 2.2

1. $x = \dfrac{y}{k}$ **3.** $w = \dfrac{A}{l}$ **5.** $h = \dfrac{V}{lw}$

7. $h = \dfrac{2A}{b}$ **9.** $h = \dfrac{3V}{\pi r^2}$ **11.** $w = \dfrac{P - 2l}{2}$

13. $C = P - a - b$ **15.** $r = \dfrac{d}{t}$ **17.** $\pi = \dfrac{A}{r^2}$

19. $p = \dfrac{I}{rt}$ **21.** $m = \dfrac{y - b}{x}$ **23.** $b = y - mx$

25. $C = \dfrac{5}{9}(F - 32)$

27. $m = \dfrac{2E}{V^2}$

29. $a = \dfrac{2y - 2V_0 t}{t^2}$

31. $h = \dfrac{2A}{B + b}$

33. $d = \dfrac{s - a}{n - 1}$

35. $a = s - (n - 1)d$

37. $n = \dfrac{2S}{a + s}$

39. $V_1 = \dfrac{QV_2 - T}{Q}$

41. $D = T - F - \dfrac{wa}{q}$ or $D = \dfrac{Tq - Fq - wa}{q}$

43. $V = \dfrac{2S + V_2 T}{T}$

45. $C = \dfrac{Q}{m(T_1 - T_2)}$

Problem Set 2.3A

1. $x < 6$

3. $x \geq 4$

5. $x < 3$

7. $x \leq 5$

9. $x > \dfrac{-2}{3}$

11. $x < 7$

13. $x \geq 2$

15. $x > 6$

17. $x < 2$

19. $y \geq -\dfrac{7}{2}$

21. $x \geq 2$

23. $y > \dfrac{3}{2}$

25. $x < \dfrac{-2}{3}$

27. $y \leq 3$

29. $n \geq -\dfrac{2}{9}$

31. $y \geq -2$

33. $n \geq 2$

35. $x < 3$

37. $x \leq 48$

39. $x < -24$

41. $x \leq -10$

43. $x \geq -4$

45. $y \leq 4$

47. $x < -6$

49. $x \geq \dfrac{2}{3}$

51. $x > \dfrac{4}{5}$

53. $n \leq \dfrac{-39}{2}$

55. $x \leq \dfrac{9}{10}$

57. $x \geq -5$

59. $n < \dfrac{9}{7}$

Problem Set 2.3B

1.

3.

5.

7.

9.

11.

13.

15.

17.

19. $x < 5$ and $x > 1$

21. $x \le \dfrac{3}{2}$ and $x > -\dfrac{7}{3}$

23. $x < 0$ and $x > 1$

No solution

25. $x \le 1$ and $x \ge 1$

27. $x \le -4$ or $x > -1$

29. $x \le -1$ and $x > 4$

No solution

Problem Set 2.4

1. $x = 2, \; -2$

3. $a = 1, \; -1$

5. $-5 < x < 5$

7. $x = 0$

9. $x = 0$

11. Impossible

13. Impossible

15. $x < -2$ or $x > 2$

17. $x \le -4$ or $x \ge 4$

19. $x \le -3$ or $x \ge 1$

21. $x \le -3$ or $x \ge 4$

23. $\dfrac{7}{5}, \; -\dfrac{11}{5}$

25. $-\dfrac{3}{5}, \; 1$

27. $-4 \le x \le 6$

29. $x < -2$ or $x > 3$

31. $-6 \le y \le 4$

33. $x < \dfrac{-1}{2}$ or $x > 2$

35. $10, -18$

37. $x \ge \dfrac{14}{3}$ or $x \le \dfrac{-2}{3}$

39. $2 \le x \le 3$

41. $\dfrac{-9}{2}, \; \dfrac{9}{2}$

43. $-16, \; -4$

45. $6 \le x \le 10$

47. $x < \dfrac{8}{5}$ or $x > \dfrac{32}{5}$

49. $x \le -12$ or $x \ge 0$

51. $1 < x < 5$

53. $x < -6$ or $x > \dfrac{2}{3}$

55. $\dfrac{-1}{2} \le x \le 1$

Problem Set 2.5

1. 6 cents

3. $3.30

5. 130 cups

7. $0.026 or 2.6 cents

9. $0.026 or 2.6 cents

11. 55 cups

13. 32 feet

15. 9.3 seconds

17. 1760 feet

Chapter 2   Review Test

1. $\dfrac{29}{3}$

2. $\dfrac{35}{8}$

3. -6

4. -3

5. 3

6. 4

7. $-\dfrac{13}{4}$

8. 4

9. $\dfrac{d}{r}$

10. $\dfrac{P-2l}{2}$

11. $\dfrac{P}{1-r}$

12. $\dfrac{2A-hB}{h}$

13. $\dfrac{8-4y}{3}$

14. $\dfrac{PV}{k}$

15. $x \le -5$

16. $y < \dfrac{1}{3}$

17. $a > \dfrac{5}{2}$

18. $x \ge -\dfrac{7}{2}$

19. $n \le \dfrac{5}{2}$

20. $x > -1$

21. $x > 3$

22. $x \le 1$

23. $-2 \le x < 5$

24. $x < -3$ or $x \ge 2$

25. $x = 6, \;\; -6$

26. $x \le -\dfrac{1}{2}$ or $x \ge 2$

27. $y > -\dfrac{5}{3}$ and $y < 3$

28. $n = -\dfrac{2}{5}, \;\; 2$

29. $x = 16, \, -4$

30. $x < \dfrac{9}{2}$ and $x > \dfrac{1}{2}$

31. $x \le -\dfrac{14}{3}$ or $x \ge \dfrac{2}{3}$

32. 42 servings

33. 11.4 seconds

■ Chapter 3

Problem Set 3.1A

1. $y = 3x - 4$

x	y
-2	-10
0	-4
1	-1
3	5
4	8

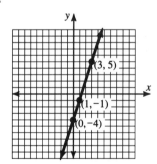

3. $2x + 3y = 6$

x	y
-6	6
-3	4
0	2
3	0
6	-2

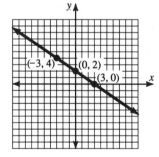

5. $y = 2x - 4$

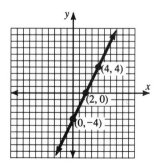

7. $y = \dfrac{3}{4}x - 2$

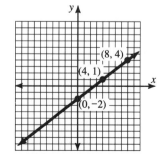

9. $y = -\dfrac{2}{3}x + 3$

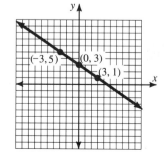

11. $y = \dfrac{3}{2}x - 1$

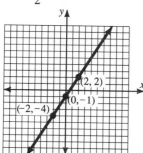

(2, 2)
(0, −1)
(−2, −4)

13. $y = -\dfrac{3}{2}x - 1$

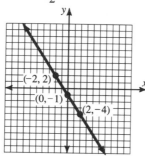

(−2, 2)
(0, −1)
(2, −4)

15. $y = -\dfrac{5}{6}x + 2$

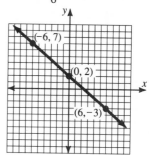

(−6, 7)
(0, 2)
(6, −3)

17. $5x - 2y = 6$

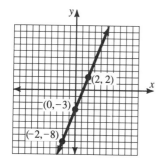

(2, 2)
(0, −3)
(−2, −8)

19. $3x + 3y = -6$

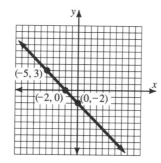

(−5, 3)
(−2, 0)
(0, −2)

21. $2x + 3y = 6$

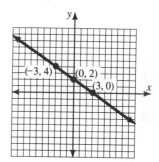

(−3, 4)
(0, 2)
(3, 0)

23. $4x - 3y = 9$

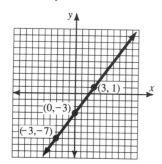

(3, 1)
(0, −3)
(−3, −7)

25. $6x - 5y = -10$

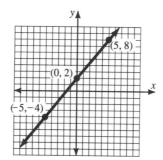

(5, 8)
(0, 2)
(−5, −4)

Problem Set 3.1B

1. $m = \dfrac{2}{5}$

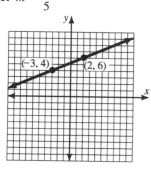

(−3, 4)
(2, 6)

3. $m = \dfrac{-9}{2}$

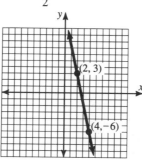

(2, 3)
(4, −6)

5. $m = \dfrac{-2}{7}$

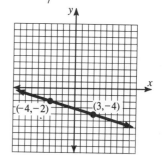

(−4, −2)
(3, −4)

7. $m = \dfrac{-1}{4}$

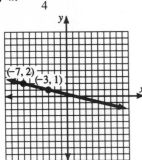

9. $m = \dfrac{11}{8}$

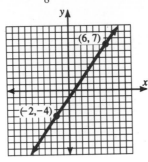

11. $m = \dfrac{-6}{5}$

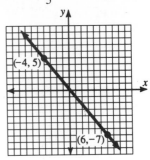

13. $m = \dfrac{10}{7}$

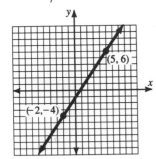

15. $m = \dfrac{-1}{3}$

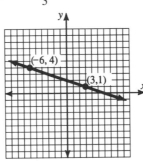

17. $m = \dfrac{3}{4}$

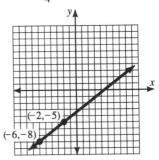

Problem Set 3.2A

1. $y = 2x + 4$

3. $y = -5x + 6$

5. $y = \dfrac{3}{4}x + 4$

7. $y = -\dfrac{4}{7}x - 5$

9. $y = 4x$

11. $y = 3x + 5$

13. $y = 4x - 2$

15. $y = -2x + 3$

17. $y = \dfrac{2}{3}x - 1$

19. $y = \dfrac{-3}{2}x - 3$

21. $y = \dfrac{3}{7}x$

23. $y = \dfrac{6}{5}x - 3$

25. $y = -2x + 3$

27. $y = -\dfrac{2}{3}x$

29. $y = -3x + 2$

31. $y = \dfrac{-6}{5}x$

Problem Set 3.2B

1. $y = \dfrac{2}{3}x + 2$

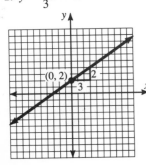

3. $y = -\dfrac{2}{3}x - 2$

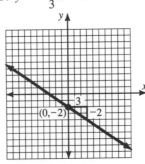

5. $y = \dfrac{3}{5}x - 2$

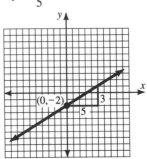

Selected Answers

7. $y = -\dfrac{5}{2}x - 3$

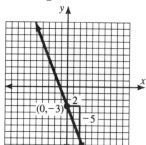

9. $-2x + 3y = 6$

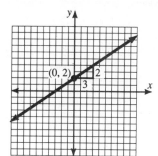

11. $-3x + 2y = -6$

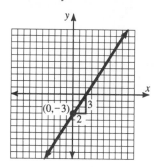

13. $y = -3x$

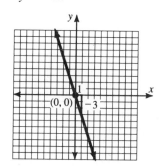

15. $4x - 3y = 12$

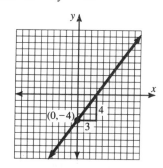

17. $3x - 4y = 12$

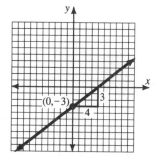

19. $-3x + 4y = 12$

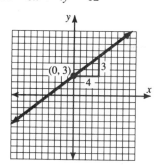

21. $7x - 4y = 8$

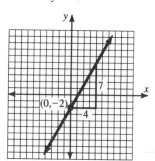

23. $y = -\dfrac{2}{3}x$

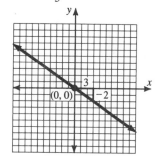

Problem Set 3.2C

1. $y = 2x - 4$

3. $y = \dfrac{3}{5}x + \dfrac{26}{5}$

5. $y = -\dfrac{4}{3}x + 1$

7. $y = -7$

9. $y = \dfrac{4}{5}x + \dfrac{26}{5}$

11. $y = -3x + 3$

13. $y = \dfrac{-5}{4}x - 4$

15. $y = \dfrac{-7}{6}x + \dfrac{29}{6}$

Problem Set 3.2D

1. $y = -\dfrac{3}{4}x + 2$

3. $y = \dfrac{12}{5}x + 6$

5. $y = \dfrac{10}{3}x + \dfrac{31}{3}$

7. $y = -x + 5$

9. $y = -x + 5$

11. $y = \dfrac{2}{5}x + 4$

13. $y = \dfrac{-5}{6}x - \dfrac{5}{2}$

15. $x = -6$

17. $y = \dfrac{5}{7}x - \dfrac{16}{7}$

Problem Set 3.3A

1. $2x + 3y = 6$

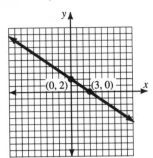

3. $3x - 2y = 6$

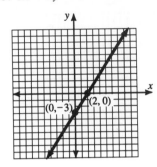

5. $2x - 5y = 10$

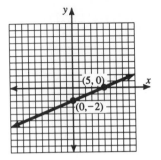

7. $-2x + 2y = 8$

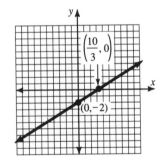

9. $3x - 2y = 8$

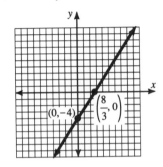

11. $3x - 4y = -4$

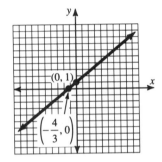

13. $3x - 5y = 10$

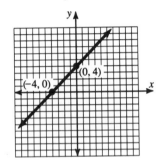

15. $3x + 7y = -14$

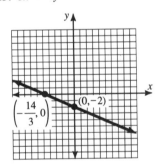

17. $3x - y = 9$

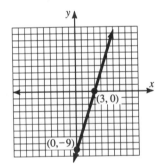

19. $x - y = 7$

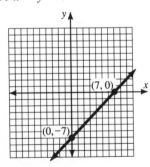

21. $3x - y = 5$

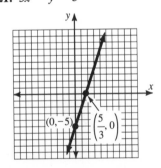

23. $-5x - y = 4$

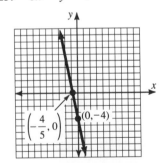

S10

Selected Answers

25. $y = 4$

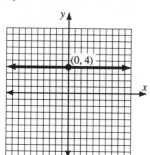

27. $x = 0$

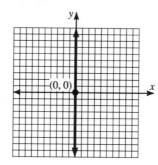

29. $x = -2$

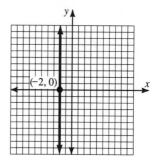

31. $x = 5$

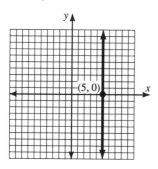

Problem Set 3.3B

1.

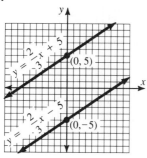

3.

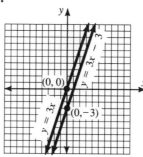

5.

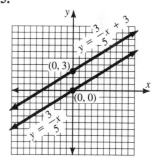

7.

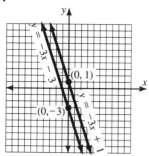

Problem Set 3.3C

1.

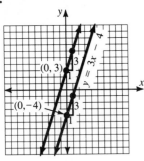

3.

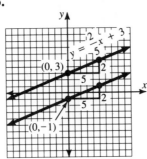

5.

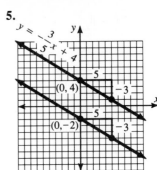

7.

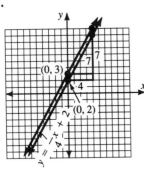

Problem Set 3.3D

1.

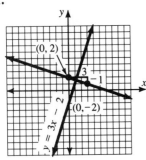

3.

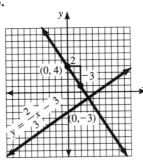

5.

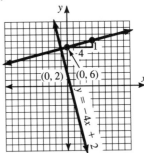

7.

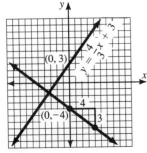

Selected Answers

Problem Set 3.3E

1. a) $y = \dfrac{2}{3}x + 8$

   b) $y = -\dfrac{3}{2}x - 5$

3. a) $y = \dfrac{1}{2}x - 4$

   b) $y = -2x + 1$

5. a) $y = \dfrac{-2}{3}x - \dfrac{10}{3}$

   b) $y = \dfrac{3}{2}x - 12$

7. a) $y = \dfrac{2}{3}x + \dfrac{10}{3}$

   b) $y = \dfrac{-3}{2}x + \dfrac{11}{2}$

9. $y = \dfrac{2}{3}x + \dfrac{4}{3}$

11. $y = \dfrac{4}{3}x + 3$

13. $y = \dfrac{3}{4}x + \dfrac{3}{2}$

15. $y = \dfrac{2}{5}x + \dfrac{6}{5}$

17. $x = -3$

19. $x = -5$

21. $y = -2$

23. $y = 4$

Problem Set 3.4A

1. $y < x$

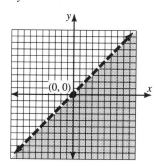

3. $y \geq x - 2$

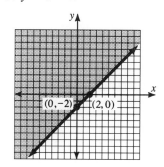

5. $y \leq x + 2$

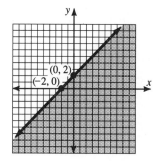

7. $y > 0$

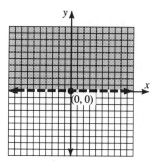

9. $y > 2x + 2$

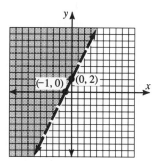

11. $y \leq 2x - 4$

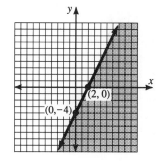

13. $y \leq \dfrac{1}{2}x$

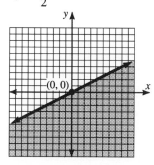

15. $y < -\dfrac{1}{2}x + 1$

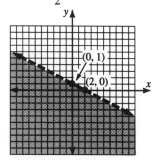

17. $y > -3x + 5$

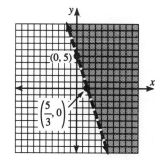

19. $x \geq -2$

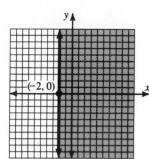

21. $y \leq -\frac{3}{4}x + 2$

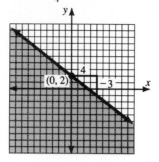

23. $y > 4 - \frac{2}{3}x$

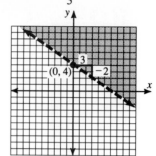

25. $y \leq -2x + 3$

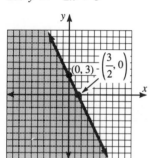

27. $y \geq 3x - 6$

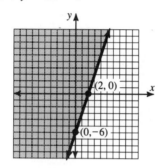

29. $y \geq \frac{2}{3}x + 1$

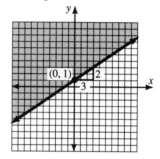

31. $y < -\frac{2}{3}x + 1$

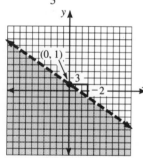

33. $x < -2$

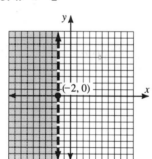

35. $x \leq 0$

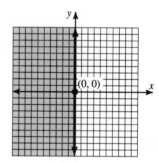

37. $y > 4x + 1$

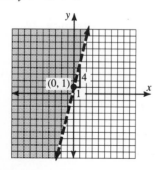

39. $y \leq -4x + 2$

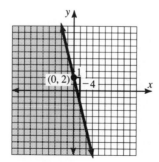

41. $y < \frac{3}{5}x$

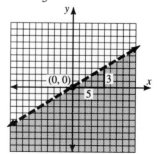

43. $y \geq -\dfrac{3}{5}x - 2$

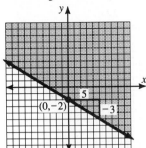

45. $y > 5 + \dfrac{5}{2}x$

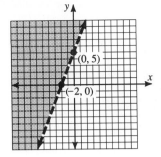

47. $y \leq -3 - \dfrac{5}{2}x$

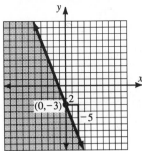

Problem Set 3.4B

1. $y = |x| + 1$

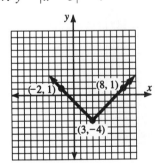

3. $y = |3x| - 5$

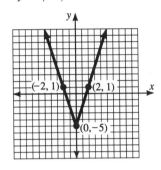

5. $y = |x - 2|$

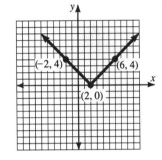

7. $y = |x - 3| - 4$

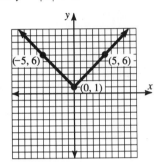

9. $y = |3x + 2|$

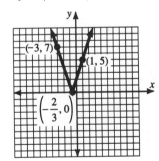

11. $y = |2x + 3| - 4$

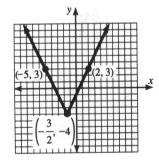

13. $y = |x + 1| - 5$

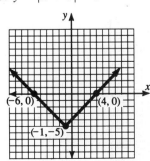

15. $y = |x + 2| - 6$

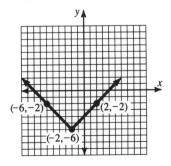

17. $y = |4x + 1| - 2$

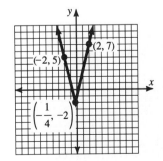

Selected Answers

19. $y = |3x + 4| - 2$

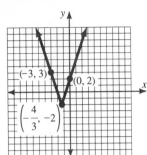

21. $y = |3x + 1| + 3$

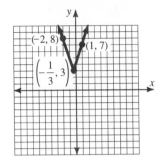

23. $y < |x - 3| + 4$

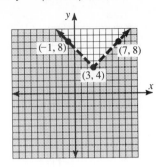

25. $y \geq |3x - 2| + 1$

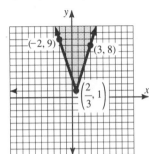

27. $y > |4x - 3| + 2$

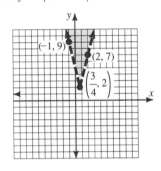

29. $y > |3x - 4| + 1$

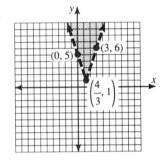

31. $y \leq |3x - 2| - 3$

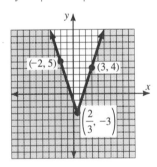

33. $y < |5x + 3| - 2$

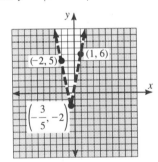

35. $y \geq |3x - 5| - 4$

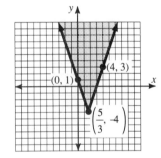

37. $y \leq |4x + 3| - 5$

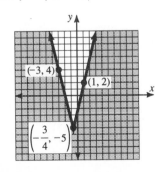

39. $y = \left| \dfrac{x}{2} - 3 \right|$

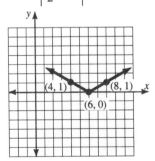

41. $y \geq \left| \dfrac{2}{3}x - 4 \right|$

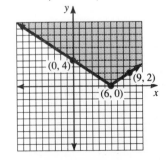

Selected Answers

Chapter 3   Review Test

1. $y = 2x + 1$

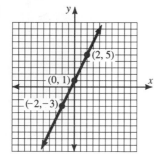

2. $3x - 4y = -8$

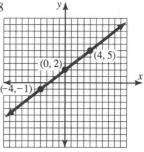

3. $m = -\dfrac{1}{3}$

4. $y = -6x + 3$

5. $y = \dfrac{2}{5}x - 4$

6. $y = 4x + 31$

7. $y = -\dfrac{4}{3}x + \dfrac{10}{3}$

8. $y = -\dfrac{9}{4}x - \dfrac{33}{4}$

9. $y = -\dfrac{9}{4}x - 3$

10. $b = -\dfrac{16}{5}; y = \dfrac{2}{5}x - \dfrac{16}{5}$

11. $y = -\dfrac{1}{2}x - \dfrac{1}{2}$

12. $y = \dfrac{2}{5}x + \dfrac{12}{5}$

13. $y = -\dfrac{3}{5}x - \dfrac{29}{5}$

14. $y = -x + 2$

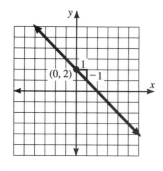

15. $3x - 5y = 15$

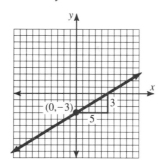

16. $3x + 4y = 12$

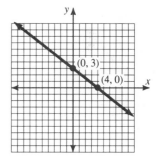

17. $2x - y = 6$

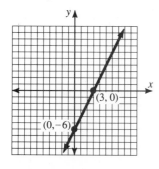

18.

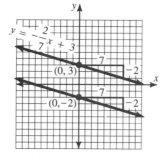

19.

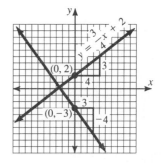

20. $y \geq -5x + 4$

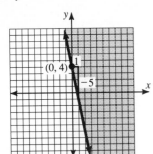

21. $x < -3$

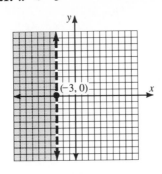

22. $y = |x + 2| - 3$

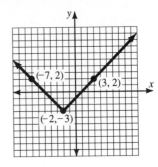

23. $y = |3x - 5| + 2$

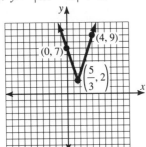

24. $y \leq |x - 3| - 4$

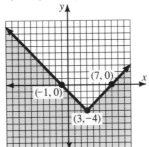

25. $y > |2x + 5| + 2$

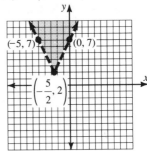

■ Chapter 4

Problem Set 4.1A
1. $(7, 1)$ **3.** Same line   **5.** $(-4, 3)$ **7.** $(-2, 7)$

9. $(5, -2)$ **11.** $(-1, 4)$ **13.** $\left(-\frac{1}{2}, \frac{1}{2}\right)$ **15.** $\left(3, \frac{1}{2}\right)$

17. $(3, 1)$ **19.** $\left(\frac{1}{2}, \frac{1}{3}\right)$ **21.** $(3, -7)$ **23.** $\left(0, -\frac{3}{4}\right)$

25. $(-3, 4)$ **27.** $\left(\frac{35}{9}, \frac{10}{3}\right)$ **29.** $(-1, 2)$

Problem Set 4.1B
1. $(-2, 5)$ **3.** $(-4, -3)$ **5.** Same line   **7.** $(-4, 1)$ **9.** No solution

11. $\left(\frac{3}{4}, -1\right)$ **13.** $(6, -3)$ **15.** No solution   **17.** $(-6, -2)$ **19.** $\left(-\frac{2}{3}, 4\right)$

21. Same line   **23.** $(-2, -2)$ **25.** $(-7, 6)$ **27.** $(3, 1)$ **29.** No solution

31. $(6, -7)$ **33.** $(-2, -1)$ **35.** $(-3, 3)$ **37.** $(-1, 2)$ **39.** $(2, -2)$

41. $(6, -4)$ **43.** $(3, -4)$ **45.** Same line   **47.** $(4, 1)$ **49.** No solution

51. $(6, 5)$ **53.** $(-6, 4)$ **55.** $(-6, 0)$ **57.** Same line   **59.** $(-4, 6)$

Problem Set 4.2A

  1. 8, 12

  3. 18, 3

  5. $-10, 6$

Problem Set 4.2B

  1. 36 dimes, 18 quarters

  5. 22 nickels, 66 dimes

  3. 1260 students, 540 adults

  7. 166 dimes, 154 quarters

Problem Set 4.2C

  1. $3\frac{1}{2}$ lbs of walnuts, $6\frac{1}{2}$ lbs of pecans

  3. 240 lbs at 72 cents, 360 lbs at 52 cents

  5. \$8000 at 8%, \$17000 at 11%

  7. \$9000 at 8%, \$7200 at 10%

  9. \$6000 at 12%, \$4000 at 10%

  11. 10 lbs at \$4.00, 6 lbs at \$6.00

Problem Set 4.3

  1. $(4, -5, 6)$

  3. $(3, -4, 2)$

  5. $(1, 1, -2)$

  7. No solution

  9. $(2, -3, 1)$

  11. $(10, 5, -2)$

  13. $(1, -2, -3)$

  15. $\left(2, \frac{1}{2}, -3\right)$

  17. $\left(-\frac{1}{3}, 2, -1\right)$

  19. $\left(4, -\frac{3}{4}, 0\right)$

  21. No Solution

  23. $\left(-6, \frac{3}{2}, 2\right)$

  25. $(10, 12, 15)$

  27. $(36, 28, 24)$

Chapter 4 Review Test

  1. $(4, 3)$

  2. $\left(-\frac{1}{2}, 2\right)$

  3. $(6, -4)$

  4. $(3, -4)$

  5. $\left(-2, \frac{1}{3}\right)$

  6. Same line

  7. $(6, -5)$

  8. $(3, 7)$

  9. $\left(\frac{2}{3}, -\frac{1}{2}\right)$

  10. 20, 34

  11. \$1600 at 8%, \$2200 at 10%

  12. 15 lbs of walnuts, 10 lbs of pecans

  13. $(-1, 2, 4)$

  14. $(12, -12, 3)$

  15. 14 nickels, 18 dimes, 7 quarters

■ Chapter 5

Problem Set 5.1A

  1. x^3

  3. $-20a^4$

  5. $-5x^5y^5$

  7. $20ab^3$

  9. $-72a^3x^3$

  11. $-7xa^2$

  13. $-2a^2x^5$

  15. $-3a^3b^4$

  17. $5ab^3$

  19. $4x^3y^5$

  21. $-32xy^2$

  23. $8x^2$

Problem Set 5.1B

1. $-20x^5$ **3.** $4x$ **5.** a^n **7.** x^{2n-2} **9.** $-\dfrac{5x}{y}$

11. $\dfrac{4x^2}{3y^3}$ **13.** x^n **15.** x^n **17.** $-4x^{11-5n}$ **19.** $-\dfrac{6x^2}{y^3}$

21. $-\dfrac{12y}{5x^3}$ **23.** $\dfrac{5x^3}{4a^2}$ **25.** $-\dfrac{6}{t^7}$ **27.** $\dfrac{a^4}{b^4}$ **29.** $\dfrac{x^6}{y^9}$

31. $\dfrac{108}{x^5 y^3}$ **33.** $32s^3$ **35.** $\dfrac{y^{12}}{x^{20}}$ **37.** $\dfrac{192 y^6 z^9}{x^4}$ **39.** $\dfrac{4a^{13} c^{12}}{27 b^9}$

41. $\dfrac{3y^9}{4x^8}$ **43.** $-\dfrac{x^2}{y^4}$ **45.** $\dfrac{4a^4}{b^{14}}$ **47.** $\dfrac{y^{14}}{4x^{14}}$ **49.** $\dfrac{64}{9x^{13}}$

51. $\dfrac{y^2}{4x^2 w^2}$ **53.** $\dfrac{-8}{a^{24} b^9}$ **55.** $\dfrac{9a^6 b^4}{4}$ **57.** $\dfrac{4(a-b)(x+y)^5}{5}$ **59.** $\dfrac{10(x-2y)^5}{9(a+b)^3}$

61. -6 **63.** 144 **65.** -4 **67.** $\dfrac{5}{4}$ **69.** 1

71. 1 **73.** $\dfrac{32}{9}$ **75.** -48 **77.** 576 **79.** 12

81. $\dfrac{1}{9}$ **83.** 4 **85.** $\dfrac{1}{4}$ **87.** 81

Problem Set 5.2A

1. 365.4 **3.** 8620 **5.** 82.65 **7.** 38600 **9.** 56000 **11.** 400

13. 59000 **15.** 80 **17.** 3869 **19.** 10.8 **21.** 71.0 **23.** 29010

25. 10^2 **27.** 10^4 **29.** 10^3 **31.** 10^3 **33.** 10^2 **35.** 10^1

37. 10^1 **39.** 10^4 **41.** 10^3 **43.** 10^4 **45.** 10^1 **47.** 10^3

Problem Set 5.2B

1. 6700 **3.** 0.0018 **5.** 47.68 **7.** -0.00005 **9.** 21000

11. 0.003769 **13.** 0.000386 **15.** 492.4 **17.** 23950 **19.** 1740.3

21. 0.000090092 **23.** 0.00029089 **25.** 6.54×10^{-1} **27.** 8.56×10^2 **29.** 8.79×10^{-1}

31. 8.59×10^5 **33.** 6.5×10^{-6} **35.** -3.91×10^2 **37.** -4×10^6 **39.** 1×10^{-7}

41. 9.28×10^1 **43.** 5.16×10^{-1} **45.** -8.649×10^3 **47.** 5.31×10^{-5} **49.** 3.000456×10^6

51. 1×10^6 **53.** 5×10^{-5} **55.** -7.21×10^{-4} **57.** 5.2×10^{-9} **59.** 4.488×10^{11}

61. 4.50×10^{-10} **63.** 1.4×10^2 **65.** 6.2×10^{-6} **67.** 4.32×10^{-10} **69.** 3.8318×10^{14}

71. 1.23×10^{-2} **73.** 3.04×10^{-3} **75.** 2.6×10^{-9} **77.** 2.08×10^7 **79.** 805 students

Problem Set 5.3A

1. $-56a^2 x^2 y$ **3.** $-3x^5 y^3$ **5.** $-32a^5 b^2 x^2$

7. $5a^3 x^6 y^3$ **9.** $6xm^2 - 30xm$ **11.** $2a^4 + 16a^3$

13. $x^2 + yx - 6y^2$ **15.** $18a^2 - 18ab + 4b^2$ **17.** $-20x^2 y^2 + 28xy^4$

19. $-27a^7 x^2 - 12a^3 x^5$ **21.** $15x^2 - 23xy + 4y^2$ **23.** $24a^2 + 62ab + 35b^2$

25. $15a^2 + 2a - 24$

31. $10x^3 - x^2 - 39x + 27$

37. $24y^3 - 35y^2 - 44y + 6$

43. $6a^4 - 4a^3 - 47a^2 + 63a - 21$

47. $21x^4 + 19x^3 - 7x^2 + 19x - 4$

27. $x^3 - 5x^2 + 7x - 3$

33. $12y^3 + 18y^2 - 40y + 14$

39. $24a^3 + 38a^2 - 9a - 20$

45. $6a^4 + 11a^3 - 23a^2 + 34a - 28$

29. $2x^3 + 9x^2 - 11x + 3$

35. $6x^3 - 7x^2 - 37x + 28$

41. $x^4 + 6x^3 - 22x + 15$

Problem Set 5.3B

1. $x^2 - x - 6$

7. $4a^2 - 17a + 4$

13. $6x^2 - 5x - 4$

19. $20x^2 + 23x - 7$

25. $2x^2 - xy - y^2$

31. $6a^2 - ab - 12b^2$

37. $30a^2 - 59ab + 28b^2$

3. $x^2 + x - 12$

9. $8x^2 - 10x + 3$

15. $21a^2 + 44a + 15$

21. $12x^2 + 55x + 63$

27. $20x^2 - xy - 12y^2$

33. $6a^2 - 17ab + 12b^2$

39. $20a^2 + 37ab + 15b^2$

5. $2x^2 - 7x + 3$

11. $30x^2 - 47x + 14$

17. $42y^2 - 51y + 15$

23. $15x^2 - 67x + 72$

29. $18r^2 - 9rs - 35s^2$

35. $12x^2 - xy - 6y^2$

41. $42x^2 - 13xy - 40y^2$

Problem Set 5.3C

1. $x^2 + 6x + 9$

7. $x^2 + 2xy + y^2$

13. $a^2 + 4ab + 4b^2$

19. $9x^2 - 25$

25. $9x^2 + 12xy + 4y^2$

31. $x^2 + 14xy + 49y^2$

37. $25x^2 - 4y^4$

43. $x^4 + 6x^2y^2 + 9y^4$

3. $x^2 - 9$

9. $x^2 - y^2$

15. $9a^2 - 4$

21. $9r^2 - 30r + 25$

27. $9x^2 - 4y^2$

33. $9x^4 - 6x^2y + y^2$

39. $4x^4 - 12x^2 + 9$

45. $36r^4 - 49t^4$

5. $4a^2 - 4a + 1$

11. $4a^2 + 4ab + b^2$

17. $9x^2 - 24x + 16$

23. $4x^2 - 49$

29. $9a^2 - 16b^2$

35. $16x^4 + 24x^2y + 9y^2$

41. $16x^4 - 9y^4$

47. $64a^4 - 144a^2b + 81b^2$

Problem Set 5.4A

1. $4x + 12$

7. $9ab + 3a$

13. $2a^2 + 3a^2b - 6$

19. $1 - 6ay + \dfrac{7a^2}{y}$

25. $\dfrac{3ab}{2} - 3b + \dfrac{4}{ab}$

31. $2xy - 3 + \dfrac{1}{2xy^2}$

37. $\dfrac{2y}{x} - 3x - \dfrac{3}{2xy^3}$

3. $x^2 + 3x$

9. $2x^2y + 3x + \dfrac{1}{y}$

15. $\dfrac{a}{5} - \dfrac{1}{b} - \dfrac{2a^2}{b^2}$

21. $\dfrac{b^2}{a} - \dfrac{3}{a} + \dfrac{b^2}{5a^2}$

27. $-x + 3xy - \dfrac{5y}{x}$

33. $4 - 3ab + \dfrac{a^2}{5b}$

39. $4b - 3a + \dfrac{3}{4ab^2}$

5. $8x - 5 + \dfrac{2}{x}$

11. $7a^2 - 9ab - 8b^2$

17. $\dfrac{x}{y} - \dfrac{2x^2}{y} + 3$

23. $\dfrac{1}{y} - \dfrac{3x}{2} + \dfrac{2y}{x}$

29. $2ab - 3 - \dfrac{1}{2b}$

35. $-\dfrac{2a}{b^2} + 3ab - a$

41. $\dfrac{7xy^2}{3} - 3x + \dfrac{11x^2}{3y}$

43. $-\dfrac{a}{2b} + 2a - \dfrac{5}{6a}$

45. $3x - \dfrac{5}{2xy} + \dfrac{3}{5x^2y}$

47. $4a - \dfrac{a^4b^2}{2} + \dfrac{3b}{4a}$

Problem Set 5.4B

1. $x + 5$

3. $2x + 3$

5. $2x - 3$

7. $3x - 6$

9. $3x + 4 + \dfrac{5}{2x + 5}$

11. $6x + 5 + \dfrac{-4}{7x - 3}$

13. $3x + 4$

15. $5x - 6$

17. $x + 3 + \dfrac{-1}{4x - 5}$

19. $x - 6$

21. $2x + 5$

23. $3x - 2 + \dfrac{2}{5x - 4}$

25. $x^2 - 2x + 3$

27. $x^2 + 4x - 5$

29. $x^2 + 3x - 5$

31. $2x^2 + 3x - 4$

33. $4x^2 - 3x + 4 + \dfrac{2}{x - 5}$

35. $2x^2 - 3x + 4 + \dfrac{-2}{2x - 3}$

37. $x^2 - 2x + 8$

39. $2x^3 + 3x^2 - x + \dfrac{-4}{2x - 3}$

41. $2x^2 + 1 + \dfrac{2}{5x - 2}$

43. $2x^2 - 3x + 5$

45. $2x^2 + 4x - 3$

47. $3x^2 - 5x - 4 + \dfrac{2}{5x - 4}$

Problem Set 5.5

1. $x + 3$

3. $x + 9 + \dfrac{-15}{x + 3}$

5. $x^2 + 2x - 3$

7. $3x^2 - 2x + 4 - \dfrac{16}{x + 2}$

9. $4x^2 - 3x + 4 - \dfrac{1}{x + 1}$

11. $x^2 + 2x + 7 + \dfrac{12}{x - 2}$

13. $3x^2 - 2x + 3 + \dfrac{-24}{x + 4}$

15. $3x^2 - x + 2 + \dfrac{-1}{x - 2}$

17. $x^3 - x + 1$

19. $3x^3 - 1$

21. $4x^3 - x + \dfrac{3}{x + 5}$

23. $3x^3 + 2x^2 + x - 3$

25. $2x^3 + 2x^2 - 2x - 1 + \dfrac{2}{x - 5}$

27. $4x^3 - x^2 + 2x - 2 + \dfrac{-5}{x - 1}$

29. $x^3 - 3x^2 - x + 1 + \dfrac{-4}{x + 3}$

31. $3x^3 - x^2 + 2x + 3 + \dfrac{-7}{x + 2}$

33. $x^2 + 4x + 16$

35. $x^4 - x^3 + x^2 - x + 1$

37. $2x^4 - 2x^3 + 2x^2 + \dfrac{5}{x + 1}$

39. $x^5 + 3x^4 + x^3 + 3x^2 - x - 3 - \dfrac{3}{x - 3}$

41. $x^3 - 2x^2 + 4x - 8$

43. $x^5 + 2x^4 + 4x^3 + 8x^2 + 16x + 32$

45. $2x^5 + 2x^4 - x^3 + 3x^2 + x - 4 + \dfrac{-4}{x - 1}$

47. $x^5 - x^4 - 3x^3 - 17x^2 - 51x - 163 + \dfrac{-488}{x - 3}$

Chapter 5 Review Test

1. $8ab^2$

2. $\dfrac{4}{x^6}$

3. $-\dfrac{72a^6}{b^8}$

4. $-8x^3y^4$

5. $\dfrac{x^3y^{12}}{c^{15}}$

6. $-4(2x - 1)^2(x + 2y)^2$

7. -4

8. 8

9. 8

10. -2

11. 5.642×10^5

12. 3.1×10^{-4}

13. 4.0×10^{-2}

14. 7.81921×10^2

15. 30000

16. -2150

17. -0.004071

18. 0.00065

19. $8ya^3 - 24ay$

20. $-3x^4y^2 + 3xy^5$

21. $20s^2 - 32s - 16$

22. $4x^2 - 12xy^2 + 9y^4$

23. $36x^2 - 49y^6$

24. $3x^3 - 7x^2 + 14x - 4$

25. $3a^4 - 2a^3 - 10a^2 + 14a - 5$

26. $x^6 + x^3y^3 - 2y^6$

27. $4 - 6x$

28. $4x - \dfrac{3}{5} + \dfrac{2}{x^3}$

29. $\dfrac{x}{y} - \dfrac{2}{x} + 3y^2$

30. $3x - 4$

31. $3x^2 - 5x$

32. $4x^2 - 7 + \dfrac{-3}{5x + 6}$

33. $4x^2 - 3x + \dfrac{6}{2x - 3}$

34. $x - 3 - \dfrac{17}{x - 3}$

35. $x^2 + 8x - 3$

36. $x^3 - 2x^2 + 3x - 3 + \dfrac{6}{x + 2}$

■ Chapter 6

Problem Set 6.1A

1. $2(x + 2)$

3. $2(3a - 5)$

5. $x(5x + 6)$

7. $4x(x - 1)$

9. $y(y^2 - 2y - 5)$

11. $4x(2x^2 - 3)$

13. $3a^2(4a - 5)$

15. $7a^2(3 - 2a^3)$

17. $9x^3(x - 3)$

19. $x(x^2 - 2x + 3)$

21. $3x(x^2 - 2x + 3)$

23. $2a(2a^2 - a + 3)$

25. $2x(2x^2 - x + 5)$

27. $7y(y^3 - 2y - 3)$

29. $b(x^2 + 3x - 5)$

31. $2ay^2(y^2 + 3y - 5)$

33. $3x(2x^2 + 3x - 4)$

35. $-2y^2(4y^3 - 6y^2 + 3)$

37. $ax^2(x^2 - x + 2)$

39. $3a^2y^3(2y^3 - 4y + 5)$

Problem Set 6.1B

1. $(2, 3)$

3. $(3, -2)$

5. $(-3, -5)$

7. $(4, -3)$

9. $(-6, 3)$

11. $(x - 8)(x - 3)$

13. $(b - 12)(b - 2)$

15. $(x - 4)(x - 8)$

17. $(x - 15)(x + 2)$

19. $(x + 10)(x + 2)$

21. $(x - 6)(x + 4)$

23. $(x + 9)(x - 8)$

25. $(x - 14)(x + 3)$

27. $(b - 13)(b + 5)$

29. $(z + 11)(z - 7)$

31. $(y - 14)(y - 3)$

33. $(x - 11)(x + 5)$

35. $(a - 13)(a - 2)$

37. $(x - 7)(x - 4)$

39. $(x - 4)(x + 9)$

41. $(x - 9)(x + 3)$

43. $(x + 2)(x + 13)$

45. $(x - 4)(x - 9)$

47. $(a - 19)(a + 2)$

49. $(x - 2)(x + 15)$

51. $(x + 3)(x + 12)$

53. $(x + 8)(x - 5)$

55. $(x - 4)(x - 11)$

57. $(x + 5)(x + 10)$

59. $(x - 10)(x + 6)$

61. $(x + 8)(x - 6)$

63. $(x + 4)(x + 16)$

65. $(x - 6)(x - 8)$

Problem Set 6.2A

1. $(2x - 1)(x + 1)$ **3.** $(3x + 1)(x - 2)$ **5.** $(7y - 5)(y - 2)$ **7.** $(5x + 1)(x + 8)$

9. $(3x - 1)(x - 2)$ **11.** $(3a + 2)(a + 7)$ **13.** $(3b + 2)(b - 7)$ **15.** $(2x + 5)(x + 4)$

17. $(3x + 1)(x + 1)$ **19.** $(5x - 3)(x + 10)$ **21.** $(7a - 2)(a + 2)$ **23.** $(2x - 5)(x - 3)$

25. $(5x + 2)(x - 3)$ **27.** $(7x - 3)(x + 4)$ **29.** $(3x - 4)(x + 3)$ **31.** $(2x + 7)(x - 3)$

33. $(5x + 6)(x - 4)$ **35.** $(7x - 5)(x + 3)$ **37.** $(2x + 5)(x - 6)$ **39.** $(5x + 3)(x - 6)$

41. $(7x + 3)(x + 6)$

Problem Set 6.2B

1. $(3x + 2)(2x - 1)$ **3.** $(3x - 2)(2x - 1)$ **5.** $(3x + 1)(2x - 3)$ **7.** $(5x - 1)(2x - 5)$

9. $(8x - 3)(x - 1)$ **11.** $(3x - 5)(3x + 1)$ **13.** $(7x - 5)(x - 1)$ **15.** $(7x + 5)(x + 1)$

17. $(3a + 5)(4a - 1)$ **19.** $(12a + 5)(a + 1)$ **21.** $(6a - 1)(2a - 5)$ **23.** $(5x + 2)(2x + 1)$

25. $(4x + 3)(5x + 1)$ **27.** $(4x + 7)(3x + 1)$ **29.** $(4x + 11)(2x + 1)$ **31.** $(5x + 2)(3x - 1)$

33. $(7x - 3)(5x + 1)$ **35.** $(4x + 7)(5x - 1)$ **37.** $(4x + 11)(5x - 1)$ **39.** $(3x - 7)(4x - 1)$

41. $(6x - 5)(4x - 1)$

Problem Set 6.2C

1. $(3x + 2)(2x + 5)$ **3.** $(3x - 2)(2x + 5)$ **5.** $(3x - 4)(2x + 7)$ **7.** $(4x - 3)(x + 5)$

9. $(2T - 3)(2t + 5)$ **11.** $(8x - 3)(3x + 7)$ **13.** $(4a + 3)(a - 6)$ **15.** $(8x - 7)(x + 3)$

17. $(6x - 1)(4x + 3)$ **19.** $(20x - 1)(x - 6)$ **21.** $(4a + 5)(2a - 3)$ **23.** $(2x + 5)(2x - 7)$

25. $(2x + 5)(3x + 4)$ **27.** $(2x - 3)(4x - 5)$ **29.** $(3x - 4)(3x + 2)$ **31.** $(3x - 7)(4x + 3)$

33. $(8x - 5)(2x - 3)$ **35.** $(16x - 5)(x - 3)$ **37.** $(20x + 3)(x - 5)$ **39.** $(8x + 5)(3x + 2)$

41. $(4x + 3)(6x + 5)$

Problem Set 6.2D

1. $(2x + 3)(x + 1)$ **3.** $-1(2x + 1)(x - 3)$ **5.** $6(2x - 1)(x - 1)$

7. $-1(10x + 1)(x - 5)$ **9.** $a^2(10x + 3)(2x + 5)$ **11.** $4a(8x + 3)(x + 1)$

13. $-3a(4x - 5)(2x + 3)$ **15.** $(3y - 1)(3y + 5)$ **17.** $-4a^2(3x - 1)(x + 2)$

19. $3(3x - 2)(3x + 4)$ **21.** $-5(2y + 3)(5y - 6)$ **23.** $2x^2(3x - 4)(4x + 5)$

25. $2ay(6y - 1)(3y + 5)$ **27.** $-a^2y(3y + 7)(2y - 3)$ **29.** $3a^2y^2(4a - 1)(3a - 2)$

31. $-4a^2y^3(2a + 5)(3a - 4)$ **33.** $-a^2(3x - 7)(4x + 5)$ **35.** $4x^2y(3y + 4)(3y - 5)$

Problem Set 6.2E

1. $(x + 3)(x - 2)$ **3.** $(x - 2)(x - 3)$ **5.** $(2x + 1)(x - 1)$ **7.** $(4x - 5)(x - 2)$

9. $(2x - 3)(3x + 4)$ **11.** $(3x - 1)(x + 2)$ **13.** $(5x + 1)(2x + 3)$ **15.** $(6x - 5)(2x + 1)$

17. $(6x + 1)(2x - 1)$ **19.** $(4x + 5)(3x - 5)$ **21.** $(6x - 5)(2x - 3)$

Problem Set 6.2F

1. $(3x + 1)(x + 2)$ **3.** $(3x + 2)(x + 1)$ **5.** $(3x - 2)(x + 1)$ **7.** $(3x + 1)(x - 2)$

9. $(4x + 1)(2x + 3)$ **11.** $(4x + 1)(2x - 3)$ **13.** $(4x + 3)(2x + 1)$ **15.** $(4x + 3)(2x - 1)$

     Selected Answers

17. $(7y - 10)(y - 1)$ **19.** $(7y + 10)(y - 1)$ **21.** $(7y - 5)(y - 2)$ **23.** $(3x - 1)(x - 8)$

25. $(3b + 2)(b - 7)$ **27.** $(3x + 1)(x - 6)$ **29.** $(3b - 2)(b + 7)$ **31.** $(3x - 4)(2x + 5)$

33. $(6y - 1)(y + 4)$ **35.** $(7x + 5)(x + 2)$ **37.** $(8a + 1)(a - 3)$ **39.** $(3x - 2)(2x + 5)$

41. $(6x - 7)(x + 2)$ **43.** $(5x - 3)(x - 4)$ **45.** $(5x - 2)(x - 4)$ **47.** $(3x + 4)(3x - 2)$

49. $(3x + 2)(x + 5)$ **51.** $(4x - 1)(x - 6)$ **53.** $(2x + 3)(3x + 4)$ **55.** $(4x + 3)(2x - 3)$

57. $(8x + 5)(x - 3)$ **59.** $(5x + 4)(x + 3)$ **61.** $(7x + 3)(x - 2)$ **63.** $(4a + 7)(2a - 3)$

Problem Set 6.3A

1. $(x + 2)^2$ **3.** $(x - 4)^2$ **5.** $(x + 7)^2$ **7.** $(3x - 1)^2$ **9.** $(2x + 5)^2$

11. $(x + y)^2$ **13.** $(3x + 2y)^2$ **15.** $(xy - 2z)^2$ **17.** $(a - 3)^2$ **19.** $(2x + 1)^2$

21. $(2x + 7)^2$ **23.** $(x + 2y)^2$ **25.** $(4a + b)^2$ **27.** $(5x - 2y)^2$ **29.** $(x^2y - 3a^2)^2$

Problem Set 6.3B

1. $(x - 2)(x + 2)$ **3.** $(x - 6)(x + 6)$ **5.** $(3x - 2)(3x + 2)$ **7.** $(y + 7)(y - 7)$

9. $(4x + 1)(4x - 1)$ **11.** $(2x - 5)(2x + 5)$ **13.** $(x - y)(x + y)$ **15.** $(x - 4y)(x + 4y)$

17. $(5x - 2y)(5x + 2y)$ **19.** $(4a + b)(4a - b)$ **21.** $(3x + 5y)(3x - 5y)$ **23.** $(4a + 5b)(4a - 5b)$

25. $(a^2b - 2)(a^2b + 2)$ **27.** $(xy - rs)(xy + rs)$

29. $(x^3y + 5)(x^3y - 5)$ **31.** $(3ab + 2x^2y^2)(3ab - 2x^2y^2)$

33. $(2x + y)(2x - y)(4x^2 + y^2)$ **35.** $(a - b)(a + b)(a^2 + b^2)(a^4 + b^4)$

37. $(a + b^2)(a - b^2)(a^2 + b^4)$ **39.** $(x - 2y)(x + 2y)(x^2 + 4y^2)$

41. $(a^{2n} - b^{3n})(a^{2n} + b^{3n})$ **43.** $(2xy - w^2)(2xy + w^2)(4x^2y^2 + w^4)$

45. $(x^{3n} + y^{4n})(x^{3n} - y^{4n})$ **47.** $(a^4 + 3b^2c^2)(a^4 - 3b^2c^2)$

Problem Set 6.3C

1. $(y + 3)(y^2 - 3y + 9)$ **3.** $(3x - 2)(9x^2 + 6x + 4)$

5. $(3a + 2)(9a^2 - 6a + 4)$ **7.** $(3x + 4)(9x^2 - 12x + 16)$

9. $(a + 2b)(a^2 - 2ab + 4b^2)$ **11.** $(5a - cd)(25a^2 + 5acd + c^2d^2)$

13. $(x + 3y)(x^2 - 3xy + 9y^2)$ **15.** $(xy - 4a)(x^2y^2 + 4axy + 16a^2)$

17. $(4y^2 + 5x^2)(16y^4 - 20x^2y^2 + 25x^4)$ **19.** $(2x - y)(4x^2 + 2xy + y^2)(2x + y)(4x^2 - 2xy + y^2)$

21. $(4x^2 + y^2)(16x^4 - 4x^2y^2 + y^4)$ **23.** $(ab - c^2)(a^2b^2 + abc^2 + c^4)$

25. $(4a^2 + b^2)(16a^4 - 4a^2b^2 + b^4)$ **27.** $(3a^2 + 4b^2)(9a^4 - 12a^2b^2 + 16b^4)$

29. $(4a^2 - 5x^2y^2)(16a^4 + 20a^2x^2y^2 + 25x^4y^4)$

31. $(xy - 2a^2)(x^2y^2 + 2a^2xy + 4a^4)$ **33.** $(3x^2y + 2ab^2)(9x^4y^2 - 6ab^2x^2y + 4a^2b^4)$

35. $(2a + w^2y^2)(4a^2 - 2aw^2y^2 + w^4y^4)$ **37.** $[(ab)^2 + (2c)^2][(ab)^4 - (2abc)^2 + (2c)^4]$

39. $(x + 0.5)(x^2 - 0.5x + 0.25)$ **41.** $(x^2 - 3y)(x^4 + 3x^2y + 9y^2)$

43. $(ab^2 + 3c^2)(a^2b^4 - 3ab^2c^2 + 9c^4)$ **45.** $(x^2 - 0.1)(x^4 + 0.1x^2 + 0.01)$

47. $(x^2 + 0.5y)(x^4 - 0.5x^2y + 0.25y^2)$ **49.** $(0.4 - 3x)(0.16 + 1.2x + 9x^2)$

Problem Set 6.4A

1. $(y^2 + 1)(y^2 + 2)$ **3.** $(x - 1)(x + 1)(x - 2)(x + 2)$

5. $(3x^2 + 2)(2x^2 - 1)$

9. $(x - 1)(x^2 + x + 1)(x - 2)(x^2 + 2x + 4)$

13. $(2x - 3)(2x + 3)(4x^2 + 9)$

17. $(x + 2)(x - 2)(x^2 + 5)$

21. $(x + 3)(x - 3)(2x^2 + 1)$

25. $(2x + 3)(2x - 3)(x + 2)(x - 2)$

29. $(3x - 2)^2(3x + 2)^2$

33. $(2x + 3)(x + 2)$

37. $(x + 1)(x + 4)(x - 1)(x + 6)$

41. $(x - 2)(x - 3)(x - 6)(x + 1)$

45. $(3x^{-1} + 2)(x^{-1} - 1)$

7. $(x + 1)(x - 1)(x^2 - 8)$

11. $(x - 2)^2(x + 2)^2$

15. $(4x^3 + 1)(2x^3 - 3)$

19. $(2x^3 + 3)(x^3 - 4)$

23. $(x + 1)(x^2 - x + 1)(3x^3 - 2)$

27. $(2x + 1)(4x^2 - 2x + 1)(3x^3 + 8)$

31. $(2x - 1)(4x^2 + 2x + 1)(x - 2)(x^2 + 2x + 4)$

35. $3(x - 2)(9x - 5)$

39. $2(2x^2 + 1)(x^2 + 2)$

43. $(3x - 4)(x + 2)(3x - 1)(x + 1)$

47. $(5x^{-1} - 2)(2x^{-1} + 3)$

Problem Set 6.4B

1. $(y - 2)(x - 4)$

7. $(a - 3)(b + 4)$

13. $(x + y)(c - d)(c + d)$

19. $(a - 2)(x + y)(x^2 - xy + y^2)$

23. $(x - 5)(a - b)(a^2 + ab + b^2)$

27. $(2a - b)(4a^2 + 2ab + b^2)(2x + y)(2x - y)$

31. $(a^2 + b^2)(x - 2)(x + 2)$

35. $(a + 3)(a - 3)(x^2 + y^2)$

39. $(x + 2 - 3b)(x + 2 + 3b)$

43. $(4x - a + b)(4x + a - b)$

47. $(4x + a + 2b)(4x - a - 2b)$

3. $(x - y)(a + 4)$

9. $(a - b)(x + 4)$

15. $(2x + y)(r - s)(r + s)$

5. $r(5r - 2s)(6 + r)$

11. $x(3x - y)(4x - 3)$

17. $(a + b)(x - 2y)(x + 2y)$

21. $(x + y)(x^2 - xy + y^2)(a - b)(a + b)$

25. $(a - 4)(2x - y)(4x^2 + 2xy + y^2)$

29. $(3a + 4)(x^2 + y^2)$

33. $(2a + b)(3x^2 + y^2)$

37. $(a + 2b)(a - 2b)(x + 2)(x - 2)$

41. $(x - y - 3a)(x - y + 3a)$

45. $(2x - 1 - 2a)(2x - 1 + 2a)$

Problem Set 6.4C

1. $3a(3x + 1)(x - 2)$

5. $8x^2(4x - 3)(4x + 3)$

9. $3x^2(2x + 3)(x - 4)$

13. $6x^2(3x + 2)(3x - 2)$

17. $5x^2(3x - 4)(2x + 3)$

21. $4x^3(3x^2 + 4)(x + 2)$

25. $5x^3y(3x^2 + 2y)(9x^4 - 6x^2y + 4y^2)$

29. $4x^2y(3x + 2y^2)(2x - y)$

33. $9ax^2(7x^2 + 7x - 6)$

37. $4x^2(x + 2)(x^2 - 2)(x^2 - 2x + 4)$

41. $5x^2y(2x + 3y^2)(3x - y^2)$

45. $3ab(x - 3)^2(x^2 + 3x + 9)$

3. $4(x + 2)(x - 2)$

7. $(3x - 2y^2)(3x + 2y^2)(9x^2 + 4y^4)$

11. $4(x + 4)(x - 4)$

15. $(2x^2 + 3y)(2x^2 - 3y)(4x^4 + 9y^2)$

19. $6x^2y^2(3x - 4y)(2x + 5y)$

23. $3xy^3(8x^3 - 27y^5)$

27. $5xy^2(4x^3 + 1)(x - 3)$

31. $3b^3(x - 2)^2(x^2 + 2x + 4)$

35. $7tx^3(2x - 3)(3x + 1)$

39. $3x^2(x^3 - 27y^4)$

43. $(2x - y^2)(2x + y^2))(3x + y)(9x^2 - 3xy + y^2)$

47. $6xy^2(2x - y)(x^3 - 4y^3)$

Chapter 6   Review Test

1. $2ax^3(3ax^2 - 4x + 6a^2)$ **2.** $-6x(x^3 + 2x^2 - 3)$ **3.** $(x + 4)(x - 3)$

4. $ax^2(x - 6)(x - 4)$ **5.** $(2x + 3)(x + 5)$ **6.** $a^2b(3a - 2)(2a + 1)$

7. $(3x + 4)(2x - 3)$ **8.** $ax^2(5x - 6)(2x + 3)$ **9.** $(4x - 5)(4x + 5)$

10. $3x(3x - 4)(3x + 4)$ **11.** $(2x + 5)(2x + 5)$ **12.** $(4x^2 - 5)(4x^2 - 5)$

13. $x^2(3x - 4y)^2$ **14.** $y^2(2x^2 - 5y)^2$ **15.** $(2x - 3)(4x^2 + 6x + 9)$

16. $(5x - 4y^2)(25x^2 + 20xy^2 + 16y^4)$ **17.** $(x + 1)(x - 1)(x + 3)(x - 3)$

18. $(x - 2y)(x + 2y)(x^2 + 4y^2)$ **19.** $(x - 1)(x^2 + x + 1)(x + 2)(x^2 - 2x + 4)$

20. $(x - 3y)(x^2 + 3xy + 9y^2)(x + y)(x^2 - xy + y^2)$ **21.** $x(x + 5)$

22. $(4x - 3)(2x + 3)$ **23.** $(x + 1)(x + 4)(x + 6)(x - 1)$

24. $(x - 1)(x - 2)(x + 1)(x - 4)$ **25.** $(x - 2y)(a + 2b)$

26. $(a - 3b)(x + 7)$ **27.** $(x - 3)(x + 3)(x + 1)$

28. $(2x + 3)(x - 2)(x + 2)$

■ Chapter 7

Problem Set 7.1A

1. $\dfrac{(x + 4)}{4}$ **3.** $-\dfrac{1}{2}$ **5.** $\dfrac{x - y}{x + y}$ **7.** $\dfrac{1}{x}$ **9.** $\dfrac{x^2}{y}$

11. $-\dfrac{a^2}{b}$ **13.** $\dfrac{4x}{y^2}$ **15.** $-\dfrac{2y^2}{x}$ **17.** $\dfrac{x - 6}{x + 5}$ **19.** $\dfrac{(3x + y)}{(2x - 3y)}$

21. $\dfrac{x - 4}{x + 4}$ **23.** $\dfrac{2x + y}{x + 2y}$ **25.** $\dfrac{x^2 - xy + y^2}{x(x - y)}$ **27.** $\dfrac{(x + y)}{(a - 3)}$ **29.** $\dfrac{y(2x + 5y)}{x(3x + y)}$

31. $\dfrac{x(x^2 + xy + y^2)}{y(x + y)}$ **33.** $\dfrac{ya(x - y)}{a - y}$ **35.** $a + b$

Problem Set 7.1B

1. $-\dfrac{1}{xy}$ **3.** $-\dfrac{8n^3}{m^2}$ **5.** $-\dfrac{a}{2}$ **7.** -1 **9.** $b - a$

11. $-\dfrac{a}{a + 4}$ **13.** 2 **15.** $\dfrac{a^2}{x}$ **17.** $\dfrac{(a + b)}{b}$ **19.** $\dfrac{x - 3}{4x(x - 12)}$

21. $\dfrac{(x^2 + 2x + 4)}{(x + 3)}$ **23.** $\dfrac{-x(y + x)}{(y^2 + x^2)}$ **25.** $\dfrac{(x - 4)}{5}$ **27.** $-\dfrac{x}{3}$ **29.** $-\dfrac{1}{(2x - y)}$

31. $(x + 2)$ **33.** $\dfrac{2(9a^2 + 12ab + 16b^2)}{(4a - b)}$ **35.** $\dfrac{x}{2x + y}$ **37.** $\dfrac{x + 3}{y}$ **39.** $\dfrac{x + 2y}{a - b}$

41. $y - 2$ **43.** $\dfrac{1}{4}$ **45.** $\dfrac{x + 4}{x^2(5 - x)}$ **47.** $\dfrac{3x}{y}$ **49.** $\dfrac{2x - y}{x - 2y}$

51. $a + 1$

Problem Set 7.1C

1. $\dfrac{y - x}{y}$

3. $-\dfrac{1}{x^2}$

5. $\dfrac{6x + 1}{6(x + 1)}$

7. $-\dfrac{1}{y^2}$

9. $-\dfrac{3}{ax}$

11. $\dfrac{2(3x + 1)}{9(x + 1)}$

13. $-\dfrac{y}{x}$

15. $\dfrac{(x + 5)}{x^2 - 5}$

17. $\dfrac{(x + 3)}{2}$

19. 2

21. $\dfrac{x}{2}$

23. $\dfrac{a(2b - a)}{2(2a - b)}$

25. $-\dfrac{y^3}{x^4}$

27. $\dfrac{(x + 1)(2x + 1)}{(x + 2)(x + 3)}$

29. $\dfrac{x - 3}{x - 2}$

31. $\dfrac{x + y}{2y - x}$

33. $\dfrac{4a^2 + 2ab + b^2}{2a + b}$

35. $\dfrac{(x - 2)(x + 1)}{x + 2}$

37. $y^2(y - 2)$

39. $\dfrac{(a - b)(a - 2b)}{2ab(a + 2b)}$

41. $-\dfrac{3(x + 1)}{4(x - 2)}$

43. $\dfrac{xy^2}{x - y}$

45. $\dfrac{(x + 1)^2}{(x + 2)}$

47. $\dfrac{25a^2 + 15ab + 9b^2}{3a + b}$

49. $\dfrac{a(a + 1)}{a - 1}$

Problem Set 7.2A

1. $\dfrac{9}{x + 3}$

3. 2

5. 1

7. 1

9. $\dfrac{2}{a - 1}$

11. $\dfrac{3}{x + 3}$

13. $\dfrac{2x}{x - 3}$

15. $\dfrac{1}{x + 4}$

17. $\dfrac{1}{x - 3}$

19. $\dfrac{1}{3x + 2}$

Problem Set 7.2B

1. $\dfrac{7y - 6x}{x^2 y}$

3. $\dfrac{x + 4}{3x(x - 2)}$

5. $\dfrac{4}{x - 4}$

7. $\dfrac{15x - 4}{3x - 1}$

9. $\dfrac{x + 4}{2(x + 2)}$

11. $\dfrac{3x - 5}{(5 - x)(5 + x)}$

13. $\dfrac{2x + 8}{3x + 4}$

15. $\dfrac{3(y - 4x)}{4y(y - 4)}$

17. $\dfrac{6}{x - 4}$

19. $\dfrac{2}{(x^2 + 2x + 4)(x + 4)}$

21. $\dfrac{9}{x(x - 3y)}$

23. $\dfrac{3m - 2}{(m + 2)^2(m - 2)}$

25. $\dfrac{2}{5x}$

27. $\dfrac{3}{x - 4}$

29. $\dfrac{-1}{2(x + 3)}$

31. $\dfrac{x^2 + 7x - 1}{(x - 1)(x + 4)(x^2 + x + 1)}$

33. $\dfrac{3x + 5y}{x(x - 3y)(x - y)}$

35. $\dfrac{4x + 3}{3(x - 1)}$

37. $\dfrac{(2x + 5)(x + 5)}{(x - 5)(x + 10)(x^2 + 5x + 25)}$

39. $\dfrac{1}{(x + 1)}$

41. $\dfrac{2(x^2 + 2x - 12)}{x(x - 2)}$

43. $\dfrac{3x^2 + 7x - 19}{3(x + 3)(x - 1)}$

45. $\dfrac{x - 9}{2(x - 3)(x + 3)}$

47. $\dfrac{3x^2 + 11x + 18}{(x + 5)(x - 3)(x^2 + 3x + 9)}$

49. $\dfrac{x^2 - 14x + 8}{x^2 - 5x + 4}$

51. $\dfrac{-3x^2 + 21x - 12}{x(x - 4)}$

Problem Set 7.3

1. $\dfrac{6}{5}$ **3.** $\dfrac{3}{4}$ **5.** $\dfrac{7}{6}$ **7.** $\dfrac{61}{46}$

9. $\dfrac{2(1+4x)}{15}$ **11.** $\dfrac{1}{a-1}$ **13.** $(x+1)(x-1)$ **15.** $\dfrac{y}{x-y}$

17. $\dfrac{12y+3}{6y-2}$ **19.** $a+2$ **21.** $\dfrac{4+6x}{5}$ **23.** $\dfrac{b}{a+b}$

25. $\dfrac{1}{2}$ **27.** $\dfrac{y+x}{xy}$ **29.** $\dfrac{4(x+1)}{x-6}$ **31.** $\dfrac{x+3}{x-3}$

33. $\dfrac{x^2}{y^2(x^2-xy+y^2)}$ **35.** $\dfrac{x-2}{x-3}$ **37.** $-\dfrac{1}{4}$ **39.** $\dfrac{3x+3}{x-15}$

41. $-\dfrac{(ab^2+a^2b)}{2a^2-2b^2+2b}$ **43.** $\dfrac{x-2y}{x+2y}$ **45.** $\dfrac{x+4}{x+5}$ **47.** $\dfrac{x-4}{x+3}$

Problem Set 7.4

1. 2   **3.** -3 **5.** 5   **7.** 9   **9.** 3   **11.** -3

13. $-\dfrac{1}{6}$ **15.** -6 **17.** 19   **19.** 4   **21.** No solution   **23.** 2

25. -6 **27.** 4   **29.** 5   **31.** No solution   **33.** -8 **35.** 6

37. No solution   **39.** 6   **41.** -3 **43.** 2   **45.** $\dfrac{5}{3}$ **47.** No solution

49. 1   **51.** 3   **53.** 2   **55.** $-\dfrac{1}{2}$ **57.** 4   **59.** No solution

Problem Set 7.5

1. 7   **3.** $\dfrac{8}{14}$ **5.** 8, 24   **7.** $\dfrac{9}{12}$ **9.** $\dfrac{7}{11}$

11. $\dfrac{7}{11}$ **13.** 3.6   **15.** 6.7 min   or $6\dfrac{2}{3}$ min   **17.** 13.3   **19.** $28\dfrac{1}{8}$ min

Chapter 7   Review Test

1. $\dfrac{x-8}{x}$ **2.** $\dfrac{2(2x+5y)}{4x+y}$ **3.** $\dfrac{9x^2-6xy+4y^2}{3x+2y}$ **4.** $\dfrac{x+2}{x}$

5. $\dfrac{1}{2}$ **6.** $x-2y$ **7.** $\dfrac{1}{4y-3x}$ **8.** $\dfrac{(x-1)^2}{2x-1}$

9. $\dfrac{1}{2a^2}$ **10.** $\dfrac{1}{x}$ **11.** $\dfrac{3}{(y+6)(y^2+3y+9)}$ **12.** $\dfrac{1}{x+1}$

13. $\dfrac{5}{4(x-10)}$ **14.** $\dfrac{1}{x-2}$ **15.** $\dfrac{b}{a-b}$ **16.** $\dfrac{x}{y}$

17. $\dfrac{x-4}{x+3}$ **18.** $x=\dfrac{7}{9}$ **19.** $y=6$ **20.** $x=2$ **21.** $\dfrac{8}{24}$ **22.** $1\dfrac{7}{8}$ hrs

Chapter 8

Problem Set 8.1A

1. 7 **3.** 2 **5.** 4 **7.** $\dfrac{1}{16}$ **9.** $\dfrac{1}{81}$ **11.** 243

13. $\dfrac{1}{125}$ **15.** $\dfrac{1}{4}$ **17.** $\dfrac{1}{27}$ **19.** 8 **21.** 64 **23.** $\dfrac{1}{5}$

25. $\dfrac{1}{2187}$ **27.** $\dfrac{1}{16}$ **29.** $a^{7/6}$ **31.** 1 **33.** $24a^{3/2}$ **35.** $\dfrac{-32}{x^{2/3}y^{3/4}}$

37. $\dfrac{1}{x^{5/2}}$ **39.** $\dfrac{1}{a}$ **41.** $x^{4/15}$ **43.** $-12x^{11/12}$ **45.** $\dfrac{1}{x^{1/6}}$ **47.** $x^{1/6}y^{1/12}$

49. $\dfrac{a^3}{b^2}$ **51.** ab^3 **53.** $\dfrac{3x^{1/3}}{2y^{1/4}}$ **55.** $\dfrac{8y^{1/2}}{x^{3/8}}$ **57.** $\dfrac{b^3y^2}{a^8}$ **59.** $\dfrac{b^{8/3}}{a^{29/12}}$

61. $\dfrac{x^9}{y^8}$ **63.** $\dfrac{1}{ab^3}$ **65.** $\dfrac{x^{2/3}}{4y^2}$ **67.** $\dfrac{9x^4}{y^2}$ **69.** $\dfrac{1}{a^8b^{10}}$ **71.** $\dfrac{a^{30}x^6}{y^5}$

Problem Set 8.1B

1. 16 **3.** 32 **5.** 3 **7.** 2 **9.** 4 **11.** -1

13. -2 **15.** 2 **17.** -81 **19.** -2 **21.** $-\dfrac{1}{25}$ **23.** 1

25. Impossible **27.** 3 **29.** $-\dfrac{1}{2}$ **31.** -8 **33.** 4 **35.** -4

37. 2 **39.** -2 **41.** 9 **43.** -9 **45.** 3 **47.** -3

Problem Set 8.2

1. x^2 **3.** x **5.** $a^2b\sqrt[3]{b}$ **7.** $3xy\sqrt{2y}$

9. $4x^2y^3\sqrt{2y}$ **11.** $-5a^2\sqrt{3b}$ **13.** $-5a^2b^3\sqrt{2a}$ **15.** $2x\sqrt[3]{2xy^2}$

17. $2xy\sqrt[4]{2xy^2}$ **19.** $-2ab\sqrt[5]{2ab^2}$ **21.** $3a^2b^5\sqrt{3a}$ **23.** $2xy^2\sqrt[4]{4x^3y}$

25. $-2xy\sqrt[5]{2y}$ **27.** $\sqrt[3]{x^2-4}$ **29.** $x\sqrt[4]{8x^2}$ **31.** $5xy\sqrt[3]{3x^2y}$

33. $5x^3\sqrt{2x}$ **35.** $4x^2y\sqrt[3]{y^2}$ **37.** $5x^2y^2\sqrt[4]{3}$ **39.** $-2ab\sqrt[3]{3a^2b^4}$

41. $\dfrac{2x^4}{y^2}\sqrt{x}$ **43.** $\dfrac{3x}{2y}$ **45.** $\dfrac{7x^3\sqrt{6abx}}{3b^2y^2}$ **47.** $\dfrac{2xy\sqrt[3]{75abxy^2}}{5a^2b}$

49. $\dfrac{3x^2y\sqrt[4]{250ab^3x}}{5ab^2}$ **51.** $\dfrac{-b\sqrt[3]{6x^2a^2y^2}}{2a^2y^3}$ **53.** $\dfrac{a\sqrt[4]{27abxy^3}}{2y^2x}$ **55.** $\dfrac{5ax\sqrt[3]{9x^2b^2y^2}}{3y^2b}$

57. $\dfrac{5}{3ab}$ **61.** $\dfrac{\sqrt[4]{4x^2y^3}}{x}$ **63.** $\dfrac{\sqrt[5]{80x^4y^3}}{4x}$

65. $\dfrac{\sqrt{10a}}{14a^2b^3}$ **67.** $\dfrac{-3a\sqrt[3]{50axy^2}}{5x^2y}$ **69.** $-\dfrac{\sqrt[4]{500a^2b^3x^3}}{5bx}$ **71.** $\dfrac{a\sqrt[5]{3240a^2b^4y^4}}{6b^2y^2}$

Problem Set 8.3

1. $11\sqrt{5}$

7. $3x\sqrt[3]{x} + x$

13. $x\sqrt[3]{3x^2}$

19. $17x^3\sqrt{5x}$

25. $14x^2\sqrt{x}$

31. $-4a\sqrt[4]{a^3} - 6a\sqrt[4]{a}$

37. $2x\sqrt[3]{2x}$

43. $6x\sqrt{6x}$

3. $3\sqrt[3]{2x^2} + 2x\sqrt{2}$

9. $\sqrt{5x}$

15. $(x+1)x\sqrt[5]{2x}$

21. $4\sqrt{2a} + \sqrt{5a}$

27. $-10x^2\sqrt{5x}$

33. $\dfrac{31\sqrt[3]{2a}}{2}$

39. $\dfrac{19a^2\sqrt[5]{a^2}}{3x}$

45. $33a^3\sqrt[3]{5a^2}$

5. $5x\sqrt[3]{2x^2}$

11. $17x\sqrt{5x} - 2\sqrt[3]{5x}$

17. $x^2\sqrt[4]{3x^3} + 4x^2\sqrt[5]{3x^3}$

23. $4x\sqrt[4]{x}$

29. $x^2\sqrt[5]{x^2}$

35. $3x^2\sqrt{x}$

41. $3x\sqrt[4]{3x^3} + x\sqrt[4]{6x^3}$

47. $\dfrac{11x\sqrt[5]{3x^4}}{6}$

Problem Set 8.4

1. $3 - 3\sqrt{2}$

5. $5\sqrt[3]{x} - x$

9. $3ay\sqrt{2} + 6y\sqrt{a}$

13. $6 + \sqrt{3x} - 3x$

17. $x - 5\sqrt[3]{x^2} + 5\sqrt[3]{x} - 25$

21. $\sqrt[3]{4x^2} - 2\sqrt[3]{2xy} + \sqrt[3]{y^2}$

25. $2x - 7$

29. $3x + 2\sqrt{3xy} + y$

33. $100 - 28\sqrt{5x} - 24x$

37. $\dfrac{7 - \sqrt{5}}{44}$

41. $2\sqrt{5} - 2\sqrt{2}$

45. $\dfrac{15\sqrt{3} - 9\sqrt{5}}{2}$

49. $\dfrac{2\sqrt{y} + 12}{y - 36}$

53. $\dfrac{9 - 3\sqrt{3}}{2}$

57. $\dfrac{10\sqrt{6} + 2\sqrt{15} + 5\sqrt{2} + \sqrt{5}}{55}$

61. $\dfrac{x + 2\sqrt{xy} + y}{x - y}$

65. $\dfrac{a\sqrt{2} + 2\sqrt{3ab}}{2a - 12b}$

69. $\dfrac{x\sqrt{2} - \sqrt{xy} - y\sqrt{2}}{2x - y}$

3. $5\sqrt{5}\,a - 5a\sqrt{2}$

7. $5x\sqrt{2} - 5x\sqrt{3}$

11. $\sqrt[3]{3x^2} - x$

15. $3x - 2\sqrt{3xy} + y$

19. $x\sqrt[4]{x^2} - 4$

23. $x\sqrt[4]{9} + 10\sqrt[4]{3x^2} + 25$

27. $x - 4\sqrt[3]{x^2} + 5\sqrt[3]{x} - 20$

31. $30x - 23\sqrt{xy} - 14y$

35. $x\sqrt[3]{9x} - 2\sqrt[3]{3ax^2} + \sqrt[3]{a^2}$

39. $\dfrac{3\sqrt{x} - 9}{x - 9}$

43. $\dfrac{5\sqrt{2} + 2\sqrt{5}}{3}$

47. $-\dfrac{(3\sqrt{6} + 15)}{19}$

51. $2\sqrt{6} + 2\sqrt{3}$

55. $\dfrac{25\sqrt{3} + 15\sqrt{5}}{2}$

59. $\dfrac{a - \sqrt{ab}}{a - b}$

63. $-\dfrac{(5\sqrt{2} + 3\sqrt{5} + 15\sqrt{6} + 9\sqrt{15})}{130}$

67. $\dfrac{3x\sqrt{3} + 3\sqrt{xy} - \sqrt{3xy} - y}{3x - y}$

1. -3

2. -2

3. Undefined

4. 4

5. 32

6. $x^{13/12}$

7. $\dfrac{x^{1/2}}{y^{8/3}}$

8. $a^{17/4}$

9. $-3x^{13/3}y^{9/4}$

10. 2

11. $\dfrac{1}{3}$

12. $4a^2\sqrt{3a}$

13. $2ab\sqrt[4]{2ab^2}$

14. $-3ab^2\sqrt[3]{2a^2}$

15. $2ab^2\sqrt{2}$

16. $\dfrac{2x\sqrt[3]{6y^2}}{3y}$

17. $\dfrac{x^2}{2a}$

18. $\dfrac{y^2\sqrt{3y}}{x^2}$

19. $5\sqrt{3} - 4\sqrt{2}$

20. $(1-x)\sqrt{3x}$

21. $2x^2\sqrt{3x}$

22. $\dfrac{x\sqrt[3]{2x} + \sqrt[3]{3x}}{x}$

23. $5 - 5\sqrt{2}$

24. $5x\sqrt{2} - 2x\sqrt{15}$

25. $y\sqrt[3]{4} - 2$

26. $2x - 2\sqrt[3]{3x^2}$

27. $3x + 2\sqrt{3x} - 48$

28. $3x - 2\sqrt{15xy} + 5y$

29. $4 + 2\sqrt{3y} - 18y$

30. $2(\sqrt{6} + \sqrt{5})$

31. $\dfrac{\sqrt{6} + \sqrt{2}}{2}$

32. $\dfrac{36\sqrt{2} + 16\sqrt{3}}{19}$

33. $\dfrac{6x + 2\sqrt{xy}}{9x - y}$

34. $\sqrt{10} + 2\sqrt{2}$

35. $\dfrac{\sqrt{10} + 5 - 2\sqrt{5} - 5\sqrt{2}}{3}$

36. $\dfrac{3x + 4\sqrt{2xy} + 2y}{x - 2y}$

■ Cumulative Review   Chapters 1–8

1. $(a + b) \cdot 3$

2. $16x^2 - 24xy + 9y^2$

3. $5a^4 - 18a^3 - 6a^2 + 14a - 3$

4. $\dfrac{-(x-5)(x+2)}{x^2 + 2x + 4}$

5. $\dfrac{(x-3y)^2}{(3x-y)(x^2 - 3xy + 9y^2)}$

6. -13

7. -14

8. -5

9. 15

10. 42

11. $\dfrac{29}{10}$

12. $\dfrac{43}{13}$

13. $\dfrac{2s - p}{s}$

14. 3

15. $\dfrac{3}{2}$

16. $a > 1$

17. $x < -18$

18. $-1 < x < 4$

19. $y \geq 7$ or $y \leq -3$

20. $m = \dfrac{-4}{3}$

21. $y = \dfrac{-3}{4}x - 4$

22. $y = -x - 4$

23. $y = \dfrac{3}{2}x - \dfrac{23}{2}$

24. $y = \dfrac{4}{5}x + \dfrac{22}{5}$

25.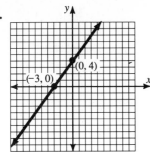

26.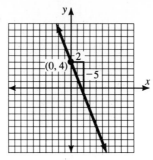

27. $x \geq \dfrac{3}{2}$

28. $-\dfrac{1}{3} < x < 5$

29. $\left(-5, \dfrac{2}{3}\right)$

30. $(-4, 5)$

31. $(3, -1, 2)$

32. 12, 8

33. 16, 64, 24

34. 7.5 hrs

35. $\dfrac{3}{7}$

36. $\dfrac{-108x^3}{y^{13}}$

37. $\dfrac{x^6 z^{12}}{y^{21}}$

38. $-8(3x - 1)^5(2x + 5)$

39. -16

40. $\dfrac{a^{\frac{21}{4}}}{b^3}$

41. $\dfrac{2y}{x}\sqrt{xy}$

42. $\dfrac{y^3\sqrt{6x}}{x^2}$

43. $\dfrac{x\sqrt{2x} + 6x\sqrt{3x}}{3}$

44. $\dfrac{3x + 4}{4(x + 2)}$

45. $\dfrac{2x^3}{3y^3} - 2y + \dfrac{3x}{2y}$

46. $8x^4 + 6x^3 + 4x^2 + 3x + 2 + \dfrac{2}{3x - 2}$

47. $3x^3 + 2x^2 - 4x + 3 + \dfrac{-4}{x + 2}$

48. $ax(2x - 7)(4x + 5)$

49. $(x^2 - 3y^3)(x^2 + 3y^3)$

50. $(x - 2)(x + 1)(x^2 + 2x + 4)(x^2 - x + 1)$

51. $(x + 4)(x - 2)(x + 3)(x - 1)$

52. $(x + 3 - y)(x + 3 + y)$

53. $(x - 4)(x + 4)(x + 2)$

54. $\dfrac{(y + 6)}{2(y - 6)(y - 4)}$

55. $\dfrac{x}{x + 5}$

56. $\dfrac{-y}{x + y}$

57. $\dfrac{x - 5}{x + 1}$

58. $4 - 2x\sqrt[3]{4}$

59. $2x - 2\sqrt{6xy} + 3y$

60. $y - 3x$

61. $\dfrac{2\sqrt{5} + 5\sqrt{2} - \sqrt{10} - 5}{3}$

■ Chapter 9

Problem Set 9.1

1. $3i$

3. $-6i$

5. $-3\sqrt{2}\,i$

7. $5\sqrt{3}\,i$

9. $6\sqrt{2}\,i$

11. $4\sqrt{2}\,i$

13. $-3\sqrt{3}\,i$

15. $-4\sqrt{5}\,i$

17. $x = 4$
$y = -3$

19. $x = 5$
$y = -\dfrac{1}{3}$

21. $x = 4$
$y = \dfrac{1}{5}$

23. $x = -2$
$y = -11$

25. $x = -\dfrac{9}{5}$
$y = 2$

27. $x = 4$
$y = 4$

29. $x = 3$
$y = 2$

31. $x = 2$
$y = -4$
33. i
35. i
37. $-i$
39. i

41. 1
43. i
45. $-2 + 8i$
47. $10 - \sqrt{2}\, i$
49. $-5 - 5i$

51. $-2 + 4i$
53. $26 - 9i$
55. $4 + i$
57. $24 + 21i$
59. $35 + 35i$

61. $11 - \sqrt{5}\, i$
63. $-24 - 48i$
65. $42 - 24i$
67. $15 + 9\sqrt{7}\, i$
69. $3 - 8i$

71. $-4i$
73. $\dfrac{1}{3} + \dfrac{\sqrt{3}}{6} i$
75. $\dfrac{24}{13} + \dfrac{36}{13} i$
77. $\dfrac{1 + 4i}{2}$
79. $\dfrac{1}{5} - \dfrac{2\sqrt{6}}{5} i$

81. $\dfrac{-4\sqrt{2} + 8i}{3}$
83. $8 - \dfrac{2\sqrt{3}}{3} i$
85. $7 - \sqrt{2}\, i$
87. $2 - \dfrac{3}{2} i$
89. $14 - 7i$

91. $\dfrac{75}{41} + \dfrac{60}{41} i$
93. $-\dfrac{54}{17} + \dfrac{90}{17} i$
95. $-\dfrac{11}{89} - \dfrac{71}{89} i$
97. $\dfrac{21}{29} - \dfrac{20}{29} i$
99. $\dfrac{3}{13} - \dfrac{4\sqrt{10}}{13} i$

Problem Set 9.2

1. 1, 5
3. $-4, \dfrac{1}{2}$
5. $\dfrac{3}{2}, \dfrac{1}{4}$
7. $-4, 4$
9. $-3, 3$
11. $-2, 6$

13. $\dfrac{4}{3}, 6$
15. $-\dfrac{5}{2}, \dfrac{1}{5}$
17. $\dfrac{5}{4}$
19. $0, \dfrac{3}{2}$
21. $-3, \dfrac{1}{5}$
23. $\dfrac{5}{3}, -\dfrac{3}{5}$

25. $-\dfrac{3}{4}, 5$
27. $-\dfrac{4}{5}, \dfrac{5}{4}$
29. $2, \dfrac{5}{4}$
31. $3, -\dfrac{1}{4}$
33. $-\dfrac{7}{5}, -2$
35. $-\dfrac{5}{3}, -\dfrac{1}{3}$

37. 5, 13
39. 12, 24
41. 10
43. 35, 56
45. 6
47. 2

49. 12
51. 4
53. 10
55. 784 feet

Problem Set 9.3A

1. $7, -7$
3. $\pm 2\sqrt{3}$
5. $4, -8$
7. $0, 10$
9. $9, 5$

11. $2, -8$
13. $-3, 7$
15. $-11, 1$
17. $1, -15$
19. $0, -4$

21. $0, 18$
23. $9, 13$
25. $-8 \pm 2\sqrt{2}$
27. $2 \pm 3\sqrt{2}$
29. $15, -5$

31. $-4 \pm 5\sqrt{3}$
33. $-11 \pm 2\sqrt{11}$
35. $7 \pm \sqrt{7}$
37. $7 \pm 3\sqrt{2}$
39. $-3 \pm 2\sqrt{6}$

41. $-5 \pm 5\sqrt{2}$
43. $-9 \pm 3\sqrt{5}$
45. $10 \pm 5\sqrt{3}$
47. $12 \pm 2\sqrt{31}$

Problem Set 9.3B

1. 4
3. 16
5. 81
7. $\dfrac{1}{4}$
9. $\dfrac{25}{4}$
11. $\dfrac{121}{4}$

13. 1
15. 49
17. $\dfrac{49}{4}$
19. $\dfrac{121}{4}$
21. $\dfrac{289}{4}$
23. $\dfrac{169}{4}$

Problem Set 9.3C

1. 4, 2
3. $6, -2$
5. $2, -7$
7. $4, -7$
9. $-5, 3$

11. $-5, -4$
13. 3, 7
15. $-2, 9$
17. $0, -4$
19. $2, -2$

21. $\pm 2\sqrt{2}$
23. $\pm 2i$
25. $\pm \sqrt{6}$
27. $\pm \sqrt{7}$
29. $-5, 0$

31. $\pm \sqrt{6}\, i$
33. $\dfrac{-1}{3}, 2$
35. $\dfrac{11 \pm \sqrt{217}}{8}$
37. $\dfrac{3}{2}, -\dfrac{2}{3}$
39. $\dfrac{3 \pm \sqrt{15}}{2}$

41. $\dfrac{-1 \pm \sqrt{61}}{10}$ **43.** $\dfrac{2 \pm \sqrt{13}}{3}$ **45.** $\dfrac{3 \pm \sqrt{41}}{8}$ **47.** $\dfrac{-2 \pm 3\sqrt{2}}{2}$ **49.** $\dfrac{-3 \pm 2\sqrt{3}\,i}{3}$

51. $\dfrac{3 \pm 2\sqrt{3}}{2}$ **53.** $\dfrac{7 \pm \sqrt{17}}{4}$ **55.** $\dfrac{-9 \pm \sqrt{21}}{6}$ **57.** $\dfrac{5 \pm \sqrt{55}\,i}{8}$ **59.** $\dfrac{2 \pm \sqrt{5}\,i}{3}$

Problem Set 9.4

1. $-3, 2$ **3.** $-\dfrac{1}{2}, 3$ **5.** $-\dfrac{4}{3}, \dfrac{1}{2}$ **7.** $0, -2$ **9.** $0, \dfrac{-4}{9}$

11. $-2, 4$ **13.** $-12, 1$ **15.** $-\dfrac{3}{2}, \dfrac{5}{2}$ **17.** $-\dfrac{3}{2}, \dfrac{5}{3}$ **19.** $0, 3$

21. $\pm\dfrac{2\sqrt{3}}{3}$ **23.** $\pm\dfrac{3i\sqrt{2}}{2}$ **25.** $\pm\dfrac{2\sqrt{6}}{3}$ **27.** $\pm\dfrac{3}{2}i$ **29.** $\dfrac{-2 \pm \sqrt{10}}{2}$

31. $\dfrac{2 \pm \sqrt{7}}{2}$ **33.** $\dfrac{1 \pm 2i\sqrt{5}}{3}$ **35.** $\dfrac{1 \pm i\sqrt{17}}{6}$ **37.** $\dfrac{1 \pm \sqrt{13}}{4}$ **39.** $\dfrac{3 \pm \sqrt{15}}{3}$

41. $\dfrac{-3 \pm \sqrt{7}\,i}{8}$ **43.** $\dfrac{-5 \pm \sqrt{73}}{4}$ **45.** $\dfrac{1 \pm 3i}{5}$ **47.** $\dfrac{1}{3}, 1$ **49.** 1 Solution

51. 1 Solution   **53.** 2 Real   **55.** 2 Real   **57.** 2 Complex   **59.** 2 Real

61. 2 Complex   **63.** 1 Solution   **65.** 2 Real   **67.** 2 Real   **69.** 2 Complex

71. 1 Solution

Problem Set 9.5

1. $0.333, -7.000$ **3.** $-0.927, 1.727$ **5.** 0.15 **7.** $2.404, 67.596$

9. $-0.051, 26.051$ **11.** $0.129 \pm 1.157i$ **13.** 0.378 **15.** $0.433 \pm 0.588i$

17. $0.973 \pm 1.236i$ **19.** $-0.419, 1.178$ **21.** $-0.093, 390.926$ **23.** $-2.015, 0.840$

25. $0.049, 3.04$ **27.** $-1.964, 1.970$ **29.** $-3.267, 1.054$

Chapter 9   Review Test

1. $5i$ **2.** $9i$ **3.** $3\sqrt{2}\,i$ **4.** $5\sqrt{3}\,i$

5. $-i$ **6.** -1 **7.** i **8.** 1

9. $x = 3, y = -3$ **10.** $x = 15, y = -3$ **11.** $-1 - 2i$ **12.** $-2 + 6i$

13. $46 + 48i$ **14.** $-33 - 14\sqrt{3}\,i$ **15.** $2 - 3i$ **16.** $1 - 2i$

17. $\dfrac{-2 + 23i}{13}$ **18.** $\dfrac{4 - 10\sqrt{2}\,i}{9}$ **19.** $x = \pm 3$ **20.** $x = \pm 4$

21. $-\dfrac{3}{2}, 6$ **22.** $-\dfrac{5}{2}, \dfrac{4}{3}$ **23.** $-\dfrac{3}{2}, \dfrac{5}{2}$ **24.** 12 in $\times$ 16 in

25. $9, 11$ **26.** ± 4 **27.** $\pm 3\sqrt{2}\,i$ **28.** $-3, 11$

29. $-5 \pm 3\sqrt{2}$ **30.** $-2, 9$ **31.** $-2, \dfrac{1}{3}$ **32.** $\dfrac{2 \pm \sqrt{6}\,i}{2}$

33. $-5, 2$ **34.** $-\dfrac{5}{2}, 3$ **35.** $\dfrac{2 \pm \sqrt{13}}{3}$ **36.** $\dfrac{5 \pm i\sqrt{7}}{8}$

37. 1 Solution   **38.** 2 Real   **39.** 2 Complex   **40.** 2 Complex

41. $-0.2, -4.8$ **42.** 1.125 **43.** $-0.39 \pm 1.35i$ **44.** $0.7726, -2.1133$

Problem Set 10.1

1. 2

3. −1 does not check
4 is a solution

5. 4

7. 0 does not check
5 is a solution

9. −2

11. −4

13. −2

15. 1, −2

17. 5

19. 9

21. −4 does not check
9 is a solution

23. −4 does not check
$\dfrac{1}{2}$ is a solution

25. 8 is a solution
−2 does not check

27. 8 is a solution
$-\dfrac{2}{3}$ does not check

29. 0 does not check
8 is a solution

31. 6

33. 15 does not check
−1 is a solution

35. −3 does not check
$\dfrac{12}{5}$ is a solution

37. −2 does not check
13 is a solution

39. 7 is a solution

41. −2 is a solution
138 does not check

43. 7 is a solution
$-\dfrac{1}{9}$ does not check

45. −3 is a solution
137 does not check

47. 9 is a solution
$-\dfrac{7}{9}$ does not check

Problem Set 10.2

1. $\pm 2, \pm\sqrt{5}$

3. $\pm 3, \pm 2$

5. $\pm 3i, \pm\sqrt{5}$

7. $\pm 2, \pm\sqrt{3}i$

9. 64, 8

11. $\dfrac{27}{8}, -125$

13. $\pm 3, \pm\sqrt{3}$

15. $\pm 1, \pm 2\sqrt{3}$

17. $\pm 2\sqrt{5}, \pm\sqrt{3}\,i$

19. $\pm\sqrt{5}, \pm\dfrac{1}{3}i$

21. $\dfrac{1}{3}$

23. $-\dfrac{1}{4}$

25. $2 \pm i, 4, 0$

27. $0, -8, -4 \pm i$

29. $1, \dfrac{1}{2}$

31. $\pm 2, \pm\dfrac{1}{2}$

33. $-\dfrac{1}{2}, \dfrac{1}{6}$

35. $\pm\dfrac{1}{3}, \pm 2$

37. $-\dfrac{1}{2}, 2, -\dfrac{3}{2}, 3$

39. $-\dfrac{5}{3}, -1, \dfrac{4}{3}, 2$

41. 16, 256

43. 81, 256

45. $\dfrac{1}{625}, \dfrac{81}{16}$

47. $-\dfrac{27}{8}, \dfrac{64}{27}$

Problem Set 10.3

1. $s = Kt$

3. $G = Kh^2$

5. $y = \dfrac{Kx}{w^2}$

7. $L = \dfrac{Kmn}{q}$

9. $d = Kt^2$

11. $A = kr^2$

13. $V = kr^2 \cdot h$

15. $V = kr^2 \cdot h$

17. $k = \dfrac{1}{2}$, 14 sq in

19. $K = 2\pi$, 30π inches

21. $4.9\,\dfrac{m}{\sec}$

23. 6 ft

25. $\dfrac{\sqrt{10}}{80}$ in

27. $K = 4\pi$, $144\,\pi$ in^2

29. $420\,\dfrac{\text{lbs}}{\text{sq ft}}$

31. 9 lb

33. 50 mph

35. 28 lb

Problem Set 10.4

1. 416 miles

3. 240 mph

5. 3 hrs 20 min

7. 46, 54 mph

9. 7 mph

11. 60 mph on 1st part
50 mph on 2nd part

13. 600 mph

15. 3 mph

17. 120 mph and
180 mph

19. $4\dfrac{8}{13}$ mph

Chapter 10 Review Test

1. $x = 4$

2. $\dfrac{25}{4}$ doesn't check
3 is a solution

3. 1 doesn't check
7 is a solution

4. -2 doesn't check
8 is a solution

5. -1 doesn't check
11 is a solution

6. 15, -1 doesn't check

7. $\pm 2, \pm 5$

8. $\pm \dfrac{3\sqrt{2}}{2}, \pm i$

9. $-8, \dfrac{27}{8}$

10. 3, -5

11. $-\dfrac{3}{2}, \dfrac{1}{4}$

12. $\pm 2, \pm \dfrac{\sqrt{6}}{2}$

13. 256, 81

14. $\dfrac{1}{3}, -2, \dfrac{4}{3}, -3$

15. 288π cm^3

16. 119.4 ft

17. 225 lb

18. 30 miles

19. 5 hrs 20 min

20. 2 mph

■ Chapter 11

Problem Set 11.1

1. a) 5
b) $\left(3, \dfrac{1}{2}\right)$

3. a) 13
b) $\left(\dfrac{1}{2}, -2\right)$

5. a) $5\sqrt{2}$
b) $\left(\dfrac{1}{2}, -\dfrac{3}{2}\right)$

7. a) $\dfrac{5}{2}$
b) $\left(\dfrac{9}{4}, \dfrac{7}{2}\right)$

9. a) $4\sqrt{5}$
b) $(-1, 4)$

11. a) $5\sqrt{2}$
b) $\left(-\dfrac{7}{2}, \dfrac{11}{2}\right)$

13. a) 10
b) $(4, -6)$

15. a) $\dfrac{29}{3}$
b) $(0, 2)$

17. $P = 12$

19. $P = 24 + 6\sqrt{2}$

21. $15 + 5\sqrt{5}$

23. $14 + \sqrt{34}$

25. $3\sqrt{13}$

27. $2\sqrt{14}$

29. 8

31. $2\sqrt{29}$

33. $20\sqrt{13}$

35. $140 + 20\sqrt{7}$

37. $5\sqrt{73}$

39. $\sqrt{41}$

Problem Set 11.2

1. $x^2 + y^2 = 64$

3. $x^2 + 4x + y^2 - 6y - 23 = 0$

5. $x^2 - 2x + y^2 - 15 = 0$

7. $x^2 + 10x + y^2 + 12y + 52 = 0$

9. $x^2 + y^2 - 6x + 4y - 12 = 0$

11. $x^2 + y^2 + 8x + 18y + 93 = 0$

13. $x^2 + y^2 + 16x - 8y + 64 = 0$

15. $x^2 + y^2 - 10y + 16 = 0$

17. $(x - 3)^2 + (y - 4)^2 = 9$

19. $(x + 2)^2 + (y - 3)^2 = 16$

21. $x^2 + (y + 2)^2 = 25$

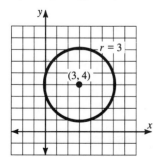

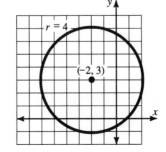

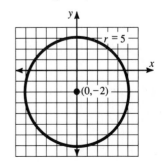

23. $(x + 2)^2 + (y - 4)^2 = 4$

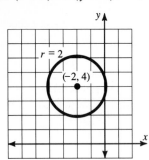

25. $(x - 2)^2 + y^2 = 16$

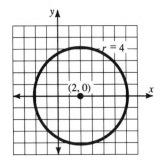

27. $(x - 4)^2 + (y + 2)^2 = 25$

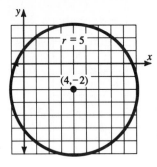

29. $x^2 + y^2 - 4x - 6y + 9 = 0$

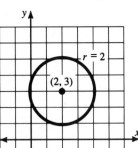

31. $x^2 + y^2 - 4x - 12 = 0$

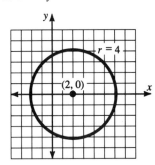

33. $x^2 + y^2 + 2x + 8y - 8 = 0$

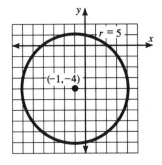

35. $x^2 + y^2 - 2x - 2y - 6 = 0$

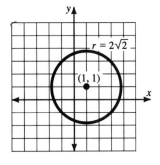

37. $x^2 + y^2 + 2x - 2y - 7 = 0$

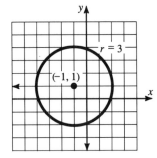

39. $x^2 + y^2 + 4x - 6y + 4 = 0$

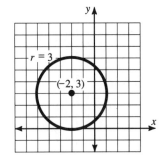

41. $x^2 + y^2 + 6x + 6y + 2 = 0$

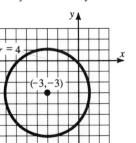

43. $x^2 + y^2 - 10x = 0$

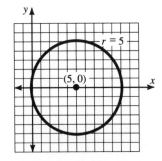

45. $x^2 + y^2 - 8x - 12y + 51 = 0$

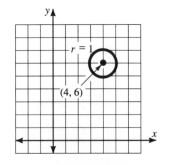

47. $x^2 + y^2 - 2x - 6y - 10 = 0$ **49.** $x^2 + y^2 \geq 25$ **51.** $(x - 2)^2 + y^2 < 36$

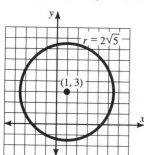

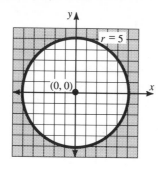

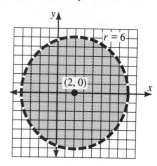

53. $x^2 + y^2 + 6x - 4y + 4 > 0$ **55.** $x^2 + y^2 \leq 36$ **57.** $x^2 + (y - 4)^2 > 16$

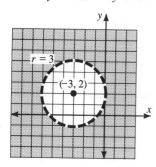

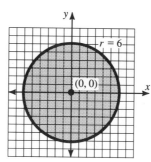

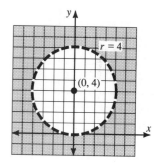

59. $x^2 + y^2 + 4x - 8y + 4 < 0$

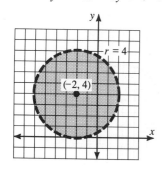

Problem Set 11.3

1. $x'^2 + y'^2 = 25$
$(x - 3)^2 + (y - 6)^2 = 25$

3. $x'^2 + y'^2 = 5$
$(x + 6)^2 + (y + 2)^2 = 5$

5. $x'^2 + y'^2 = 9$
$(x + 7)^2 + (y - 4)^2 = 9$

7. $x'^2 + y'^2 = 25$
$(x - 8)^2 + (y - 1)^2 = 25$

9. $x'^2 + y'^2 = 6$
$(x - 2)^2 + y^2 = 6$

Selected Answers

S39

11. $x^2 + y^2 + 2x + 6y + 6 = 0$ **13.** $x^2 + y^2 - 6x - 4y + 7 = 0$ **15.** $x^2 + y^2 - 8x - 4y + 11 = 0$

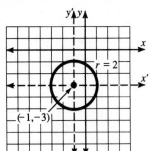

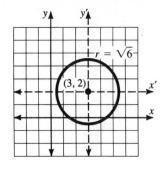

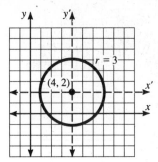

17. $x^2 + y^2 + 10x - 4y + 25 = 0$

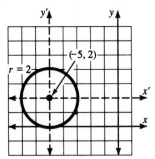

19. $x^2 + y^2 + 12x - 8y + 51 = 0$

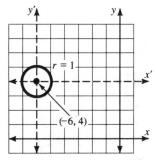

Problem Set 11.4

1. $y = 3x^2$

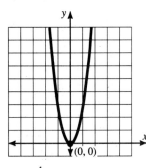

3. $y = \dfrac{1}{2}x^2$

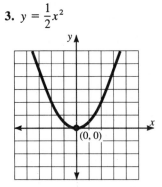

5. $y = -4x^2$

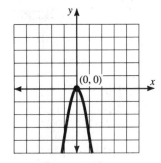

7. $y = \dfrac{1}{3}x^2$

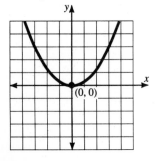

9. $y = -2x^2$

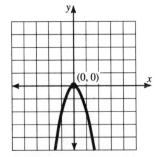

11. $y = \dfrac{1}{4}x^2$

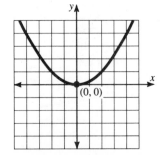

13. $y = x^2 + 4$

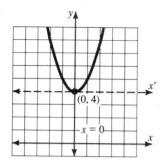

15. $y = -x^2 + 3$

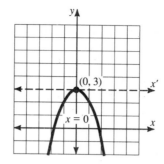

17. $x = -2y^2 + 3$

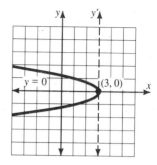

19. $y = -\frac{1}{2}x^2 + 4$

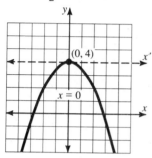

21. $y = -2(x + 3)^2$

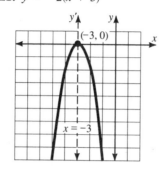

23. $y = -x^2 - 2$

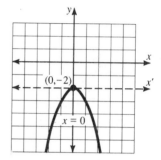

25. $y = -2x^2 - 1$

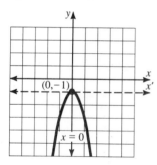

27. $x = -\frac{1}{2}y^2 + 4$

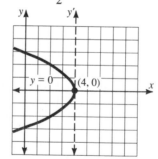

29. $y = \frac{1}{2}(x - 4)^2$

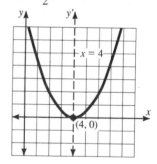

31. $x = 2(y + 1)^2$

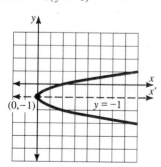

33. $y = 3(x - 1)^2$

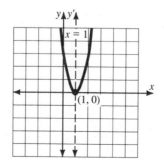

35. $y = -(x + 3)^2 + 2$

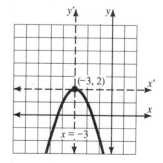

37. $y - 3 = -(x + 3)^2$

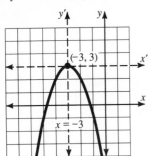

39. $y + 1 = 3(x - 2)^2$

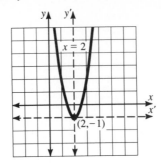

41. $x - 5 = -2(y + 1)^2$

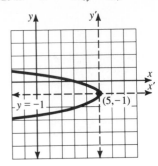

43. $x - 3 = \frac{1}{2}(y + 2)^2$

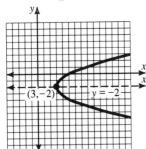

45. $y = -(x + 3)^2 + 2$

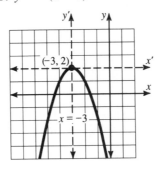

47. $y + 2 = 3(x - 2)^2$

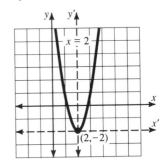

49. $x = -\frac{1}{2}(y + 1)^2 - 1$

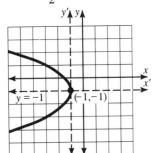

51. $y + 2 = \frac{1}{2}(x + 4)^2$

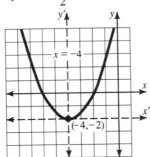

53. $x + 4 = 3(y + 1)^2$

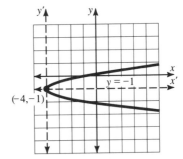

Problem Set 11.5

1. $y < x^2 + 1$

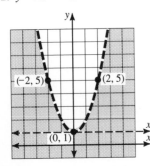

3. $y \geq (x - 2)^2$

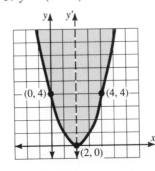

5. $y + 2 \leq (x + 1)^2$

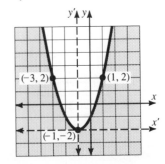

7. $x > y^2 - 1$

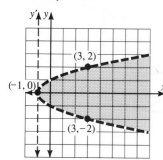

9. $y \geq 2(x + 1)^2$

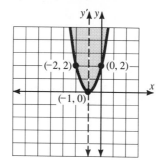

11. $x^2 + y^2 > 4$

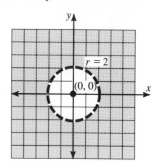

13. $y + 3 < (x - 2)^2$

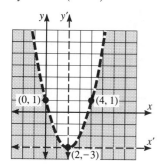

15. $x^2 + y^2 \leq 9$

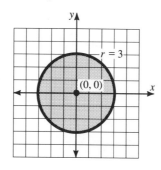

17. $x + 2 > (y + 1)^2$

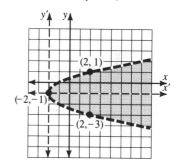

19. $(x + 2)^2 + (y - 3)^2 \geq 4$

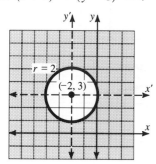

21. $y - 2 \leq -\dfrac{1}{2}(x + 1)^2$

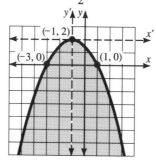

23. $x^2 + y^2 + 6x - 2y - 6 < 0$

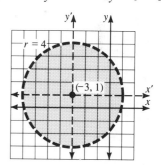

Chapter 11   Review Test

1. $3\sqrt{2}$ **2.** 9   **3.** $2\sqrt{34}$ **4.** 10   **5.** $(1, 1)$

6. $(1, 4)$ **7.** $18 + 2\sqrt{41}$ **8.** $2\sqrt{41} + 18$ **9.** 5   **10.** $4\sqrt{58}$

11. $x^2 + 8x + y^2 - 10y + 32 = 0$ **12.** $x^2 + y^2 + 6y - 55 = 0$

13. center: $(1, -2)$ radius: 2

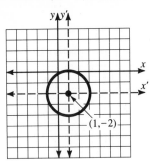

14. center: $(-5, 2)$ radius: $2\sqrt{2}$

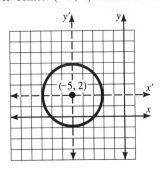

15. center: $(4, -3)$ radius: 4

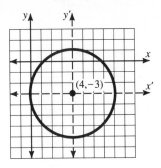

16. center: $(-5, 1)$ radius: 3

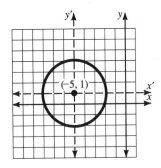

17. $x'^2 + y'^2 = 16$
$(x + 5)^2 + (y - 4)^2 = 16$

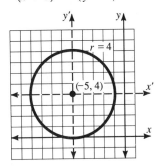

18. $x'^2 + y'^2 = 7$
$(x - 3)^2 + (y + 6)^2 = 7$

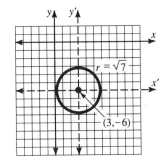

19. $(x - 3)^2 + (y + 4)^2 = 16$

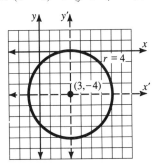

20. $(x + 2)^2 + (y - 3)^2 = 8$

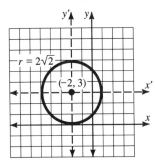

21. $y = 3x^2$

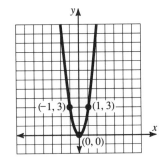

22. $x = -\dfrac{1}{2}y^2$

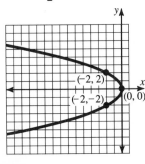

23. $x = 2y^2 - 3$

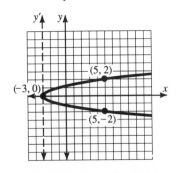

24. $y = 2(x + 3)^2$

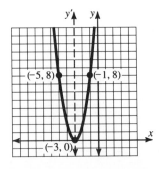

25. $y - 3 = \frac{1}{2}(x + 4)^2$

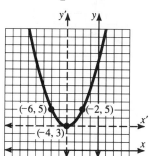

26. $y + 2 = -3(x - 4)^2$

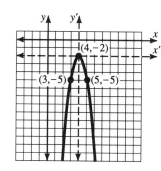

27. $(x - 2)^2 + (y + 1)^2 \geq 9$

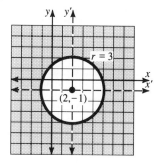

28. $x^2 + y^2 + 2x - 6y - 6 < 0$

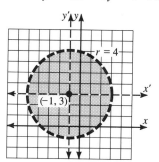

29. $y - 1 < (x + 2)^2$

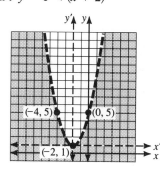

30. $x \geq y^2 - 2$

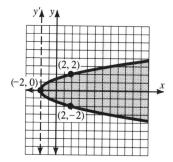

■ Chapter 12

Problem Set 12.1A

1. $\dfrac{x^2}{1} + \dfrac{y^2}{4} = 1$

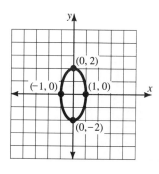

3. $\dfrac{x^2}{16} + \dfrac{y^2}{4} = 1$

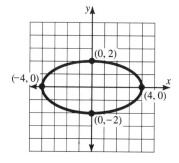

5. $\dfrac{x^2}{4} + \dfrac{y^2}{25} = 1$

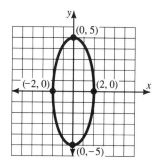

7. $\dfrac{x^2}{16} + \dfrac{y^2}{4} = 1$

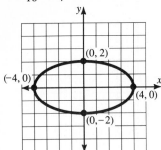

9. $x^2 + 4y^2 = 36$

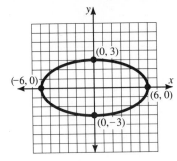

11. $16x^2 + 25y^2 = 400$

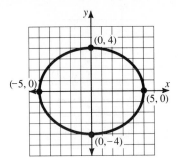

13. $4x^2 + y^2 = 36$

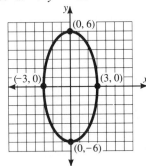

15. $x^2 + 9y^2 = 36$

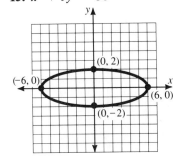

Problem Set 12.1B

1. $\dfrac{(x+1)^2}{9} + \dfrac{(y-2)^2}{16} = 1$

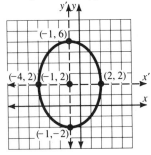

3. $\dfrac{(x-2)^2}{1} + \dfrac{(y-4)^2}{4} = 1$

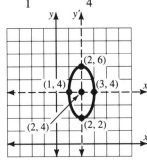

5. $\dfrac{x^2}{25} + \dfrac{(y+4)^2}{16} = 1$

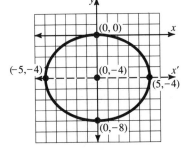

7. $\dfrac{(x-2)^2}{1} + \dfrac{(y+3)^2}{9} = 1$

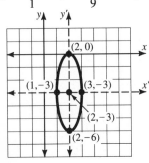

9. $\dfrac{(x-4)^2}{16} + \dfrac{y^2}{4} = 1$

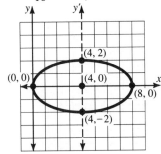

11. $\dfrac{(x+2)^2}{16} + \dfrac{(y-1)^2}{1} = 1$

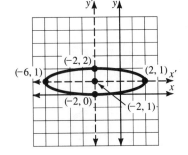

Selected Answers

13. $4x^2 + 9y^2 - 18y - 27 = 0$ **15.** $4x^2 + y^2 + 16x - 6y + 9 = 0$ **17.** $9x^2 + 4y^2 - 36x - 32y - 44 = 0$

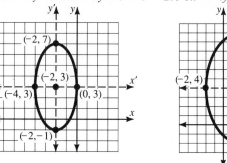

19. $9x^2 + y^2 - 36x < 0$ **21.** $x^2 + 4y^2 + 6x - 32y + 69 \geq 0$ **23.** $9x^2 + 4y^2 + 36x + 16y + 16 \leq 0$

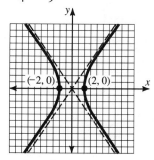
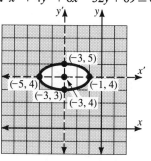
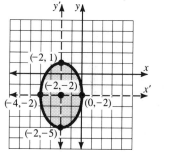

Problem Set 12.2

1. $\dfrac{x^2}{4} - \dfrac{y^2}{9} = 1$ **3.** $\dfrac{y^2}{4} - \dfrac{x^2}{25} = 1$ **5.** $\dfrac{x^2}{4} - \dfrac{y^2}{4} = 1$

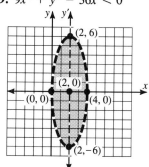
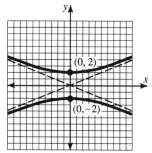
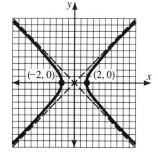

7. $\dfrac{x^2}{9} - \dfrac{y^2}{4} = 1$ **9.** $\dfrac{y^2}{25} - \dfrac{x^2}{16} = 0$ **11.** $\dfrac{y^2}{16} - \dfrac{x^2}{4} = 1$

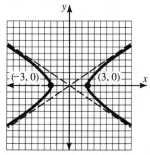
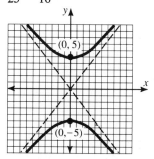
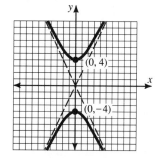

13. $x^2 - 4y^2 = 16$

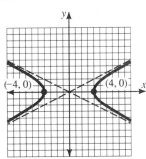

15. $4x^2 - 9y^2 = 36$

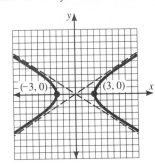

17. $\dfrac{(y - 3)^2}{4} - \dfrac{(x - 4)^2}{9} = 1$

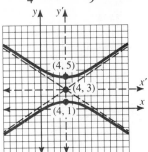

19. $\dfrac{(x - 2)^2}{25} - \dfrac{(y + 3)^2}{16} = 1$

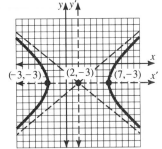

21. $\dfrac{(x + 4)^2}{9} - \dfrac{(y - 3)^2}{9} = 1$

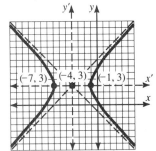

23. $\dfrac{(x + 2)^2}{4} - \dfrac{y^2}{18} = 1$

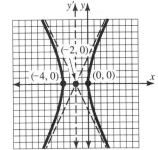

25. $\dfrac{(x + 5)^2}{20} - \dfrac{(y + 2)^2}{9} = 1$

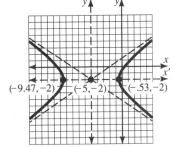

27. $\dfrac{(x - 2)^2}{4} - \dfrac{(y + 1)^2}{9} = 1$

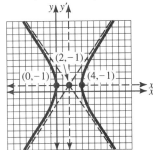

29. $\dfrac{(y - 4)^2}{4} - \dfrac{(x - 2)^2}{4} = 1$

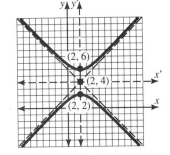

31. $\dfrac{(x - 1)^2}{9} - \dfrac{y^2}{4} = 1$

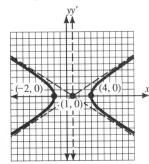

33. $\dfrac{(y + 2)^2}{9} - \dfrac{x^2}{25} = 1$

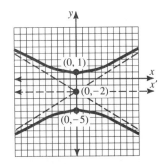

35. $\dfrac{(x + 2)^2}{4} - \dfrac{(y + 2)^2}{8} = 1$

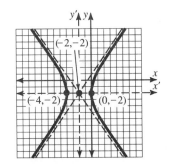

37. $4y^2 - 24y - x^2 - 4x - 4 = 0$

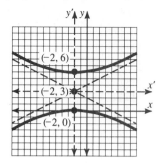

39. $16x^2 - 25y^2 - 150y - 625 = 0$

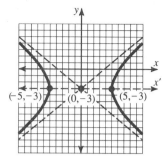

41. $4x^2 - 3y^2 + 8x + 30y - 119 < 0$

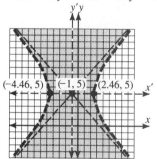

43. $9y^2 - 4x^2 - 8x - 54y + 41 \geq 0$

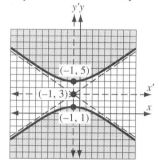

45. $9x^2 - 16y^2 + 36x - 32y - 124 \leq 0$

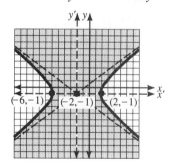

47. $y^2 - x^2 + 8x + 2y - 24 > 0$

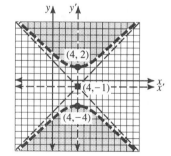

Problem Set 12.3A

1. circle

3. ellipse

5. hyperbola

7. ellipse

9. circle

11. parabola

13. circle

15. ellipse

17. hyperbola

19. straight line

21. parabola

23. ellipse

25. $xy = 6$

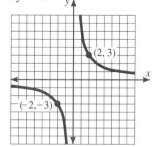

27. $xy = -5$

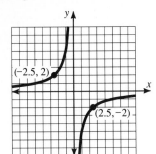

$(-2.5, 2)$

$(2.5, -2)$

29. $2xy = -5$

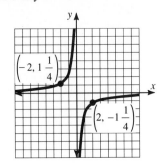

$\left(-2, 1\frac{1}{4}\right)$

$\left(2, -1\frac{1}{4}\right)$

31. $6xy = 9$

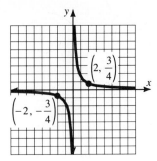

$\left(2, \frac{3}{4}\right)$

$\left(-2, -\frac{3}{4}\right)$

Problem Set 12.3B

1. $(4, 3)$, $(3, 4)$ **3.** $(1, 2)$, $(-1, -2)$ **5.** $(-6, 24)$, $(2, 8)$ **7.** $(2, 1)$, $(-2, -1)$

9. $(2, 1)$, $(-2, -1)$ **11.** $(-2, 6)$, $(1, 3)$ **13.** No real solution   **15.** $(3, 4)$, $(-4, -3)$

17. $(-4, -3)$ **19.** $(1, 2)$, $\left(-\frac{5}{3}, -\frac{2}{3}\right)$ **21.** $(-1, -2)$, $(1, 2)$ **23.** $(2, 5)$, $(-1, -7)$

25. $(2, 3)$, $(-3, -2)$ **27.** $(-1, -7)$, $(7, 1)$ **29.** No solution   **31.** $\left(\frac{5}{2}, 2\right)$, $\left(-\frac{5}{2}, -2\right)$

33. $(0, 8)$, $(4, 0)$ **35.** $(3, -2)$, $(2, -3)$

Problem Set 12.4A

1.

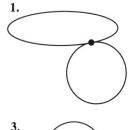

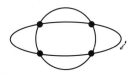

3.

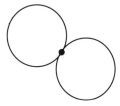

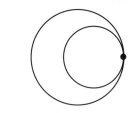

5.

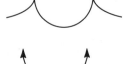

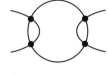

7.

9.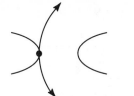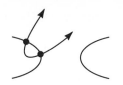

Problem Set 12.4B

1. $(-1, 0)$, $(1, 0)$ **3.** $(\pm\sqrt{6}\,i, 4)$, $(\pm 3, -1)$ **5.** $(\pm 1, 0)$ **7.** $(0, 3)$

9. $(4, 1)$, $(-4, -1)$, $(1, 4)$, $(-1, -4)$ **11.** $(\pm 1, \pm 2.83)$ **13.** $(\pm 2, \pm 1)$ **15.** $(\pm 3, \pm 2)$

17. $\left(-\dfrac{2}{3}, \dfrac{4}{3}\right)$, $(1, 3)$ **19.** $(\pm 1, \pm\sqrt{2})$ **21.** $(\pm\sqrt{3}, 2)$ **23.** $(\pm 2, \pm\sqrt{2})$

25. $(3, 0)$ **27.** $(3, \pm 2\sqrt{2})$, $(-1, 0)$ **29.** $(-\sqrt{3}, \sqrt{3})$, $(\sqrt{3}, -\sqrt{3})$, $\left(\sqrt{6}, -\dfrac{\sqrt{6}}{2}\right)$, $\left(-\sqrt{6}, \dfrac{\sqrt{6}}{2}\right)$

31. $(-4, 32)$, $\left(\dfrac{1}{2}, \dfrac{1}{2}\right)$ **33.** $(4, -3)$, $(-4, 3)$ **35.** $(0, -2)$, $\left(\pm\dfrac{\sqrt{30}}{3}, \dfrac{4}{3}\right)$

Chapter 12   Review Test

1. $\dfrac{x^2}{16} + \dfrac{y^2}{1} = 1$ **2.** $9x^2 + y^2 = 36$ **3.** $4x^2 + 5y^2 = 80$

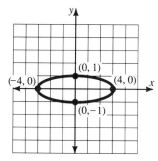 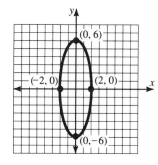 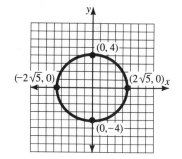

4. $\dfrac{(x+3)^2}{25} + \dfrac{(y-2)^2}{36} = 1$ **5.** $4x^2 + 9y^2 - 32x + 18y + 37 \geq 0$

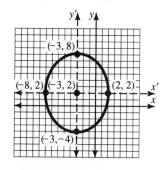 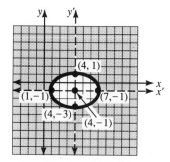

6. $16x^2 + 9y^2 + 64x + 90y + 145 = 0$ **7.** $\dfrac{x^2}{25} - \dfrac{y^2}{16} = 1$ **8.** $9y^2 - 4x^2 = 36$

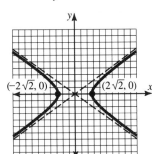

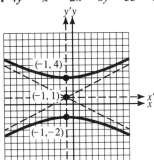

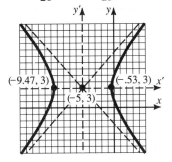

9. $x^2 - 2y^2 = 8$ **10.** $4y^2 - x^2 - 2x - 8y - 33 = 0$ **11.** $\dfrac{(x+5)^2}{20} - \dfrac{(y-3)^2}{25} = 1$

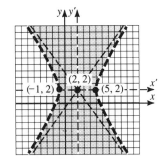

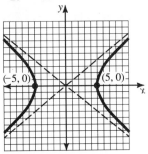

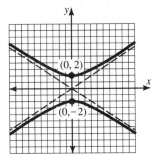

12. $16x^2 - 9y^2 - 64x + 36y - 116 < 0$

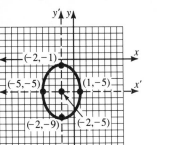

13. circle   **14.** ellipse   **15.** hyperbola   **16.** parabola

17. $(6, -2), (-2, 6)$ **18.** $(2, 1), (0.66, -1.66)$ **19.** $(1, -1), (-5, -7)$ **20.** $(-1, 2), (3, 22)$

21. $(\pm 6, \pm 2\sqrt{3})$ **22.** $(\pm 2, \pm \sqrt{2})$ **23.** $(2, 3), (-2, -3),$ **24.** $(\pm 2, 3)$
                                                              $(3, 2), (-3, -2)$

■ Chapter 13

Problem Set 13.1A

1. 4	**3.** -2	**5.** 6	**7.** -6	**9.** -9	**11.** 0	**13.** 9
15. 7	**17.** 2	**19.** 6	**21.** 10	**23.** 6	**25.** -3	**27.** -6

29. 2   **31.** $-\dfrac{17}{2}$ **33.** $3a + 5$ **35.** $3h$ **37.** $\dfrac{4}{3}a + 7$ **39.** $\dfrac{4}{3}h$

41.
$$f(-4) = 4(-4) + 2 = -14 \quad (-4, -14)$$
$$f(0) \;\;= 4(0) \;\;\; + 2 = 2 \quad (0, 2)$$
$$f(-6) = 4(-6) + 2 = -22 \quad (-6, -22)$$
$$f(2) \;\;= 4(2) \;\;\; + 2 = 10 \quad (2, 10)$$
$$f(1) \;\;= 4(1) \;\;\; + 2 = 6 \quad (1, 6)$$

43.
$$g(-6) = \dfrac{1}{2}(-6) - 4 = -7 \quad (-6, -7)$$
$$g(-4) = \dfrac{1}{2}(-4) - 4 = -6 \quad (-4, -6)$$
$$g(0) \;\;= \dfrac{1}{2}(0) \;\;\; - 4 = -4 \quad (0, -4)$$
$$g(2) \;\;= \dfrac{1}{2}(2) \;\;\; - 4 = -3 \quad (2, -3)$$
$$g(8) \;\;= \dfrac{1}{2}(8) \;\;\; - 4 = 0 \quad (8, 0)$$

45.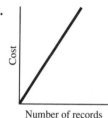

47.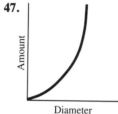

49.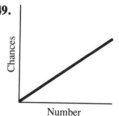

51.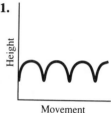

Problem Set 13.1B

1. a) Domain: $\{1, 3, 5, -2\}$
   Range: $\{2, -6, -4, 1\}$
   b) is a function
   c) is one to one

3. a) Domain: $\{2, 5, 4, -2\}$
   Range: $\{-1, 6, 0, -2\}$
   b) is a function
   c) is one to one

5. a) Domain: $\{-3, -2, -1\}$
   Range: $\{-4, -3, -2\}$
   b) is a function
   c) is one to one

7. a) Domain: $\{-2, -5, 2, 3, 6\}$
   Range: $\{3, 6, 4, -2\}$
   b) is a function
   c) is not one to one

9. a) Domain: $\{-6, -2, 0, 2, 8\}$
   Range: $\{1, 4, 7, 8\}$
   b) is a function
   c) is not one to one

11. a) Domain: $\{-1, 0, 3, 5\}$
   Range: $\{-4, 2, 7, 9\}$
   b) is a function
   c) is one to one

13. a) Domain: $\{-3, 0, 3, 5\}$
   Range: $\{0, 2, 4, 6\}$
   b) is a function
   c) is one to one

15. a) Domain: $\{-1, 2\}$
   Range: $\{2, 3, 4, 8\}$
   b) is not a function
   c) is not one to one

17. a) is a function
   b) is not one to one

19. a) is a function
   b) is not one to one

21. a) is a function
   b) is not one to one

23. a) is not a function
   b) is not one to one

25. a) is not a function
   b) is not one to one

27. a) is a function
   b) is one to one

29. a) Domain: $x \in R$
   Range: $0 \le y$
   Note: Assume that graph
   touches x-axis
   b) is not one to one

31. a) Domain: $0 \le x$
   Range: $y \in R$
   b) is not one to one

33. a) Domain: $x \in R$
Range: $y \le -2$ or $y \ge 2$
b) is not one to one

35. a) Domain: $x \in R$
Range: $y \ge 2$
b) is not one to one

37. a) Domain: $-5 \le x \le 5$
Range: $-4 \le y \le 4$
b) is not one to one

39. a) Domain: $x \in R$
Range: $y \ge 0$
b) is not one to one

41. a) Domain: $x \in R$
Range: $y \ge 2$
b) is not one to one

43. a) Domain: $x \le -4$ or $x \ge 4$
Range: $y \in R$
b) is not one to one

45. a) Domain: $0 \le x$
Range: $y \in R$
b) is not one to one

47. a) Domain: $x \ge -1$
Range: $y \in R$
b) is not one to one

49. a) Domain; $x \in R$
Range: $y \in R$
b) is one to one

51. a) Domain: $x \in R$

Range: $y \in R$
b) is one to one

53. a) $y = \dfrac{8 - 3x}{2}$
b) is a function
c) is a function

55. a) $y = \sqrt[3]{x + 3}$
b) is a function
c) is a function

57. a) $y = \pm\sqrt{1 - 4x^2}$

b) is not a function
c) is not a function

59. a) $y = \dfrac{3x - 1}{4}$

b) is a function
c) is a function

61. a) $y = \pm\sqrt{x + 4}$

b) is a function
c) is not a function

63. a) $y = \sqrt[3]{x - 6}$
b) is a function
c) is a function

65.

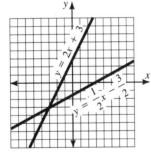

67.

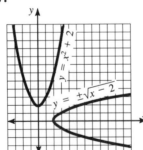

69.

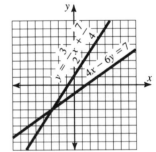

71.

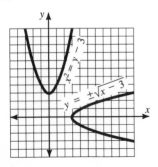

Problem Set 13.2

1. $y = 4^x$

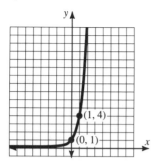

3. $y = \left(\dfrac{3}{4}\right)^x$

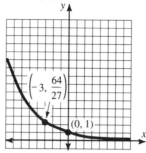

5. $y = 2^{2x}$

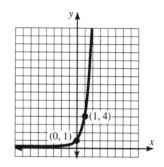

7. $y = 5^x$

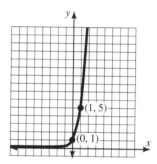

9. $y = \left(\dfrac{2}{5}\right)^x$

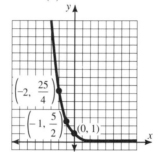

11. $y = 3^{2x}$

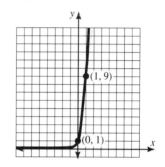

13. $y = 2^x + 2$

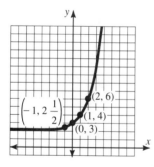

15. $y = 2^{x-1}$

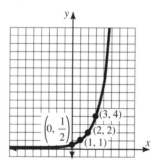

17. $y = 2^{2x-1}$

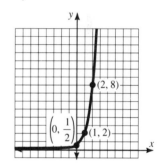

19. $y = 2^x - 3$

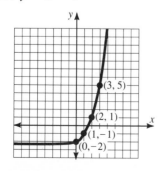

21. $y = 2^{x-2}$

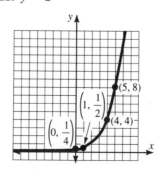

23. $y = 3^{-x+2}$

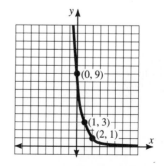

25. 4   **27.** 3   **29.** $\dfrac{5}{3}$ **31.** -4 **33.** -4 **35.** 3   **37.** 1   **39.** $\dfrac{4}{3}$ **41.** $\dfrac{2}{3}$

43. -4 **45.** 3   **47.** 2   **49.** 4   **51.** 1   **53.** $\dfrac{3}{2}$ **55.** 2   **57.** 2   **59.** -3

61. 1573.72   **63.** 2881.85   **65.** 3909.86   **67.** \$2539.00   **69.** 4527.29   **71.** \$3225.44

Problem Set 13.3A

1. $4 = \log_2 16$ **3.** $-5 = \log_2 \dfrac{1}{32}$ **5.** $3 = \log_{\frac{3}{2}} \dfrac{27}{8}$ **7.** $-3 = \log_{10} 0.001$

9. $-4 = \log_3 \dfrac{1}{81}$ **11.** $-2 = \log_{\frac{3}{2}} \dfrac{4}{9}$ **13.** $2^7 = 128$ **15.** $10^{-2} = 0.01$

17. $5^{-3} = \dfrac{1}{125}$ **19.** $2^6 = 64$ **21.** $10^{-4} = 0.0001$ **23.** $3^{-4} = \dfrac{1}{81}$

25.

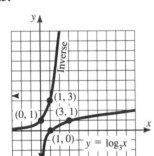

27.

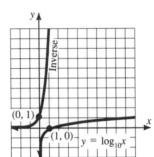

29.

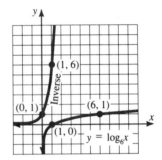

31.

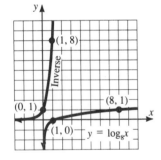

33. -2 **35.** 4   **37.** -3 **39.** 10

41. -3 **43.** -3 **45.** 3   **47.** -4

49. -4 **51.** 4   **53.** 3   **55.** 27

57. 27   **59.** $\dfrac{1}{2}$ **61.** $\dfrac{1}{16}$ **63.** 5

65. -5 **67.** $\dfrac{27}{8}$ **69.** 4   **71.** $\dfrac{1}{6}$

Problem Set 13.3B

1. 8   **3.** 9   **5.** 2   **7.** 7   **9.** 11

11. 2   **13.** 2   **15.** 5   **17.** 3   **19.** 2

21. 17   **23.** 2

25. $\log_a 3 + \log_a x + \log_a y$

27. $\log_a 8 + \log_a x + 2\log_a y - \log_a z$

29. $\log_a 7 + 2\log_a x + \log_a y$

31. $\log_a 18 + 2\log_a x + \dfrac{1}{3}\log_a y - 2\log_a z$

33. $\log_a 12 + \dfrac{1}{2}\log_a x + \log_a y - \log_a 5 - 3\log_a z$

35. $\dfrac{1}{2}[2\log_b x + 3\log_b y - 4\log_b z]$

37. $\dfrac{1}{4}[\log_b 7 + 5\log_b x + 4\log_b y - \log_b 6 - 2\log_b z]$

39. $\log_b 16 + \dfrac{1}{2}\log_b x + 3\log_b y - \log_b 7 - \dfrac{1}{3}\log_b z$

41. $\dfrac{3}{4}\log_b x + \dfrac{2}{5}\log_b y - \dfrac{1}{3}\log_b z$

43. $\dfrac{1}{3}[\log_b 5 + 3\log_b x + 7\log_b y - \log_b 2 - 2\log_b z]$

45. $\log_b \dfrac{\sqrt[3]{x^2}}{y}$

47. $\log_a \dfrac{x^4}{y^2}$

49. $\log_a x^9 y^2$

51. $\log_a \dfrac{16x^2}{27y^6}$

53. $\log_a z \sqrt[3]{\dfrac{x^2}{y}}$

55. $\log_a x^{5/4} y^{3/8}$

57. $\log_a \dfrac{x}{y^{2/3}}$

59. $\log_a x^{10/3} y$

Problem Set 13.4

1. 2000   **3.** 0.6961   **5.** 0.3445   **7.** 0.7830   **9.** 0.93055   **11.** 12.0204

13. 0.95024   **15.** 0.9241   **17.** 26.2956   **19.** 4.7265   **21.** 1123.4   **23.** 1.9773

25. 2.4308   **27.** 2.469   **29.** 14.1165   **31.** 14.7373   **33.** 13.5463   **35.** 6.3736

Chapter 13   Review Test

1. a) Domain: $\{-3, -2, 5, 3\}$
    Range: $\{2, 6, 4\}$
  b) is a function
  c) is not one to one

2. a) Domain: $\{0, -2, 8, 9\}$
    Range: $\{6, 0, 7, 2, 3\}$
  b) is not a function
  c) is not one to one

3. a) is a function
  b) is one to one

4. a) is a function
  b) is not one to one

5. a) is not a function
  b) is not one to one

6. a) Domain: $-4 \le x \le 4$
    Range: $-2 \le y \le 2$
  b) is not one to one

7. a) Domain: $x \in R$
    Range: $y \le 0$
  b) is not one to one

8. a) Domain: $x \in R$
    Range: $y \in R$
  b) is one to one

9. a) Domain: $x \le -5$ or $x \ge 5$
    Range: $y \in R$
  b) is not one to one

10. a) Domain: $-2 \le x \le 2$
    Range: $-4 \le y \le 4$
  b) is not one to one

11. a) Domain: $x \in R$
    Range: $4 \le y$
  b) is not one to one

12. a) $y = \dfrac{2}{3}x + \dfrac{8}{3}$
  b) is a function
  c) is a function

13. a) $y = \sqrt[3]{x} + 2$
  b) is a function
  c) is a function

14. a) $y = \pm\dfrac{2}{3}\sqrt{9 - x^2}$
  b) is not a function
  c) is not a function

15.

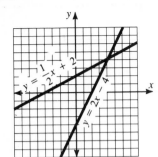

16.

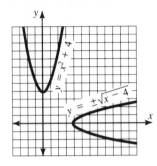

17. $y = 2^{3x}$

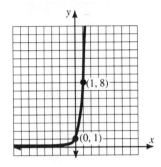

18. $y = 2^{2x+3}$

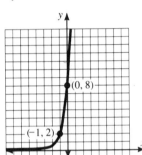

19. $y = 2^{2x} + 3$

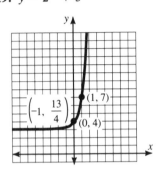

20. $y = 3^{-2x}$

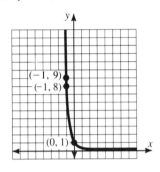

21. 5

22. 2

23. 3

24. $\dfrac{5}{6}$

25. 886.47

26. 2257.64

27. 3076.45

28. $\log_3 81 = 4$

29. $\log_{10} 0.001 = -3$

30. $\dfrac{1}{8} = 2^{-3}$

31. $0.01 = 10^{-2}$

32. $y = \log_4 x$

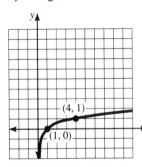

33. 2

34. -5

35. 100

36. $\dfrac{5}{3}$

37. 2

38. $\dfrac{1}{81}$

39. 2

40. 3

41. 6

42. $\log_a 3 + \log_a 5 + \dfrac{1}{2}\log_a x + 3\log_a y$

43. $\dfrac{1}{3}(\log_a 6 + 2\log_a x + \log_a y - \log_a 4 - 3\log_a z)$

44. $\log_a \sqrt[3]{\dfrac{x}{y^4}}$

45. $\log_a \sqrt{xy}$

46. 119.8

47. 0.3489

48. 0.6999

Problem Set 14.1

1. 4   **3.** 5.13   **5.** 3   **7.** 1.58   **9.** 2.63

11. 3.41   **13.** 0.78   **15.** 0.19   **17.** 3.01   **19.** $-4, -1$

21. $-1.33, 0$ **23.** $-4, \dfrac{3}{2}$ **25.** $-2, 3$ **27.** $\dfrac{-5 \pm \sqrt{17}}{2}$ **29.** $-4, \dfrac{1}{2}$

31. $\dfrac{-3}{2}, 4$ **33.** $-4, 1$ **35.** $\dfrac{3}{4}$ **37.** $\dfrac{2}{3}$, reject -2 **39.** $\dfrac{1}{4}$

41. 2, 5   **43.** 8   **45.** 2, reject -5 **47.** 25, reject -4 **49.** 3, reject -1

Problem Set 14.2

1. 0.1386, 14.87%   **3.** $t = 10$ yrs   **5.** 3.78 yrs   **7.** 5.9 billion

9. 2806   **11.** 0.1133   **13.** 4.19 years   **15.** 9.43 lbs/sq in

17. 12.91 lbs/sq in   **19.** 11.47 lbs/sq in   **21.** 5.03 lbs/sq in   **23.** 80%

25. 51%   **27.** $3566.96   **29.** $3630.71   **31.** 3643.79

33. 3644.24   **35.** 9.2 years   **37.** 19.2   **39.** 9.5 hrs

41. $A = 5000e^{0.924t}$; $t = 2.5$ hrs

Problem Set 14.3

1. pH $= 12.6$ **3.** $H^+ = 3.98 \times 10^{-10}$

5. pH $= 2.99$ **7.** $t = 2397.3$ or approximately 2400 yrs

9. $A = 1.25$, 0.63 gr   **11.** a) $t = 13412$ or approximately 13400 yrs

                 b) $t = 19188$ or approximately 19200 yrs

13. 1000 times   **15.** $L = 40$ db

17. 3.6×10^{10} ft lb   **19.** 1.042×10^{14}

21. 9.6   **23.** $10^{1.5}$ or approximately 32 times

Chapter 14   Review Test

1. 3.81   **2.** 2.01   **3.** $-\dfrac{1}{2}, -4$ **4.** $\dfrac{3}{4}, -1$

5. 2   **6.** 5, -1 **7.** $-\dfrac{2}{3}$, reject -4 **8.** $\dfrac{5}{2}$, reject -2

9. 0.116, 12%   **10.** 1576   **11.** approx. 330000   **12.** $A = 25{,}937$; 174,494
                                            approx. 896000

13. 8.44   **14.** 7.01   **15.** 10.9, 0.74 or 74%   **16.** $A = 16936$, 35853.51

17. a. 10836.67   **18.** 4×10^{-7} **19.** -1.08 **20.** a. 8.41 g
    b. 10932.71   b. 7.07 g
                                                                      c. 1.77 g

21. $t = 55.5$;   **22.** 1519.3   **23.** 500 times   **24.** 60 db
    $A = 19.9, 19.5, 15.6$

25. 2.9×10^{11} **26.** 7   **27.** $10^{3.4}$ or 2512 times

■ Chapter 15

Problem Set 15.1

1. 5   **3.** 22   **5.** -15 **7.** 28   **9.** -84 **11.** -10

13. 38   **15.** 14   **17.** 0   **19.** -3 **21.** 11   **23.** $\dfrac{7}{3}$

25. 0.044   **27.** -0.158 **29.** 0   **31.** -2 **33.** $\dfrac{1}{12}$ **35.** 0.232

Problem Set 15.2

1. $(-1, 2)$ **3.** $(-2, -5)$ **5.** $(-4, 5)$ **7.** No solution   **9.** $\left(\dfrac{2}{3}, -\dfrac{1}{2}\right)$

11. $(3, -2)$ **13.** $(5, -6)$ **15.** $\left(\dfrac{1}{2}, \dfrac{1}{3}\right)$ **17.** $\left(\dfrac{2}{3}, \dfrac{1}{4}\right)$ **19.** No solution

21. Same line   **23.** $\left(\dfrac{5}{4}, \dfrac{4}{3}\right)$ **25.** $(-1, -1)$ **27.** $(15, 12)$ **29.** No solution

31. $\left(-3, \dfrac{1}{3}\right)$ **33.** $(4, 12)$ **35.** 5, 8

Problem Set 15.3

1. -11 **3.** -41 **5.** 5   **7.** 24   **9.** -16 **11.** 0

13. -74 **15.** 9   **17.** 0   **19.** 90   **21.** 0   **23.** -2

25. -74 **27.** 9   **29.** 137

Problem Set 15.4

1. $(2, -1, 3)$ **3.** $(-3, 4, 1)$ **5.** $\left(\dfrac{1}{3}, -\dfrac{1}{2}, 0\right)$ **7.** $\left(2, -1, \dfrac{1}{2}\right)$ **9.** No solution

11. $(-1, 0, -2)$ **13.** $(-1, 1, 2)$ **15.** $(-2, 4, 0)$ **17.** $(0, 4, -1)$ **19.** $\left(-3, \dfrac{2}{3}, 0\right)$

21. $\left(0, \dfrac{1}{2}, -3\right)$ **23.** $\left(3, 0, -\dfrac{1}{2}\right)$ **25.** $(4, 6, 8)$ **27.** $4000 at 8%, $6000 at 10%, $8000 at 12%

Chapter 15 Review Test

1. 22   **2.** 0   **3.** -2 **4.** $(-6, 5)$ **5.** No solution   **6.** $\left(\dfrac{1}{2}, -3\right)$

7. $(3, -2)$ **8.** 39   **9.** 0   **10.** -15 **11.** $(-1, 3, -2)$ **12.** $\left(\dfrac{1}{2}, \dfrac{1}{3}, -\dfrac{1}{4}\right)$

13. $(3, -2, 4)$ **14.** 15 feet by 7 feet   **15.** 15 nickels, 20 dimes, 18 quarters

■ Chapter 16

Problem Set 16.1

1. 3, 4, 5, 6, 7   **3.** $\dfrac{1}{2}, 1, \dfrac{3}{2}, 2, \dfrac{5}{2}$ **5.** 3, 9, 19, 33, 51   **7.** 13, 11, 9, 7, 5

9. $1, \dfrac{1}{2}, \dfrac{1}{3}, \dfrac{1}{4}, \dfrac{1}{5}$ **11.** $\dfrac{1}{4}, \dfrac{1}{6}, \dfrac{1}{8}, \dfrac{1}{10}, \dfrac{1}{12}$ **13.** $2, \dfrac{5}{3}, 2, \dfrac{17}{7}, \dfrac{26}{9}$ **15.** $-1, 2, -3, 4, -5$

17. $a_n = 2n - 1$ **19.** $a_n = \dfrac{1}{n^2}$ **21.** $a_n = \dfrac{2}{n + 1}$ **23.** $a_n = 2n^2$

25. $a_n = n^2 + 1$ **27.** $a_n = n(n + 1)$ or: $a_n = n^2 + n$

Problem Set 16.2

1. 2   **3.** -4 **5.** $a_n = 2n - 1$ **7.** $a_n = 20 - 5n$ **9.** 66   **11.** 700

13. 500   **15.** 25   **17.** -22.4 **19.** 950   **21.** 830   **23.** 1320

25. -35 **27.** 295

Problem Set 16.3

1. 2, 6, 18, 54, 162   **3.** $10, 5, \dfrac{5}{2}, \dfrac{5}{4}, \dfrac{5}{8}, \dfrac{5}{16}, \dfrac{5}{32}$ **5.** 3, -6, 12, -24, 48   **7.** 2

9. $\dfrac{1}{2}$ **11.** -2 **13.** $a_n = 1 \cdot 5^{n-1}$ **15.** $a_n = 64\left(\dfrac{1}{2}\right)^{n-1}$

17. 510   **19.** 65535   **21.** 1.9980469   **23.** 2046

25. 1111

Problem Set 16.4A

1. 6   **3.** 2   **5.** 1   **7.** 362880   **9.** 42   **11.** 99

13. 7084   **15.** $\dfrac{22!}{18!}$ **17.** $\dfrac{200!}{9!}$ **19.** $\dfrac{n!}{(n - 5)!}$

Problem Set 16.4B

1. 1, 7, 21, 35, 35, 21, 7, 1   **3.** $a^3 + 3a^2b + 3ab^2 + b^3$

5. $p^4 - 4p^3q + 6p^2q^2 - 4pq^3 + q^4$ **7.** $a^4 + 4a^3b + 6a^2b^2 + 4ab^3 + b^4$

9. $x^5 - 5x^4y + 10x^3y^2 - 10x^2y^3 + 5xy^4 - y^5$

11. $84a^6b^3$ **13.** $243x^5$ **15.** $-20x^3y^3$ **17.** $2835x^4y^3$ **19.** $(125970)(7^{12})(4^8)x^{12}y^8$

Chapter 16 Review Test

1. $\frac{1}{2}, \frac{2}{5}, \frac{3}{10}, \frac{4}{17}, \frac{5}{26}$

2. $-\frac{1}{3}, \frac{1}{2}, -\frac{3}{5}, \frac{2}{3}, -\frac{5}{7}$

3. $a_n = n^2 - 1$

4. $a_n = \frac{(-1)^n}{2(n+1)}$

5. $a_n = 3n - 2$

6. $a_n = 4n - 17$

7. -11

8. -31

9. 4

10. 44

11. $5, -10, 20, -40, 80, -160$

12. $27, 18, 12, 8, \frac{16}{3}, \frac{32}{9}$

13. -2

14. $\frac{3}{4}$

15. $a_n = 3^n$

16. 3280

17. $\frac{31}{16}$

18. $22724.64

19. 101, 467.67

20. 720

21. 40320

22. 1320

23. 627

24. $x^5 + 10x^4y + 40x^3y^2 + 80x^2y^3 + 80xy^4 + 32y^5$

25. $-4320x^3y^3$

26. $1792x^3y^5$

■ Cumulative Review   Chapters 1–16

1. 50

2. -2

3. 2

4. $2a^2b$

5. $\frac{72y^7}{x^{12}}$

6. $\frac{4b^4}{a^{12}}$

7. $\frac{x^{1/2}}{y^{15/4}}$

8. -18

9. $2\sqrt{6}\,i$

10. $-i$

11. $12a^4 + 15a^3 - 19a^2 + 42a - 20$

12. $9a^2 - 4b^2$

13. $3x^3 + 2x^2 - 2 + \frac{-3}{5x - 2}$

14. $\frac{3x}{x^2 + 3x + 9}$

15. $\frac{a - b}{2a - 3}$

16. $\frac{-6y - 4}{(y + 4)(y - 4)(y - 1)}$

17. $\frac{a + 6}{a - 1}$

18. $\frac{x^2}{x + 2}$

19. $7\sqrt{3x} - 2x\sqrt{3x}$

20. $2x - 2\sqrt{2x} - 24$

21. $3\sqrt{2} + 2\sqrt{3}$

22. $-1 + 4i$

23. $-21 - 7\sqrt{3}\,i$

24. $\frac{1}{25} - \frac{18}{25}i$

25. $(4x - 3)(2x + 5)$

26. $(5x + 2y)^2$

27. $(4x^4 + y^2)(2x^2 + y)(2x^2 - y)$

28. $(x^2 - 2y)(x^4 + 2x^2y + 4y^2)$

29. $(x - 3)^2(x + 3)^2$

30. $(x - 3)(x + 1)(x - 4)(x + 2)$

31. $(2x - 5)(y + 6)$

32. 2

33. 3

34. $d = \frac{2n - ac}{a}$

35. $-9 \le x \le 3$

36. $x \ge 4$ or $x \le \frac{4}{3}$

37. -3

38. $\frac{2}{3}, -4$

39. $-\frac{5}{2}, 3$

40. $\frac{1 \pm \sqrt{11}\,i}{4}$

41. 5; 13 doesn't check

42. $\pm\sqrt{3}$, ± 2

43. $\dfrac{3}{2}$, $-\dfrac{1}{4}$

44. 2

45. 2

46. 1.6789

47. 6, reject $\dfrac{5}{3}$

48. $(3, -2)$

49. $(-1, 2, -3)$

50. $(-8, -6)$, $(6, 8)$

51. $(1, 2)$, $(1, -2)$, $(-1, 2)$, $(-1, -2)$

52. $y = -\dfrac{4}{5}x - 2$

53. $y = -\dfrac{4}{3}x - \dfrac{11}{3}$

54. $y = \dfrac{5}{2}x - 13$

55. 3.75×10^{-9}

56. $2x^3 - x^2 - 4x + 12 + \dfrac{-20}{x + 3}$

57. $2\sqrt{41}$

58. $\left(2, \dfrac{5}{2}\right)$

59. $C:(3, -5)$; $r = 6$

60. a. $x'^2 + y'^2 = 36$
b. $(x - 3)^2 + (y + 5)^2 = 36$

61. $\dfrac{1}{4}[\log_a 8 + 3\log_a x + \log_a y - 2\log_a z]$

62. 0.04290

63. $x \le \dfrac{13}{4}$

64. $x \le -1$ or $x > 1$

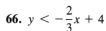

65. $3x - 4y = 12$

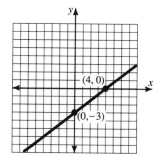

66. $y < -\dfrac{2}{3}x + 4$

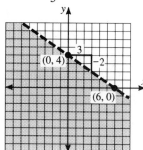

67. $y - 2 = -2(x + 3)^2$

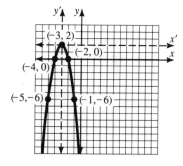

68. $9x^2 + 4y^2 - 36x + 8y + 4 = 0$

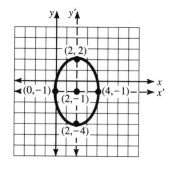

69. $\dfrac{(x - 1)^2}{16} - \dfrac{(y + 2)^2}{9} = 1$

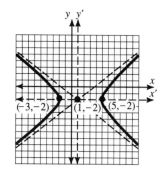

70. 8, 5

71. 6 nickels, 12 dimes, 9 quarters

72. 2 hrs 24 min

73. 400 ohms

74. 3 mph　　**75.** a) 8811.71　b) 8954.25　c) 9030.55　d) 9104.30　e) 9108.68　f) 9110.59

76. 3.98　　**77.** 10.18 lb/in^2　**78.** 12.2%; 9.5 yrs　**79.** 10.8 hrs　　**80.** 5.13×10^8 yrs

81. 7636 yrs old　　**82.** 2.83 grams　**83.** 100,000 times　**84.** $10^{-5.5}$ watts/meter^2　**85.** 100 times

86. -11　　**87.** -2　　**88.** $(-5, -4)$　**89.** $(2, 2, 1)$

90. 800 at 10%, 1200 at 12%　　**91.** first -6, second 4, third 10

92. $-\dfrac{1}{3}, 1, -\dfrac{9}{5}, \dfrac{8}{3}, -\dfrac{25}{7}$　　**93.** $a_n = 4n - 26$　**94.** 51　　**95.** -11

96. 6　　**97.** $81, -54, 36, -24, 16, -\dfrac{32}{3}$　　**98.** $a_n = 16\left(-\dfrac{1}{2}\right)^{n-1}$　**99.** $-2\dfrac{5}{8}$

or　$a_n = -32\left(-\dfrac{1}{2}\right)^{n}$

100. 7752　　**101.** $32x^5 - 80x^2y + 80x^3y^2 - 40x^2y^3 + 10xy^4 - y^5$　　**102.** $-4320x^3y^3$

Radicant, 324
Radioactive material, half-life of, 555–56, 572
Radiocarbon dating, 562–64, 572
Radius, 423
Range, of relation, 502, 537
Rational expressions
  addition and subtraction of, 284–85, 286–90
  division of, 280–81
  equality of, 270
  evaluating, 30–31
  multiplication of, 269–75, 277, 279–80, 309
  simplification of, 309
Rationalizing the denominator, 327–30
Rational numbers, 21, 52, 337
  addition of, 34, 36, 53
  additive inverse of, 37
  division for, 22–23
  division of, 25
  multiplication of, 22
  operations with, 21
  product of, 52
  subtraction of, 34, 36, 38, 53
Rectangle
  area of, 41
  perimeter of, 40
Rectangular solid, volume of, 45
Reflexive property of equality, 57, 94
Relation, 502
  domain of, 502, 537
  inverse of a, 506–9, 537
  range of, 502, 537
Richter, Dr. C. F., 564
Richter Scale, 564
Right triangle, hypotenuse of the, 416

Scientific notation, 195, 197, 227
  conversion from, to decimal notation, 197–99, 227
Second degree equation, 491
  systems of equations with, 491–92
  equations solving, 484–85, 486–89
  identification of, 476–77
  systems of equations involving a line and, 479–83
Semi-major axis, of ellipse, 456

Sequences, 595–601, 627
  general term of, 627
  term of, 627
Set, 2, 51
  equal, 2–3
  membership of, 2
Signed numbers
  addition of, 51
  division of, 52
  multiplication of, 52
  subtraction of, 51
Simultaneous solution, 144–45, 182
  for systems of three equations, 182
Slope, 103–8
Slope-intercept form of, equation of a line, 109–10, 139
Slope-intercept method, graphing of lines by, 111–12
Slope of a line, 139
Solution set, 58, 94
Sound, measurement of, 567–69, 573
Special products, 207–9
Square, of a binomial, 210, 362–63
Square root, 314–16, 337
Substitution, solving systems of equations by, 154–55, 182
Substitution property of equality, 57, 94
Subtraction, 9–10
  of algebraic fractions, 286–90
  of fractions, 284–85, 310
  of integers, 9
  of radical expressions, 331–32
  of rational numbers, 34, 36, 53
  of signed numbers, 11, 51
  with unlike denominators, 34–35
Sum and difference, product of, 211, 227
Sum of two cubes, factoring, 267
Symmetric property of equality, 57, 58, 94
Symmetry, axis of, 437, 477
Synthetic division, 185, 222–24, 227
Systems of equation
  addition method for solving, 182
  involving a line and a second

degree equation, 479–83
  involving second degree equations, 491–92
  simultaneous solution of, 182
  solving, 144–52
    using Cramer's rule, 589–90
    by substitution, 154–55, 182
  with three equations, 175–80

Tautology, 152
Term, 13
  of a sequence, 595
Transitive property
  of equality, 57, 58, 94
  of inequality, 94
Translation equations, 434–35, 452
Translations, 430–33
Triangle, area of right, 42
Trichotomy axiom, 94
Trinomial, 203, 227
  factoring, 230–31
Trinomial square, factoring, 267

Value, absolute, 51
Variable, 3, 51
  solving equations for specified, 65–68, 94
  use of, for problem solving, 406–7
Variation, 396
  direct, 396–98, 400
  inverse, 400
  joint, 401
Vertical line test, 501–2
Visualization, use of, for problem solving, 407–11
Volume, 44–45, 399
  of a cylinder, 45, 46
  of a rectangular solid, 45

Whole numbers, 5, 51

x-axis, 98, 138

y-axis, 98, 138
y-intercept, 109, 139

Zero, 51
  in addition, 5
  as an exponent, 185–86
  multiplication by, 13–14, 52
  product equal to, 352–57, 382
Zero exponent, 226